Shop Manual for

Automotive Brake Systems

Shop Manual for Automotive Brake Systems

Lane Eichhorn
Parkland College
Champaign, IL

Drew Corinchock

Jack Erjavec
Series Advisor
Columbus State Community College
Columbus, Ohio

Delmar Publishers

I(T)P™ An International Thomson Publishing Company

Albany • Bonn • Boston • Cincinnati • Detroit • London • Madrid • Melbourne
Mexico City • New York • Pacific Grove • Paris • San Francisco • Singapore • Tokyo
Toronto • Washington

NOTICE TO THE READER

Publisher does not warrant or guarantee any of the products described herein or perform any independent analysis in connection with any of the product information contained herein. Publisher does not assume, and expressly disclaims, any obligation to obtain and include information other than that provided to it by the manufacturer.

The reader is expressly warned to consider and adopt all safety precautions that might be indicated by the activities described herein and to avoid all potential hazards. By following the instructions contained herein, the reader willingly assumes all risks in connection with such instructions.

The publisher makes no representations or warranties of any kind, including but not limited to, the warranties of fitness for particular purpose or merchantability, nor are any such representations implied with respect to the material set forth herein, and the publisher takes no responsibility with respect to such material. The publisher shall not be liable for any special, consequential, or exemplary damages resulting, in whole or in part, from the readers' use of, or reliance upon, this material.

COVER PHOTO: Courtesy of Ford Motor Company
PHOTO SEQUENCES: Photography by Jeff Hinckley

Portions of materials contained herein have been reprinted with permission of General Motors Corporation, Service Technology Group.

DELMAR STAFF
Senior Administrative Editor: Vernon Anthony
Developmental Editor: Catherine Eads
Project Editor: Eleanor Isenhart
Production Coordinator: Karen Smith
Art/Design Coordinator: Heather Brown

Printed in the United States of America

For information, contact:

Delmar Publishers
3 Columbia Circle, Box 15015
Albany, New York 12212-5015

International Thomson Publishing Europe
Berkshire House 168-173
High Holborn
London, WC1V7AA
England

Thomas Nelson Australia
102 Dodds Street
South Melbourne, 3205
Victoria, Australia

Nelson Canada
1120 Birchmont Road
Scarborough, Ontario
Canada M1K 5G4

International Thomson Editors
Campos Eliseos 385, Piso 7
Col Polanco
11560 Mexico DF Mexico

International Thomson Publishing GmbH
Königswinterer Strasse 418
53227 Bonn
Germany

International Thomson Publishing Asia
221 Henderson Road
#05-10 Henderson Building
Singapore 0315

International Thomson Publishing Japan
Hirakawacho Kyowa Building, 3F
2-2-1 Hirakawacho
Chiyoda-ku, Tokyo 102
Japan

8 9 10 XXX 02 01 00 99

Library of Congress Cataloging-in-Publication Data

Eichhorn, Lane.
Automotive brake systems / Lane Eichhorn, Drew Corinchock, Jack Erjavec.
p. cm. — (Today's technician)
Includes index.
Contents: [1] Classroom manual — [2] Shop manual.
ISBN 0-8273-6188-2
1. Automobiles — Brakes. I. Corinchock, John A. II. Erjavec, Jack. III. Title. IV. Series.
TL269.E37 1995
629.24'6 — dc20 95-9362
CIP

CONTENTS

Photo Sequences

PREFACE

Unlike yesterday's mechanic, the technician of today and for the future must know the underlying theory of all automotive systems and be able to service and maintain those systems. Today's technician must also know how these individual systems interact with each other. Standards and expectations have been set for today's technician, and these must be met in order to keep the world's automobiles running efficiently and safely.

The *Today's Technician* series, by Delmar Publishers, features textbooks that cover all mechanical and electrical systems of automobiles and light trucks. Principal titles correspond with the eight major areas of ASE (National Institute for Automotive Service Excellence) certification. Additional titles include remedial skills and theories common to all of the certification areas and advanced or specialized subject areas that reflect the latest technological trends.

Each title is divided into two manuals: a Classroom Manual and a Shop Manual. Dividing the material into two manuals provides the reader with the information needed to begin a successful career as an automotive technician without interrupting the learning process by mixing cognitive and performance-based learning objectives.

Each Classroom Manual contains the principles of operation for each system and subsystem. It also discusses the design variations used by different manufacturers. The Classroom Manual is organized to build upon basic facts and theories. The primary objective of this manual is to allow the reader to gain an understanding of how each system and subsystem operates. This understanding is necessary to diagnose the complex automobile systems.

The understanding acquired by using the Classroom Manual is required for competence in the skill areas covered in the Shop Manual. All of the high priority skills, as identified by ASE, are explained in the Shop Manual. The Shop Manual also includes step-by-step instructions for diagnostic and repair procedures. Photo Sequences are used to illustrate many of the common service procedures. Other common procedures are listed and are accompanied with fine-line drawings and photographs that allow the reader to visualize and conceptualize the finest details of the procedure. The Shop Manual also contains the reasons for performing the procedures, as well as when that particular service is appropriate.

The two manuals are designed to be used together and are arranged in corresponding chapters. Not only are the chapters in the manuals linked together, the contents of the chapters are also linked. Both manuals contain clear and thoughtfully selected illustrations. Many of the illustrations are original drawings or photos prepared for inclusion in this series. This means that the art is a vital part of each manual.

The page layout is designed to include information that would otherwise break up the flow of information presented to the reader. The main body of the text includes all of the "need-to-know" information and illustrations. In the side margins are many of the special features of the series. Items such as definitions of new terms, common trade jargon, tools lists, and cross-references are placed in the margin, out of the normal flow of information so as not to interrupt the thought process of the reader. Each manual in this series is organized in a like manner and contains the same features.

Jack Erjavec, Series Advisor

Classroom Manual

To stress the importance of safe work habits, the Classroom Manual dedicates one full chapter to safety. Included in this chapter are common safety practices, safety equipment, and safe handling of hazardous materials and wastes. This includes information on MSDS sheets and OSHA regulations. Other features of this manual include:

Cognitive Objectives

These objectives define the contents of the chapter and define what the student should have learned upon completion of the chapter.
Each topic is divided into small units to promote easier understanding and learning.

Marginal Notes

New terms are pulled out and defined. Common trade jargon also appears in the margin and gives some of the common terms used for components. This allows the reader to speak and understand the language of the trade, especially when conversing with an experienced technician.

Cautions and Warnings

Throughout the text, cautions are given to alert the reader to potentially hazardous materials or unsafe conditions. Warnings are also given to advise the student of things that can go wrong if instructions are not followed or if a nonacceptable part or tool is used.

References to the Shop Manual

Reference to the appropriate page in the Shop Manual is given whenever necessary. Although the chapters of the two manuals are synchronized, material covered in other chapters of the Shop Manual may be fundamental to the topic discussed in the Classroom Manual.

A Bit of History

This feature gives the student a sense of the evolution of the automobile. This feature not only contains nice-to-know information, but also should spark some interest in the subject matter.

Summaries

Each chapter concludes with summary statements that contain the important topics of the chapter. These are designed to help the reader review the contents.

Terms to Know

A list of new terms appears next to the Summary. Definitions for these terms can be found in the Glossary at the end of the manual.

Review Questions

Short answer essay, fill-in-the-blank, and multiple-choice type questions follow each chapter. These questions are designed to accurately assess the student's competence in the stated objectives at the beginning of the chapter.

Shop Manual

To stress the importance of safe work habits, the Shop Manual also dedicates one full chapter to safety. Other important features of this manual include:

Performance Objectives

These objectives define the contents of the chapter and define what the student should have learned upon completion of the chapter. These objectives also correspond with the list of required tasks for ASE certification. *Each ASE task is addressed.*

Although this textbook is not designed to simply prepare someone for the certification exams, it is organized around the ASE task list. These tasks are defined generically when the procedure is commonly followed and specifically when the procedure is unique for specific vehicle models. Imported and domestic model automobiles and light trucks are included in the procedures.

Photo Sequences

Many procedures are illustrated in detailed Photo Sequences. These detailed photographs show the students what to expect when they perform particular procedures. They also can provide a student a familiarity with a system or type of equipment, which the school may not have.

Marginal Notes

Page numbers for cross-referencing appear in the margin. Some of the common terms used for components, and other bits of information, also appear in the margin. This provides an understanding of the language of the trade and helps when conversing with an experienced technician.

Cautions and Warnings

Throughout the text, cautions are given to alert the reader to potentially hazardous materials or unsafe conditions. Warnings are also given to advise the student of things that can go wrong if instructions are not followed or if a nonacceptable part or tool is used.

References to the Classroom Manual

Reference to the appropriate page in the Classroom Manual is given whenever necessary. Although the chapters of the two manuals are synchronized, material covered in other chapters of the Classroom Manual may be fundamental to the topic discussed in the Shop Manual.

Customer Care

This feature highlights those little things a technician can do or say to enhance customer relations.

Tools Lists

Each chapter begins with a list of the Basic Tools needed to perform the tasks included in the chapter. Whenever a Special Tool is required to complete a task, it is listed in the margin next to the procedure.

Service Tips

Whenever a special procedure is appropriate, it is described in the text. These tips are generally those things commonly done by experienced technicians.

Case Studies

Case Studies concentrate on the ability to properly diagnose the systems. Each chapter ends with a case study in which a vehicle has a problem, and the logic used by a technician to solve the problem is explained.

Terms to Know

Terms in this list can be found in the Glossary at the end of the manual.

Diagnostic Chart

Chapters include detailed diagnostic charts linked with the appropriate ASE task. These charts list common problems and most probable causes. They also list a page reference in the Classroom Manual for better understanding of the system's operation and a page reference in the Shop Manual for details on the procedure necessary for correcting the problem.

ASE Style Review Questions

Each chapter contains ASE style review questions that reflect the performance objectives listed at the beginning of the chapter. These questions can be used to review the chapter as well as to prepare for the ASE certification exam.

Instructor's Guide

The Instructor's Guide is provided free of charge as part of the *Today's Technician Series* of automotive technology textbooks. It contains Lecture Outlines, Answers to Review Questions, Pretest and Test Bank including ASE style questions.

Classroom Manager

The complete ancillary package is designed to aid the instructor with classroom preparation and provide tools to measure student performance. For an affordable price, this comprehensive package contains:

Instructor's Guide
200 Transparency Masters
Answers to Review Questions
Lecture Outlines and Lecture Notes
Printed and Computerized Test Bank
Laboratory Worksheets and Practicals

Reviewers

Allan Foster
Portage Lakes Career Center
Greensburg, OH

David Hewitt
Northern Alberta Institute of Technology
Edmonton, AB CANADA

Thomas G. Ihlenfeldt
MoTech Education Center
Livonia, MI

Royal L. Johnson
Hennepinn Technical College
Eden Prairie, MN

Lawrence F. Meicher
Madison Area Technical College
Madison, WI

Wayne E. Musser
Harrisburg Area Community College
Harrisburg, PA

Scott Schiefe
Milwaukee Area Technical College
Milwaukee, WI

Richard Slinkard
Jefferson College
Hillsboro, MO

CHAPTER 1

Brake Service Tools and Equipment

Upon completion and review of this chapter, you should be able to:

- ❑ List the basic units of measure for length and volume in the metric and USCS systems.
- ❑ Identify and describe the purpose and use of hand tools commonly found in a basic service technician's tool kit.
- ❑ Identify and describe the purpose and use of commonly used power tools.
- ❑ Identify and use the major measuring tools and instruments used in brake service work.
- ❑ Identify and use the major electrical test tools used in ABS diagnostic work.
- ❑ Explain the procedure for measuring with both an inside and outside micrometer.
- ❑ Identify and describe the purpose and use of specialty tools used in brake service work.
- ❑ Describe asbestos containment equipment commonly used in the shop.

Selection, Storage, and Care of Tools

Brake system service requires the use of a wide variety of tools. Many of these tools are the common hand and power tools used in all types of automotive service. Other tools are more specialized and are used only for specific brake system repairs and procedures. Your measuring tools are particularly important in brake service work. Brake service involves the precision measurement of rotors, brake drums, and other components where measurements as small as one-ten-thousandth of an inch can determine the serviceability of a part.

To work quickly, safely, and efficiently, you must have the right tools for the job. Professional service technicians use quality tools and keep them clean, organized, and within reach. Select tools made of alloy steel that is heat treated for strength and durability. The tool's finish should be smooth for easy cleaning. The tool should feel comfortable and balanced in your hand.

A roll cabinet (Figure 1-1), tool chest, and smaller tote tray are standard items used to store and carry your tool set. Store cutting tools such as files, chisels, and drills in separate drawers to avoid damaging the cutting edges. Keep the most frequently used tools at hand, and keep tool sets such as wrenches, sockets, and drill bits together. Leaving a tool on a workbench is the first step in losing it. Make it a habit to return frequently used tools, such as screwdrivers, to your tool chest or tote tray after each use. Keep delicate measuring tools, such as micrometers, in their protective cases and store in a clean, dry area. After each use, clean the tool using a lightly oiled lintfree cloth.

Never dip a precision measuring tool, such as a micrometer, into a cleaning solvent unless you have time to dismantle and hand-dry each part. And never use shop compressed air to clean a precision measuring tool.

Measuring Systems

Two different measurement systems are currently used in the United States: the US Customary System (USCS) and the international system (SI) or metric system (Figure 1-2). Domestic auto manufacturers are slowly changing over to the simpler, more popular metric system. But during the changeover, cars made in the United States may have both US Customary and metric fasteners and specifications. Japan, Korea, Germany, Sweden, and most other car-producing countries use the metric system; so the import vehicles you work on will require metric-sized wrenches and sockets. Fortunately, vehicle specifications and tightening torques are normally listed in both meters and inches, so tools such as micrometers and torque wrenches can be based on either system.

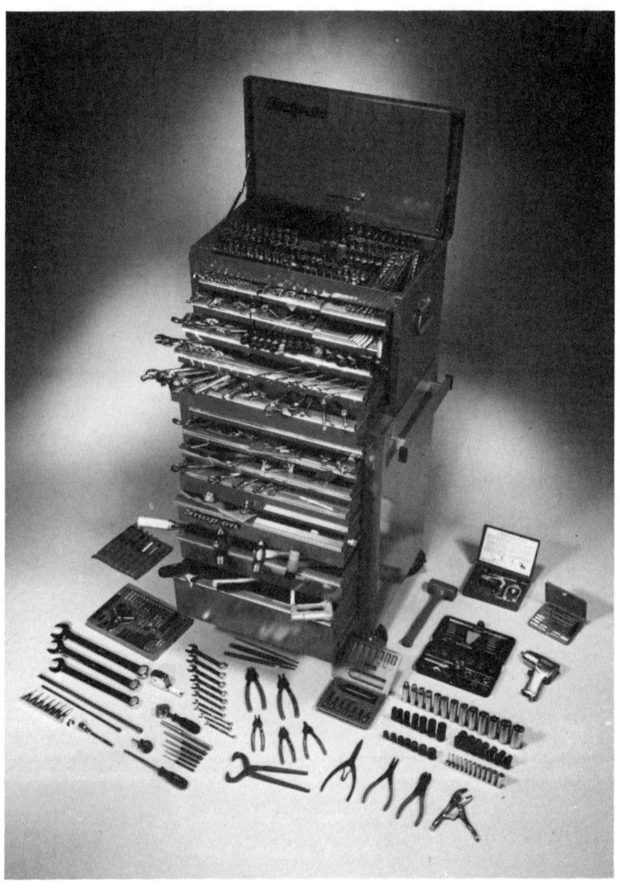

Figure 1-1 Professional set of automotive service technician hand tools (Courtesy of Snap-on Tools Corporation)

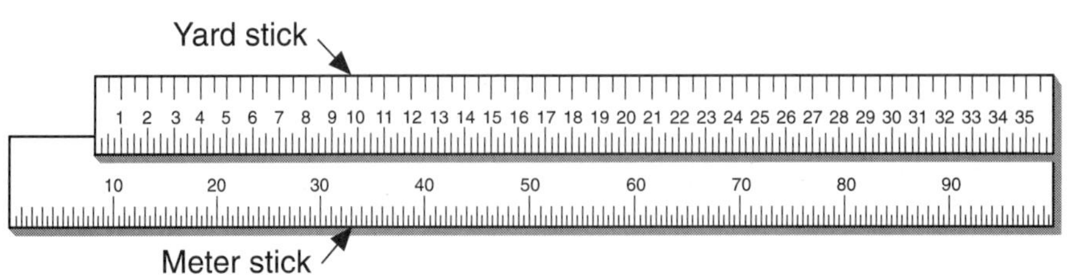

Figure 1-2 Comparison of a US Customary yard and the metric system meter

US Customary System

In the US Customary System, the basic unit of linear measurements for automotive work is the inch. The inch can be broken down into fractions, such as those used to designate wrench or socket sizes (1/4, 5/16, 1/2, 9/16, and so on).

For component size and tolerance measurements, the inch is commonly broken down by tenths, hundredths, and even thousands. When an inch is divided in this way, the measurement is written down using a decimal point. A single digit to the right of the decimal point indicates tenths of an inch (0.1 in.). Tenths of an inch can be further divided by ten into hundredths of an inch, which is written using two digits to the right of the decimal point (0.01 in.). The division after hundredths is thousandths (0.001 in.), followed by ten-thousandths (0.0001 in.).

Metric System

The basic unit of linear measurement in the metric system is the meter. For exact measurements, the meter is divided into units of ten. One-tenth of a meter is called a decimeter (dm), one-hundredth of a meter a centimeter (cm), and one-thousandth of a meter a millimeter (mm). One dm is written as 0.1 m, 1 cm as 0.01 m, and 1 mm as 0.001 m. The millimeter is the most common metric measurement used in automotive work. Metric wrenches are sized by millimeter; e.g., 10 mm, 12 mm, and so on.

WARNING: Never use a metric wrench or socket on a US Customary bolt or nut, or a Customary wrench on a metric bolt. The wrench or socket will always be slightly oversized and will slip off or strip the nut.

Common Hand Tools

At the beginning of each chapter in this manual, a list of the basic tools needed to perform the work covered is given in the margin. Any special tools required to complete a particular job will be listed in the margin next to the procedure and included in the procedure. Both basic and specialized brake tools are discussed in this chapter.

Wrenches

To work on today's brake systems, you should have complete sets of both USCS and metric wrenches. Most older domestic cars are assembled using bolts and nuts that require wrenches of USCS sizes. USCS or "standard" wrenches are most commonly sized in increments of 1/16 of an inch. Most import and newer domestic cars require metric-sized wrenches. These wrenches are sized in increments of 1 mm.

Wrenches are either boxed or open-ended. A box wrench completely encircles a nut or bolt head (Figure 1-3). It is less likely to slip and cause damage or injury. But when clearances are very tight, it is often not possible to place a box wrench around the nut or bolt.

Box wrenches are available in either 6 or 12 points. Twelve-point wrenches allow you to work in tighter areas than 6-point wrenches.

Open-ended wrenches solve that problem. They have open squared ends and grasp only two of the nut's four flats (Figure 1-4). They are more likely to slip than box wrenches but are the only alternative in a tight spot. An open-end wrench is normally the best tool for turning the nut down or holding a bolt head.

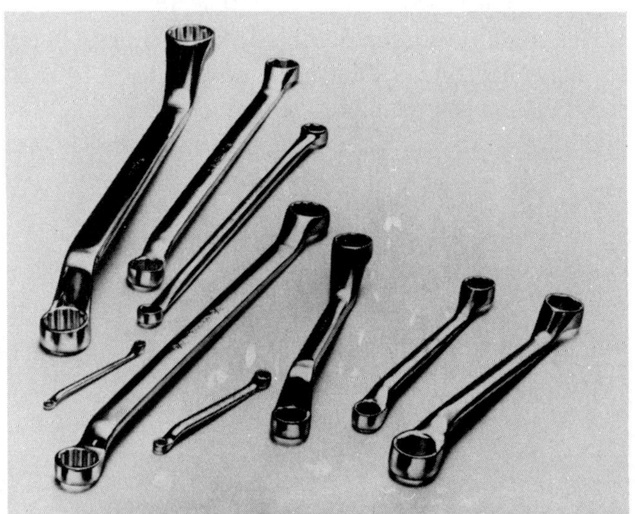

Figure 1-3 Assortment of box-end wrenches (Courtesy of Snap-on Tools Corporation)

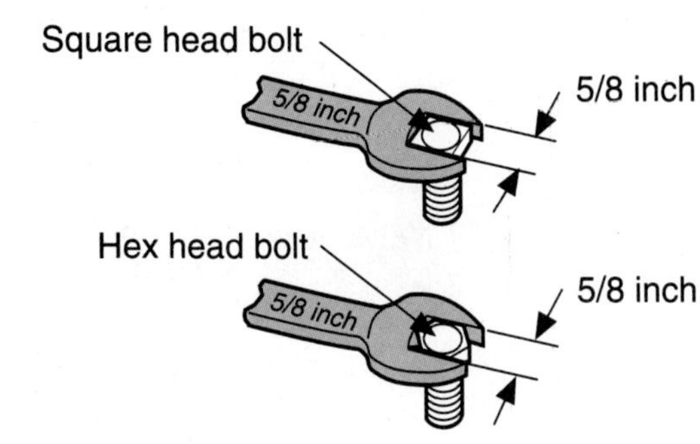

Figure 1-4 An open-end wrench grips only two faces of a bolt or nut head.

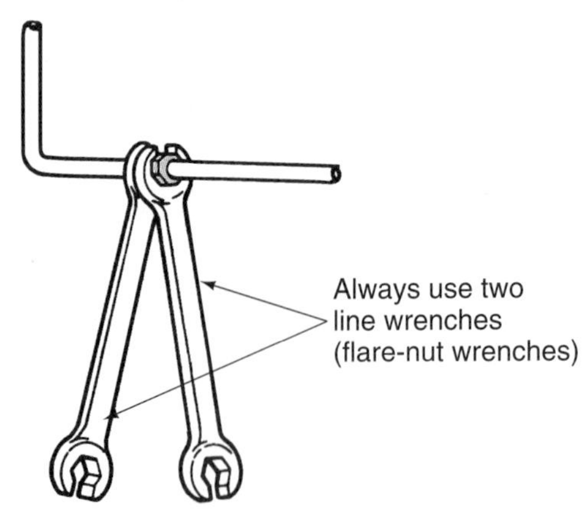

Figure 1-5 Using flare nut wrenches to loosen a flare fitting or bleeder screw fitting (Courtesy of Raybestos Division, Brake Systems Inc.)

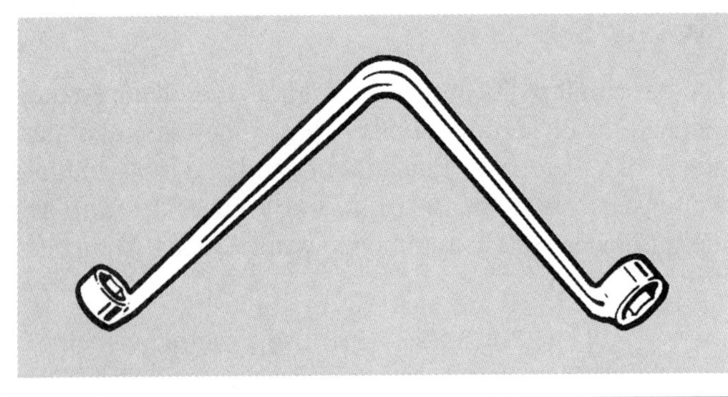

Figure 1-6 Special brake bleeder valve wrench (Courtesy of U.S. Navy)

To reduce cost and storage space, it is recommended that you have a set of combination wrenches that have an open-end wrench on one end and a box wrench on the other. Both ends are sized the same and can be used interchangeably on the same nut or bolt.

Flare Nut (Line) and Bleeder Screw Wrenches

To loosen or tighten brake line or tubing fittings, special flare nut wrenches should be used. Using open-ended wrenches on these fittings will tend to round the corners of the nut, which are typically made of soft metal and can distort easily. Flare nut wrenches surround the nut and provide a better grip on the fitting. They also have a section cut out so that the wrench can be slipped around the brake line and dropped over the flare nut (Figure 1-5).

In brake work, a special bleeder wrench may be used for opening bleed screws (Figure 1-6). Some service manuals specify a certain size box-end wrench for bleeder screw removal.

Socket Wrenches

Socket wrenches consist of two parts: the socket, which fits over the nut or bolt, and the handle or racket, which is used to drive the socket (Figure 1-7). Your tool set should have a full set of both USCS and metric sockets combined with a ratchet handle and several extensions (Figure 1-8). A 6-point has stronger walls and improved grip on a bolt compared to a normal 12-point socket. Six-point sockets are mostly used on fasteners that are rusted or rounded. Eight-point sockets are

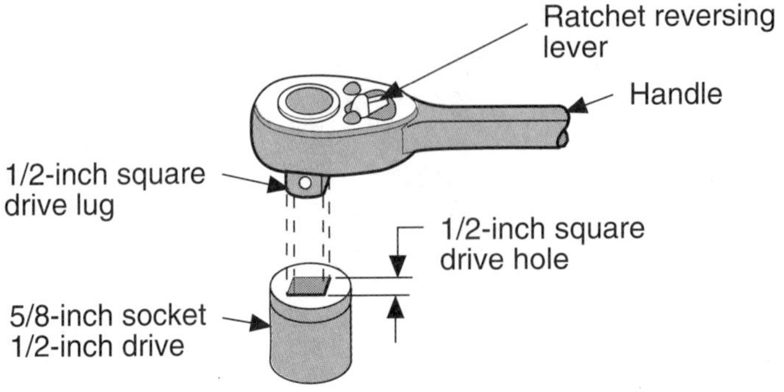

Figure 1-7 Parts of a socket wrench

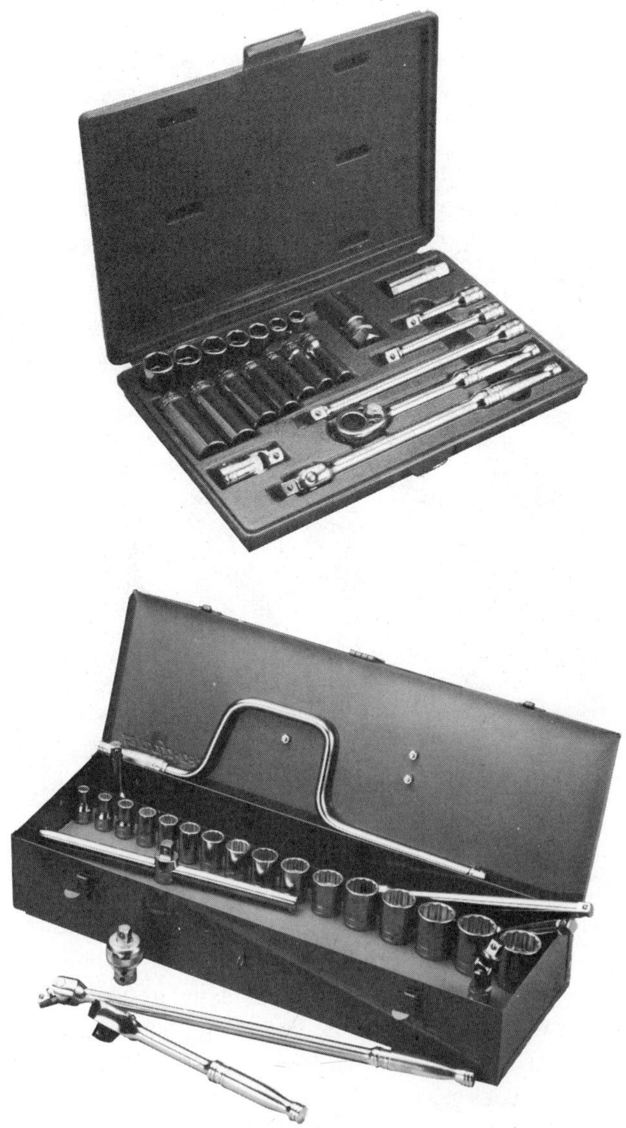

Figure 1-8 Typical 3/8-inch and 1/2-inch socket wrench sets (Courtesy of Snap-on Tools Corporation)

available to use on square nuts or square-headed bolts. Most brake work can be handled using standard depth sockets, although deep sockets are useful in other areas of automotive work. Swivel sockets allow you to turn a fastener at an angle. The square driving hole in the base of the socket that the ratchet attaches to can measure 3/8 inch, 1/4 inch or 1/2 inch.

WARNING: Deep-well sockets are also good for reaching nuts and bolts in limited access areas. Deep-well sockets should not be used when a regular-size socket will work. The longer socket develops more twist torque and tends to slip off the fastener.

Ratchets or drivers are available in many lengths and styles. Extensions allow you to put the handle in the best position while working. They range from 1 inch to 3 feet long.

To convert from foot-pounds to newton-meters, simply multiply by 1.356. Divide newton-meters by 1.356 to convert to foot-pounds.

Torque Wrenches

Torque wrenches measure the torque or twisting force applied to a fastener. Many fasteners such as caliper mounting bolts, bleeder screws, brake hose-to-caliper fasteners, and others must be tightened to meet a torque specification that is expressed in foot-pounds (USCS) or newton-meters

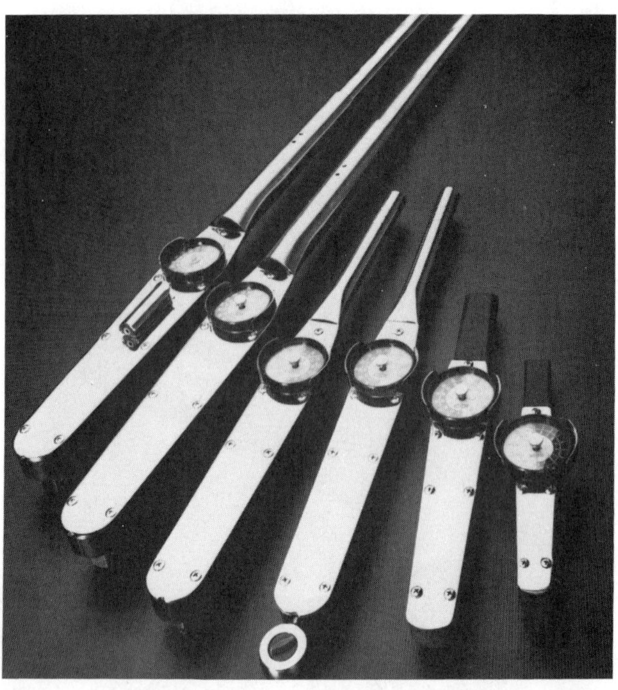

Figure 1-9 Dial-type torque wrench

(metric). A foot-pound is the work or pressure accomplished by a force of 1 pound through a distance of 1 foot. A newton-meter is the work or pressure accomplished by a force of 1 kilogram through a distance of 1 meter.

For automotive brake work, most torque requirements are measured in foot-pounds or newton-meters. Some bleeder screws may require the use of an inch-pound torque wrench.

Torque wrenches are available with 1/4-, 3/8-, and 1/2-inch drives. There are several types of torque wrenches. Common designs include the dial (Figure 1-9), beam (Figure 1-10), breakover (Figure 1-11), and the digital readout types. These types all have a scale that measures turning effort as the fastener is tightened. With a breakover torque wrench, the desired torque is dialed in.

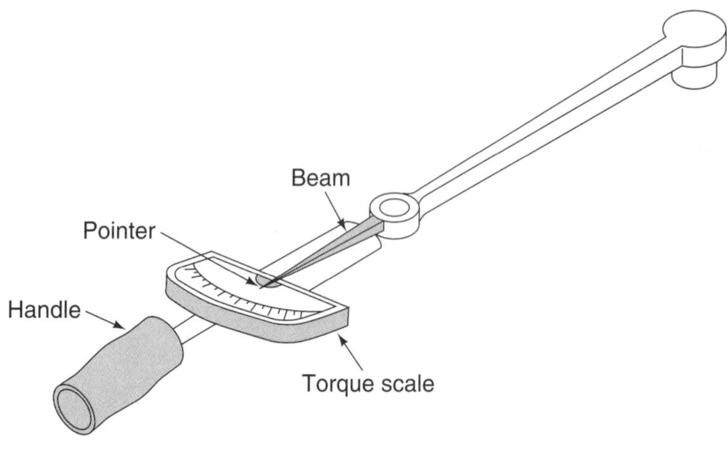

Figure 1-10 Beam-type torque wrench

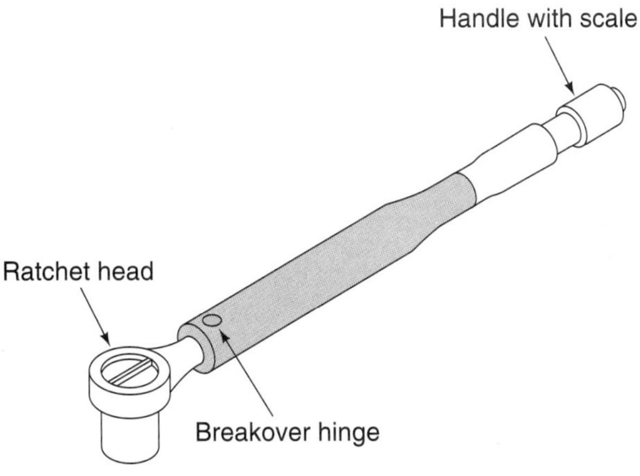

Figure 1-11 Breakover hinge torque wrench

The wrench then makes an audible click when you have reached the correct force. For accurate torque readings, the threads on the bolt and hole must be clean. Unless the service manual cautions against its use, a high-temperature lubricant should be applied to the threads and to the area where the nut or capscrew head contacts the part. Whenever possible, pull on the wrench rather than pushing on it.

Fasteners

While not true "tools," bolts and fasteners are part of everyday automotive work and you must understand their characteristics and grading system. Figure 1-12 lists (a) US Customary and (b) metric system bolt terminology. The SAE Grade and strength markings for bolts used in automotive and other industrial assemblies are shown in Table 1-1. Service manuals also list important fastener information. Remember, using an incorrect fastener or a fastener of poor quality can result in dangerous failures and personal injury.

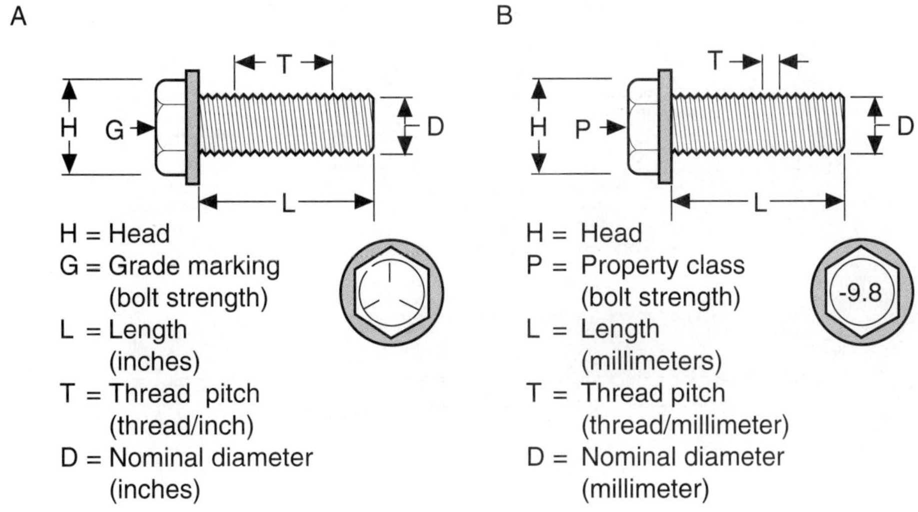

Figure 1-12 (A) US Customary and (B) metric bolt terminology

Table 1-1 STANDARD BOLT STRENGTH MARKINGS

SAE Grade Markings	(no lines)	(3 lines)	(4 lines)	(5 lines)	(6 lines)
DEFINITION	No lines: unmarked indeterminate quality SAE grades 0-1-2	3 Lines: common commercial quality Automotive and AN bolts SAE grade 5	4 Lines; medium commercial quality Automotive and AN bolts SAE grade 6	5 Lines: rarely used SAE grade 7	6 Lines: best commercial quality NAS and aircraft screws SAE grade 8
MATERIAL	Low carbon steel	Med. carbon steel tempered	Med. carbon steel quenched and tempered	Med. carbon alloy steel	Med. carbon alloy steel quenched and tempered
TENSILE STRENGTH	65,000 psi	120,000 psi	140,000 psi	140,000 psi	150,000 psi

Screwdrivers

The most commonly used screwdrivers are the flat blade and Phillips types. A blade-style fits into a straight slot in the head of the screw (Figure 1-13A).

WARNING: A screwdriver should never be used as a chisel, punch, or pry bar. Screwdrivers are not made to withstand blows or bending pressures. In some brake service procedures, it is acceptable to use a screwdriver to help lift off dust boots and sealing rings as long as the work is done carefully and the tool is not put under stress.

A Phillips screwdriver has a cross point tip. The cross point has four surfaces that fit into a similar pattern cut into the head of the screw (Figure 13B). The increased tip-to-screw head contact makes the Phillips less likely to slip out and increases gripping power and stability.

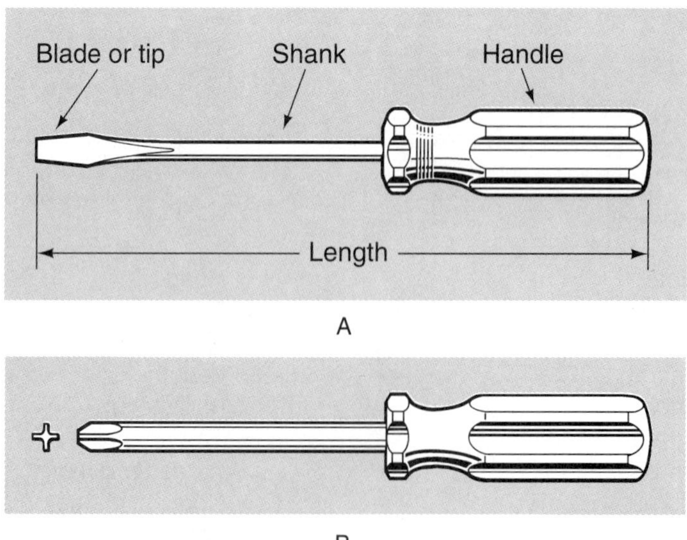

Figure 1-13 (A) Flat blade and (B) Phillips screwdrivers (Courtesy of U.S. Navy)

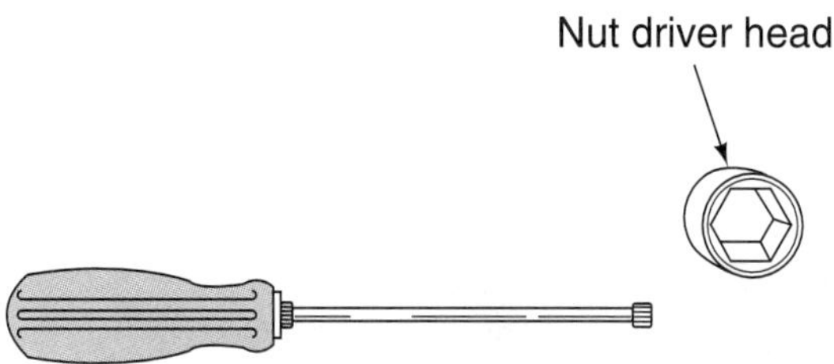

Figure 1-14 Nutdrivers combine the ease of a screwdriver with the holding power of a socket for removing small diameter nuts and bolts.

Your tool set should include both flat blade and Phillips drivers in lengths from 2-inch "stubbies" to 12-inch screwdrivers.

Nutdrivers

Nutdrivers combine the ease of use of a screwdriver with the superior gripping power of a socket (Figure 1-14). They are normally used on small nuts and bolts and are particularly suited for reaching hard to get at locations. A set of nutdrivers, while not an absolute necessity, is a handy addition to a basic tool set.

Pliers

The two most commonly used pliers are combination or slip-joint pliers 8 or 9 inches long, and diagonal cutters 7 inches long. Interlocking slip-joint pliers are a general gripping and pulling tool for jobs such as pulling out caliper guide pins and removing brake hose lock clips (Figure 1-15). The slip joint can be set to two jaw opening settings.

Diagonal cutters are used to cut wire and remove cotter pins (Figure 1-16). Other plier designs are also used in servicing a vehicle's brake system. These include slip-joint, needle-nose, adjustable joint, and snap ring pliers (Figure 1-17).

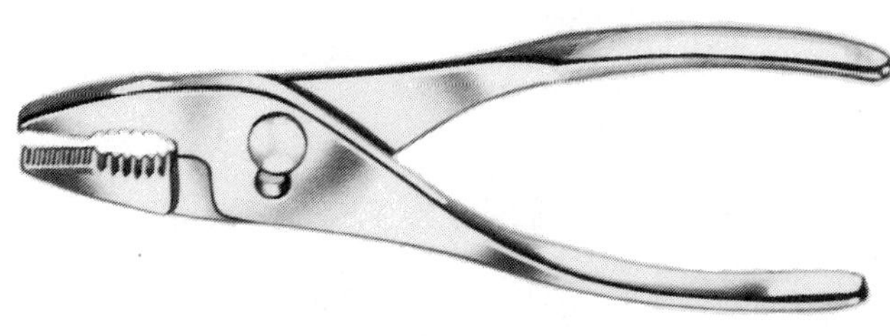

Figure 1-15 Combination or slip-joint piers. (Courtesy of Snap-on Tools Corporation)

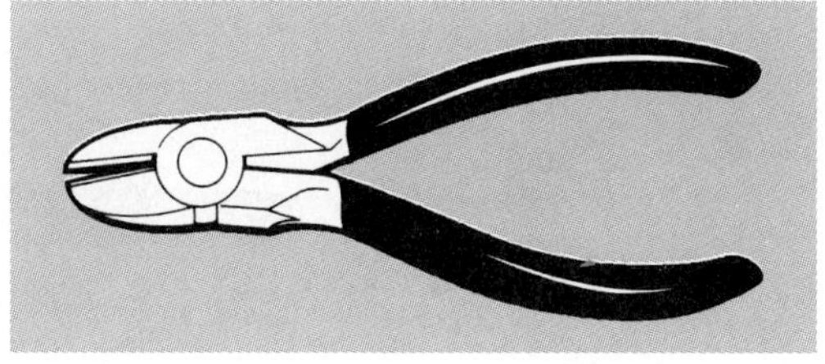

Figure 1-16 Diagonal cutters (Courtesy of U.S. Navy)

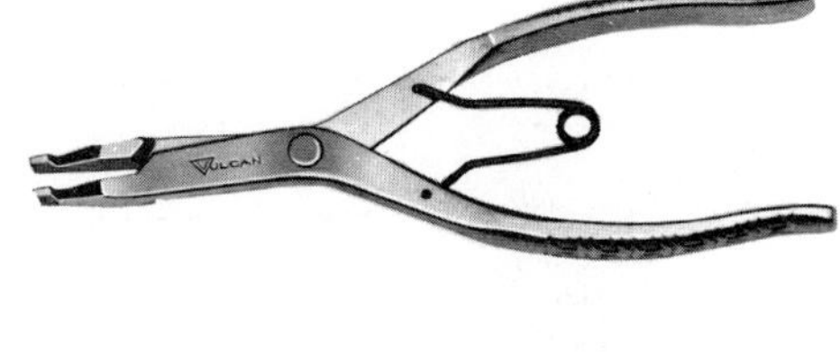

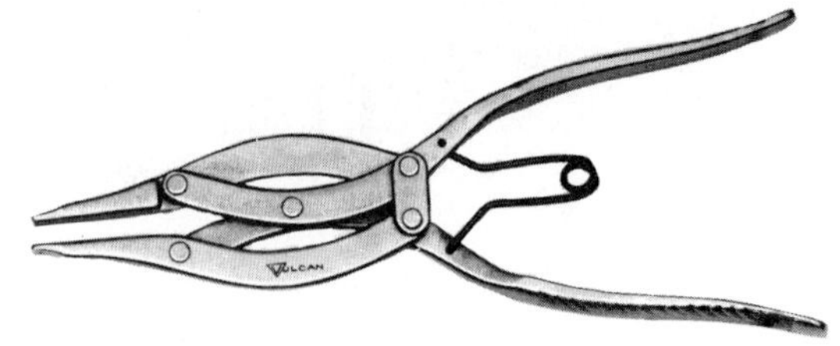

Figure 1-17 Snap ring pliers

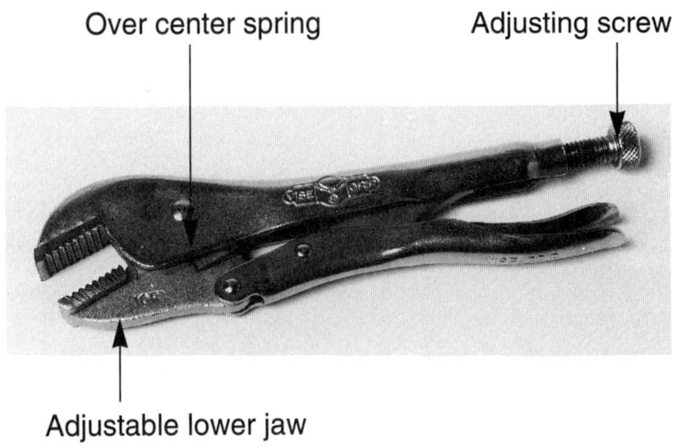

Figure 1-18 Vise grip pliers

Vise grip pliers are indispensable for most automotive work (Figure 1-18). They are excellent for gripping small parts or can serve as your "third hand" in unique service situations.

Hammers and Mallets

In brake work, a ball-peen hammer is used to tap off pads, bend over outer pad ears, or drive punches or drifts when tapping out caliper retaining keys, mounting pins, and so on. A 12- to 16-ounce ball-peen hammer is best for brake work (Figure 1-19). You should also have a plastic, leather, or brass-faced mallet (Figure 1-20). Always use a soft mallet when tapping parts apart or aligning parts together. A soft-faced mallet is also best when striking dust boot installer tools and the like, as it will not damage the part.

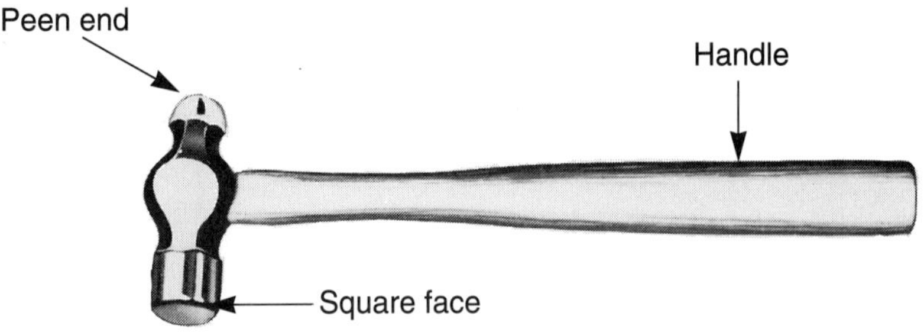

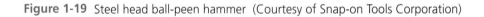

Figure 1-19 Steel head ball-peen hammer (Courtesy of Snap-on Tools Corporation)

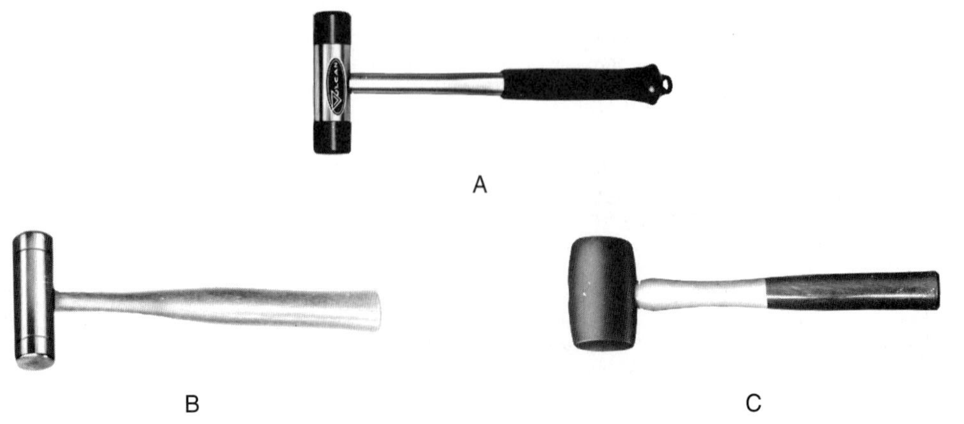

Figure 1-20 Soft-faced mallets: (A) plastic, (B) brass, and (C) rubber

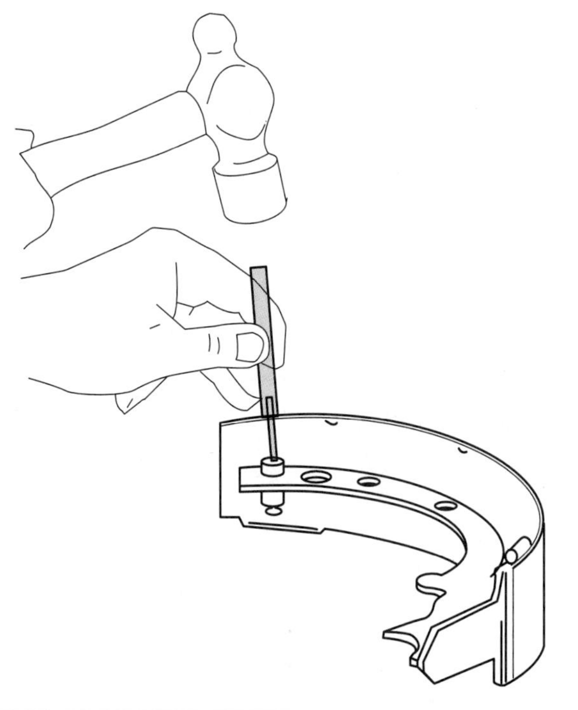

Figure 1-21 Using a drift to drive out lever pins on a brake shoe (Courtesy of Raybestos Division, Brake Systems Inc.)

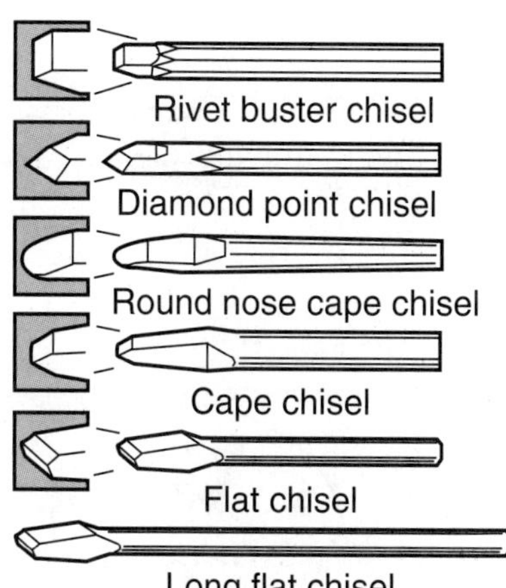

Figure 1-22 Chisel head types

Punches and Chisels

Drift punches are used to remove drift and roll pins. Brass drifts should be used whenever you are concerned about possible damage to the pin or surface surrounding the pin (Figure 1-21). Tapered punches are used to line up bolt holes. Chisels are used to cut off rivet heads, bolts, and rusted nuts. Your tool set should include flat, cape, round-nose cape, and diamond-point chisels (Figure 1-22).

Files

Files are used to clean metal surfaces, remove metal, and deburr parts. The most commonly used files are the half-round and flat with either single-cut or double-cut designs (Figure 1-23). A single-cut file has its cutting grooves lined up diagonally across the face of the file. The cutting grooves of a double-cut file run diagonally in both directions across the face. Double-cut files are considered first cut or roughening files because they can remove large amounts of metal. Single-cut files are considered finishing files because they remove small amounts of metal.

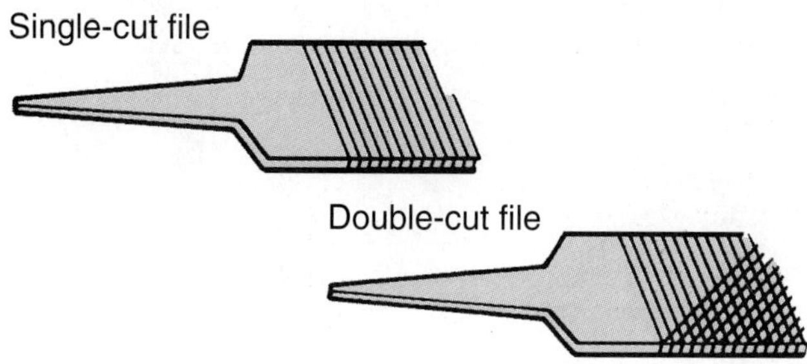

Figure 1-23 Common file designs

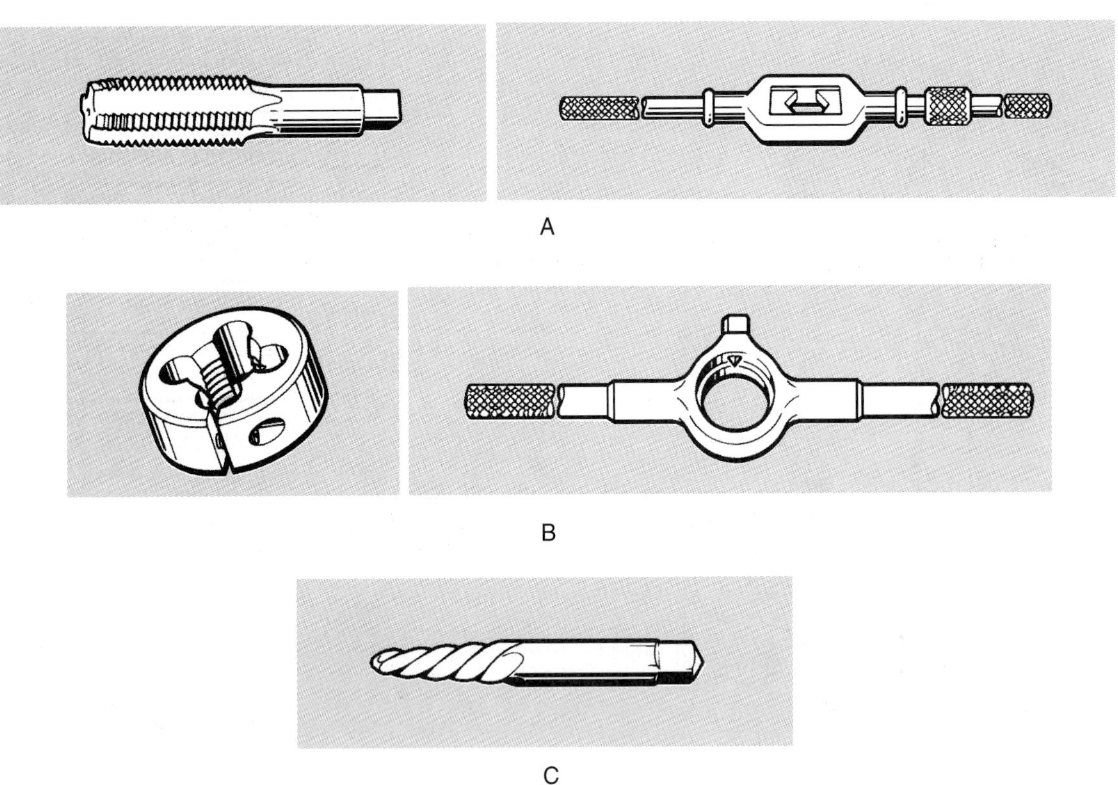

Figure 1-24 (A) Tap and tap wrench, (B) die and die stock, and (C) screw extractor (Courtesy of U.S. Navy)

Rotary files are chucked in a pneumatic or electric drill and are useful for cleaning blind holes or recesses.

Taps and Dies

Taps are used for cutting internal threads in a bore. Dies are used to cut external threads on bolts and screws when a replacement fastener is not handy. Invest in two sets of taps and dies: one USCS and one metric (Figure 1-24).

Vise

A bench-mounted vise is needed to secure master cylinders for bench bleeding. It is also used to hold calipers, wheel cylinders, and other small-to-medium sized components (Figure 1-25).

CUSTOMER CARE: Remind your customers to periodically check the brake fluid level in their master cylinder reservoir, particularly on older vehicles that may develop slow leaks. Leaking fluid is a sign of trouble. Stress that service is needed whenever this problem is noticed.

Special Tools

Many tools are designed for a specific purpose. Car manufacturers and specialty tool companies work closely together to design and manufacture special tools required to repair cars. Most special tools required for brake service work are listed in the appropriate service manuals. Special tools

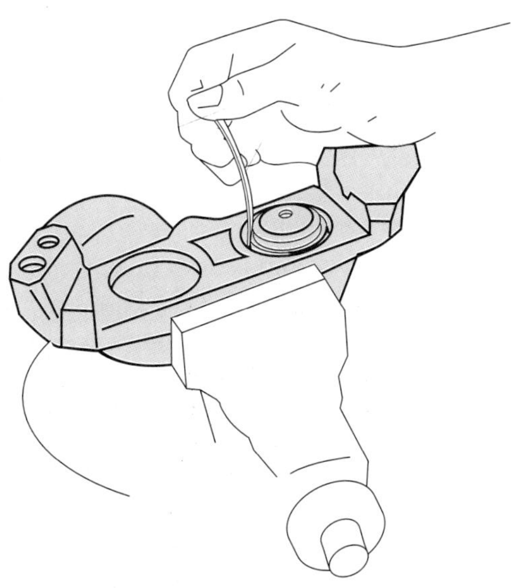

Figure 1-25 A bench-mounted vise is ideal for holding small and medium-sized parts and assemblies. (Courtesy of Raybestos Division, Brake Systems Inc.)

needed for the procedures are discussed in this manual. Examples of special tools used in brake work include:

- Brake spring pliers
- Drum micrometer
- Bleeder wrench
- Push rod height gauge
- Tandem vacuum booster separating tool
- Boot seal installer
- Power piston seal protector
- Brake pedal effort gauge
- Bleeder adapters
- Tube cutters
- Double-flare forming tool
- ISO flare forming tool
- Rotor micrometer
- Master cylinder bleeder tubes
- Line and port plugs
- Bench bleeding syringe
- Drum brake star wheel adjustment tool
- Brake shoe adjusting gauge
- Drum brake adjusting tool
- Dust boot seating tool
- Sleeve installer
- Return-spring tool

Some brake specialty tools are shown (Figure 1-26).

Figure 1-26 Examples of brake service specialty tools

Miscellaneous Tools and Equipment

The following tools and items are regularly used in brake system service work.

- Dowels for pushing plunger in master cylinder
- Clear jar or containers for bleeding system
- C-clamp for forcing brake piston back into bore
- Siphon to remove brake fluid from master cylinder
- 80-grit sandpaper
- Denatured alcohol for cleaning parts
- Stud cutter for removing studs on drum
- Press for installing new stud on drum
- Wood dowels to remove wheel bearing grease seals, force out bushings, and so on
- Wire for tying up calipers
- Clamps to hold pads in place

Power Tools

Most shops prefer pneumatic tools over electric tools because they produce more torque, weigh less, and require less maintenance.

Power tools, whether electric or pneumatic (air), save time and energy. In most cases, power tools are typically provided by the shop.

CAUTION: Carelessness or mishandling of power tools can cause serious injury. Make sure you know how to operate a tool before using it.

Impact Wrenches

An impact wrench hammers or impacts a nut or bolt loose or tight (Figure 1-27). Removing and installing lug nuts on wheels is an excellent job for an impact wrench. Light-duty impact wrenches are available in three drive sizes—1/4 in., 3/8 in., and 1/2 in.—and two heavy-duty sizes—3/4 in. and 1 in.

WARNING: Do not use an impact wrench to tighten critical parts or parts that may be damaged by the hammering force of a wrench.

Air Ratchet

Air ratchets are used for general disassembly or reassembly work. Because they turn sockets without a jarring impact force, air ratchets can be used on most parts and with ordinary sockets. Air ratchets typically have a 3/8-in. drive.

Keep in mind that air ratchets are not torque sensitive. After snugging the fastener down with an air ratchet, a torque wrench must be used to set the final fastener tightness.

Figure 1-27 Impact wrench

Figure 1-28 Typical brake lathe (Courtesy of PMC Automotive Brake Equipment Division)

Brake Lathes

Brake lathes are used for turning disc brake rotors or drum brake drums (Figure 1-28). Turning involves cutting away very small amounts of material to restore the surface of the rotor or drum.

Lifting Tools

Lifting tools are necessary tools for most brake system repair procedures. These tools are usually provided by the shop. Correct operating and safety procedures should always be followed when using lifting tools.

Jacks

Jacks are used to raise a vehicle off the ground. For shop use two basic designs are used—hydraulic and pneumatic. The most popular jack is the hydraulic floor jack, which is classified by the weight it can lift: 1-1/2, 2, 2-1/2 tons, and so on. A hydraulic floor jack is operated by pumping the handle up and down (Figure 1-29). The second type of portable floor jack used operates on compressed air. Pneumatic jacks are operated by controlling air pressure to the jack.

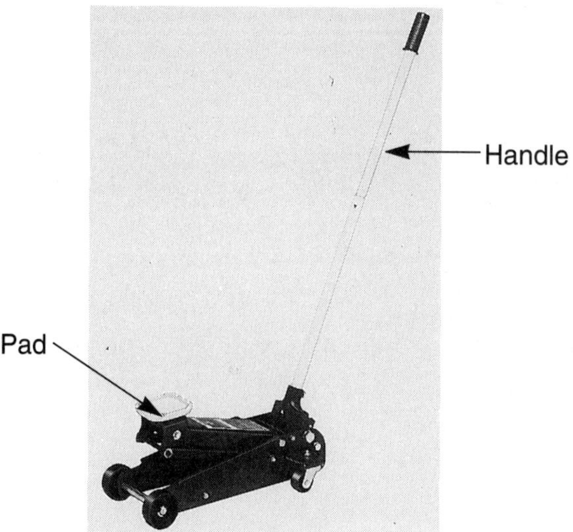

Figure 1-29 Typical hydraulic floor jack (Courtesy of Lincoln St. Louis)

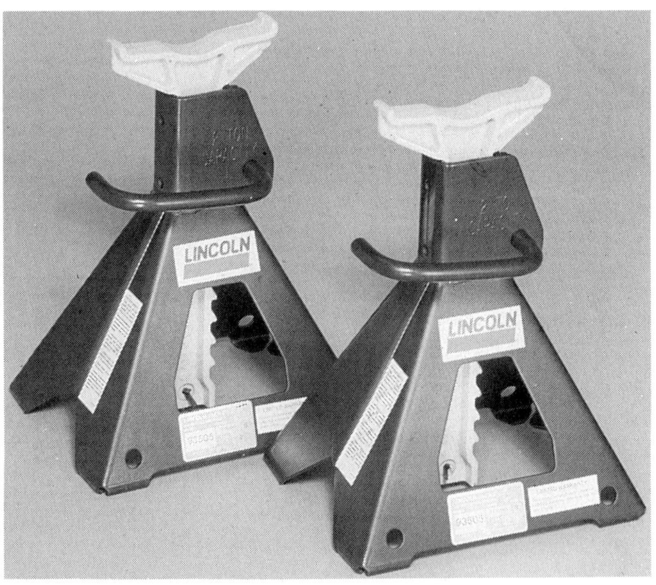

Figure 1-30 Jack safety stands (Courtesy of Lincoln St. Louis)

Safety Stands

Whenever a vehicle is raised by a jack, it must be supported by safety stands (Figure 1-30). Never work under a car with only a jack supporting it; always use safety stands. The hydraulic seals of a floor jack can let go and allow the vehicle to drop.

Check in the vehicle service manuals for the proper locations for positioning the jack and the safety stands. Always follow these guidelines.

Hydraulic Lift

When used correctly, a hydraulic floor lift is the safest, most convenient lifting tool. It allows you to raise the vehicle high enough to work comfortably and quickly (Figure 1-31). Various safety features prevent a hydraulic lift from dropping if a seal does leak or if air pressure is lost. Before lifting a vehicle, make sure the lift is positioned correctly.

Measuring Tools

Many of the procedures involved in brake work require exact measurements of parts and clearances. Accurate measurements require the use of precision measuring tools capable of measuring to the thousandth of an inch. For example, the acceptable lateral runout of a brake rotor may be as little as 0.004 in. The minimum acceptable thickness variation may be as small as 0.0005 in. Tools for measuring such small increments are delicate instruments and should be handled with great care. Never strike, pry, drop, or force these tools. Clean them before and after every use.

SERVICE TIP: Periodically check all measuring tools against known good equipment or tool standards. This ensures they are working properly and giving accurate measurements.

Many different measuring devices are used by automotive technicians. This chapter covers only those commonly used to service brake systems.

Steel Ruler or Tape Measure

A steel ruler or tape measure is needed to measure brake pedal travel.

Figure 1-31 Twin-post hydraulic floor lift

Micrometer

Micrometers are used to measure the outside diameter of an object or the inside diameter of a bore (Figure 1-32). Disc brake rotor micrometers are used to measure rotor thickness. The throats on these micrometers are deeper than on standard micrometers to accommodate the diameter of the

Rotor micrometers are equipped with a pointed anvil to make it easier to check rotor wear and measure the depth of any grooves in the rotor surface.

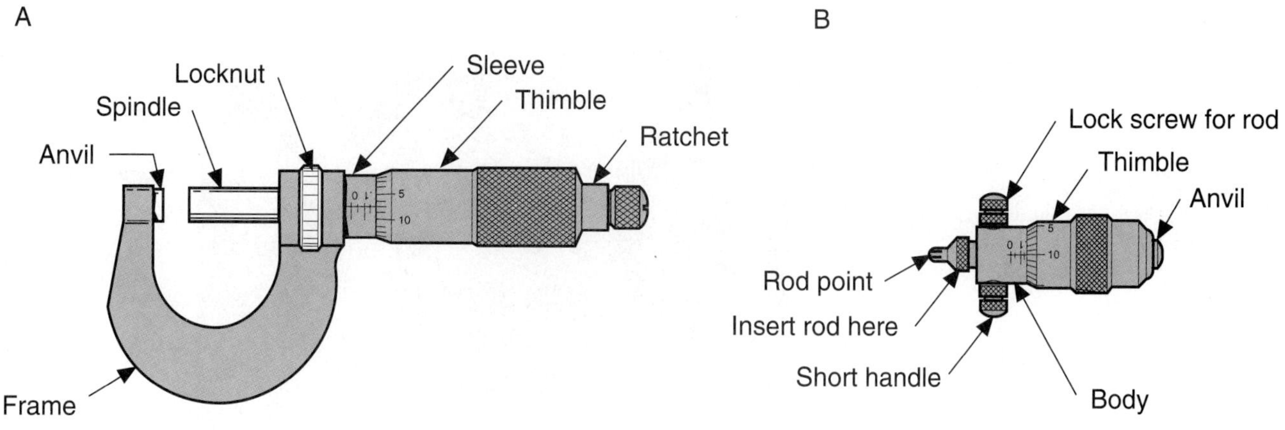

Figure 1-32 Major components of (A) an outside and (B) an inside micrometer

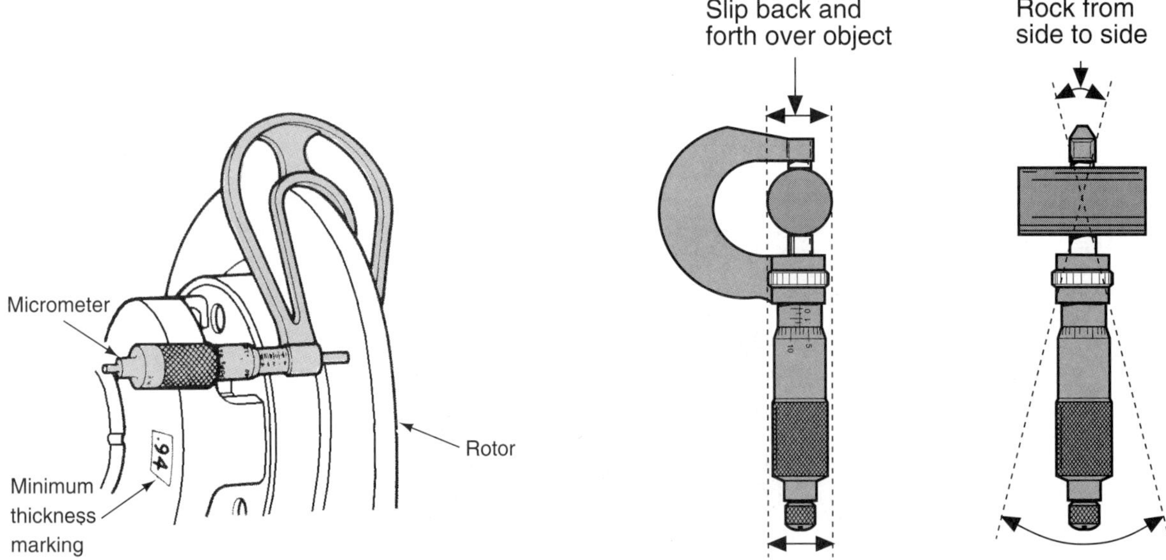

Figure 1-33 Brake rotor micrometer

Figure 1-34 Using a standard outside micrometer

rotor. The anvils are pointed (Figure 1-33). The major components and markings of a micrometer include the frame, anvil, spindle, locknut, sleeve, sleeve numbers, sleeve long line, thimble marks, thimble, and ratchet. Micrometers are calibrated in either inch or metric graduations and are available in a range of sizes.

To measure small objects with an outside micrometer, open the jaws of the tool and slip the object between the spindle and the anvil. While holding the object against the anvil, turn the pressure on the thimble to allow the object to just fit between the anvil and the spindle (Figure 1-34). The object should slip through with only a very slight resistance. When a satisfactory feel is reached, lock the micrometer. Each graduation on the sleeve represents 0.025 inch, so begin reading the measurement by counting the visible lines on the sleeve and multiply that number by 0.025 (Figure 1-35). The graduations on the thimble assembly define the area between the lines on the sleeve; therefore, the number indicated on the thimble should be added to the measurement shown on the sleeve. This sum is the outside measurement of the object.

To measure larger objects such as a disc brake rotor, hold the frame of the micrometer and slip it over the object. Continue to slip the micrometer over the object until you feel a very light resistance, while at the same time rocking the tool from side to side to make certain the spindle cannot be closed any further. Then, lock the micrometer and take a measurement reading.

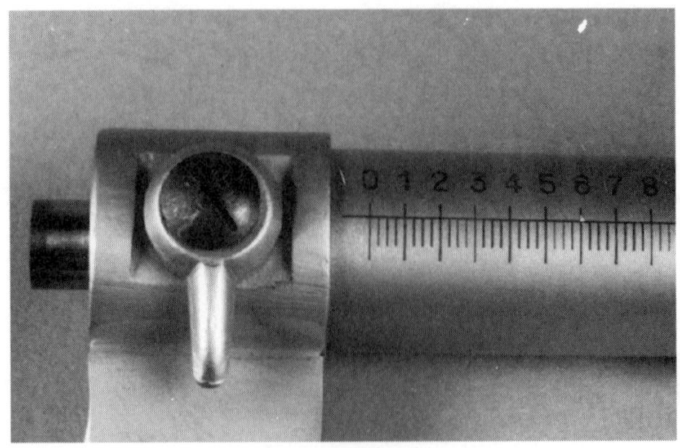

Figure 1-35 Reading the micrometer scale

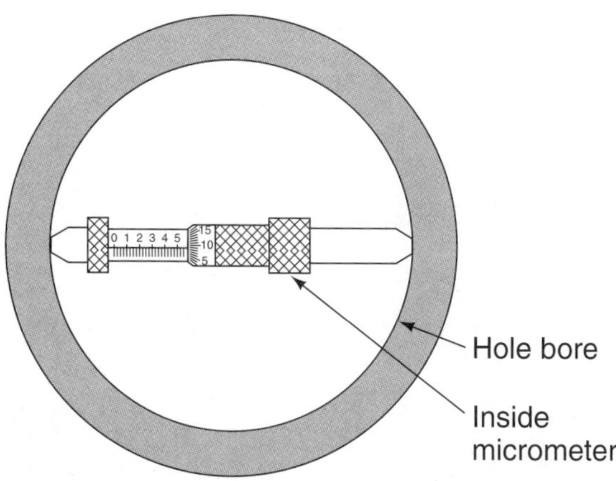

Figure 1-36 Using a standard inside micrometer

Inside micrometers also have a micrometer body, thimble, sleeve, and ratchet. They have interchangeable spindles of various lengths to use in holes of different sizes. To use this type of micrometer, select a spindle that will just fit into the bore when it is threaded to the thimble and sleeve. Then, extend the thimble until the spindle touches one side of the bore and the thimble end touches the other side (Figure 1-36). Read the measurement in the same way as you would read an outside micrometer. It is good practice to take measurements at two or three different locations as a check for bore out-of-roundness.

Drum Micrometer

A drum micrometer is used to measure the inside diameter of a brake drum (Figure 1-37).

SERVICE TIP: Like all tools, measuring tools should only be used for the purpose for which they were designed. Some instruments are not accurate enough for very precise measurements; others are too accurate to be practical for less critical measurements.

Precision guages are available to calibrate drum micrometers. Calibration ensures accuracy to the thousandths of an inch.

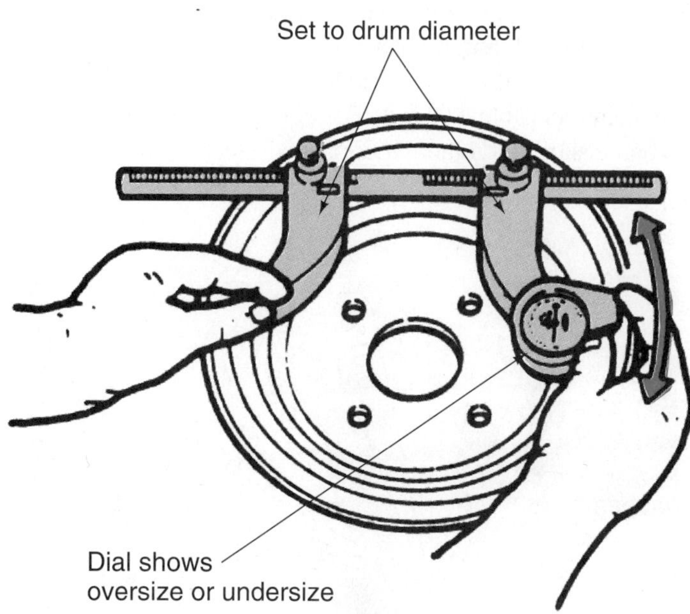

Figure 1-37 Using a drum micrometer to measure inside drum diameter (Courtesy of Raybestos Division, Brake Systems Inc.)

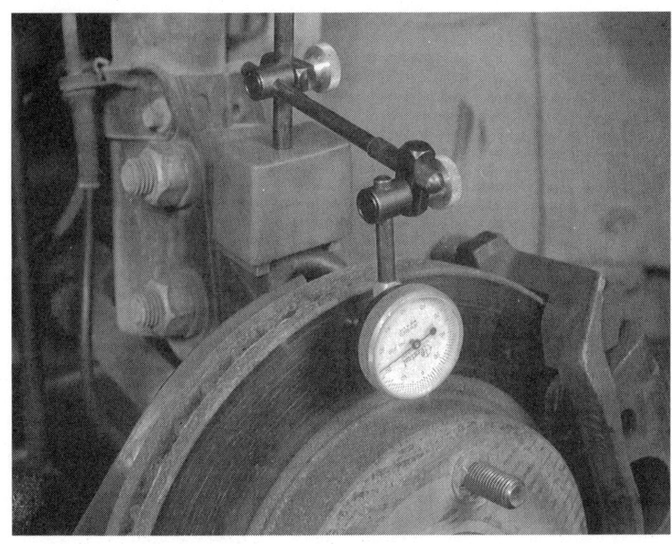

Figure 1-38 Using a dial indicator to measure lateral runout on a disc brake rotor

Dial Indicator

A dial indicator is used to measure variations or movements along the surface of an object. Dial indicators are normally calibrated in 0.001-in. increments. Metric indicators are also available and measure in 0.01-mm increments. The most common use for dial indicators in brake work is suring lateral runout in brake rotors (Figure 1-38). A dial indicator uses a spring-loaded rod that contacts the part being measured. Once the rod is in contact and the indicator is locked into position, the dial gauge is set to zero. The rotor is then rotated. All changes or movements will show on the indicator.

Feeler Gauge

A feeler gauge is a thin strip of metal of a known thickness. A feeler gauge set is a collection of these strips, each with a slightly different thickness (Figure 1-39). A steel feeler gauge set usually contains strips of 0.002- to 0.010-in. thicknesses (in increments of 0.001 in.) and strips of 0.012- to 0.024-in. thicknesses (in increments of 0.002 in.). Feeler gauges are used to check piston-to-wheel cylinder bore, drum to brake shoe, and other types of small clearances down to the thousandths of an inch.

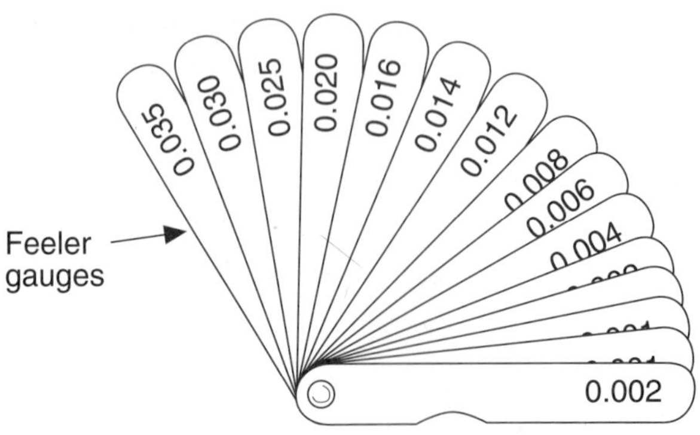

Figure 1-39 Feeler gauge set

Figure 1-40 Using a depth gauge to check lining thickness

Depth Gauge

A simple depth gauge is used to check the lining thickness of drum brake shoes (Figure 1-40).

Electronic Test Equipment

With the increasing use of antilock braking systems, electronic controls have become part of the brake systems you service. For troubleshooting these systems, you will need two major pieces of electrical test equipment—a circuit tester and a digital multimeter.

Circuit Testers

Circuit testers or test lights are used to identify shorted and open electrical circuits. Low-voltage testers are used to troubleshoot 6- to 12-volt circuits commonly used in electronic control systems. A circuit tester looks like a stubby ice pick. Its handle is transparent and contains a light bulb. A probe extends from one end of the handle and a ground clip and wire extend from the other end. When the ground clip is attached to a good ground and the probe is touched to a live connector, the bulb in the handle lights up. If the bulb does not light up, voltage is not available at the connector (Figure 1-41).

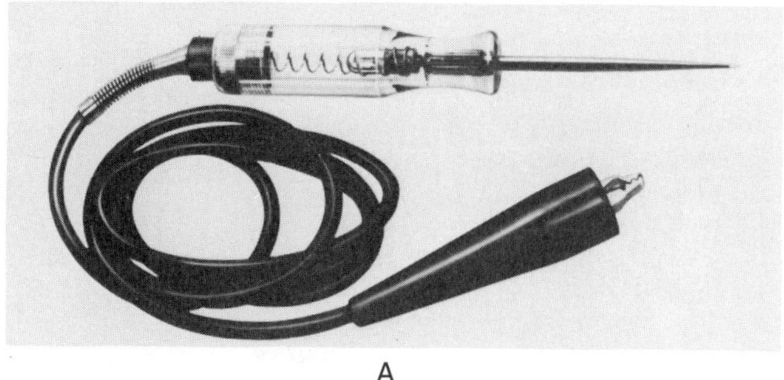

A

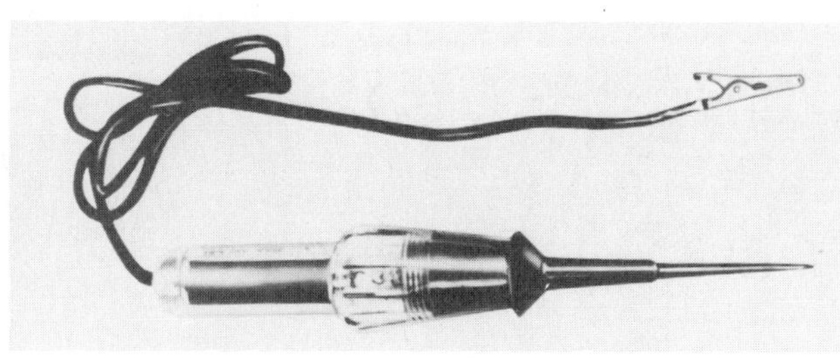

B

Figure 1-41 (A) Circuit tester or test light and (B) self-powered or continuity tester (Courtesy of MAC Tools)

Don't confuse an ordinary test lamp with a self-powered circuit tester. The first is used to test for power in a circuit. The second is only connected to unpowered circuits.

A self-powered circuit tester is called a continuity tester. It is used with the power off in the circuit being tested. It looks like a regular test light, except that it has a small internal battery. When the ground clip is attached to the ground terminal of a component and the probe is touched to the feed wire, the light will be illuminated if there is continuity in the circuit. If an open circuit exists, the light will not come on.

Multimeters

The multimeter is one of the most versatile tools used in diagnosing engine performance and electrical systems. The most common multimeter is the volt/ohm 1 milliamp meter (VOM) (Figure 1-42). As its name implies, the VOM tests voltage, resistance, and current.

Top-of-the line multimeters are multifunctional. Most test dc, ac, volts, ohms, and amperes. Several test ranges are usually provided for each of these functions. More advanced multimeters may also test engine revolutions per minute (rpm), ignition dwell, diode connection, distributor conditions, frequency, and even temperature.

Multimeters are available with either digital or analog displays. Many electronic component tests require very precise voltage measurement results. Digital multimeters (DMM) provide this accuracy by measuring volts, ohms, or amperes in tenths and hundredths of a unit. Some digital meters have multiple test ranges that must be manually selected. Others are autoranging. A typical multimeter may have 18 test ranges. It may measure dc voltage readings as small as 1/10 millivolt or as high as 1,000 volts.

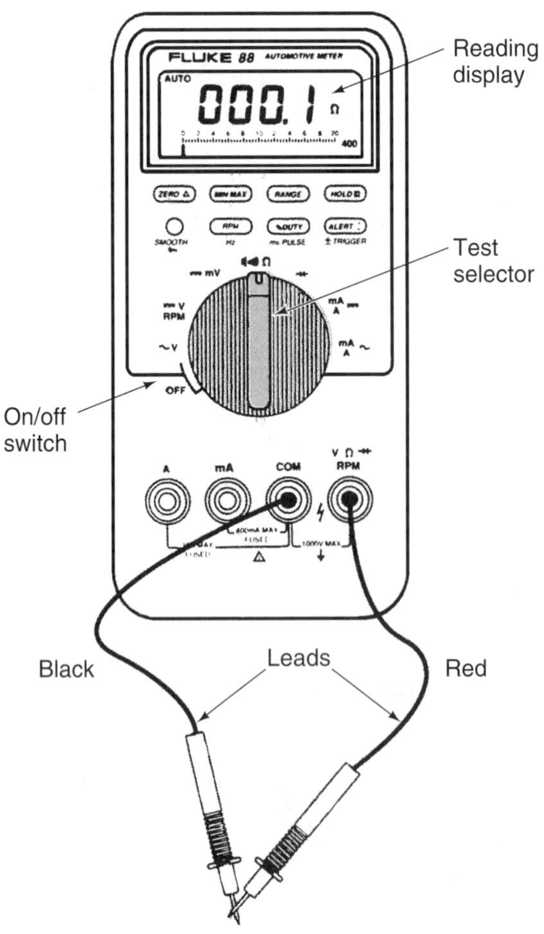

Figure 1-42 Typical multimeter for checking voltage, resistance, and current (Courtesy of Fluke 88)

Analog meters use a needle and scale to display readings and are not as accurate as digital readout equipment. Another problem with analog meters is their low internal resistance (input impedance). The low input impedance allows too much current to flow through circuits; therefore, analog meters should not be used on delicate electronic devices.

Digital meters, on the other hand, have a high input impedance, usually at least 10 megohms (10 million ohms). Metered voltage for resistance tests is well below 5 volts, reducing the risk of damage to sensitive components and delicate computer circuits.

One of the major problems with circuit checking is the vehicle's wiring harness. The introduction of *breakout boxes* allows the technician to check voltage and resistance readings between specific points within the harness.

Scan Tools

Electronic scan tools (Figure 1-43) are invaluable in the servicing of most modern antilock brake systems. The scan tool plugs into the brake system's control computer and allows the technician to read out stored or current digital trouble codes that help pinpoint system problems. Many scan tools also are capable of activating system functions for testing individual components or capturing a "snapshot" of the system in operation during a test drive.

Pressure Bleeders

Pressure bleeding is the fastest method of bleeding a brake system for two reasons. The master cylinder does not have to be refilled several times, and the job can be done by one person.

A pressure bleeder is a tank separated into two sections by a flexible diaphragm (Figure 1-44). The top section is filled with brake fluid. Compressed air is fed into the bottom section, and as the air pushes on the diaphragm, the brake fluid above it is also pressurized.

A supply hose runs from the top of the tank to the master cylinder. The hose is connected to the master cylinder by means of an adapter fitting that fits over the reservoir, taking the place of the reservoir cap. These adapters exist in different configurations, for the different types of reservoirs, including the plastic reservoirs on some of the newer vehicles.

The pressurized brake fluid flows into the master cylinder and out through the brake lines, quickly forcing all air pockets out of the lines.

Pressure bleeding is often recommended for bleeding antilock brake systems.

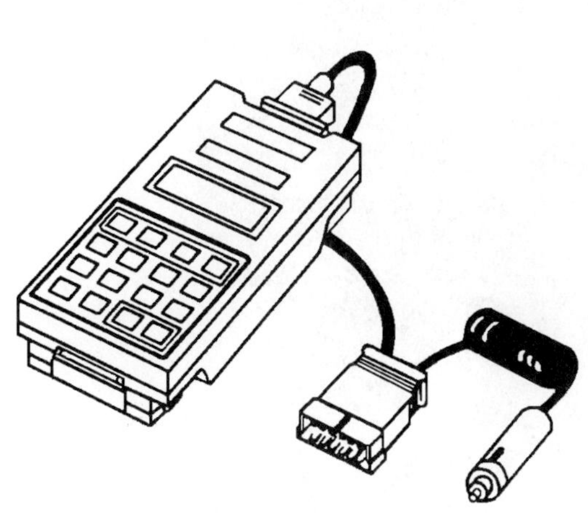

Figure 1-43 TECH 1 diagnostic scan tool (Courtesy of General Motors Corporation)

Figure 1-44 Typical brake system pressure bleeder and adapters

SERVICE TIP: Since most brake fluids (except for silicone) have a tendency to absorb moisture from the air, always keep containers tightly capped. It is better to buy smaller containers of brake fluid and keep them sealed until needed. Taking these steps to minimize water in the brake fluid will help reduce corrosion and keep the brake fluid boiling point high throughout the hydraulic system.

Cleaning Equipment and Containment Systems

The following systems and methods are used to safely contain asbestos in the workplace.

Negative Pressure Enclosure/HEPA Vacuum Systems

In this type of containment system, brake system inspection, disassembly, repair, and assembly is performed inside a tightly sealed protective enclosure that covers and contains the brake assembly (Figure 1-45). The enclosure prevents the release of asbestos fibers into the worker's breathing zone.

The enclosure is designed so that the worker can clearly see the work in progress. It is equipped with impermeable sleeves that allow you to perform the brake inspection, disassembly, repair, and reassembly. Examine the condition of the enclosure and its sleeves before beginning work. Inspect the enclosure for leaks and a tight seal.

A high efficiency particle air (HEPA) filtered vacuum is used to keep the enclosure under negative pressure as work is being performed. Because particles cannot escape the enclosure, compressed air may be used to remove asbestos fibers or particles from brake parts. The HEPA vacuum can also be used to loosen the asbestos-containing residue from the brake clutch parts. Once the asbestos is loose, draw it out of the enclosure using the vacuum port. The dust is then trapped in the vacuum cleaner's filter.

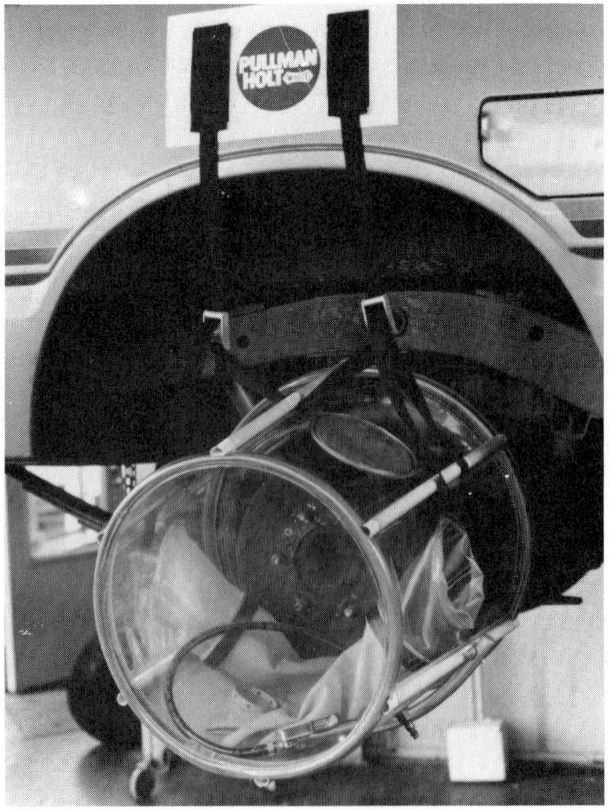

Figure 1-45 Negative pressure enclosure for containing asbestos dust (Courtesy of Pullman, White, Holt, Inc.)

When the vacuum cleaner filter is full, spray it with a fine mist of water, then remove and place it immediately in an impermeable container. Label as follows:

DANGER. CONTAINS ASBESTOS FIBERS. AVOID CREATING DUST. CANCER AND LUNG DISEASE HAZARD.

Asbestos waste must be collected, recycled, and disposed of in sealed impermeable bags or other closed, impermeable containers. Any spills or release of asbestos-containing waste material from inside of the enclosure or vacuum hose or vacuum filter should be immediately cleaned up using vacuuming or wet cleaning methods.

Low-Pressure Wet Cleaning Systems

Low-pressure wet cleaning systems wash asbestos from the brake assembly, catching the contaminated cleaning agent in a catch basin (Figure 1-46). Place a catch basin under the brake assembly, positioning it to avoid splashes and spills. The reservoir contains water with an organic solvent or wetting agent. In order to prevent the asbestos-containing brake dust from becoming airborne, control the flow of liquid so that the brake assembly is gently flooded.

Allow the liquid solution to flow between the brake drum and brake support before the drum is removed. Once the brake drum is removed, thoroughly wet the wheel hub and back of the assembly to suppress dust. Wash the brake support plate, brake shoes, and brake components used to attach the brake shoes before removing the old shoes.

Some wet cleaning equipment uses a filter. When the filter is full, first spray it with a fine mist of water. Remove the filter and place it in an impermeable container. Label and dispose of the container as described earlier.

CAUTION: Never use dry brushing during low-pressure/wet cleaning operations.

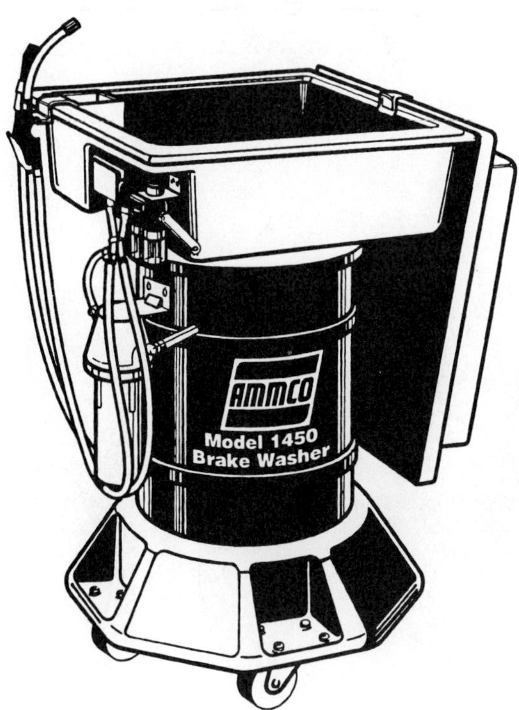

Figure 1-46 Low-pressure spray cleaning system with catch basin (Courtesy of Ammco Tools, Inc.)

Wet Cleaning Method

The wet cleaning method of containing asbestos dust is the simplest and easiest to use. But it must be performed correctly to provide protection. First, thoroughly wet the brake parts using a spray bottle, hose nozzle, or other implement that creates a fine mist of water or cleaning solution. Once the components are completely wet, wipe them clean with a cloth.

Place the cloth in a correctly labeled, impermeable container and properly dispose of it. The cloth can also be professionally laundered by a service equipped to handle asbestos-laden materials and reused. See Photo Sequence 1-1.

CAUTION: Never use dry brushing during wet method operations.

Photo Sequence 1
Typical Procedure for Using Enclosure Equipment for Cleaning Brake Drums

P1-1 Check that the hose is securely fastened to the HEPA vacuum and to the brake enclosure.

P1-2 Check that the vacuum container clips and seals are in position and working properly.

P1-3 Remove the wheel.

P1-4 Turn on the vacuum cleaner.

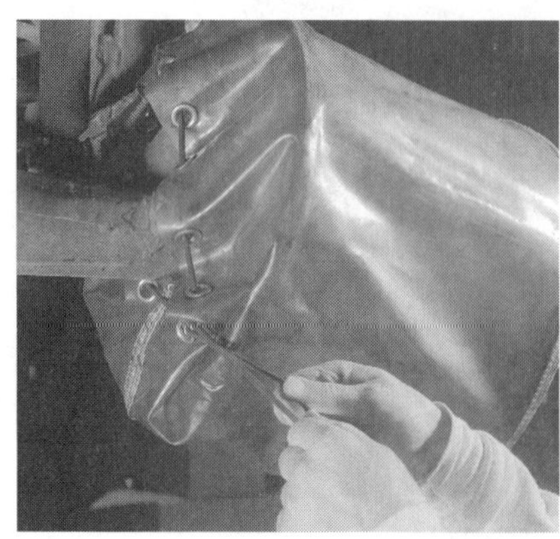

P1-5 Place the enclosure over the drum. It must form a tight seal behind the backing plate.

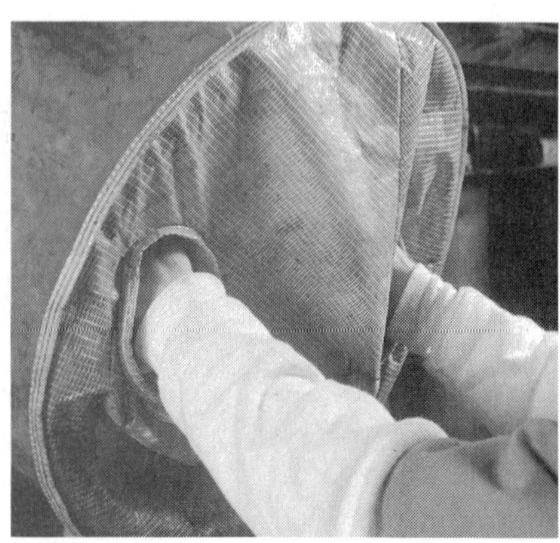

P1-6 Place your hands into the attached rubber gloves.

Photo Sequence 1
Typical Procedure for Using Enclosure Equipment for Cleaning Brake Drums (continued)

P1-7 Remove the brake drum.

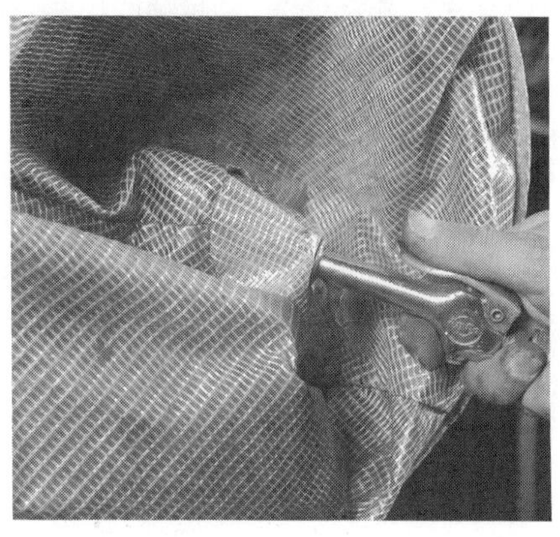

P1-8 Blow dust off the drum and brake assembly using the air gun attachment inside the enclosure. Blow dust off of all inside surfaces of the enclosure, directing it toward the vacuum exit.

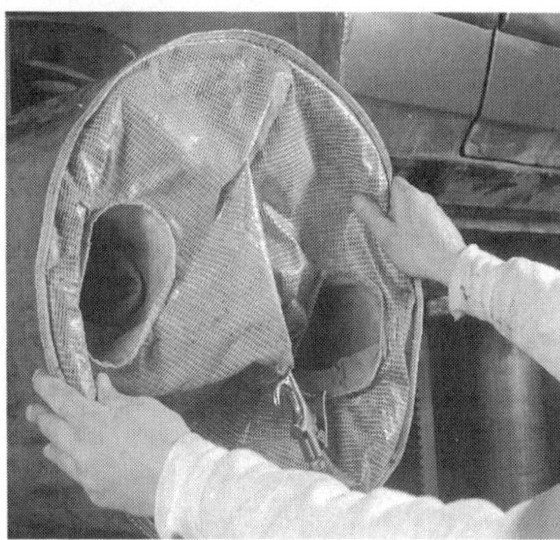

P1-9 Remove the enclosure.

P1-10 Turn off the vacuum. Then proceed with the drum brake inspection.

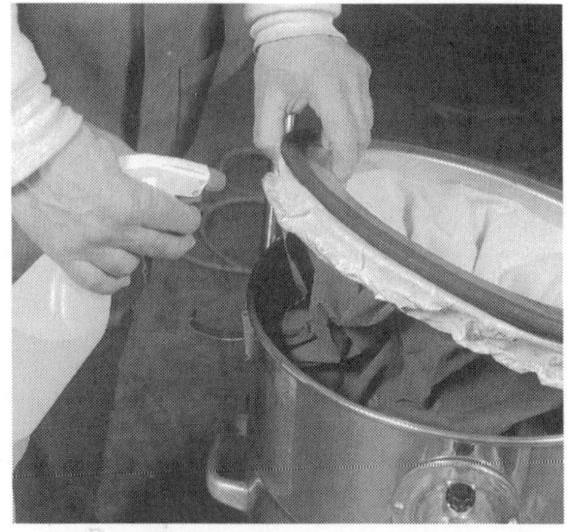

P1-11 When the vacuum cleaner filter is full, spray it with a fine mist of water, then remove and place it immediately in an impermeable container. Label the cleaner as follows: DANGER. CONTAINS ASBESTOS FIBERS. AVOID CREATING DUST. CANCER AND LUNG DISEASE HAZARD.

Vacuum Cleaning Equipment

Several types of vacuum cleaning systems are available to control asbestos in the shop. The vacuum system must be equipped with a high-efficiency HEPA filter designed to handle asbestos dust (Figure 1-47). After vacuum cleaning, any remaining dust should be wiped from components using a damp cloth.

Because they contain asbestos fibers, the vacuum cleaner bags and any cloth used in asbestos cleanup are classified as hazardous material. Such hazardous materials must be disposed of in accordance with the Occupational Safety and Health Administration (OSHA) regulations.

Figure 1-47 Commercial vacuum system with HEPA filter for handling asbestos cleanup

Figure 1-48 Typical shop manual

Always wear your respirator when removing vacuum cleaner bags or handling asbestos-contaminated waste. Seal the cleaner bags and wipes in heavy-mill plastic bags. Label the bags as follows:

> **DANGER.** CONTAINS ASBESTOS FIBERS. AVOID CREATING DUST. CANCER AND LUNG DISEASE HAZARD.

Shop Manuals

Shop manuals are essential for safe, complete brake service (Figure 1-48). They are needed to obtain desired specifications on torque settings and critical measurements such as drum and rotor discard limits. Manuals also provide drawings and photographs that show where and how to perform all major service procedures on that particular vehicle.

Special tools or instruments are also listed and shown when they are required. Precautions are given to prevent injury or damage to parts.

Most automobile manufacturers publish a service manual or set of manuals for each model and year of their cars. These manuals provide the best and most complete information for those cars. Various editions are available, covering different ranges of model years for both domestic and import cars. Brake part manufacturers often publish service manuals and service literature independently of vehicle manufacturers.

Although the manuals from different publishers vary in presentation and arrangement of topics, all service manuals are easy to use after you become familiar with their organization. Most shop manuals are divided into a number of sections, each of which covers different aspects of the vehicle. The beginning sections deal with each different vehicle system in detail, including diagnostic, service, and overhaul procedures. Each section has an index indicating more specific areas of information.

Throughout this book, you will be told to refer to the appropriate shop manual to find the correct procedures and specifications. Although the brake systems of all automobiles function in much the same way, there are many variations in design, particularly with regard to antilock brakes. Each design has its own set of repair and diagnostic procedures. It is very important that you always follow the recommendations of the manufacturer to identify and repair problems. Make using vehicle service manuals a habit. The benefits include increased productivity, less rework, and safer working conditions.

CASE STUDY

The shop's senior service technician has just replaced a run of brake tubing between the master cylinder and the right rear drum assembly. Leaks in the line were causing a loss of braking power. He shows the old line to the two new technicians in the shop. Can they tell him the source of the problem? One technician notices several kinks in the line at points where the steel tubing is bent. "And what caused the kinks?" the senior technician asks. The young technicians shrug. "Laziness," says the senior technician. "Whoever installed this line was too lazy to find and use a tubing bender. The bends were made by hand and the lines were kinked, just a little, but just enough to make them fail at half their service life." Always use the special tools available to increase the quality of your work and ensure your personal safety.

Terms to Know

Flare nut wrench
Foot-pound
HEPA filter
Low-pressure wet cleaning
Metric measurement system
Millimeter
Negative pressure enclosure/HEPA vacuum system
Newton-meter
Pressure bleeder
Rotor micrometer
Torque
US Customary measurement system
Wet cleaning

ASE Style Review Questions

1. *Technician A* always uses flare nut wrenches to open brake lines.
   Technician B uses an open-end wrench for the same job.
   Who is correct?
   A. A only **C.** Both A and B
   B. B only **D.** Neither A nor B

2. *Technician A* states that the only danger in using a metric wrench on a US Customary size bolt is skinned knuckles if the wrench slips.
   Technician B says the head of the bolt may become rounded as well.
   Who is correct?
   A. A only **C.** Both A and B
   B. B only **D.** Neither A nor B

3. The use of torque wrenches is being discussed:
   Technician A says the bolt and hole must be clean to get an accurate torque reading.
   Technician B states that it is safer to pull the torque wrench toward you rather than push it.
   Who is correct?
   A. A only **C.** Both A and B
   B. B only **D.** Neither A nor B

4. Impact wrenches are being discussed:
   Technician A states they are an ideal tool for general assembly work because they are strong and fast.
   Technician B states they are only suited for certain types of "rough" bolt removal and installation, such as lug nuts.
   Who is correct?
   A. A only **C.** Both A and B
   B. B only **D.** Neither A nor B

5. Measuring tools are being discussed:
Technician A says a dial indicator is an ideal tool for measuring disc brake rotor thickness.
Technician B says the job is best done using a depth gauge.
Who is correct?
A. A only **C.** Both A and B
B. B only **D.** Neither A nor B

6. Bleeding brake system is being discussed:
Technician A states that a pressure bleeder is the fastest method of doing the job.
Technician B states that with a pressure bleeding outfit one technician, not two, can do the job.
Who is correct?
A. A only **C.** Both A and B
B. B only **D.** Neither A nor B

7. Cleaning asbestos from a brake system is being discussed:
Technician A says an air gun can be used to blow dust off if the worker is wearing a HEPA-type respirator.
Technician B says an air gun can only be used if the work area is enclosed in a negative pressure "bubble."
Who is correct?
A. A only **C.** Both A and B
B. B only **D.** Neither A nor B

8. *Technician A* says testing continuity in a circuit requires the use of a self-powered test light.
Technician B says digital multimeters are preferred over analog test instruments when working with computerized engine controls.
Who is correct?
A. A only **C.** Both A and B
B. B only **D.** Neither A nor B

9. *Technician A* refers to the vehicle shop manual for specific torque values when installing caliper guide pins.
Technician B refers to the shop manual to find the proper lift points of the vehicle.
Who is correct?
A. A only **C.** Both A and B
B. B only **D.** Neither A nor B

10. *Technician A* says high-pressure spray hoses are a preferred method of removing asbestos from brake parts.
Technician B states all wash-downs should be low pressure or done by hand.
Who is correct?
A. A only **C.** Both A and B
B. B only **D.** Neither A nor B

Safety

Upon completion and review of this chapter, you should be able to:

- ❑ Understand the health hazards created by asbestos dust and perform the proper tasks needed to protect yourself from harm.
- ❑ Understand the purpose of Material Safety Data Sheets and how the Right-to-Know laws affect you as an employee in the auto service industry.
- ❑ Know which types of brake systems generate high hydraulic pressures and how to safely relieve such pressure prior to service.
- ❑ Discuss how to ensure a safe work environment in the shop.
- ❑ Properly lift heavy objects.
- ❑ Extinguish the common types of fires that may occur in the shop.
- ❑ Inspect and use tools safely.
- ❑ Discuss basic safety rules and describe how common sense dictates these rules.

Carbon Monoxide

Engine exhaust contains large amounts of carbon monoxide, a deadly gas that is odorless and colorless. Never run an engine in the shop without properly venting the exhaust fumes to the outside or to a dedicated ventilation system for exhaust gas (Figure 2-1).

Basic Tools

Safety glasses or goggles

Respirator

Vacuum with HEPA filter

Wet clean system

Carbon monoxide vent system

Fire extinguisher(s)

Air Conditioning Systems and Freon Gas

When working around air conditioning lines, be careful you do not accidently open a connection. The system is filled with R-12 (freon) or other types of refrigerants that will escape as a gas and instantly freeze anything it comes in contact with, including your skin and eyes. If freon is burned by the engine or by contact with an open flame or extreme heat, dangerous phosgene gas is formed.

CAUTION: Phosgene gas is poisonous and will make you sick or fatally ill.

Figure 2-1 When running an engine in the shop, always vent the vehicle exhaust to the outdoors.

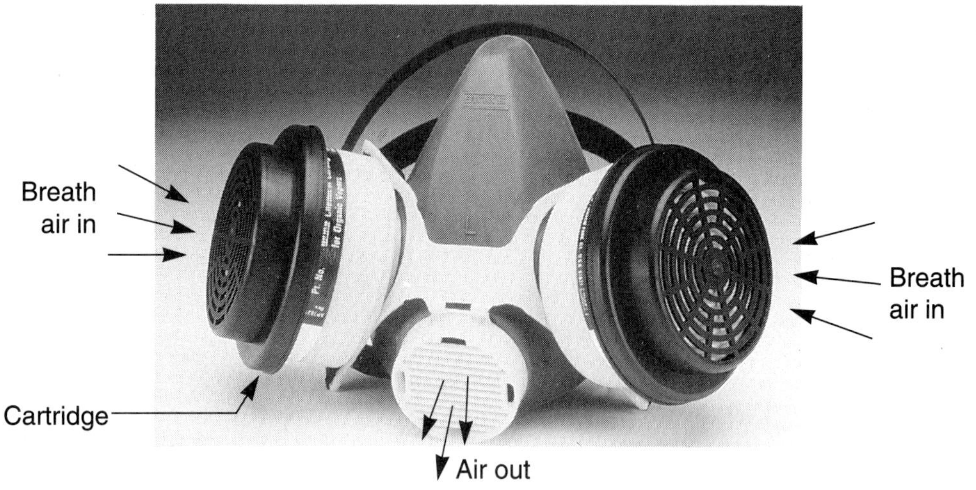

Figure 2-2 Half-mask respirator with replaceable cartridge filters (Courtesy of Mine Safety Appliance Co.)

Asbestos Dust Precautions

Many brake lining materials contain asbestos fibers. Studies have shown that exposure to asbestos poses a health hazard that may lead to cancer. The problem happens when asbestos dust becomes airborne and is inhaled into the nasal passages and lungs. Asbestos dust is created when you grind brake linings, clean brake assemblies, or perform other brake-related work. Take the proper steps to protect yourself and create a safe work environment.

Respirators

Whenever you perform brake service, wear an air-purifying respirator that is approved by the National Institute of Safety and Health (NISH). This can be a simple throw-away mask, or a mask with replaceable filter cartridges (Figure 2-2). See Photo Sequence 2 for the proper care of a respirator. Tying a bandanna or cloth over your nose and mouth will not do the job! Wear an approved respirator from the time you remove the wheels through to final assembly. Replace the mask or its filters as soon as you feel an increase in breathing effort.

The facepiece is critical to the effectiveness of any respirator. Every facepiece must seal against the face of the wearer, yet provide comfort when worn over an extended period of time. It should fit both large and small faces. A full facepiece should provide a full field of vision for the wearer, yet remain light enough to be worn over long periods of time. A facepiece must be tough and durable enough to withstand both constant exposure to chemical vapors and repeated washings in harsh detergents in hot water.

A vinyl facepiece has a plasticity that allows it to mold itself to your face. This is preferred over an elastic or rubber model that would snap back into its own shape. A vinyl facepiece is also more comfortable to wear over long periods of time.

Filters are the air-purifying elements used when particulates such as dust, mist, or metal fumes are present. Felt filters use an electrostatic charge to help trap particulates with the fibers.

High-efficiency paper filters use a fiberglass paper with fine pores to stop the passage of particulates. They are mounted in a cartridge housing to minimize breathing resistance.

Filter life depends upon the amount of particulates in the air. Change your respirator's filters as soon as you notice an increased resistance to inhaling.

Felt filters function properly only when they are installed correctly. Place the filter in the snap-off filter cap. Then install the cap over the cartridge, if used, or on the receptacle in combination with a cartridge. Do not place the filter directly into the receptacle next to the inhalation valve.

Photo Sequence 2
Typical Procedure for Caring for Your Respirator

P2-1 Check the headbands by stretching the elastic. Inspect for frayed areas, loose strands, tears, or loss of elasticity.

P2-2 Remove the filter and exhale valve covers. Carefully inspect the rubber valve discs because an aging or damaged exhaust valve can allow asbestos dust to enter the mask.

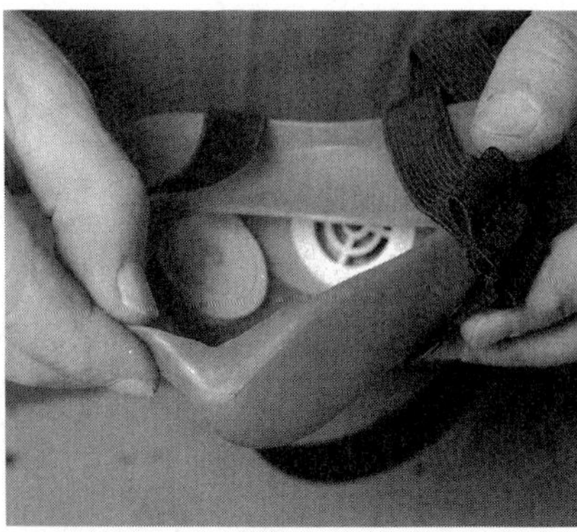

P2-3 Examine the facepiece for cuts, tears, holes, melting, stiffening, or crushing. Replace it if there is any damage or deterioration.

P2-4 Make sure the filter/cartridge housings are not cracked or scratched, and that the threads are not damaged.

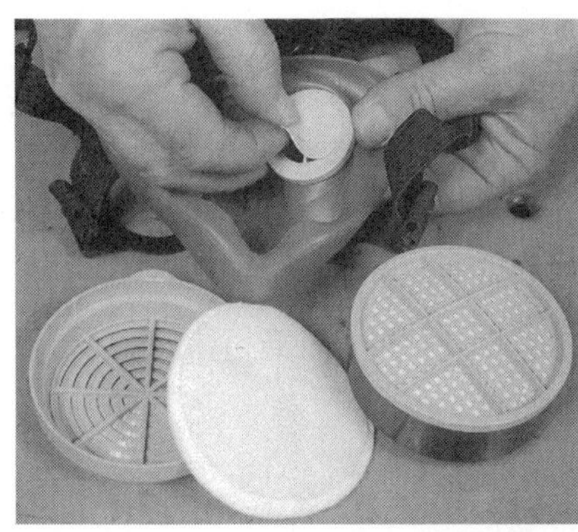

P2-5 Remove filters, cartridges, and valve flaps. Inspect for worn or deteriorating parts.

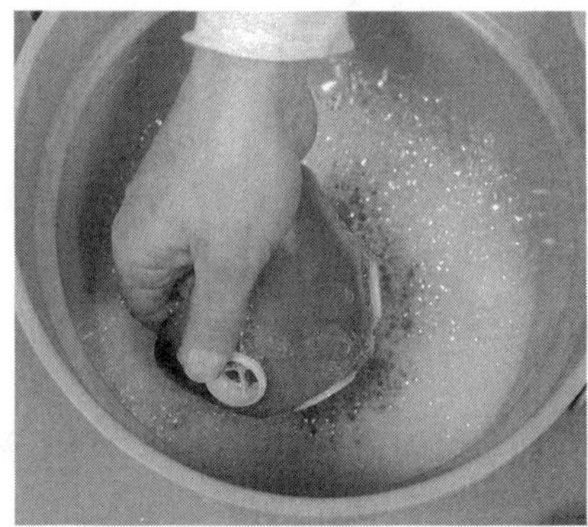

P2-6 Immerse the facepiece in a solution of germicidal detergent and water, following instructions on the detergent label. Do not use cleaning compounds that contain alcohol or other organic solvents.

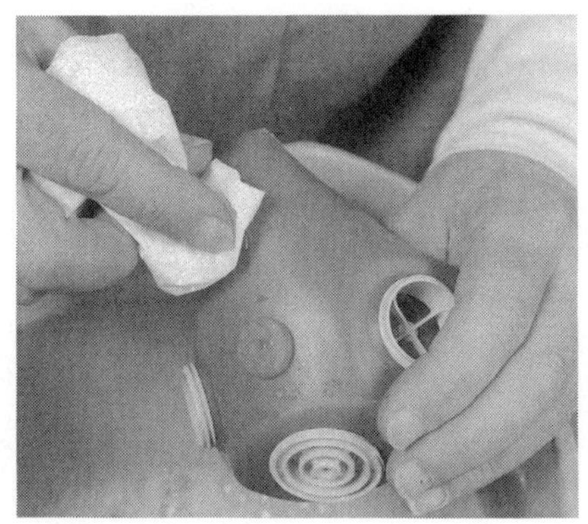

P2-7 Use a soft brush or cloth to remove dust, grease, paint, or other materials that have not lifted with soaking.

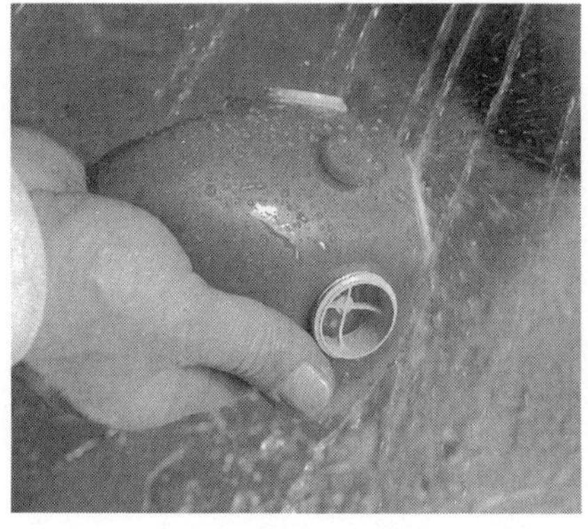

P2-8 Rinse thoroughly with clear water.

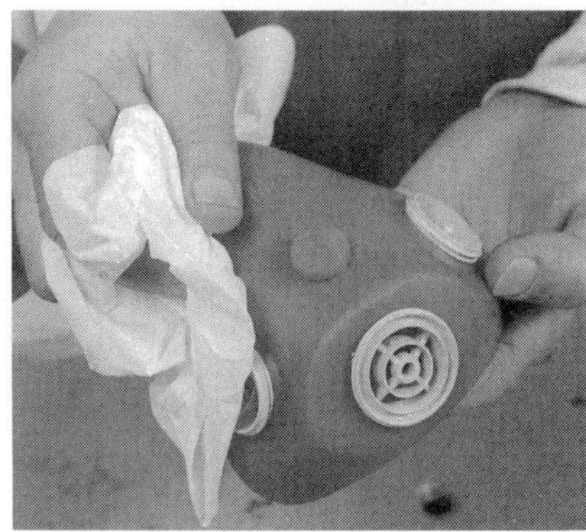

P2-9 Air or towel dry.

Photo Sequence 2
Typical Procedure for Caring for Your Respirator (continued)

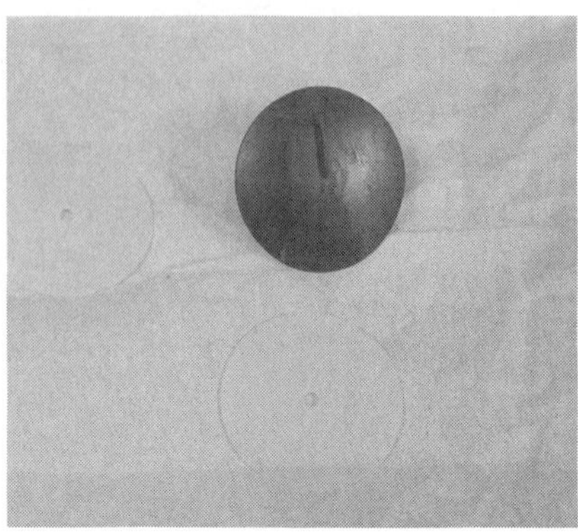

P2-10 Wash valve flaps and dry them on a flat surface. If they are deformed, replace them with new valve flaps.

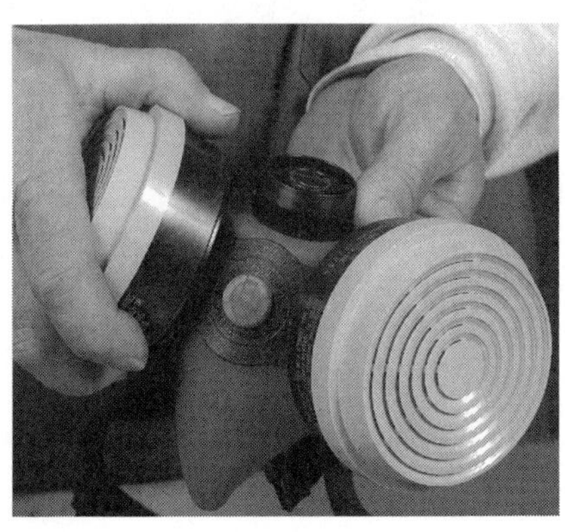

P2-11 Reassemble the respirator.

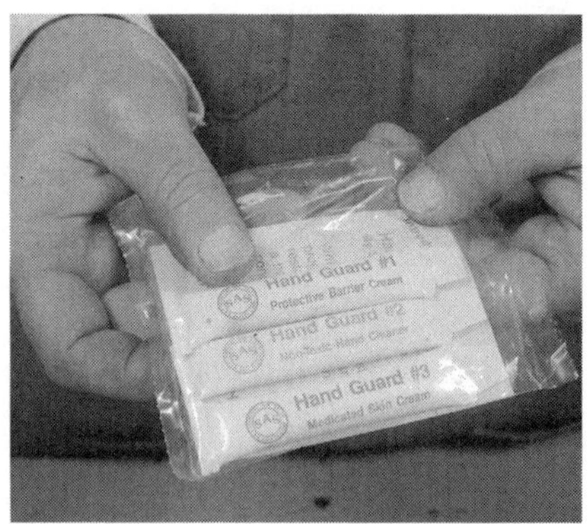

P2-12 Store your respirator in its sealed container and keep it away from heat, extreme cold, sunlight, excessive moisture, dust, and contaminating chemicals.

A respirator that fits poorly is worse than having no respirator at all as it provides a false sense of security.

To check the fit of a respirator, block the exhale valve opening with the heel of your hand or thumb (Figure 2-3). Gently exhale to build up a slight pressure in the facepiece. Any escaping air will be felt immediately. If light exhalation causes a leak, the facepiece should be adjusted and the seal rechecked.

This simple "positive pressure check" should be performed when the facepiece is first put on and at frequent intervals during the day to ensure that there is continuous protection.

Avoiding Asbestos Dust

The best way to avoid asbestos dust is to not create it. Handle brake parts carefully. Never use compressed air or dry brushing for cleaning dust from brake drums, brake backing plates, and brake assemblies. Use specialized vacuum cleaning equipment or the proper wet-cleaning procedures described in Chapter 1: Brake Service Tools and Equipment. Keep yourself and your clothing clean. Wash thoroughly before eating, and avoid eating in the shop area. If your work clothes become visibly contaminated with dust and dirt, change them before leaving work and shower or

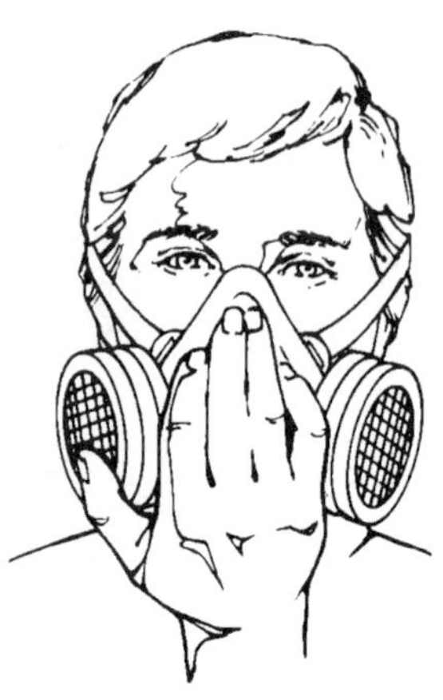

Figure 2-3 Checking respirator fit (Courtesy of Technical Innovations, Inc.)

bathe as soon as possible. Clothing that is heavily contaminated with asbestos fibers should not be washed at home. See your shop supervisor and request industrial cleaning of the garments.

When coupled with exposure to asbestos dust, the dangers associated with cigarette smoking increase dramatically. Do not smoke in the shop area.

Grinding Precautions

The risk of asbestos exposure is highest during grinding operations. Whenever possible, purchase friction materials preground and ready for installation. If grinding is necessary, wear an approved respirator. The grinding area must have its own local exhaust ventilation system to draw away the fibers generated by the cutters. The exhaust system must keep operator exposure levels below the present OSHA standards. If you have any doubts as to the efficiency of the shop's exhaust system, contact its manufacturer.

Housekeeping

In addition to cleaning brake system components, the HEPA filter-vacuum system can be used for general floor cleanup in the brake service area (Figure 2-4). Use the recommended attachments. If a vacuum unit is not available, the shop floor should be damp mopped to remove asbestos dust. Water or other dust suppressants should be applied to the floor prior to broom cleaning. Never dry broom the floor or use compressed air to blow away dust.

Material Safety Data Sheets and Right-to-Know Laws

Asbestos is not the only hazardous material found in auto service facilities. Solvents, cleaners, brake fluids, gasoline, oils, and other chemicals all present certain hazards if not handled properly. They may be flammable, emit harmful vapors, or be irritating to the eyes or skin.

All workers are protected by federal and state Right-to-Know laws concerning hazardous materials in the workplace. These laws specify that:

1. It is the responsibility of the employer to train all employees in the safe handling and disposal of all potentially hazardous chemicals in the workplace.

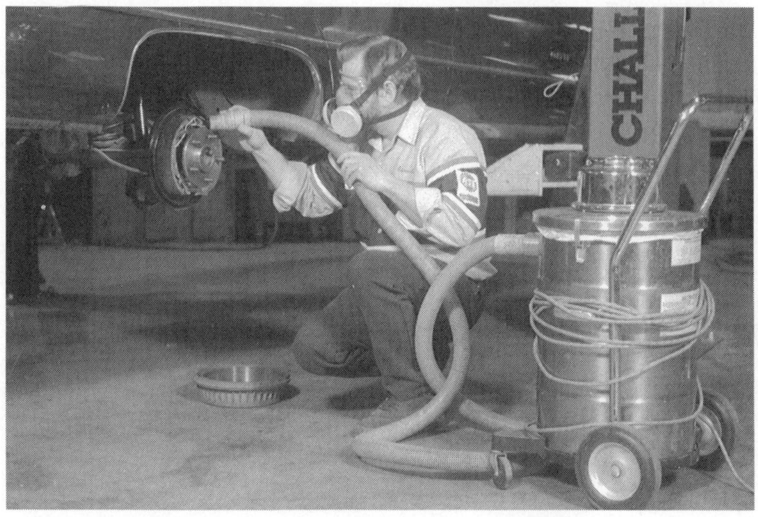

Figure 2-4 Vacuum the brake service bays with a commercial vacuum equipped with a HEPA filter.

Important information about such materials is contained on Material Safety Data Sheets (Figure 2-5). The MSDS is issued by the manufacturer of the material. It includes data regarding the material's chemical composition, ignitability, corrosiveness, reactivity, toxicity, and other distinctive characteristics. It often states recommended uses for the material and lists specific handling instructions and safety precautions that must be observed. Emergency treatments for accidental ingestion, inhalation, and eye and skin contact are given when applicable. Guidelines for cleaning up spills or responding to other emergencies are included. The Canadian equivalent to the MSDS is the workplace hazardous materials information sheet.

It is the responsibility of your employer to obtain all MSDS for the materials in the shop and make this information available to all employees. The employer must also provide formal training on the safe handling of all hazardous materials and update this training on a yearly basis.

While it is the employer's job to provide you with proper safety training and equipment, you are ultimately responsible for practicing safety in the shop.

2. Containers storing potentially hazardous materials must be properly labeled with regard to health, fire, reactivity, and handling hazards. The simplest way to ensure this is to keep materials in their original containers. If a chemical is moved into another container, it is the responsibility of the shop to see that the container is of the proper type and is

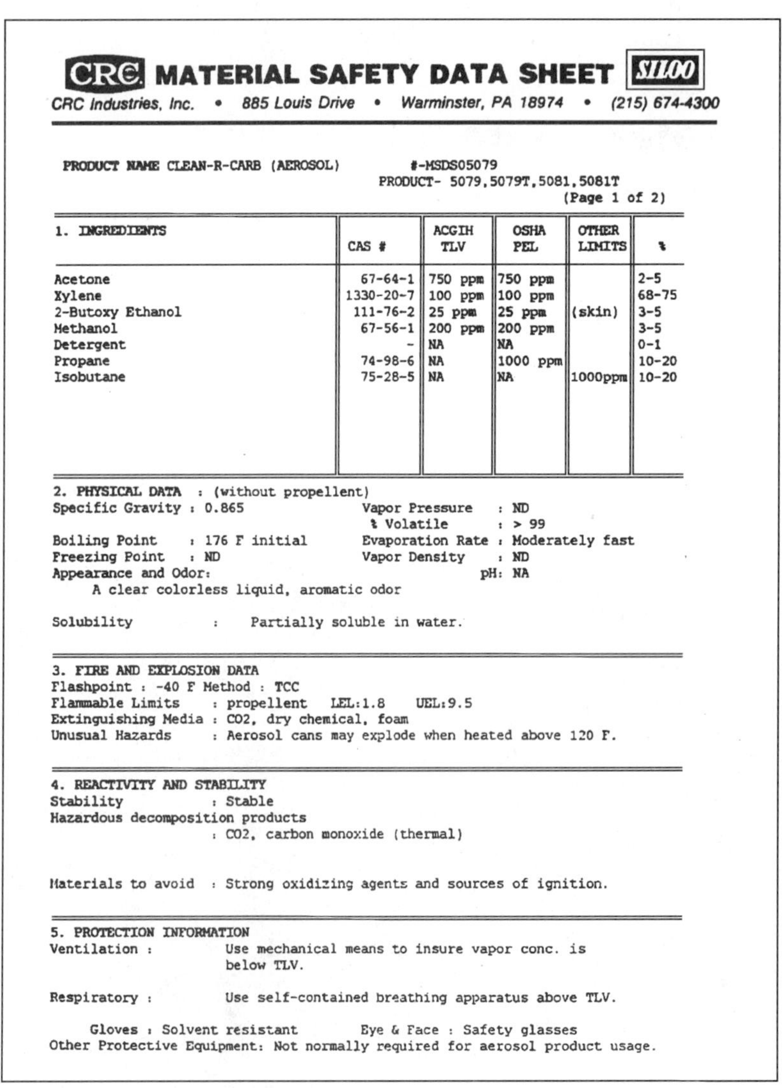

CRC MATERIAL SAFETY DATA SHEET SILOO

CRC Industries, Inc. • 885 Louis Drive • Warminster, PA 18974 • (215) 674-4300

PRODUCT NAME CLEAN-R-CARB (AEROSOL) #-MSDS05079

PRODUCT- 5079,5079T,5081,5081T

(Page 1 of 2)

1. INGREDIENTS	CAS #	ACGIH TLV	OSHA PEL	OTHER LIMITS	%
Acetone	67-64-1	750 ppm	750 ppm		2-5
Xylene	1330-20-7	100 ppm	100 ppm		68-75
2-Butoxy Ethanol	111-76-2	25 ppm	25 ppm	(skin)	3-5
Methanol	67-56-1	200 ppm	200 ppm		3-5
Detergent	-	NA	NA		0-1
Propane	74-98-6	NA	1000 ppm		10-20
Isobutane	75-28-5	NA	NA	1000ppm	10-20

2. PHYSICAL DATA : (without propellent)

Specific Gravity : 0.865 Vapor Pressure : ND

% Volatile : > 99

Boiling Point : 176 F initial Evaporation Rate : Moderately fast

Freezing Point : ND Vapor Density : ND

Appearance and Odor: pH: NA

A clear colorless liquid, aromatic odor

Solubility : Partially soluble in water.

3. FIRE AND EXPLOSION DATA

Flashpoint : -40 F Method : TCC

Flammable Limits : propellent LEL:1.8 UEL:9.5

Extinguishing Media : CO2, dry chemical, foam

Unusual Hazards : Aerosol cans may explode when heated above 120 F.

4. REACTIVITY AND STABILITY

Stability : Stable

Hazardous decomposition products

: CO2, carbon monoxide (thermal)

Materials to avoid : Strong oxidizing agents and sources of ignition.

5. PROTECTION INFORMATION

Ventilation : Use mechanical means to insure vapor conc. is below TLV.

Respiratory : Use self-contained breathing apparatus above TLV.

Gloves : Solvent resistant Eye & Face : Safety glasses

Other Protective Equipment: Not normally required for aerosol product usage.

Figure 2-5 MSDS sheet for asbestos fibers (Courtesy of CRC Industries, Inc.)

correctly labeled. Do not use materials in unmarked containers. They may not be what you think they are, or they may be contaminated.

3. The employer must maintain documentation on all hazardous materials used in the shop. The employer must provide proof of training programs, keep records of all accidents or spills, and satisfy all employee requests to review MSDS. Even if a hazardous material is phased out of use, the MSDS must be kept on file for 30 years! OSHA and other regulatory agencies are quite serious when it comes to employee safety and hazardous materials. You should be too.

A liquid's flash point is the lowest temperature at which it will give off enough flammable vapor to ignite.

Hazardous Waste Disposal

When the shop is finished using a hazardous material, it becomes hazardous waste. The Environmental Protection Agency (EPA) defines hazardous waste as solid or liquid materials that have one or more of the following characteristics:

A hazardous material is one that is flammable and/or explosive or has shown to produce adverse health effects on people during unprotected exposure during normal use.

- ❑ *Ignitability.* This includes liquids with flash points below 140°F or solids that can spontaneously ignite.
- ❑ *Corrosivity.* These are materials that dissolve metals or other materials or burn the skin on contact.
- ❑ *Reactivity.* Reactive materials include those that react violently with water or other materials. They may release cyanide gas, hydrogen sulfide gas, or similar gases when exposed to low pH acid solutions. This also includes materials that generate toxic or flammable vapors.
- ❑ *EP toxicity.* These materials leach one or more of either heavy metals in concentrations greater than 100 times primary drinking water standard concentrations.

A complete listing of hazardous wastes can be found in the Code of Federal Regulations.

CAUTION: When handling any hazardous waste material, always wear the safety equipment as specified in the MSDS. In many cases, this includes full eye protection, chemical resistant gloves, and a respirator (Figure 2-6).

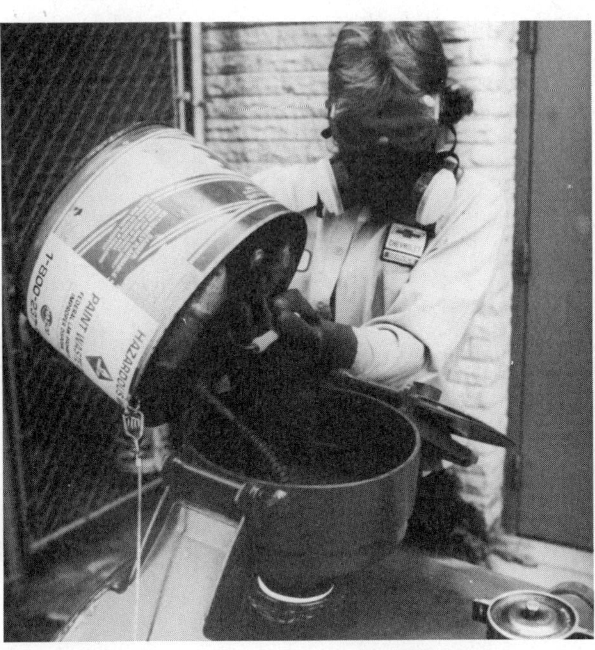

Figure 2-6 Wear proper safety equipment when handling hazardous materials such as cleaning solvents. (Courtesy of DuPont Co.)

Hydraulic Pressure

Many antilock brake systems generate extremely high brake fluid pressures that typically range between 2,000 and 2,600 psi. Failure to fully depressurize the hydraulic accumulator of an antilock brake system before servicing any brake system components could result in severe personal injury from high-pressure brake fluid escaping from a service connection. Follow the exact shop manual procedure for the vehicle being serviced. Following is a typical depressurizing procedure.

1. Disconnect the negative battery cable (Figure 2-7A).
2. Keep the ignition key in the off position (Figure 2-7B).
3. Pump the brake pedal a minimum of 40 times using about 50 pounds of pedal force (Figure 2-7C).
4. Continue pumping until you feel a definite increase in pedal pressure. Pump the pedal a few additional times to ensure removal of hydraulic pressure from the system.
5. You can now proceed with system service.

Hydraulic Power-Assisted Brakes

While not as high as those found in ABS systems, pressures within brake systems equipped with hydraulic power-assist units can range upwards of 700 psi. These units rely on pressures generated by the vehicle's power steering pump. Like ABS systems, hydraulic-assisted brake systems use an accumulator to store brake fluid under pressure. The method of relieving this pressure is also similar to that for ABS systems. With the ignition off or the battery positive terminal disconnected, pump the brake pedal approximately 10 times with a pedal force of approximately 50 to 75 pounds.

Safety in the Shop

Personal safety and accident prevention must be the top priority in the shop. The potential for serious injury exists in many areas of the shop. Exhaust fumes are poisonous. Fuels, cleaning solvents, and battery acid are highly flammable and/or explosive. High pressures inside components such as the hydraulic accumulator of antilock brake systems, pressurized fuel lines, and battery cases can

A

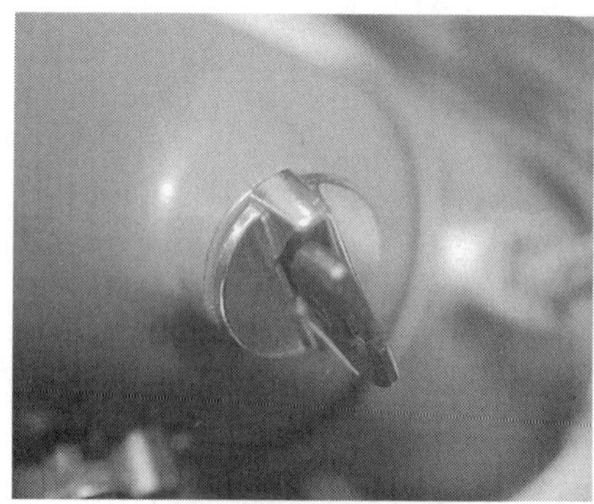

B

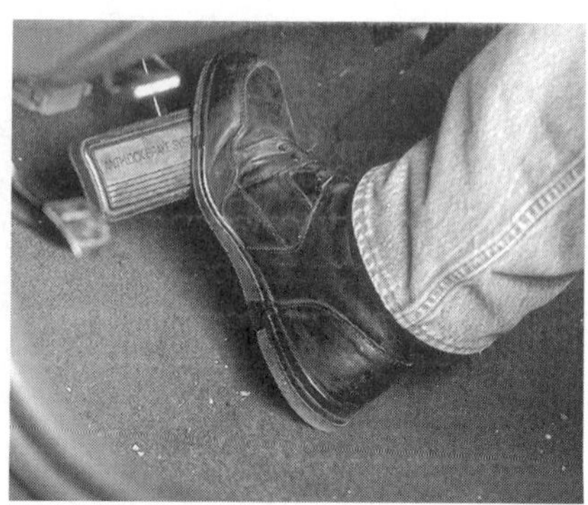

C

Figure 2-7 Steps in relieving high hydraulic pressures in antilock braking systems: (A) disconnect negative battery cable, (B) keep ignition in off position, and (C) pump pedal firmly until there is an increase in pedal pressure.

spray liquids or acids if pressure is not properly relieved. When released, small retaining clips and other pressure-held hardware can fly off at amazing speeds.

CUSTOMER CARE: Auto repair work draws customers like a magnet. But don't be tempted to look the other way if a customer casually comes into the service bay. Set strict rules that keep nonworkers away from service areas.

Many components are heavy or have sharp flanges or edges. Carelessness with hoists, jacks, and jack stands can lead to disaster. Many tools are potentially dangerous if mishandled or operated with guards. Engine block and exhaust system parts become hot enough to cause severe burns.

Lifting and Carrying

Knowing the right way to lift heavy objects is important (Figure 2-8). Even small or midsize parts such as a rotor or brake drum can wrench your back or pull an arm muscle if mishandled. When lifting any object:

1. Place your feet close to the load, spreading them slightly for better balance.
2. Keep your back and elbows straight and bend at the knees until your hands reach the load.
3. Grasp the object firmly. Keep the object close to your body and lift by straightening your legs.
4. Turn your entire body when changing direction. Don't simply twist at the waist.
5. When placing the object on a shelf, do not bend forward. Place the edge of the load on the surface of the shelf and slide it forward. When setting an object on the floor, lower it by bending your knees and keeping your back straight. Bending forward strains your back muscles.

CAUTION: A box soaked with grease, brake fluid, or other liquid has lost much of its strength. You may pick up the box only to have the part exit the bottom and find the floor or worse, your foot.

Safe Work Areas

Familiarize yourself with the way the shop is laid out. Find out where fire extinguishers, first-aid kits, eye wash stations, and other safety items are located. Be sure you know the types of fire each extinguisher can handle. Know how to operate the extinguishers. The same holds true for the eye wash stations and other first-aid equipment. Memorize the route to the nearest exit in the event of a major fire or hazardous material spill.

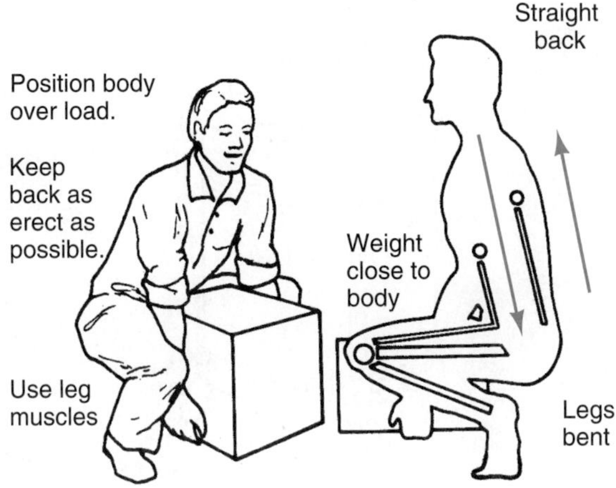

Figure 2-8 Safe lifting involves using your leg muscles and keeping your back straight.

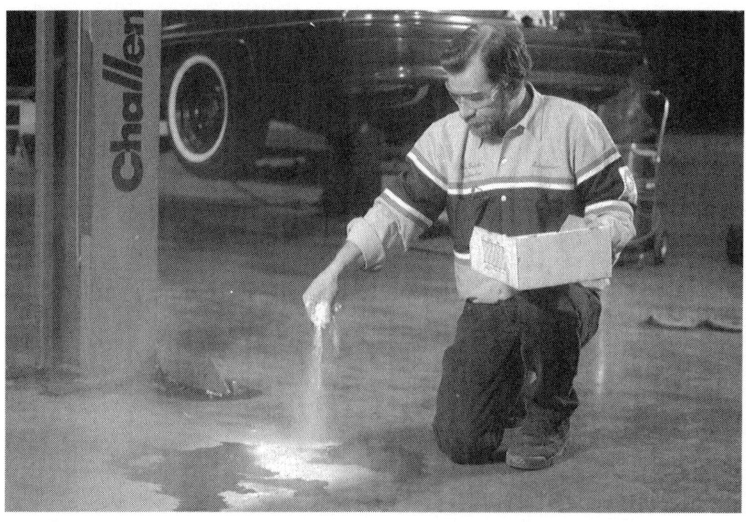

Figure 2-9 Using a commercial absorbent to soak up a spill

Ask if there are certain stalls reserved for brake work or other special jobs. Abide by these rules. Take note of all posted instructions. No smoking signs, special instructions for shop tools and equipment, and danger zone warnings are all posted to help the shop run smoothly and safely.

Housekeeping

Good housekeeping is a safety issue. A cluttered shop is a dangerous shop. You are responsible for keeping your work area and the rest of the shop clean and safe. A clean, well-organized work area will also help you become a better technician. Organization and neatness will be reflected in your work. You'll spend more time doing work and less time searching for that lost tool or misplaced part.

All surfaces should be kept clean, dry, and orderly. Any oil, coolant, or grease on the floor can cause slips that could result in serious injuries. To clean up oil or brake fluid spills, be sure to use a commercial oil absorbent (Figure 2-9). Dirty or oily rags must be stored in a sealed metal container and disposed of properly (Figure 2-10). Keep all water off the floor; remember that water is a conductor of electricity. A serious shock hazard will result if a live wire happens to fall into a puddle in which a person is standing.

Figure 2-10 Store rags soaked in flammable liquids in sealed waste cans designed for that job.

Figure 2-11 Vehicle raised on jack or safety stands

When you raise a vehicle with a hand jack, always set the car down on safety stands and remove the jack (Figure 2-11). Don't leave the jack handle sticking out from under the car where someone can trip over it. The same holds true for creepers. When not in use, stand the creeper on end against a wall. Pushing it completely under the vehicle gets it out of the way, but it is easy to forget that it is there and drive over it once the job is done.

Air hoses and power extension cords should be neatly coiled and hung. Don't leave a tangled mess in walkways or on the shop floor. Keep all exits open. A blocked exit violates fire codes and leaves the shop liable to legal action if people become trapped in a fire or dangerous situation.

Fire Hazards and Prevention

There are four basic types of fires (Table 2-1). Class A fires involve the burning of wood, paper, and other "ordinary" materials. Class B fires involve flammable liquids such as gasoline, diesel fuel, solvents, grease, and oil. Class C fires are electrical fires, while class D fires involve a unique group of flammable metals.

One such metal is magnesium, which is used to make high-performance, lightweight wheels. It is also a common ingredient in many aerial fireworks, so subjecting a "mag" to high heat can result in a large unexpected "sparkler."

Extinguishing agents are designed to handle one or more class of fire. The extinguisher is normally marked with a symbol or letter designating the fire classes it is effective on (Figure 2-12). Extinguishers charged with pressurized water should be used only on class A fires. Foam-charged extinguishers are suitable for class A and B fires. The foam should be sprayed directly on the fire to smother it. A number of extinguishing agents are suitable for class B and C fires. These include dry chemical charges, carbon dioxide, and halon gas. As noted in Table 2-1, working with carbon dioxide or halon gas requires special safety concerns.

Multipurpose dry chemical extinguishers capable of handling class A, B, and C fires are quite popular. With any type of dry chemical extinguisher, the spray should be directed at the base of the fire to suffocate the flames. Class D fires require a special dry-powder extinguishing agent.

The shop should be equipped with multiple extinguishers capable of handling all fire classifications. Know their locations and how to operate them. Fire extinguishers require periodic inspection and recharging by a qualified technician. If the inspection tag on the extinguisher has lapsed, notify the shop supervisor. In the event of a fire, you will have to react quickly and you are

Small under-the-hood fires have destroyed many vehicles and buildings and have cost some service technicians their lives. Know what to do to protect yourself and your fellow employees.

Table 2-1 GUIDE TO FIRE EXTINGUISHER SELECTION

	Class of Fire	Typical Fuel Involved	Type of Extinguisher
Class A Fires (green)	**For Ordinary Combustibles** Put out a class A fire by lowering its temperature or by coating the burning combustibles.	Wood Paper Cloth Rubber Plastics Rubbish Upholstery	Water*[1] Foam* Multipurpose dry chemical[4]
Class B Fires (red)	**For Flammable Liquids** Put out a class B fire by smothering it. Use an extinguisher that gives a blanketing, flame-interrupting effect; cover whole flaming liquid surface.	Gasoline Oil Grease Paint Lighter fluid	Foam* Carbon dioxide[5] Halogenated agent[6] Standard dry chemical[2] Purple K dry chemical[3] Multipurpose dry chemical[4]
Class C Fires (blue)	**For Electrical Equipment** Put out a class C fire by shutting off power as quickly as possible and by always using a nonconducting extinguishing agent to prevent electric shock.	Motors Appliances Wiring Fuse boxes Switchboards	Carbon dioxide[5] Halogenated agent[6] Standard dry chemical[2] Purple K dry chemical[3] Multipurpose dry chemical[4]
Class D Fires (yellow)	**For Combustible Metals** Put out a class D fire of metal chips, turnings, or shaving by smothering or coating with a specially designed extinguishing agent.	Aluminum Magnesium Potassium Sodium Titanium Zirconium	Dry powder extinguishers and agents only

*Cartilage-operated water, foam, and soda-acid types of extinguishers are no longer manufactured. These extinguishers should be removed from service when they become due for their next hydrostatic pressure test.

Notes:

(1) Freeze in low temperatures unless treated with antifreeze solution, usually weighs over 20 pounds, and is heavier than any other extinguisher mentioned.

(2) Also called ordinary or regular dry chemical (solution bicarbonate).

(3) Has the greatest initial fire-stopping power of the extinguishers mentioned for class B fires. Be sure to clean residue immediately after using the extinguisher so sprayed surfaces will not be damaged (potassium bicarbonate).

(4) The only extinguishers that fight A, B, and C classes of fires. However, they should not be used on fires in liquefied fat or oil of appreciable depth. Be sure to clean residue immediately after using the extinguisher so sprayed surfaces will not be damaged (ammonium phosphates).

(5) Use with caution in unventilated, confined spaces.

(6) May cause injury to the operator if the extinguishing agent (a gas) or the gases produced when the agent is applied to a fire is inhaled.

counting on the extinguisher to do its job. If there is not a fire extinguisher handy, a blanket or fender cover may be used to smother a small fire. Don't count on this method to handle larger fires. Evacuate the building and call the local fire department. Keep fire department and other emergency response telephone numbers posted in several areas around the shop. If your clothing catches fire, drop to the floor and roll to smother the flames.

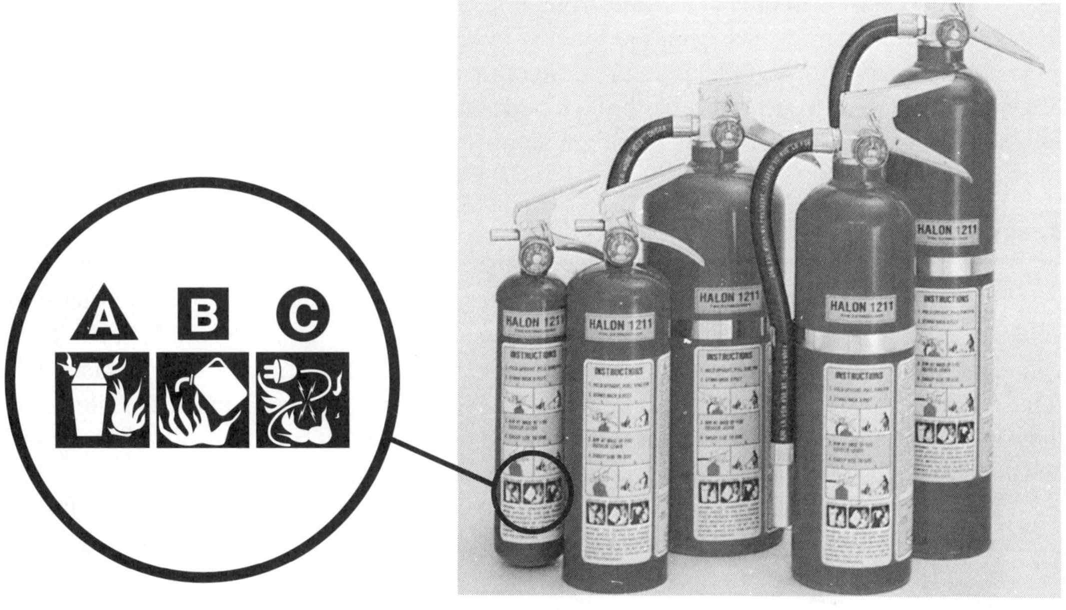

Figure 2-12 The fire extinguisher label shows what types of fire it will extinguish.

Accidents

Make sure you are aware of the location and contents of the shop's first-aid kit. There should be an eye wash station in the shop so that you can rinse your eyes thoroughly should you get hydraulic fluid, battery acid, asbestos dust, or other irritants in your eyes (Figure 2-13). Find out if there is a resident nurse in the shop or at the school, and locate the nurse's office. If there are specific first-aid rules in your school or shop, find out what they are and abide by them.

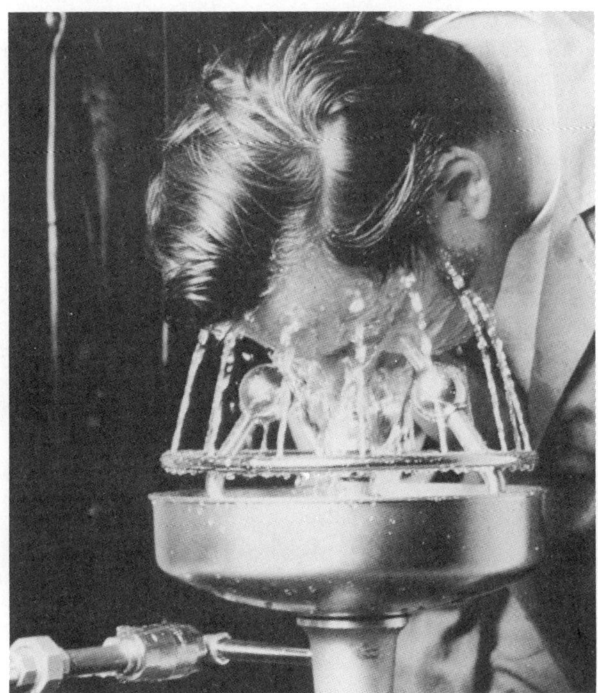

Figure 2-13 Eye wash station for flushing contaminants from your eyes (Courtesy of Western Emergency Equipment)

If someone is overcome by carbon monoxide, move the person to fresh air immediately. Immediately cool burns by rinsing in cold water or by applying an ice pack. To stop bleeding from a deep cut or puncture wound, apply pressure on or around the wound and get medical help. Never move someone you suspect has broken bones or a back injury unless the person is in danger from another hazard such as fire or carbon monoxide gas. Call for medical assistance.

Battery Safety

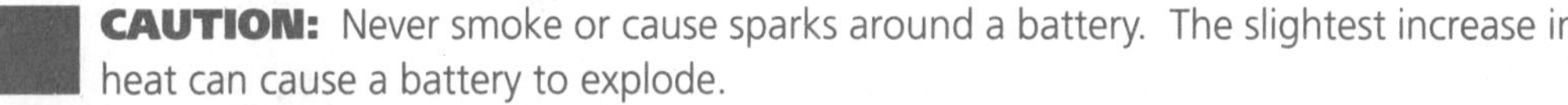

Disconnect the vehicle battery before you disconnect any electrical wire or component such as an antilock brake system controller. This prevents the possibility of a fire or electrical shock. It also eliminates the possibility of an accidental short, which can ruin electronic components or wire runs. To properly disconnect the battery, first disconnect the negative or ground cable. Removing the battery's ground eliminates the possibility that an electric circuit can accidently be completed and energized. Once the ground is disconnected, remove the battery's positive cable.

When reconnecting the battery, reverse the procedure. Connect the positive cable to the battery's positive terminal. Then connect the negative or ground cable to the negative terminal.

CAUTION: Never smoke or cause sparks around a battery. The slightest increase in heat can cause a battery to explode.

CUSTOMER CARE: To keep your customer happy, always record the stations set on the radio before disconnecting the battery. Cutting power to the radio will erase the programmed station settings from the radio's memory. Reset the stations and the dashboard clock after the battery has been reconnected.

Jump Starting

To avoid hurting yourself on the car's electrical system, carefully follow these steps when using jumper cables and a booster battery to start an engine (Figure 2-14). Do not follow this procedure if the drained battery is a sealed battery equipped with a charge indicator lamp that indicates the need for battery replacement.

1. If the batteries are equipped with vent caps, remove them. On sealed batteries do not attempt to pry off the top.
2. Cover the vent holes with a wet cloth.
3. Position the vehicles so they do not touch one another.
4. Turn off head lights, heater fans, and all electrical accessories on both cars.
5. Connect the positive (+) terminal of the booster battery to the positive terminal of the drained battery using the positive jumper cable.

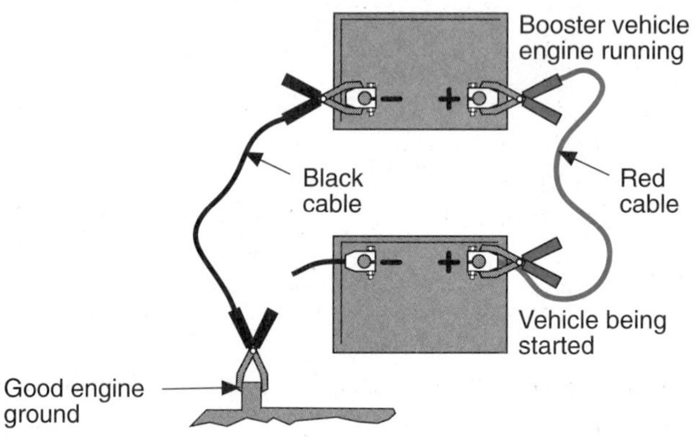

Figure 2-14 Proper connections for jump starting a vehicle

6. Connect one end of the negative jumper cable to the negative (-) terminal of the booster battery.
7. Connect the second end of the negative jumper cable to a known good ground on the car being "jumped." The vehicle frame or engine block are good examples. NEVER CONNECT THIS CABLE TO THE NEGATIVE TERMINAL OF THE DRAINED BATTERY.
8. Start the engine of the booster car.
9. Start the engine of the car being jumped.
10. After the engine of the jumped car starts, first disconnect the negative jumper cable of the jumped car from its good ground. Next disconnect the negative cable from the negative terminal of the booster car battery.
11. Disconnect the positive jumper cable from both positive terminals.

Personal Safety

Accidents don't just happen to other people, they can happen to you. To ensure your personal safety, think about what you are doing at all times and take these steps to protect yourself.

Always wear safety glasses, goggles, or a safety face shield when there is a possibility of dirt, metal chips, or chemicals getting in your eyes (Figure 2-15). If you wear prescription glasses, use glasses with safety lenses and side shields. If you have long hair, tie it up or keep it under a hat so it doesn't get caught in the moving parts of an engine or power tool. Always use any extra safety devices or protective equipment provided for special jobs such as grinding or lathe work.

Classroom Manual
Chapter 2, page 21

A

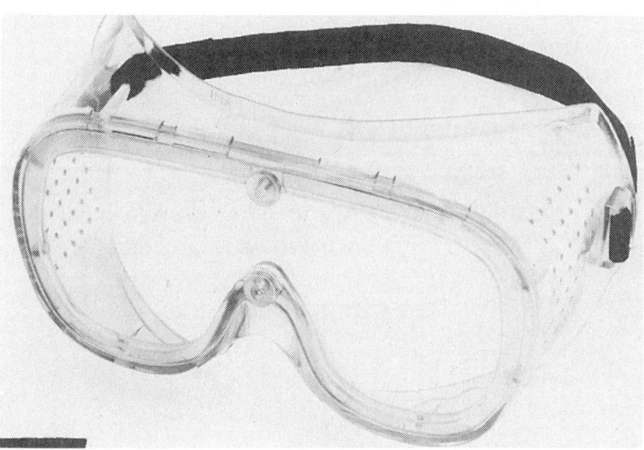

B

C

Figure 2-15 Types of eye protection: (A) safety glasses, (B) goggles, and (C) full face shield (Courtesy of Siebe North, Inc.)

A

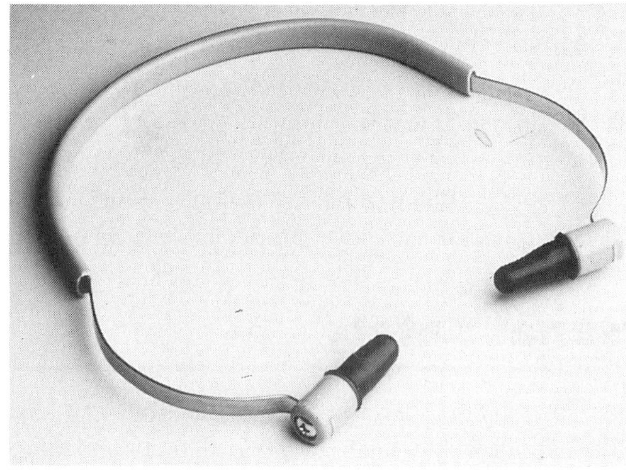

B

Figure 2-16 Types of hearing protection (Courtesy of Siebe North, Inc.)

Wear the correct clothing for the trade, such as well-fitted coveralls. Avoid wearing loose clothing or jewelry that can become caught in moving machinery. Wear heavy-duty work shoes or steel-toed safety shoes. They provide protection against flying sparks, heavy falling objects, and even some acids and solvents. A protective head covering such as a bump cap is advised when working in a pit or under an overhead hoist.

Classroom Manual
Chapter 2, page 22

The noise level in most shops is quite high. Protect your hearing by wearing earplugs or earmuffs (Figure 2-16).

Never get underneath a vehicle that does not have properly installed jack safety stands. Never play around in the shop area. To prevent serious burns, avoid contact with hot metal parts such as the radiator, exhaust manifold, tailpipe, and muffler.

Classroom Manual
Chapter 2, page 26

Move vehicles very slowly (about 2 mph) in and out of the shop, checking to see that no one is in the way. Avoid getting between a moving vehicle and a wall or other large stationary object. Never leave a running vehicle unattended. Use wheel chocks to block the wheels of any running vehicle you are working on.

Do not use drugs or alcohol before or during your work.

Tool and Equipment Safety

Always follow all operating instructions and precautions covering the use of tools and equipment. Do not use damaged tools or power equipment. Report any damage to your instructor or the shop supervisor.

Hand Tools

Many of the tools used in brake service work are general-purpose hand tools. Misuse of hand tools is the leading source of accidents in the shop. Know what a tool has been designed to do and use it properly.

1. Select the correct size and type of hand tool for the job, and only use a hand tool for the job for which it was intended.

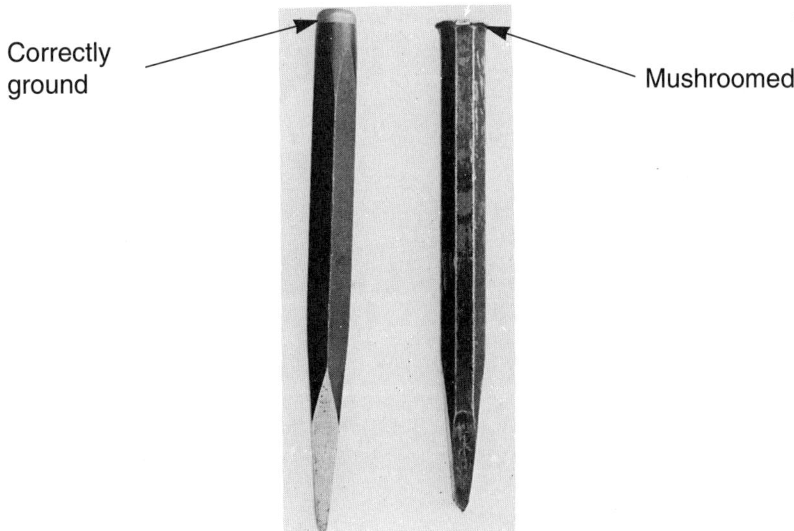

Figure 2-17 Examples of correctly ground and dangerous mushroomed heads on chisels and punches

2. Keep hand tools in good condition and store them in a safe location when not in use. Keep cutting tools properly sharpened.
3. Never put sharp or pointed tools in your pockets.
4. Secure small parts in a vise or clamp when working on them.
5. Repair or replace tools with loose or cracked handles.
6. Select a chisel with a blade at least as large as the cut to be made. Don't use chisels or punches on hardened materials such as locating pins and never use a chisel, punch, or knife as a prying tool. Excessive force can damage or snap the tool.
7. Grind off any mushroom heads (Figure 2-17), as the sharp edges could break off when struck. Keep a smooth bevel ground on the heads of all punches and chisels.
8. When using cutting tools, always cut away from your body, keeping hands and fingers behind the cutting edge. Keep handles clean and dry to ensure a firm grip.
9. Never strike a file with a hammer or use a file as a prying tool. When using a file, always stroke away from yourself (Figure 2-18) and use a file card to keep the file clean.
10. Use a screwdriver only to drive screws, never as a punch or prybar tool. Make sure that the screwdriver blade fits properly into the screw slot. An improper fit could damage both the screw slot and the screw blade. Keep the screwdriver blade perpendicular to the screw slot to minimize any slippage.

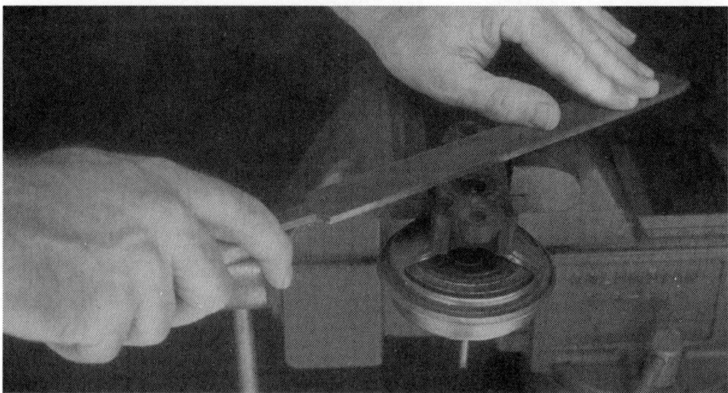

Figure 2-18 Correct method of holding a file

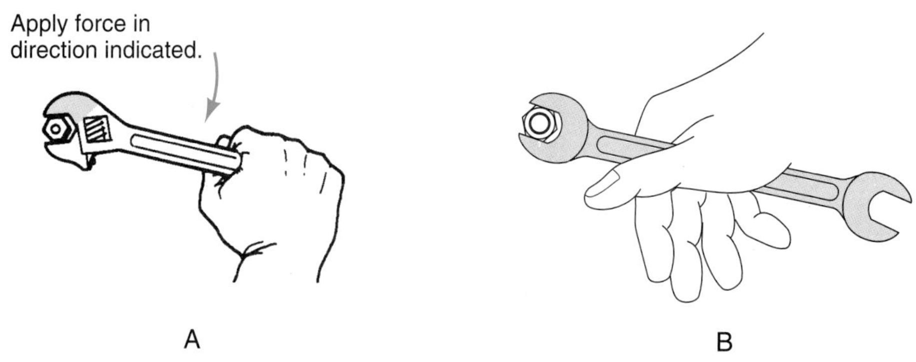
Apply force in
direction indicated.
A
B

Figure 2-19 (A) Pull on a wrench when tightening whenever possible. (B) Proper method of pushing on a wrench with the heel of the hand, fingers loose.

11. Always wear the proper eye protection when working with striking tools. Use soft-faced hammers on hardened surfaces. Never strike one hammer with another. The hammer could be damaged or chipped and the flying chips could cause injury.
12. Always use the proper size wrench for the job. A slipping wrench can damage bolt heads and nuts and cause personal injury. Use a wrench that gives a straight, clean pull (Figure 2-19). If you must push the wrench, do so with the heel of your hand, and do not wrap your fingers around the tool (Figure 2-19B). Do not cock the wrench, as this puts a strain on the points of contact, which can lead to tool failure.
13. Do not use a pipe to extend the length of a wrench. Under excessive force, the wrench or bolt can slip or break. Also, do not use a hammer with a wrench unless the wrench has been specifically designed for this purpose.
14. Replace cracked or worn wrenches. Do not attempt to straighten a bent wrench. It will only weaken it further.
15. Pliers are made for holding, pinching, squeezing, and cutting, but not for turning. Do not substitute pliers for a wrench, as the pliers can slip and damage bolt heads and nuts.
16. Always wear the proper eye protection when using pliers or cutters. Pieces of wire can fly when cut, causing injury.
17. Sockets designed for use with power, hand, or impact tools should not be interchanged as this can result in damage or injury.
18. Use a torque wrench for tightening purposes only. Never use this tool to break nuts or bolts loose.

Power Tools

Portable power tools are powered by electricity or air (pneumatic). Never operate a power tool without being properly trained in its use. Always use a tool for the job it was designed to do and keep all guards in place and in working order. General safety rules follow:

1. Always wear the proper eye and ear protection. Gloves and a face shield are required when operating air chisels or air hammers.
2. All electrical tools, unless they are double-insulated type, must be grounded. Replace damaged or frayed power cords, and do not use a two-prong electrical adapter to plug in a three-prong, grounded piece of equipment. Never use a grounded piece of equipment that has the third ground plug removed.
3. Never try to make adjustments, lubricate, or clean a power tool while it is running or plugged in. Keep all guards in place and in working order.
4. Be sure pneumatic tools and lines are attached properly.

5. Turn off and unplug all portable power tools when not in use, and return all equipment to its proper place.

When operating stationary power tools such as brake lathes and grinders:

1. Always wear eye and ear protection when operating machinery.
2. Do not operate any machine without receiving instructions on the correct operating procedures. Read the owner's manual to learn the proper applications of a tool and its limitations. Make sure all guards and shields are in place. Remove all key and adjusting wrenches before turning the tool on.
3. Give the machinery your full attention. Do not look away or talk to fellow students. Work area clean and well-lighted. Never work in damp or wet location.
4. Don't abuse electrical cords by yanking them from receptacles or running them over with vehicles or equipment.
5. Inspect equipment for any defects before using it. Make all adjustments before turning on the power. Whenever safety devices are removed to make adjustments, change blades, adapters or make repairs, turn off and unplug the tool. Lock and tag the main switch, or keep the disconnected power cord in view at all times.
6. Always wait for the machine to reach full operating speed before applying work.
7. Do not leave the power tool until it has come to a complete stop. Allow a minimum of four inches between your hands and any blades, cutters, or moving parts. Do not overreach—try to keep the proper footing and balance at all times.

Working with Cleaning Equipment

Classroom Manual
Chapter 2, page 32

Be careful when using solvents. Most are toxic, caustic, and flammable. Avoid placing your hands in solvent; wear protective gloves, if necessary. Read all manufacturer's precautions and instructions and Material Safety Data Sheets (MSDS) before using. Do not use gasoline to clean components—this is very dangerous. Gasoline vaporizes at such a rate that it can form a flammable mixture with air at temperatures as low as -50°F. Never mix solvents. One could vaporize and act as a fuse to ignite the others.

Wipe up spilled solvents promptly, and store all rags in closed, properly marked metal containers. Store all solvents either in their original containers or in approved, properly labeled containers. Finally, when using a commercially made parts washer, be sure to close the lid when you are finished.

All bolts, nuts, lock rings, and other fastening components mentioned in the manufacturer's service manual are crucial to the safe operation of the brake system. Failure to install these items could cause damage or unsafe operating conditions. Manufacturer's torque specifications must be followed.

Lift Safety

Many types of brake service require that you raise the vehicle off the floor. Always make sure the vehicle is properly placed on the lift before raising it and always use the safety locks to prevent the lift from coming down unexpectedly.

CAUTION: Check your shop manual before lifting a vehicle. Failure to use the correct lifting points is potentially dangerous and may also damage the sheet metal of the vehicle. It is wise to refer to a shop manual before lifting any vehicle to identify the proper locations for lifting (Figure 2-20).

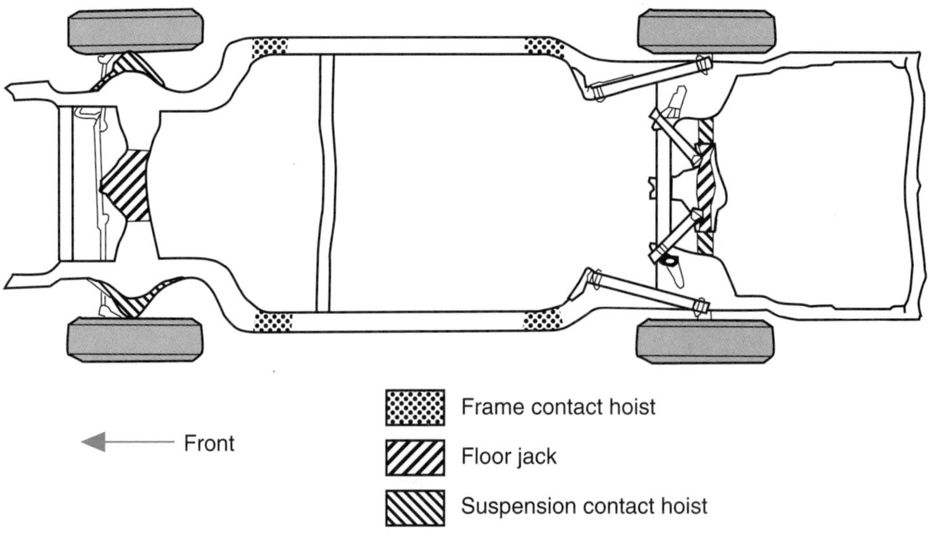

Figure 2-20 Lift point illustration for the various types of lift equipment (Courtesy of General Motors Corporation)

SERVICE TIP: To ensure safe operation of a hydraulic lift, follow these suggestions:

1. Inspect each lift daily.
2. Never operate a lift if it doesn't work right or if it has broken or damaged parts.
3. Never overload a lift. The rate capacity of the lift is listed on the manufacturer's nameplate. Never exceed that rating.
4. Always make sure the vehicle is properly positioned before making the lift.
5. Never raise a vehicle with someone in it.
6. Always keep the lift clean and clear of obstructions.
7. Before moving a vehicle over a lift, position the arms and supports to allow for free movement of the vehicle over the lift. Never drive over or hit the lift arms, adapters, or supports because this may damage the vehicle or the lift.
8. Carefully load the vehicle onto the lift and align the lift arm and contact pads with the specified lift points on the vehicle. Raise the lift until it barely supports the vehicle, then check the contact area of the lift.
9. Always lock the lift into position while working under the raised vehicle.
10. Before lowering the lift, make sure all tools and other equipment are removed from under the vehicle. Also make sure no one is standing under or near the vehicle as it descends.

CASE STUDY

A technician stands in the bay area of a busy brake service shop reviewing a work order. Suddenly she feels a sharp sting on her cheek. She inspects the surrounding area. Several feet to one side another technician is working on an older vehicle raised on a lift. He is cutting off a frozen nut using a hammer and chisel. When he pauses in his work, the injured technician asks to see the chisel he is using. It is an old chisel and the head is flat and mushroomed. A tiny piece of the chisel head has broken off when struck. From a distance of five feet, it still hit the first technician with enough force to cut her. She was not wearing goggles. If it had hit her eye, the injury would have been much worse. There are several lessons to learn from this accident. Keep even the simplest of tools, such as chisels, in good working order. Always wear eye protection in work areas, even if you yourself are not directly involved in the job. Finally, review work orders and paper work away from the work area. It can distract your attention from what is going on around you.

Terms to Know

Asbestos
Carbon monoxide
Class A, B, C, and D fires
Hazardous waste
Material Safety Data Sheet (MSDS)
OSHA
Phosgene gas
Respirator
Right-to-Know laws
Wet cleaning

ASE Style Review Questions

1. When working on a vehicle equipped with antilock brakes:
Technician A relieves system pressure by slightly opening a bleeder screw and allowing fluid to spray into a wide-mouth container.
Technician B turns the ignition off and strongly pumps the brake pedal until an increase in pedal pressure is felt.
Who is correct?
A. A only
B. B only
C. Both A and B
D. Neither A nor B

2. To clean the floor in the brake service bay:
Technician A uses the shop's industrial vacuum system.
Technician B wet mops the floor, keeping it damp at all times.
Who is correct?
A. A only
B. B only
C. Both A and B
D. Neither A nor B

3. An underhood electrical harness has shorted out and caught fire:
Technician A douses the compartment with water from a nearby cleaning bucket.
Technician B clears the area of people and returns with a class C fire extinguisher to handle the fire.
Who is correct?
A. A only
B. B only
C. Both A and B
D. Neither A nor B

4. To prevent excess asbestos dust in the shop area:
Technician A buys preground lining materials whenever possible.
Technician B always uses a grinder equipped with its own exhaust system.
Who is correct?
A. A only
B. B only
C. Both A and B
D. Neither A nor B

5. To learn more about the cleaning solvents used in the shop:
Technician A reads the MSDS from the manufacturer.
Technician B attends the shop's Right-to-Know training session.
Who is correct?
A. A only
B. B only
C. Both A and B
D. Neither A nor B

6. When removing a vehicle's ABS control unit:
Technician A relieves hydraulic pressure in the system.
Technician B disconnects the battery ground cable.
Who is correct?
A. A only
B. B only
C. Both A and B
D. Neither A nor B

7. While working on a vehicle with the engine running, a coworker feels dizzy and sick to her stomach:
Technician A tells her to sit on the floor with her head lowered and then leaves to find the school nurse.
Technician B leads her outside to fresh air and instructs another student to turn off the car engine.
Who is correct?
A. A only **C.** Both A and B
B. B only **D.** Neither A nor B

8. *Technican A* never mixes cleaning solvents in the shop.
Technician B says gasoline is a good all-purpose cleaner for greasy parts.
Who is correct?
A. A only **C.** Both A and B
B. B only **D.** Neither A nor B

9. Before raising a vehicle on a hydraulic lift:
Technician A checks the vehicle service manual for the proper lift points.
Technician B checks for proper positioning once the vehicle is driven on the lift.
Who is correct?
A. A only **C.** Both A and B
B. B only **D.** Neither A nor B

10. *Technician A* says sockets are interchangeable between pnuematic and electric power drivers.
Technician B says bent wenches are dangerous and should be straightened before using.
Who is correct?
A. A only **C.** Both A and B
B. B only **D.** Neither A nor B

Master Cylinder Service

Upon completion and review of this chapter, you should be able to:

- ❑ Perform a safe brake system test drive.
- ❑ Diagnose poor stopping, brake drag, or hard pedal caused by master cylinder problems and perform needed repairs.
- ❑ Diagnose poor stopping, brake drag, and high, low, or hard pedal conditions caused by problems in the step bore master cylinder or its internal valves. Perform needed repairs.
- ❑ Check and adjust master cylinder fluid level.
- ❑ Measure and adjust brake pedal pushrod length.
- ❑ Inspect master cylinder for leaks and defects and determine needed repairs.
- ❑ Remove and replace a master cylinder and bleed system.
- ❑ Overhaul a master cylinder.

Brake System Road Test

Basic Tools

Basic mechanic's tool set

Clean shop towel

Flare nut wrench

To operate safely, the master cylinder and other hydraulic components of a vehicle's brake system must work perfectly (Figure 3-1). Because the system uses brake fluid under pressure to apply the brake shoes or pads, leaks in the master cylinder or brake lines rob the system of pressure and can cause dangerous operating conditions. That is why the master cylinder and hydraulic system must

The crisscross hydraulic brake systems work independently for greater safety.

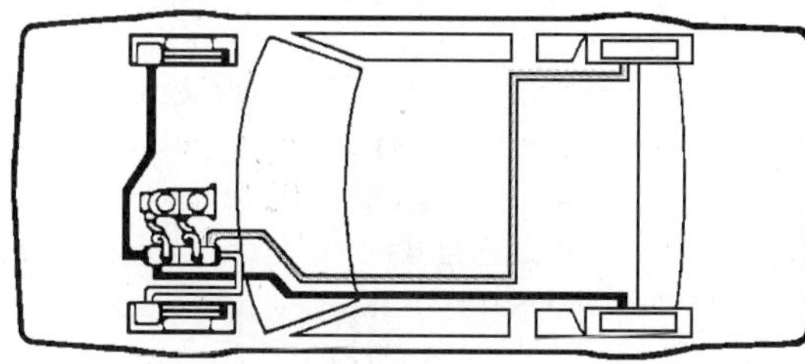

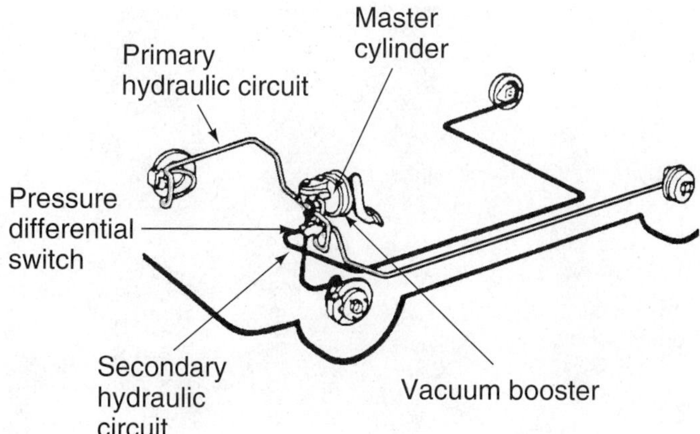

Brake line routing of the diagonally split brake system.

Figure 3-1 Components of a hydraulic brake system (Courtesy of Chrysler Corporation)

be inspected whenever the brake pads or linings are changed, or when a customer complains of poor braking power. Any problems must be corrected immediately.

Check for the following conditions that can cause poor brake performance:

Tire Problems. Worn, mismatched, or underinflated or overinflated tires will cause unequal braking.

Unequal Vehicle Loading. A heavily loaded vehicle requires more braking power. If the load is unequal from front to back or side to side the brakes may grab or pull to one side.

Wheel Misalignment. Wheels that are out of alignment may cause problems that appear brake related. For example, tires with excessive camber or caster settings will pull to one side.

If the vehicle's tires are in good shape and the wheel alignment and vehicle loading do not appear to be the problem, proceed with a brake system road test. Follow these guidelines when road testing a vehicle for brake problems:

1. Test drive the vehicle on a dry, clean, relatively smooth roadway or parking lot. Roads that are wet or grease slicked, or that have loose gravel surfaces will not allow all wheels to grip the road equally. Rough roads cause the wheels to bounce and lose contact with the gripping surface.
2. Avoid crowned roadways. They can throw the weight of the vehicle to one side, which will give an inaccurate indication of brake performance.
3. First test the vehicle at low speeds. Use both light and fairly heavy pedal pressure. If the system can safely handle it, test the vehicle at higher speeds. Avoid locking the brakes and sliding the tires. Braking heavily while turning the wheels will stop a vehicle quicker than locking the brakes.

Check the brake warning light on the instrument panel. It should light when the ignition switch is in the START position and go off when the ignition returns to the RUN position (Figure 3-2).

If the brake warning light remains on when the ignition is in the RUN position, check to see that the parking brake is fully released. If it is, the problem may be a low brake fluid level in the master cylinder. Some vehicles are equipped with a separate master cylinder fluid level warning light. If either warning light remains on, check the fluid level in the master cylinder reservoir.

Listen for unusual brake noise during the test drive. Do you hear squeals or grinding? Do the brakes grab or pull to one side? Does the brake pedal feel spongy or hard when applied? Do the brakes release promptly when you take your foot off the brake pedal?

Figure 3-2 The brake warning light should go on when the ignition switch is in the START position.

Master Cylinder Fluid Levels

Check the master cylinder fluid levels. Although normal brake lining wear may cause a slight drop in fluid level, an abnormally low or empty level in either chamber is a strong indication that there is a leak in the system. The procedure for filling the master cylinder reservoir is shown in Photo Sequence 3.

Photo Sequence 3
Typical Procedure for Filling the Master Cylinder Reservoir

CAUTION: Be careful to avoid spraying brake fluid. To protect the face, never bend directly over the reservoir.

On some antilock brake systems, the manufacturer recommends depressurizing the system before adding brake fluid. When depressurized, the reservoir level may rise slightly, giving a more accurate level reading.

P3-1 Thoroughly clean the reservoir cover before removing it to prevent dirt from entering the reservoir body.

P3-2 Remove the reservoir cover and the diaphragm.

P3-3 Check that the vent hole in the reservoir cover is open.

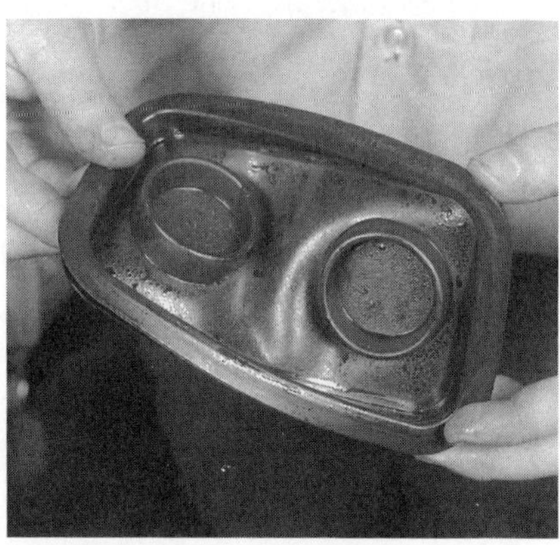

P3-4 Inspect the diaphragm for holes, tears, and other signs of damage. Replace as needed.

P3-5 Check the brake fluid level and its appearance. The fluid level should be within 1/4 inch of the top of the reservoir or at the reservoir's full level marking. The fluid should be clean. There should be no rust or other contamination. Note the level of fluid in each section of the reservoir for diagnostic purposes.

P3-6 Fill the reservoir with the recommended brake fluid. The wrong type of brake fluid, contaminated fluid, water, or mineral oil may cause the brake fluid to boil or the rubber components in the system to deteriorate.

The master cylinder replenishing port is also known as the compensating port.

When checking the master cylinder fluid level, keep in mind that the following conditions do not in themselves indicate a need for master cylinder service:

1. Unequal fluid levels in the master cylinder reservoir chambers on front disc/rear drum brake vehicles may result as fluid moves from the reservoir into the calipers to compensate for normal lining wear. Fill both chambers to the full mark.
2. A slight squirt of brake fluid from one or both master cylinder reservoir chambers when the brake pedal is applied is normal. It is caused by fluid moving through the reservoir replenishing port(s) as the master cylinder pistons move forward in the bore.
3. Light fluid turbulence in the reservoir when the brake pedal is released is the result of brake fluid returning to the master cylinder after the brakes have been released.
4. A slight trace of brake fluid on the booster shell below the master cylinder mounting flange is normal. It results from the lubricating action of the master cylinder wiping seal.

Brake Pedal Test for Fluid Leaks

Hydraulic brake system leaks can be internal and external leaks. Most internal master cylinder leaks result when the cups lose their ability to seal the piston. Brake fluid leaks past the cups internally. Sometimes it also appears as an external leak. Internal and external rubber parts wear with usage or can deteriorate with age or fluid contamination. Moisture or dirt in the hydraulic system can cause corrosion or deposits to form in the bore, resulting in the wear of the cylinder bore or its parts. Although internal leaks do not cause a loss of brake fluid, they can result in a loss of brake performance.

The brake fluid in the master cylinder reservoir will drop as brake shoes and pads wear.

When external leaks occur, the system loses brake fluid. External leaks are caused by cracks or brakes in master cylinder reservoirs, loose system connections, damaged seals, or leaking brake lines or hoses.

To check for a brake fluid leak, perform the following procedure:

1. Run the engine at idle with the transmission in neutral.
2. Depress the brake pedal and hold it down with a constant foot pressure. The pedal should remain firm and the foot pad should be at least 2 inches from the floor for manual brakes and 1 inch for power brakes (Figure 3-3).

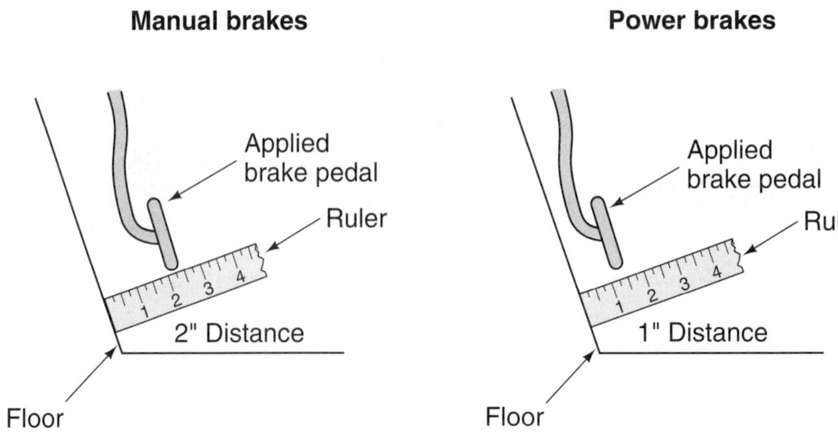

Figure 3-3 Proper floor-to-applied pedal distance is at least 2 inches for manual brakes and 1 inch for power brakes.

3. Hold the pedal depressed with medium foot pressure for about 15 seconds to make sure that the pedal does not drop under steady pressure. If the pedal drops under steady pressure, the master cylinder may have an internal leak or there may be a leak in a brake line or hose. Visually inspect the system as outlined on page 59.

Brake Pedal Mechanical Check

While you are performing the brake pedal test for system leaks, you should also check the pedal's mechanical operation.

1. Check for friction and noise by depressing and releasing the brake pedal several times (with the engine running for power brakes). Be sure the pedal movements are smooth and the pedal returns with no lag or noise.
2. Move the brake pedal from side to side. Excessive side movement indicates worn pedal mounting parts.
3. Check stop light operation by depressing and releasing the brake pedal several times. Have a coworker check that the lights come on each time the pedal is depressed and go off each time it is released (Figure 3-4).

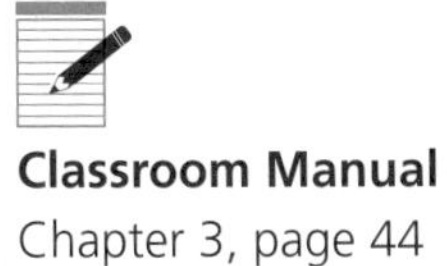

Classroom Manual
Chapter 3, page 44

Pedal Travel Test

Special Tools

Brake pedal effort gauge

Tape measure

Air in the hydraulic system causes most low pedal problems. If this is the cause, bleeding the system should solve the problem. However, low pedal can also be caused by a leak in the hydraulic system, incorrect pushrod length adjustment, a parking brake that is out of adjustment, worn brake shoes, or a brake shoe adjuster that isn't working.

When a given amount of force is applied to the pedal, brake pedal travel must not exceed a certain maximum distance. This maximum travel specification is normally around 2.5 in. (64 mm) when 100 lb. (445 N) of force is applied. The exact specifications are found in the shop manual.

Force applied to the pedal is measured using a brake effort gauge. To perform the test, proceed as follows:

1. Turn off the engine. On vehicles equipped with vacuum assist, pump the pedal until all reserve is exhausted from the booster.
2. Install the brake pedal effort gauge onto the brake pedal (Figure 3-5).

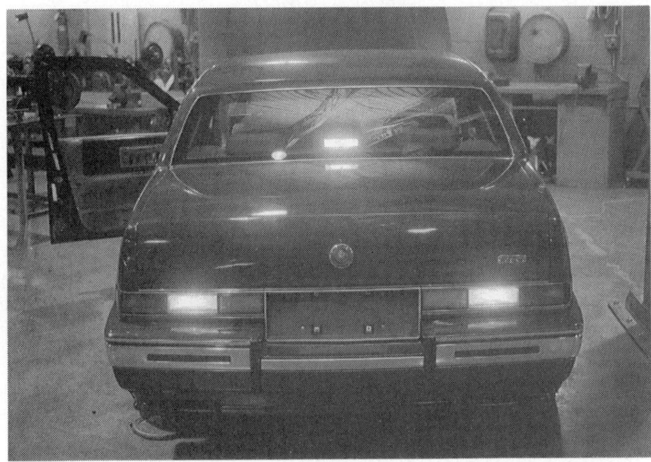

Figure 3-4 Checking brake stop light operation

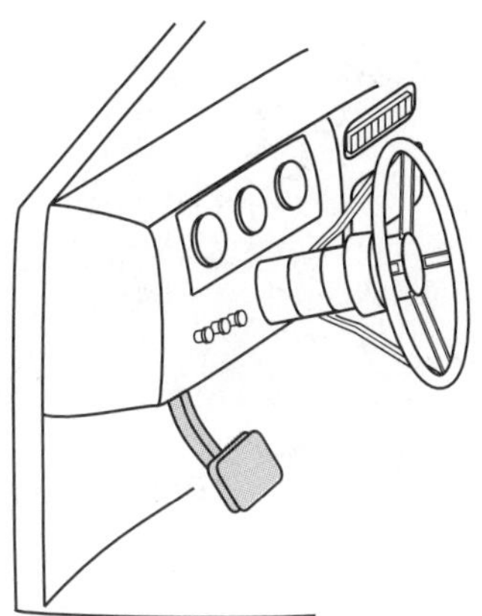

Figure 3-5 Installing the brake pedal effort gauge

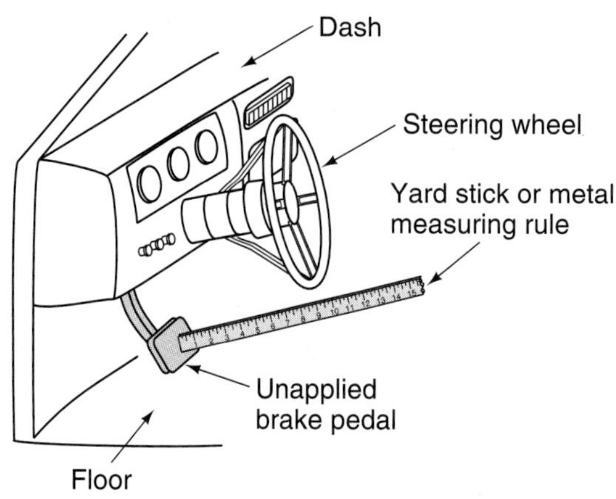

Figure 3-6 Measuring steering wheel rim to unapplied pedal distance

3. As shown (Figure 3-6), hook the lip of the tape measure over the top edge of the brake pedal and measure the distance from the pedal to the steering wheel rim.
4. Apply the brake pedal until the specified test force registers on the brake effort gauge (Figure 3-7).
5. Note the change in pedal position on the tape measure (Figure 3-8). The increased distance should not exceed the maximum specification listed in the shop manual. If it does, look for a leak in the hydraulic system and check pushrod adjustment. Remember, worn shoes, bad shoe adjusters, or a poorly adjusted parking brake can also cause excessive pedal travel.

Master Cylinder Checks

Check for cracks in the master cylinder housing. Look for drops of brake fluid around the master cylinder. A slight dampness in the area surrounding the master cylinder is considered normal and is usually no reason for concern. However, if a reservoir chamber has cracked, it may be completely empty and the surrounding area may be dry. This is because the fluid drained very quickly

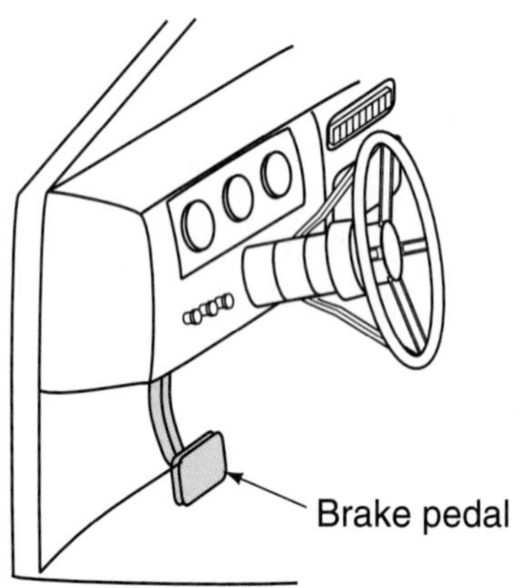

Figure 3-7 Depress the brake pedal until the specified force registers on the gauge.

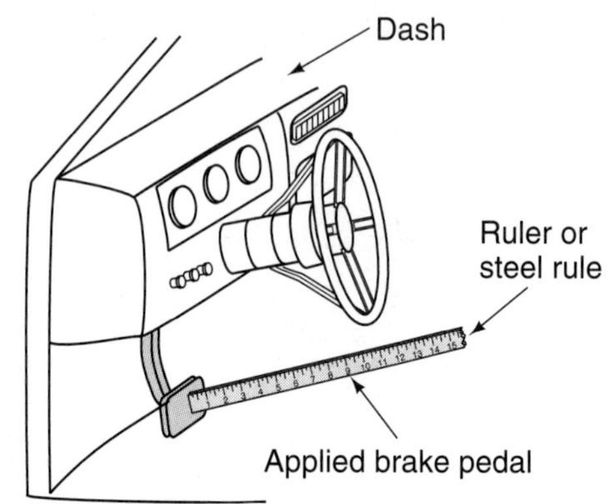

Figure 3-8 Measuring steering wheel rim-to-applied pedal distance

A

B

C

Figure 3-9 Checking for system leaks: (A) inspecting brake lines and hoses; (B) checking rear brake backing plate; (C) checking caliper area on front wheel

and has had time to evaporate or wash away. But with only one-half of the brake system operational, the test drive should reveal the loss of braking power.

Refill the master cylinder reservoir section that is empty and apply the brakes several times. Wait five to ten minutes and check for signs of leakage or fluid level drop in the reservoir.

Hydraulic System Visual Check

If the master cylinder does not appear to be leaking, raise the vehicle on the lift and visually check all brake lines, hoses, and connections (Figure 3-9). Look for brake fluid on the floor under the vehicle and at the wheels. Brake lines should not be kinked, dented, or otherwise damaged, and there should be no leakage. Brake hoses should be flexible and free of leaks, cuts, cracks, and bulges.

Backing plates at the wheels should be free of brake fluid and grease. Any parts attached to them should be tight.

Master Cylinder Tests

While brake pedal response and reservoir fluid levels are strong indicators of problems with the master cylinder or hydrualic system, other tests can be performed to help pinpoint the problem.

Test Hydraulic System for Trapped Air

The following test requires a helper to pump the brake pedal while you observe the reaction to fluid in the master cylinder.

1. Top off the master cylinder reservoir with fresh brake fluid.
2. Loosely place the gasket and cover assembly on the master cylinder. Do not tighten the cover.

CAUTION: This test may result in brake fluid bubbling or spraying out of the master cylinder reservoir. Wear safety goggles. Cover the master cylinder reservoirs with clear plastic wrap or other suitable cover to keep brake fluid off of the paint. Have water and towels ready, just in case.

3. Have your helper rapidly pump the brake pedal approximately 20 times. The brake pedal should be held down on the last application.

4. Remove the master cylinder cover and observe the fluid in the reservoirs.
5. Have your helper release the brake pedal quickly. If air is trapped in the system, pumping the brake pedal will compress it. When the brake pedal is released, the compressed air will expand. This will push brake fluid back into the master cylinder with enough force to produce bubbling or even a small geyser.
6. If air is found in the system, bleed the brake system following the shop manual's recommended sequence.

Testing for Master Cylinder Bypass

If primary piston cup seal is leaking, the fluid will bypass the seal and move between reservoirs. To test for bypass in the master cylinder, proceed as follows:

1. Remove the master cylinder cap and top off the reservoirs with fresh fluid.
2. Watch the fluid levels in the reservoirs while a helper slowly depresses the brake pedal and then quickly releases it .
3. The master cylinder is bypassing if the level of fluid rises in one reservoir while falling in the other as the brake pedal is depressed (or the reverse when the pedal is released), but the total overall level remains the same.
4. Replace or rebuild the master cylinder.

Test for Open Replenishing Ports

To test for open replenishing ports in the master cylinder, proceed as follows:

1. Remove the cover from the master cylinder. While a helper pumps the brake pedal, observe the fluid reservoirs. A small ripple or geyser should be seen in the reservoirs as the brakes are applied .
2. If no turbulence is seen, loosen the bolts securing the master cylinder to the vacuum booster about 1/4 in. and pull the cylinder forward, away from the booster. Hold it in this position and repeat step 1.
3. If turbulence (indicating compensation) now occurs, adjust the brake pedal pushrod length. If turbulence still does not occur, replace the master cylinder. Turbulence can only be seen in the front reservoir of the quick take-up master cylinders.

Testing Master Cylinder Reserve Stroke on Vacuum Boost Systems

When a low pedal or the feel of a bottomed out condition exists on a vacuum power brake-equipped vehicle, check the master cylinder reserve stroke. Use the following procedure to test for brake pedal reserve:

1. Put the transmission in either the PARK or NEUTRAL position and operate the engine at idle.
2. Apply the brake pedal until it stops moving downward or there is increased resistance to the pedal travel. A clunk might be felt as the pedal reaches this point.
3. While holding the pedal in the applied position, raise the engine speed to approximately 2,000 rpm.
4. Release the accelerator pedal, and watch the brake pedal to see if it moves downward as the engine returns to idle speed. As the entine runs down, the increased engine manifold vacuum exerts more force on the brake booster, causing the additional movement of the brake pedal. Although the brake system appears to be bottoming out, an additional stroke is available in the master cylinder.

If the diagnostic test indicates that the master cylinder is not functioning properly, the unit should be serviced, rebuilt, or replaced with a new one.

Removing the Master Cylinder (Non-ABS Systems)

To remove the master cylinder assembly from the vehicle:

1. Disconnect the negative battery cable.
2. To relieve any vacuum boost pressure, pump the pedal 15 to 20 times until a change in pedal feel is noted.
3. Use a shop towel to remove any loose dirt or grease around the master cylinder that could work its way into open lines or the vacuum boost unit.
4. Unplug the electrical connector from the fluid level switch (if so equipped).
5. Place a container under the master cylinder to catch the brake fluid that will leak from the outlet ports when the lines are disconnected.
6. Use a flare nut wrench to disconnect each brake tube line from the master cylinder. Plug the end of each tube as it is disconnected to keep dirt from entering the lines and to prevent excessive brake fluid loss.
7. Remove the nuts attaching the master cylinder to the vacuum booster unit (Figure 3-10).
8. Lift the master cylinder unit out of the vehicle. It may be necessary to insert a small prybar between the booster and the master cylinder to free the master cylinder. Do not drip brake fluid onto the vehicle paint. Before removing the master cylinder in some vehicles, the proportioning valve assembly must be slid off of the master cylinder mounting studs. In some vehicles, the vacuum valve from the booster must be removed and the pressure warning switch connector must be disconnected before lifting off the master cylinder. In vehicles that have manual (nonpower) brake system master cylinders, the pushrod must be disconnected from the brake pedal before the master cylinder can be removed.
9. Using a clean shop towel, clean the master cylinder and vacuum booster contact surfaces.

Some composite master cylinders have a float-type fluid level sensor built into the fluid reservoir or reservoir cap. This electric sensor activates the brake warning light.

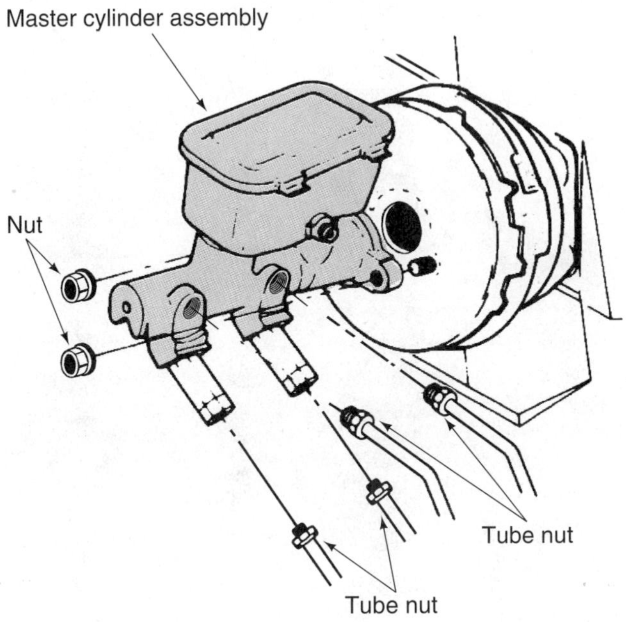

Master cylinder assembly

Nut

Tube nut

Tube nut

Figure 3-10 Removing master cylinder assembly (Courtesy of General Motors Corporation)

SERVICE TIP For many ABS-equipped vehicles, the master cylinder is part of the ABS hydraulic modulator/master cylinder assembly. This ABS assembly is removed as a unit. The master cylinder is then separated from the modulator. See Chapter 8: Antilock Brake System Service for complete details on safely removing the modulator/master cylinder assembly.

Rebuilding the Master Cylinder

Rebuilding a master cylinder involves the following major steps:

Draining the reservoirs

Completely disassembling the unit

Inspecting all parts, including the cylinder bore

Replacing all rubber seals (cups) and "O" rings

Reassémbling the unit with new seals and "O" rings

If the bore of an aluminum master cylinder is pitted or scored, it must be replaced.

The rebuild kit for the master cylinder usually contains all replacable seals, "O" rings, and retainer clips (Figure 3-11). Other components such as the reservoir body, proportioning valves, and fluid sensor switches are only replaced if they are determined to be faulty. See Chapter 5: Hydraulic Line and Valve Service for details on testing these sensors and valves. A typical rebuild procedure for a GM quick take-up master cylinder (Figure 3-12) is illustrated in Photo Sequence 4.

WARNING: Do not clean master cylinder parts with petroleum products (gasoline, kerosene, and so on). Damage to rubber seals and "O"rings will result.

Master Cylinder Reservoir Removal/Replacement

To remove a plastic reservoir body without damaging it, first secure the master cylinder in a vise. Clamp on the cylinder-body flange. This will avoid damaging the cylinder body. Insert a small pry bar between the reservoir and cylinder body and push the reservoir body away from the master

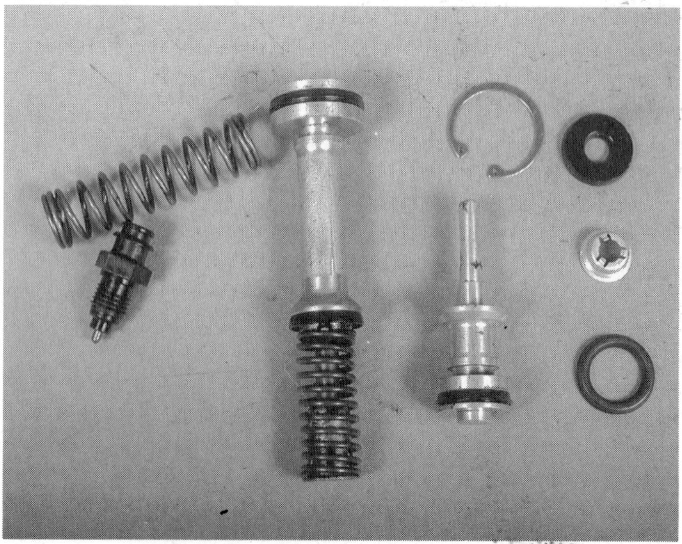

Figure 3-11 Typical master cylinder rebuild kit

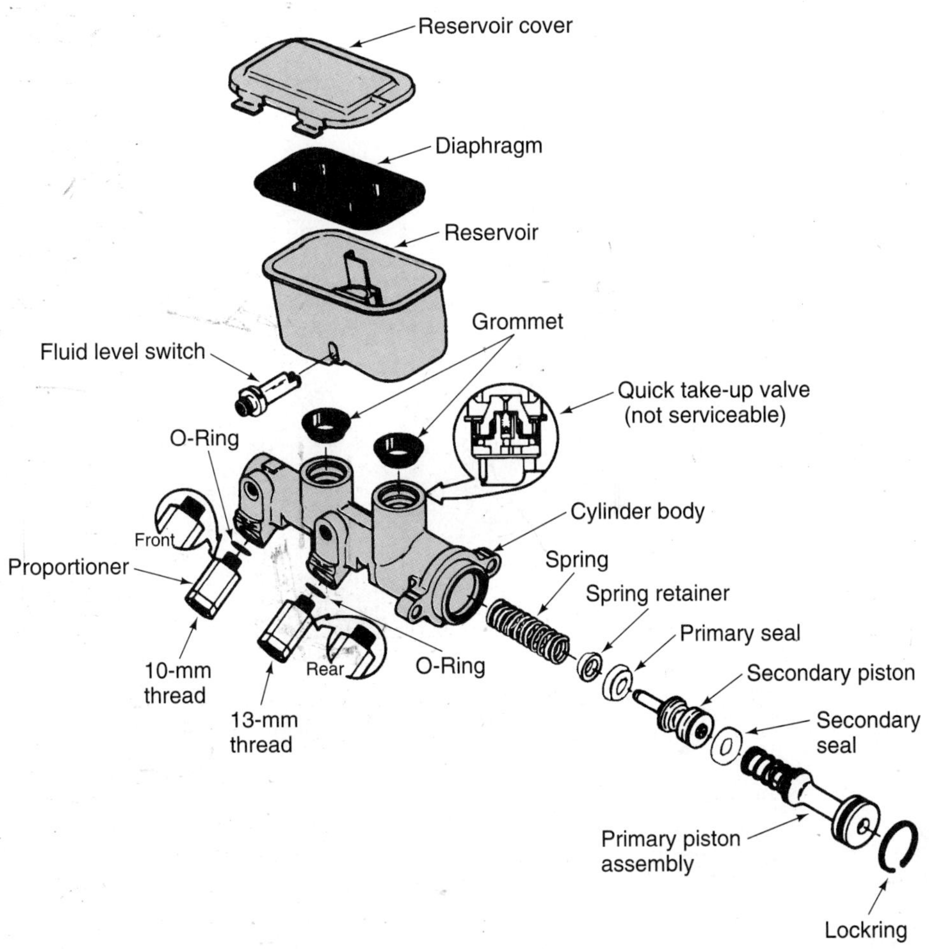

Figure 3-12 Exploded view of GM quick take-up master cylinder (Courtesy of General Motors Corporation)

Photo Sequence 4
Typical Procedure for Rebuilding the Master Cylinder

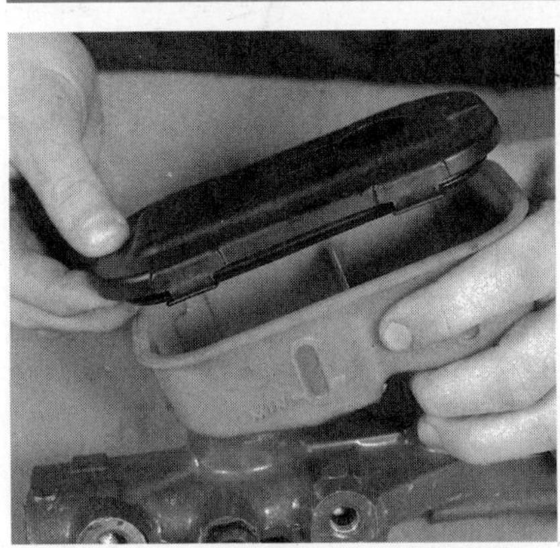

P4-1 Detach the reservoir cap and discard old brake fluid.

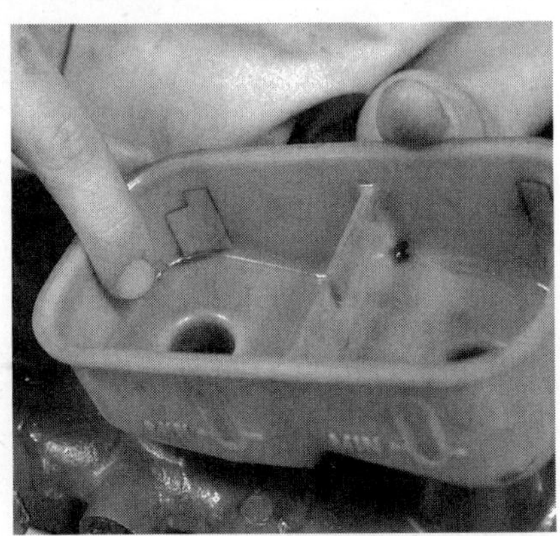

P4-2 Inspect the plastic fluid reservoir for cracks. If the reservoir is cracked or damaged, replace it.

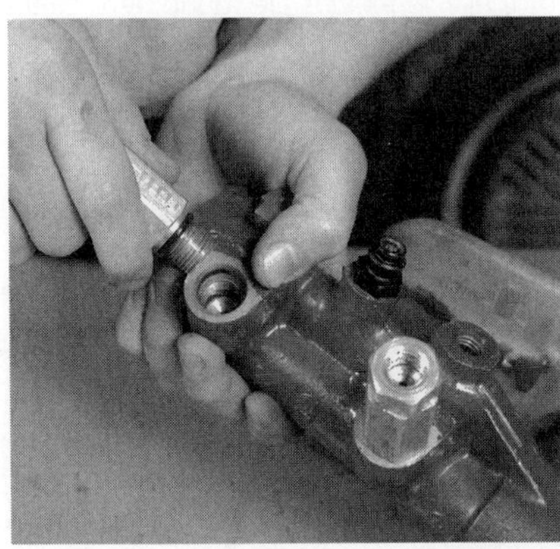

P4-3 Remove the proportioning valves from the cylinder body by screwing them out. Discard the proportioning valve "O" rings.

Photo Sequence 4
Typical Procedure for Rebuilding the Master Cylinder (continued)

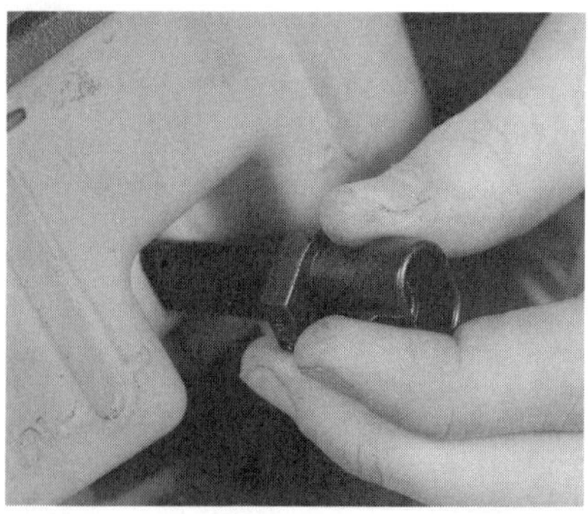

P4-4 Remove the fluid sensor light switch by unlocking the switch tabs using snap ring pliers. Discard the sensor "O" ring.

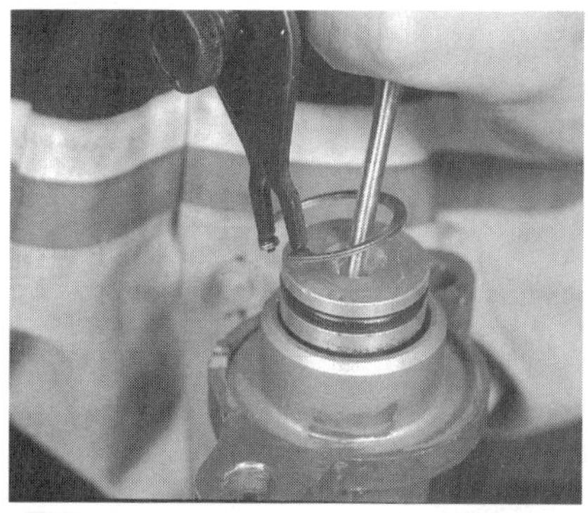

P4-5 While depressing the primary piston, remove the lock ring.

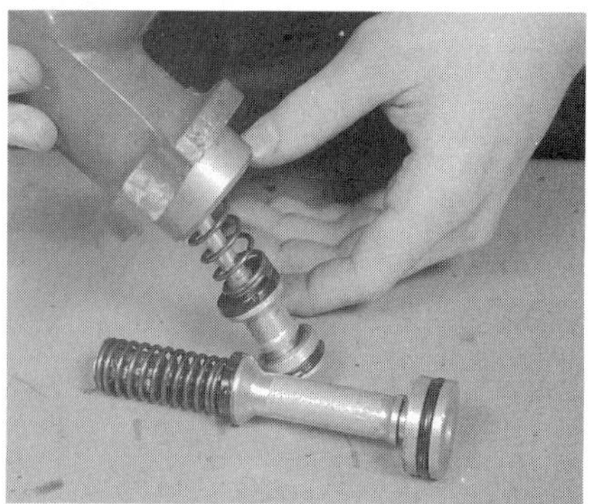

P4-6 Extract the primary and secondary piston assemblies from the bore. Discard the old rubber seals, cups, and "O" rings.

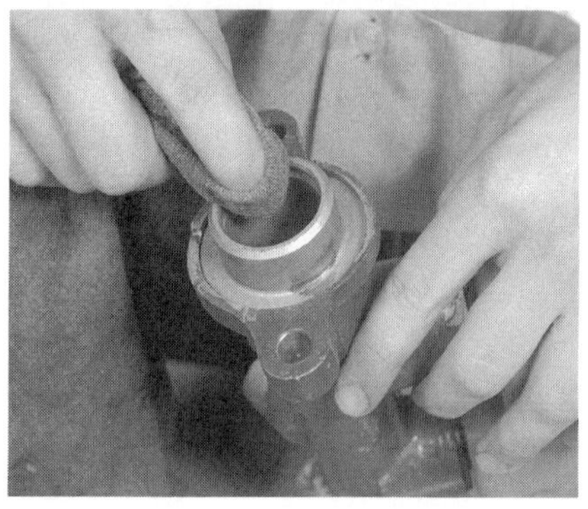

P4-7 Clean the master cylinder and related hard parts with denatured alcohol or clean brake fluid. Blow dry with clean, unlubricated, compressed air. Inspect the cylinder bore for pits, scoring, or corrosion. If any of these conditions are found, replace the master cylinder. Do **not** attempt to hone the cylinder bore.

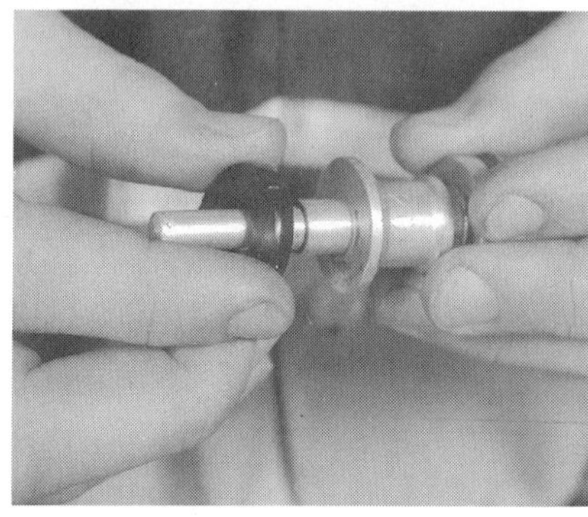

P4-8 To begin reassembly, install the new primary and secondary cup seals onto the secondary piston.

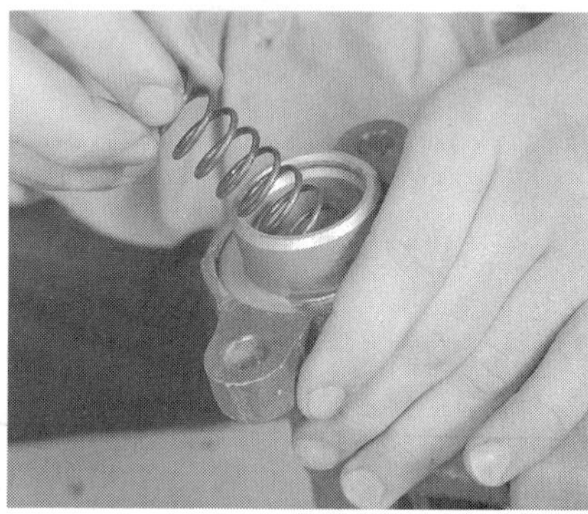

P4-9 Install the spring assembly into the secondary bore.

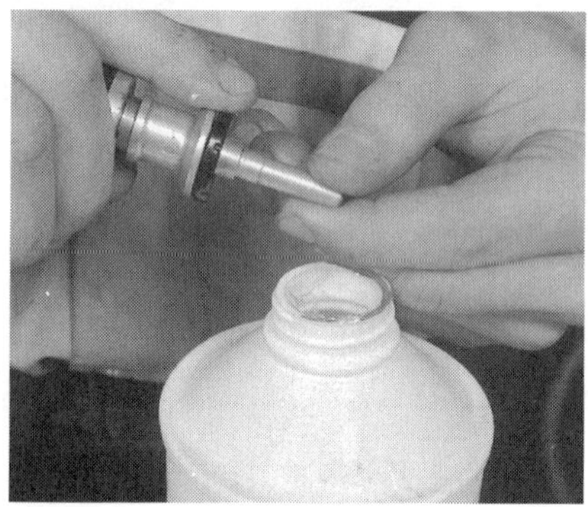

P4-10 Lubricate the secondary piston assembly with brake fluid.

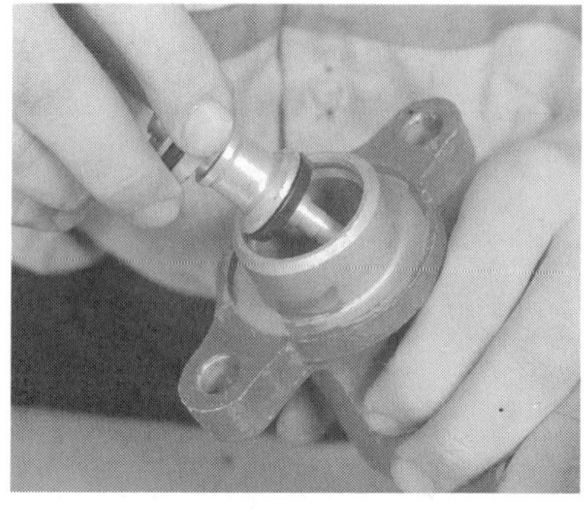

P4-11 Install the secondary piston assembly into the master cylinder bore.

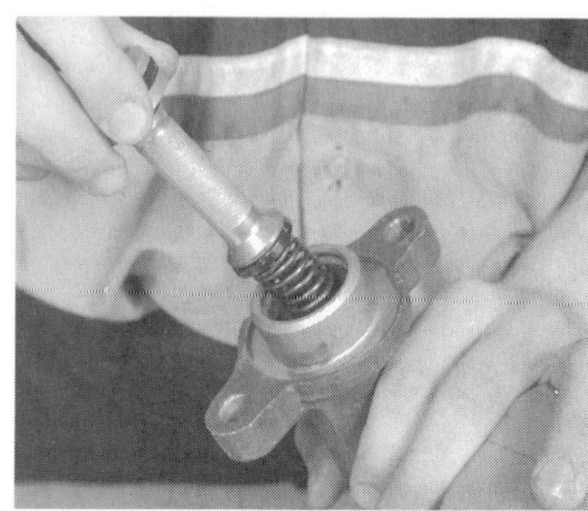

P4-12 Insert the new primary piston assembly into the master cylinder.

Photo Sequence 4
Typical Procedure for Rebuilding the Master Cylinder (continued)

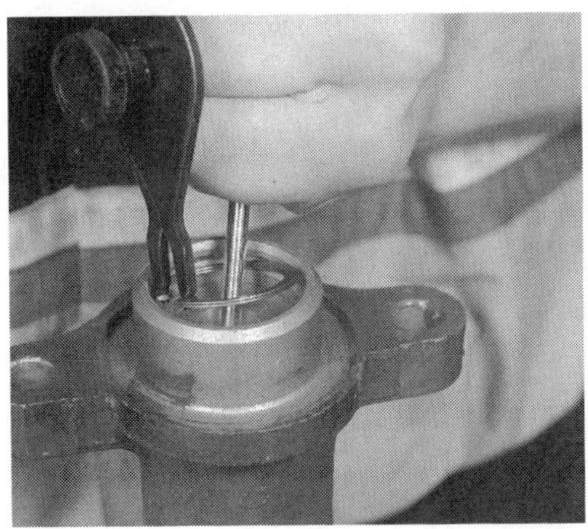

P4-13 Depress the pistons into the bore while installing the lockring retainer.

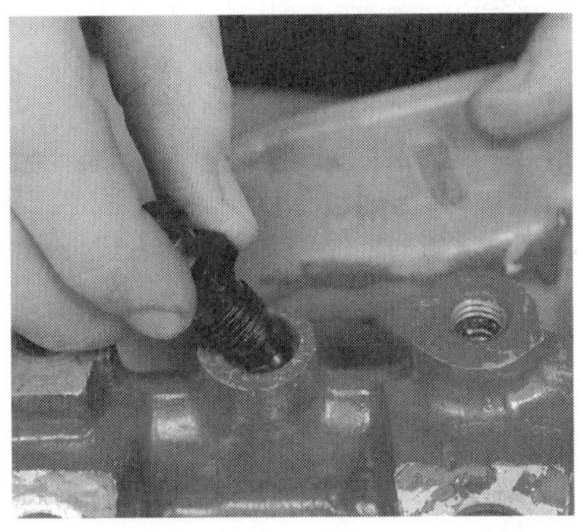

P4-14 Install the fluid level warning light switch using a new "O" ring.

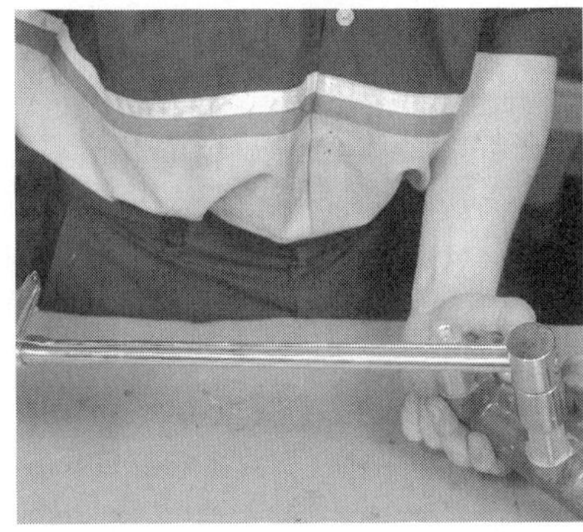

P4-15 Install the proportioning valves and new "O" rings into the master cylinder. Torque to specifications.

cylinder (Figure 3-13). Once the reservoir is free, remove and discard the rubber grommets that seal the reservoir to the cylinder body. Make sure the reservoir is not cracked or deformed. Replace it if it is.

If the reservoir is serviceable, clean it with denatured alcohol and dry it with clean, unlubricated compressed air. Using clean brake fluid, lubricate the new grommets and the bayonets on the bottom of the reservoir. To reinstall the reservoir, place the reservoir top down on a hard, flat surface, such as a workbench. Start the cylinder body onto the reservoir at an angle, working the lip of the reservoir bayonets completely through the grommets until seated. Using a steady downward force and a smooth rocking motion, press the cylinder body onto the reservoir (Figure 3-14).

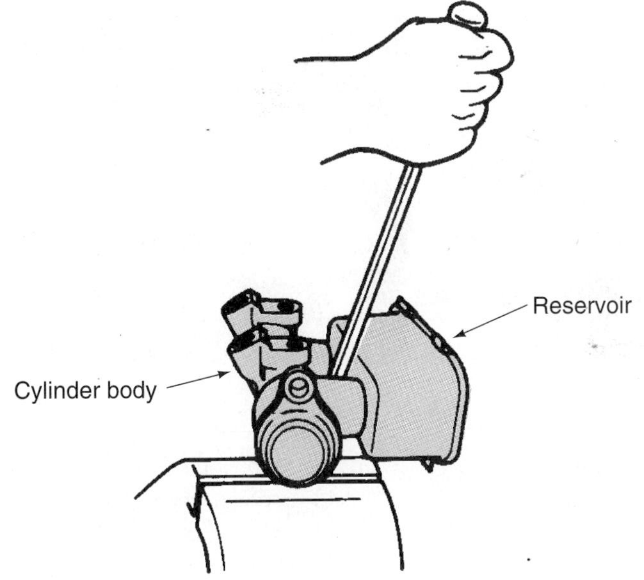

Figure 3-13 Removing the reservoir from the cylinder body on a composite master cylinder (Courtesy of General Motors Corporation)

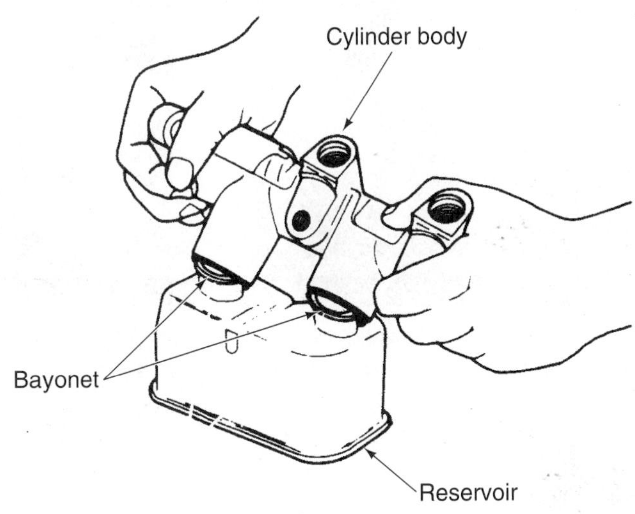

Figure 3-14 Installing reservoir from the cylinder body on a composite master cylinder (Courtesy of General Motors Corporation)

Special Tools

Vise

Bleeding kit tubes and fittings

Bleeding syringe

Port plugs

Wooden dowel

Bench Bleeding Master Cylinders

To remove all air from a new or rebuilt master cylinder, bench bleed it before installing it on the vehicle. Bench bleeding reduces the possibility of air getting into the brake lines. Proper bench bleeding is particularly important with modern dual-piston cylinders, tandem chamber reservoirs, and GM master cylinders that mount on an angle from the horizontal.

Bench bleeding involves mounting the master cylinder in a vise and forcing all air out of the unit. The most popular bench bleeding technique involves installing lengths of tubing to the cylinder ports and feeding them back into the reservoir. The cylinder pistons are manually pumped to recirculate fluid back to the reservoir in a closed loop until all air bubbles to the surface. The procedure for bench bleeding the master cylinder is shown in Photo Sequence 5.

Photo Sequence 5
Typical Procedure for Bench Bleeding the Master Cylinder

P5-1 Mount the master cylinder firmly in a vise being careful not to apply excessive pressure to the casting. Position the master cylinder so the bore is horizontal.

P5-2 Connect short lengths of tubing to the outlet ports, making sure the connections are tight.

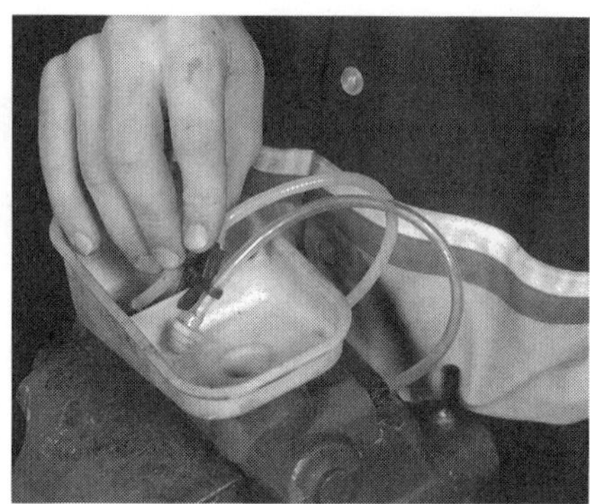

P5-3 Bend the tubing lines so that the ends are in each chamber of the master cylinder reservoir.

P5-4 Fill the reservoirs with fresh brake fluid until the level is above the ends of the tubes.

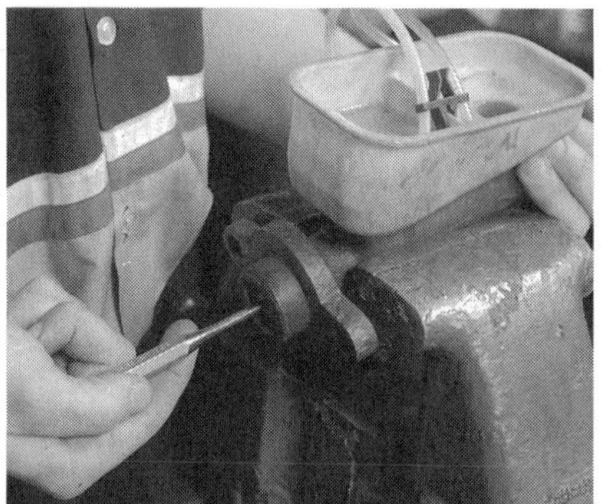

P5-5 Using a wooden dowel or the blunt end of a drift or punch, slowly push on the master cylinder's pistons until both are completely bottomed out in their bore.

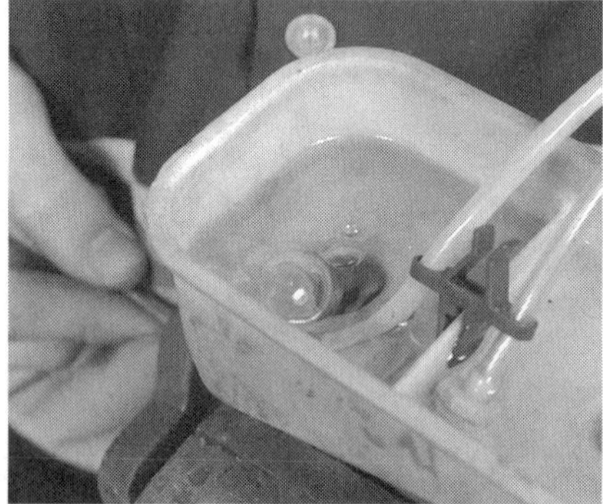

P5-6 Watch for bubbles to appear at the tube ends immersed in the fluid. Slowly release the cylinder piston and allow it to return to its original position. On quick take-up master cylinders, wait 15 seconds before pushing in the piston again. On non-quick take-up units, repeat the stroke as soon as the piston returns to its original postion. Slow piston return is normal on some master cylinders.

Photo Sequence 5
Typical Procedure for Bench Bleeding the Master Cylinder

P5-7 Pump the cylinder piston until no bubbles appear in the fluid.

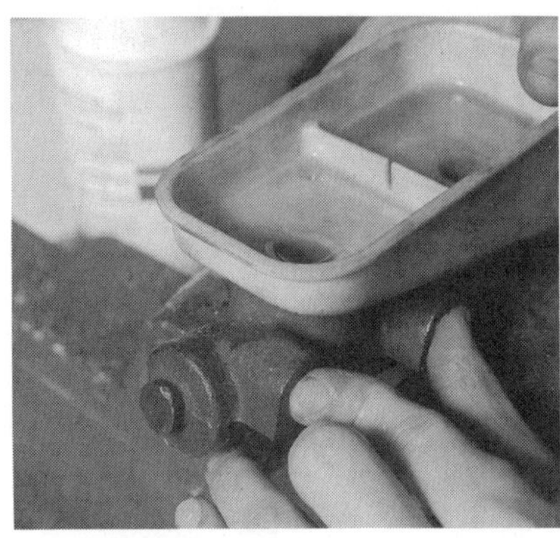

P5-8 Remove the tubes from the outlet port and plug the openings with temporary plugs or your fingers. Keep the ports covered until you install the master cylinder on the vehicle.

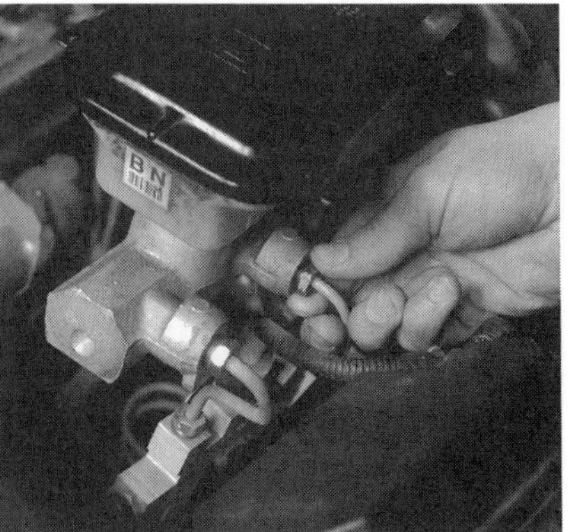

P5-9 Install the master cylinder on the vehicle. Attach the lines, but do not tighten the tube connections.

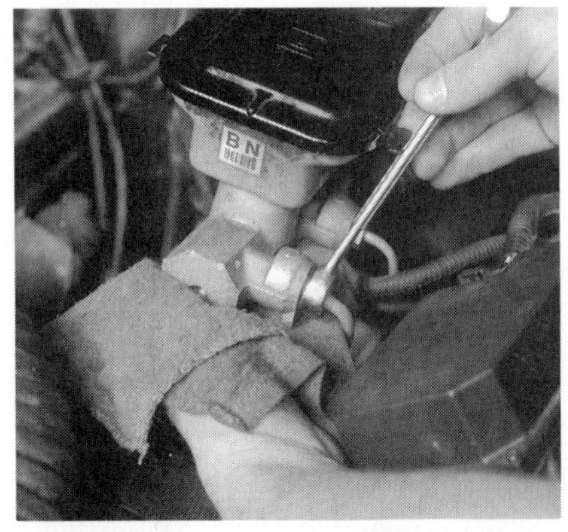

P5-10 Slowly depress the pedal several times to force out any air that might be trapped in the connections. Before releasing the pedal, tighten the nut slightly and loosen it before depressing the pedal each time. Soak up the fluid with a rag to avoid damaging the car finish.

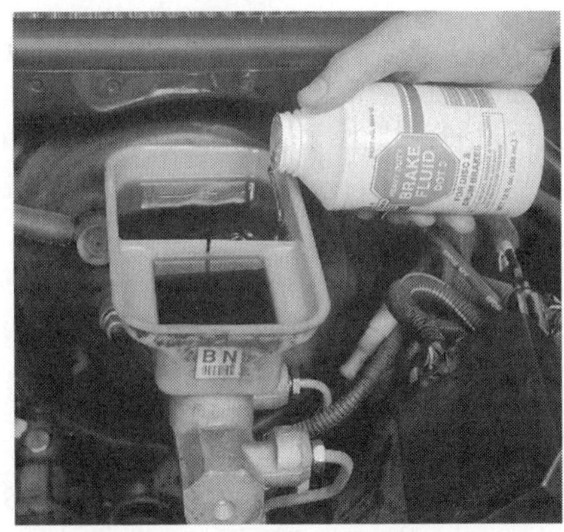

P5-11 When there are no air bubbles in the fluid, tighten the connections to the manufacturer's specifications. Make sure the master cylinder reservoirs are adequately filled with brake fluid.

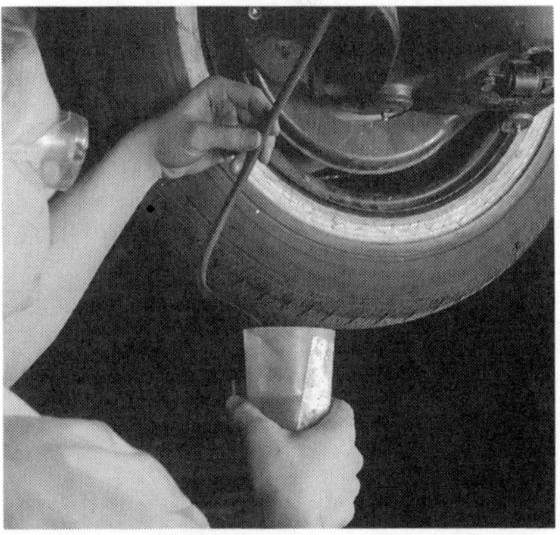

P5-12 After reinstalling the master cylinder, bleed the entire brake system on the vehicle.

Another bench bleeding procedure involves using a special bleeding syringe to draw fluid out of the reservoir, remove air from it, and inject the fluid back into the unit.

CAUTION: Avoid spraying brake fluid or making it bubble violently during the bench bleeding procedure. Do not hold the face directly above the reservoirs. Wear safety goggles.

SERVICE TIP: Bench bleeding kits are available that contain an assortment of fittings and tubing that will fit most domestic and import vehicles. You can also make your own bleeding kit using ordinary brake line hose and fittings. Always make sure the tube nuts are tightened securely to guard against air being sucked into the master cylinder on the return stroke.

SERVICE TIP: Some master cylinders can be bled on the vehicle. First, the master cylinder bore must be horizontal in its normal mounting position. There also must be sufficient working room around the cylinder. Finally, the procedure requires a helper.

Disconnect the brake lines coming out of the master cylinder and install the flexible bleeder tubes back into the reservoir chambers as in Photo Sequence 5. Follow the same basic procedure as in Photo Sequence 5, but use the brake pedal to stroke the master cylinder piston in and out.

When all air has been purged from the master cylinder, remove the tubes. Refill the reservoirs and securely install the master cylinder cap.

Another bench bleeding technique for the master cylinder utilizes a specially designed bleeder syringe (Figure 3-15). The procedure is as follows:

1. Plug the outlet holes of the master cylinder. Carefully mount it in a vise with the pushrod end slightly elevated (Figure 3-15A). Do not clamp the cylinder by the bore or exert pressure on a plastic reservoir.
2. Pour brake fluid into the master cylinder until it is half full.
3. Remove a plug from the cylinder so you can use the syringe to draw fluid out of the chamber. Completely depress the syringe plunger and place its rubber tip firmly against the outlet hole to seal it. Slowly pull back on the syringe plunger to draw fluid out of the master cylinder. Fill the syringe body about one-half full (Figure 3-15B).
4. Point the tip of the syringe upward. Slowly depress the plunger until all air is expelled (Figure 3-15C).
5. Place the tip of the syringe (with fluid) firmly against the same outlet and slowly depress the plunger to inject the fluid back into the cylinder (Figure 3-15D). Air bubbles should appear in the reservoir. When these bubbles stop, remove the syringe and plug the outlet. Repeat this procedure at the other outlet(s). Plug all outlets tightly.
6. With the pushrod end tilted downward slightly, reclamp the master cylinder in the vise.
7. Slowly slide the master cylinder pushrod back and forth about 1/8 inch until no air bubbles are seen in the reservoirs (Figure 3-15E).
8. Remount the master cylinder with the pushrod end up (Figure 3-15F). Fill the syringe with brake fluid and expel the air as in step 4. Remove one outlet plug at a time and repeat steps 4 and 5. The master cylinder is now completely bled.

Vacuum Booster Pushrod Length Check

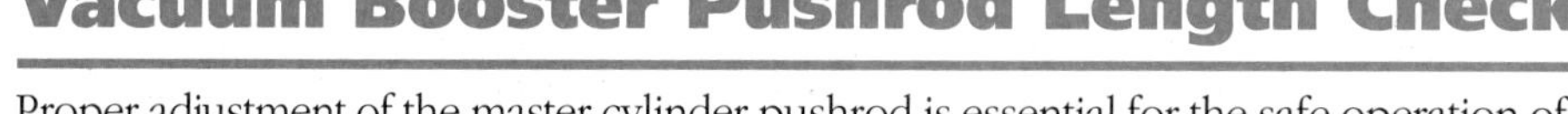

Classroom Manual
Chapter 3, page 44

Special Tool

Pushrod gauge

Proper adjustment of the master cylinder pushrod is essential for the safe operation of power (vacuum boost) brake systems. If the pushrod is adjusted too long, the master cylinder piston closes off the compensating port. This prevents the hydraulic pressure from being released, resulting in brake drag.

If the pushrod is adjusted too short, the brake pedal will be low and the pedal stroke length will be reduced, resulting in a loss of braking power. When the brakes are applied, groaning noises may be heard from the booster.

During factory assembly, the pushrod is matched to the booster. It is normally adjusted only when the vacuum booster unit or the master cylinder is serviced. Vacuum booster pushrod adjustment is usually checked with a gauge.

The pushrod gauge measures from the end of the pushrod to the power unit shell. Two basic designs of gauges are used: Bendix and Delco-Moraine. Be certain the pushrod is properly seated in the booster when making the guage check.

Bendix System Gauge

1. Disconnect the master cylinder from the vacuum booster housing, leaving the brake lines connected. Secure or tie up the master cylinder to prevent the lines from being damaged.

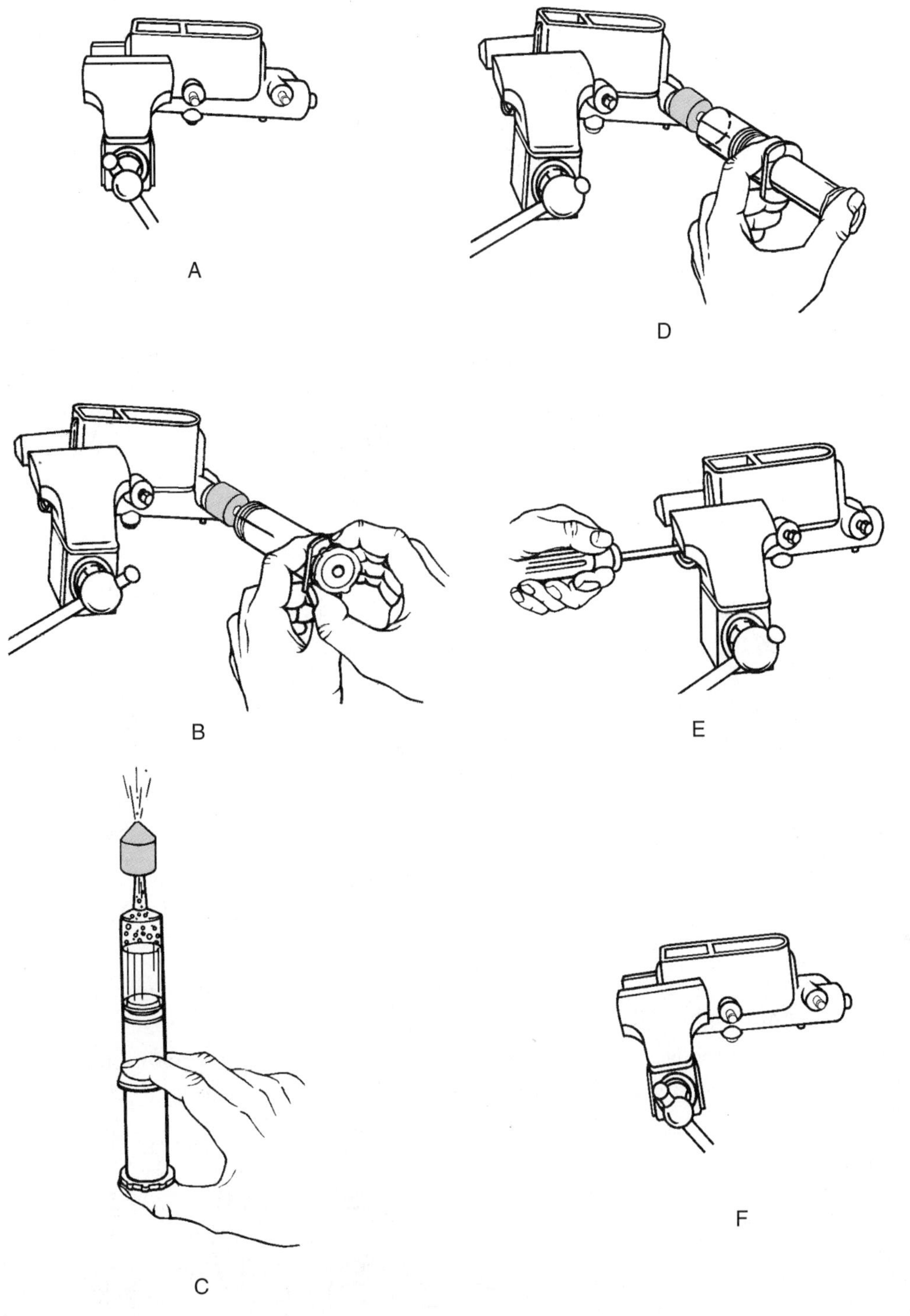

Figure 3-15 Alternate method of bench bleeding a master cylinder using a bleeding syringe

2. Start the engine and let it run at idle.
3. Place the gauge over the pushrod (Figure 3-16). When a force of about 5 lb. is applied to the pushrod, the gauge should bottom out against the booster housing.
4. If a force greater than 5 lb. is needed to seat the gauge against the housing, shorten the length of the pushrod. If the force required to seat the gauge is less than 5 lb., increase the pushrod length.
5. Reinstall the master cylinder.

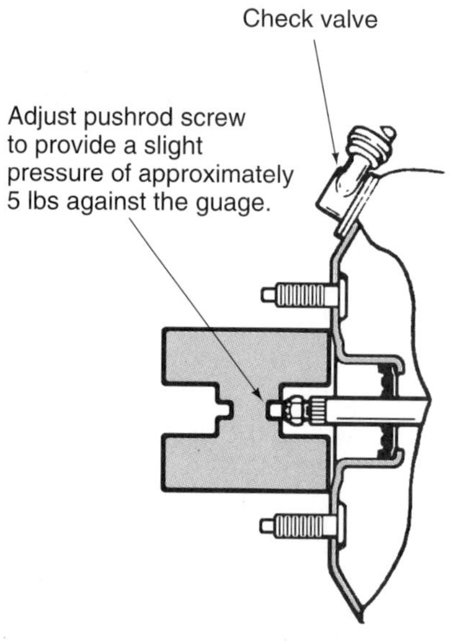

Figure 3-16 Bendix-type pushrod adjustment gauge (Courtesy of Raybestos Division, Brake Systems Inc.)

6. Remove the reservoir cover and observe the fluid while a helper rapidly pumps and releases the brake pedal. If the fluid level does not change, the pushrod is too long. Disassemble and readjust the rod length.

Delco-Moraine (GM) System Gauge

On most GM brake systems, the master cylinder pushrod length is fixed. If the pushrod length needs to be adjusted after master cylinder or booster service, a service-adjustable pushrod must be installed.

1. With the pushrod fully seated in the booster unit, position the go/no-go gauge over the pushrod (Figure 3-17).
2. If the pushrod is not within the limits of the gauge, replace the original pushrod with an adjustable replacement pushrod. Adjust the new rod to obtain the correct height.

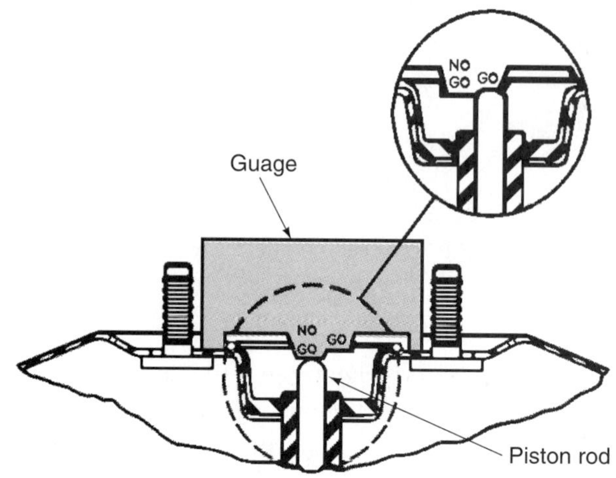

Figure 3-17 Delco-Moraine (GM) pushrod adjustment gauge (Courtesy of General Motors Corporation)

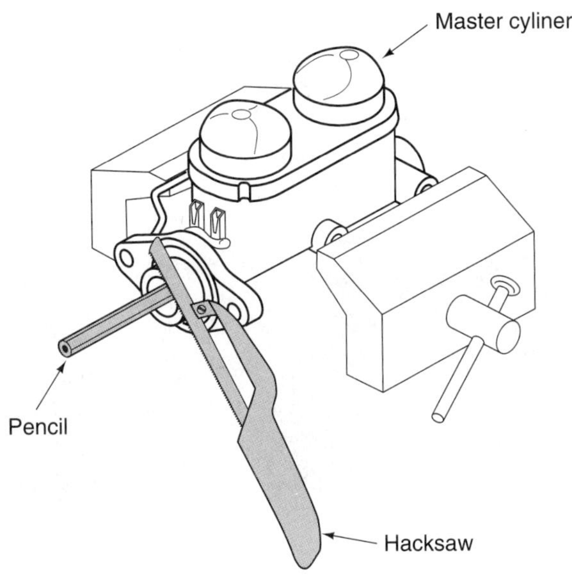

Figure 3-18 Marking pencil

3. Install the vacuum booster and check the adjustment. The master cylinder compensating port should be open with the engine running and the brake pedal released.

Alternate Measurement Method of Pushrod Adjustment

The following procedure is also used on certain models to measure and set pushrod clearance. Always refer to the specific shop manual for the exact procedure required.

1. Insert a pencil in the pushrod socket of the master cylinder. Using a hacksaw blade, mark the point on the pencil that is even with the end of the master cylinder (Figure 3-18).
2. Use a ruler to measure the length of the pencil tip to the saw mark.
3. Now measure how far the master cylinder pushrod protrudes out of the booster assembly.
4. Measure the length of the master cylinder boss (Figure 3-19). Subtract the length of the boss from the length of the pencil to obtain the correct adjusted pencil length. The difference in length between the master cylinder pushrod and the adjusted pencil length is equal to the pushrod clearance.

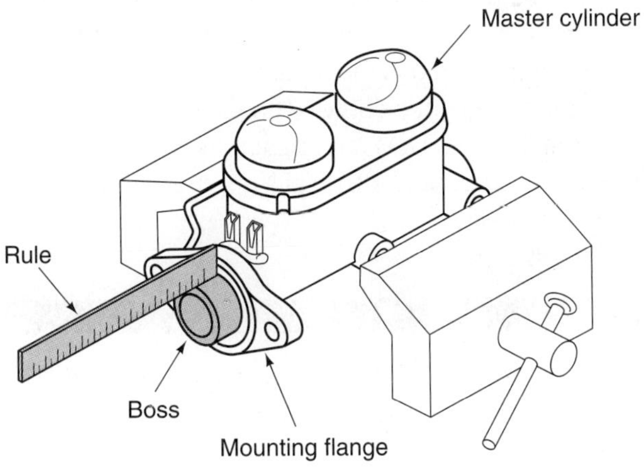

Figure 3-19 Measuring boss

Figure 3-20 Installing brake lines to master cylinder

5. Adjust the pushrod length to obtain the correct clearance. It should be approximately 0.025 in. (1 mm) shorter than the pushrod socket.

Installing a Master Cylinder (Non-ABS Systems)

Some master cylinders may trap air in the bore because the outlets are located slightly below the top of the bore. To remove this air, raise the rear of the vehicle, remove the master cylinder cover, and lightly tap the brake pedal several times.

To install a properly rebuilt and bench bled master cylinder:

1. Install the master cylinder onto the vacuum booster studs.
2. Install the retaining nuts and torque to specifications.
3. Unplug each outlet port opening and use a flare nut wrench to install the brake tube. Tighten securely (Figure 3-20).
4. Reconnect the wiring harness connector to the brake fluid level sensor connector if so equipped.
5. Reconnect the negative battery cable.
6. Properly bleed the system.

Final System Bleeding

Test drive the vehicle only after the system is bled and a good firm brake pedal is present.

Bench bleeding removes all air from fluid in the master cylinder only. After the new or rebuilt master cylinder is installed on the vehicle, the entire hydraulic system must be bled on the vehicle to remove all remaining air from the lines. It is important to use proper bleeding methods and to follow the correct bleeding sequence. Complete details on system bleeding methods are covered in Chapter 5: Hydraulic Line and Valve Service.

CASE STUDY

A customer complains of excessive pedal travel on a ten-year-old domestic sedan. The master cylinder reservoir is full. A careful test drive confirms the condition. The car is placed on the lift. Lines and hoses are in good condition and are leakfree at all connections.

The wheels are pulled and the caliper and wheel cylinders are inspected for leakage. All check out ok. The service technician now suspects problems with the master cylinder or an improperly adjusted pushrod. The first problem is more likely. A fluid bypass test confirms the problem. The technician watches the fluid levels in the reservoirs while a helper depresses the brake pedal slowly and then releases it quickly. The level in one reservoir rises, while the level in the other reservoir drops. But the total level remains the same. A leaking primary piston cup seal is allowing the fluid to bypass the seal and move between reservoirs. The technician replaces the master cylinder and the problem is solved.

Terms to Know

Bench bleeding
Brake grab
External leak
Hard pedal
Internal leak
Master cylinder reserve stroke
Pushrod adjustment
Replenishing or compensating ports

ASE Style Review Questions

1. As part of the brake system inspection and diagnosis:
Technician A checks for correct wheel alignment, inspects the tires, and notes any unbalanced loading of the vehicle.
Technician B performs a test drive on a smooth, level road, testing the brakes at various speeds.
Who is correct?
A. A only **C.** Both A and B
B. B only **D.** Neither A nor B

2. Performing a brake fluid level check is being discussed:
Technician A says unequal fluid levels in the master cylinder reservoir chambers may be caused by normal lining wear.
Technician B says a slight squirt of brake fluid from one or both master cylinder reservoir chambers when the brake pedal is applied is normal. It is caused by the fluid displacement through the reservoir replenishing port(s).
Who is correct?
A. A only **C.** Both A and B
B. B only **D.** Neither A nor B

3. Checks for brake system fluid leaks are being discussed:
Technician A states that if the brake pedal drops under steady foot pressure, it is possible the master cylinder has an internal leak or there may be an external leak in a brake line or hose.
Technician B says a slight trace of brake fluid on the booster shell below the master cylinder mounting flange is an indication of a master cylinder leak and the unit should be replaced.
Who is correct?
A. A only **C.** Both A and B
B. B only **D.** Neither A nor B

4. Brake pedal travel is being discussed:
Technician A says checking pedal travel is best learned through experience, and good technicians develop a "feel" over time for a good travel range.
Technician B says brake pedal travel is a set specification that is found in the shop manual and measured using a special gauge and tape measure.
Who is correct?
A. A only **C.** Both A and B
B. B only **D.** Neither A nor B

5. Testing for open replenishing ports on the master cylinder is being discussed:
Technician A says ripples or a small geyser should be visible in the master cylinder reservoir when the brake pedal is pumped.
Technician B says these ripples will only be visible in the front reservoir of a quick take-up master cylinder.
Who is correct?
A. A only **C.** Both A and B
B. B only **D.** Neither A nor B

6. Rebuilding master cylinders is being discussed:
Technician A automatically replaces all rubber "O" rings and seals.
Technician B inspects all rubber "O" rings and seals to determine if replacement is needed.
Who is correct?
A. A only **C.** Both A and B
B. B only **D.** Neither A nor B

7. Cleaning and inspection of master cylinder components is being discussed:
Technician A suggests cleaning the master cylinder body in the shop's solvent tank cleaning system.
Technician B says the bore of aluminum master cylinder bodies should be honed to remove any small pits or burrs that have developed.
Who is correct?
A. A only **C.** Both A and B
B. B only **D.** Neither A nor B

8. Pushrod adjustment is being discussed:
Technician A says if the pushrod is too long, it causes the master cylinder piston to close off the compensating port, resulting in brake drag.
Technician B says if the pushrod is too short, the brake pedal will be low and pedal stroke length will be reduced with a loss of braking power.
Who is correct?
A. A only **C.** Both A and B
B. B only **D.** Neither A nor B

9. Master cylinder replacement is being discussed:
Technician A says the master cylinder should be bench bled before being installed on the vehicle.
Technician B says after installing the master cylinder on the vehicle, the entire system should be bled at each wheel.
Who is correct?
A. A only **C.** Both A and B
B. B only **D.** Neither A nor B

10. Master cylinder design is being discussed:
Technician A says that on composite master cylinders the reservoir can be removed and replaced if cracked or warped.
Technician B says that on many antilock braking systems the master cylinder is part of the ABS hydraulic modulator/master cylinder assembly that is removed as a unit and then separated.
Who is correct?
A. A only **C.** Both A and B
B. B only **D.** Neither A nor B

Table 3-1 ASE TASK

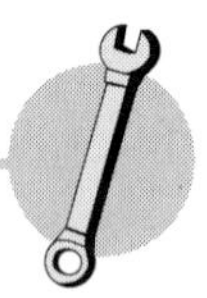

Diagnose poor stopping, brake drag, or high, low, or hard pedal caused by problems in the master cylinder and/or its internal valves.

Problem Area	Symptoms	Possible Causes	Classroom Manual	Shop Manual
BRAKES GRAB	When applied, brakes apparently modulate between on and off.	1. Insufficient fluid in master cylinder	47	55
		2. Defective master cylinder seals	48	60
		3. Cracked master cylinder casting or reservoir	49	58
		4. Leaks at calipers or wheel cylinders or in lines or at connections	96	59
		5. Air in hydraulic system	47	59

Table 3-1 ASE TASK (continued)

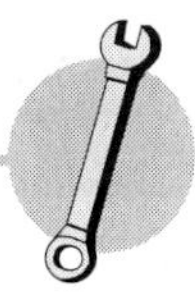

Diagnose poor stopping, brake drag, or high, low, or hard pedal caused by problems in the master cylinder and/or its internal valves.

Problem Area	Symptoms	Possible Causes	Classroom Manual	Shop Manual
HARD PEDAL	When applied, the brake pedal gives hard resistance and is difficult to depress.	1. Broken or damaged hydraulic lines	96	59
		2. Brake pedal binding on shaft	44	57
		3. Brake pedal interference in linkage	44	57
		4. Insufficient clearance between master cylinder pushrod and piston that allows pressure to build in system	44	68
		5. Plugged master cylinder compensating port	48	60
		6. Swollen primary cup in master cylinder	49	60
		7. Frozen master cylinder piston	51	63
LOW, SPONGY, OR SINKING BRAKE PEDAL	Excessive brake pedal travel. Pedal sinks to floor when constant pressure is applied.	1. Low fluid level in master cylinder	47	55
		2. Air in hydraulic system	47	59
		3. Excessive clearance between master cylinder pushrod and the master cylinder piston	44	68
		4. Bad master cylinder check valve that does not hold residual pressure in system	52	63
		5. Leaking primary cup seal in master cylinder that permits fluid bypass	55	60
		6. Leaking hydraulic system	47	59
		7. Inferior or contaminated brake fluid in system	95	112
		8. Excessive movement of brake shoes caused by a frozen starwheel adjuster	148	205
PEDAL RISING AND ALL BRAKES GRABBING **Note:** Some antilock brake systems (ABS) will create a pedal rising effect during ABS application. This is normal for these systems.	Pedal hard to apply; feels as if it is pushing back. When applied, brakes apparently modulate between on and off.	1. Blocked master cylinder compensating port caused by foreign matter, contaminated brake fluid, swollen primary cup seal, insufficient clearance between pushrod and cylinder piston, or sluggish master cylinder return	48	60
		2. Damage to internal rubber parts due to exposure to petroleum-based solvents	49	55
		3. Incorrectly installed internal master cylinder parts due to sloppy rebuild practices	50	63

Table 3-1 ASE TASK (continued)

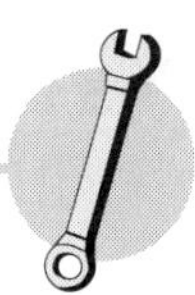

Diagnose poor stopping, brake drag, or high, low, or hard pedal caused by problems in the master cylinder and/or its internal valves.

Problem Area	Symptoms	Possible Causes	Classroom Manual	Shop Manual
BRAKES DRAGGING OR LOCKING AT ONE OR MORE WHEELS	Uneven braking or vehicle pulls to one side or nosedives when brakes are applied.	1. Defective proportioning valve or combination valve	51	63
		2. Improper drum brake adjustment	148	204
		3. Stuck parking brake cable	212	269
BRAKE FADE	System gradually loses braking power after heating up.	1. Sluggish hydraulic system release allowing the brakes to drag and heat up	51	63
		2. Hydraulic system contaminated	95	112
LOSS OF BRAKE OPERATION	Pedal easily depressed without braking effect.	1. No fluid in reservoir	47	55
		2. Air in system	47	59
		3. Bleeder screw left open	138	124
		4. Defective primary cup in master cylinder	55	60
		5. Leaks in hydraulic system	47	59

Power-Assist Unit Service

Upon completion and review of this chapter, you should be able to:

- ❑ Diagnose vacuum power-assist unit problems.
- ❑ Test pedal-free travel with and without the engine running to check power booster operation.
- ❑ Check vacuum supply to a vacuum booster unit using a vacuum gauge.
- ❑ Inspect a vacuum power booster for vacuum leaks.
- ❑ Inspect the unit's check valve for proper operation.
- ❑ Disassemble, repair, and adjust a vacuum booster unit as required to restore proper operation.
- ❑ Inspect and test a hydro-boost booster unit and accumulator for leaks and proper operation.
- ❑ Repair, adjust, or replace components on a hydro-boost unit as needed.
- ❑ Flush and bleed a hydro-boost system.

Types of Power-Assist Systems

Basic Tools

Basic mechanic's tool set

Flare nut wrench

Vacuum gauge

Two types of power-assist systems are used on today's vehicles: vacuum-boost units and hydro-boost units. Both systems multiply the force exerted on the master cylinder piston by the operator. This increases the hydraulic brake fluid pressure delivered to the wheel cylinder and caliper pistons, resulting in increased stopping performance.

Vacuum-assist power boost units (Figure 4-1) use engine vacuum and, in some cases, vacuum pressure created by an external vacuum pump to help apply the brakes. Hydro-boost power-assist units (Figure 4-2) use hydraulic pressure generated by the vehicle's power steering pump or other external pump.

Power-assisted brakes can be thought of as conventional brakes with an added power unit. When troubleshooting and servicing power brake systems, keep the two systems separate. Check

Figure 4-1 Vacuum-boost power-assist unit

Figure 4-2 Hydro-boost power-assist unit

for faults in the master cylinder and hydraulic system first. As with conventional brakes, a spongy pedal in a power brake system is caused by air in the hydraulic lines. Brake grab may be caused by grease on the brake linings. It is important to remember, however, that leaks found in power-assist units are more critical because of the higher hydraulic line pressures.

Check out all conventional components before moving on to the power-assist system.

With the exception of the master cylinder pushrod adjustment, vacuum- and hydro-boost power-assist units are not adjusted in normal service. If the unit is suspect, it is removed and replaced with a new or rebuilt unit. Complete overhaul kits are available.

SERVICE TIP: Always use all the parts contained in the vacuum-boost or hydro-boost overhaul kit. The replace or rebuild method of power-unit servicing eliminates guesswork from power brake servicing.

Vacuum-Boost Systems

Classroom Manual
Chapter 4, page 67

Vacuum-assist units (Figure 4-3) generate their application energy by opposing engine vacuum to atmospheric pressure. A piston cylinder and flexible diaphragm use this energy to provide brake assistance. Modern vacuum-assist units are vacuum-suspended units. This means the diaphragm within the power unit is suspended in a vacuum when the brakes are not applied. The diaphragm is balanced by engine vacuum on both sides until the brake pedal is depressed. When the brake pedal is pressed, an air control valve that is attached to the brake pedal pushrod opens. The air control valve admits atmospheric pressure to the back side of the diaphragm. This forces the diaphragm forward where it increases the amount of force applied to the pushrod and master cylinder piston.

Single and Tandem Units

The power piston contains and operates the vacuum and air valves that control the diaphragm movement.

Vacuum-boost units use one or two diaphragms, but most are single-diaphragm units. Single-diaphragm units are larger in diameter than dual- or tandem-diaphragm vacuum boosters.

Vacuum Check Valve

Check valves are found in all vacuum boosters. The check valve is located between the engine manifold and the vacuum-boost unit. Vacuum can reach the unit through the one-way check valve, but it

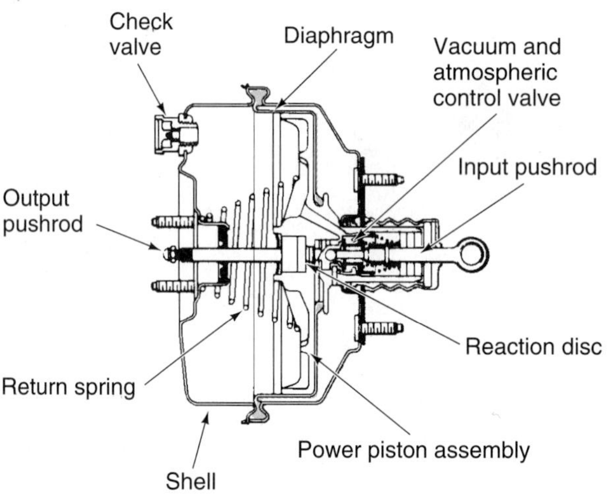

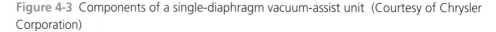

Figure 4-3 Components of a single-diaphragm vacuum-assist unit (Courtesy of Chrysler Corporation)

cannot leak back past the valve. As a result, vacuum is maintained inside of the vacuum booster even after the engine is turned off.

Diagnosing Vacuum Booster Problems

Insufficient manifold vacuum, leaking or collapsed vacuum lines, punctured diaphragms, or leaky piston seals can all result in weak power-unit operation. A steady hiss when the brake is held down indicates a leak that can cause poor power-unit operation. Hard brake pedal is usually the first signal that the unit is on the way to complete failure.

If brakes do not release completely, they may have a tight or misaligned connection between the power unit and the brake linkage. If the pedal-to-booster linkage appears in good condition, loosen the connection between the master cylinder and the brake booster. If the brakes release, the trouble is in the power unit. A piston, diaphragm, or bellows return spring may be broken. If the brakes do not release when the master cylinder is removed from the booster, the brake line may be restricted or there may be a problem in the regular hydraulic circuit.

If the brakes grab, look for common causes such as greasy linings or scored drums before checking out the power unit. If the problem is in the power unit, it may be a damaged reaction control. The reaction control is normally made up of a diaphragm, spring, and valving that tends to resist pedal action. This feature is designed to give the brake pedal "feel" to the operator.

Basic Vacuum-Boost Operational Test

To conduct a basic operating test of the vacuum-boost unit:

1. Turn off the engine.
2. Repeatedly pump the brake pedal to remove all residual vacuum from the booster unit.
3. Hold the brake pedal down firmly and start the engine.
4. If the system is working correctly, the pedal should move downward a short distance, then stop. Only a small amount of pressure should be needed to hold down the pedal.
5. If you do not get the results outlined in step 4, perform the other tests.

Vacuum Supply Tests

If the unit is giving weak braking assistance or no assistance at all, there may be a problem with the vacuum supply to the unit. Vacuum boost efficiency is affected by loose or kinked vacuum lines and/or clogged air intake filters. Another factor may be the check valve, which retains some vacuum in the system when the engine is off. This valve can be checked with a vacuum gauge to determine if it is restricted or stuck open or closed.

Many late model vehicles use a vacuum reservoir to maintain a steady source of vacuum that does not fluctuate with engine speed.

SERVICE TIP: In newer emission-controlled engines, the manifold vacuum generated at idle may be lower than the specification listed in the shop manual.

See Photo Sequence 6 for a typical procedure for vacuum booster testing.

SERVICE TIP: Do not mistake normal booster breathing for a vacuum leak. When the pedal is moved, air rushing through the filter installed in the rubber boot of the booster's input pushrod causes a slight breathing sound, which is normal.

Fluid Loss Test

If the fluid level in the master cylinder reservoir level is low, but there is no sign of an external leak, remove the vacuum line that runs between the intake manifold and the power booster. Inspect it carefully for signs of brake fluid. If you find evidence of brake fluid, the master cylinder secondary seal may be leaking. The master cylinder requires rebuilding. See Chapter 3: Master Cylinder Service for details.

Photo Sequence 6
Typical Procedure for Vacuum Booster Testing

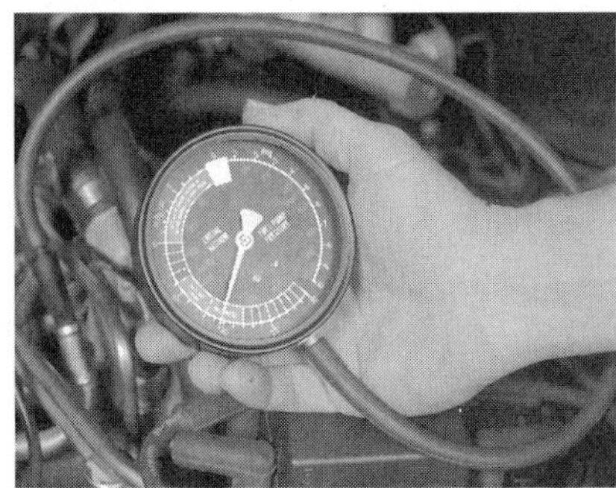

P6-1 With the engine idling, attach a vacuum guage to an intake manifold port. Any reading below 14 in. Hg of vacuum may indicate an engine problem.

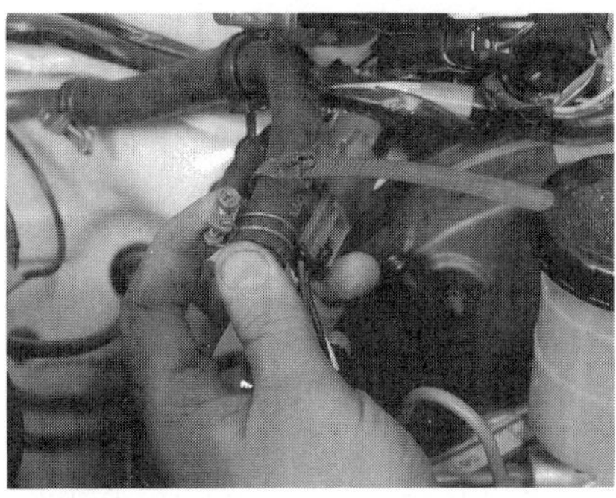

P6-2 Disconnect the vacuum tube or hose that runs from the intake manifold port to the booster unit, and quickly place your thumb over it before the engine stalls. You should feel strong vacuum.

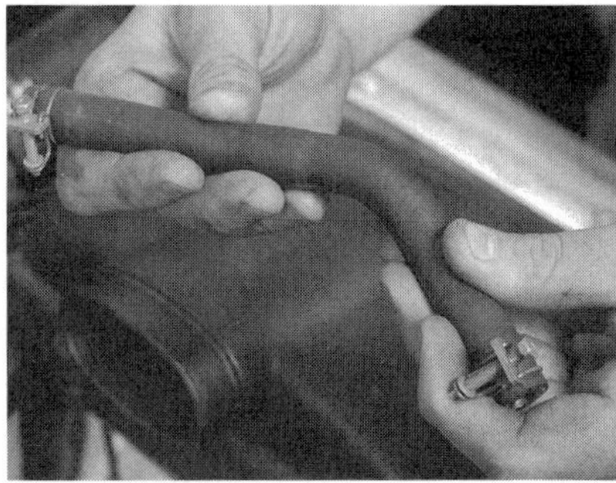

P6-3 If you do not feel a strong vacuum in step 2, shut off the engine, remove the tube and see if it is collapsed, crimped, or clogged. Replace it if needed.

P6-4 To test the operation of the vacuum check valve, shut off the engine and wait 5 minutes. Apply the brakes. There should be power assist on at least one pedal stroke. If there is no power assist on the first application, the check valve is leaking.

P6-5 Remove the check valve from the booster body.

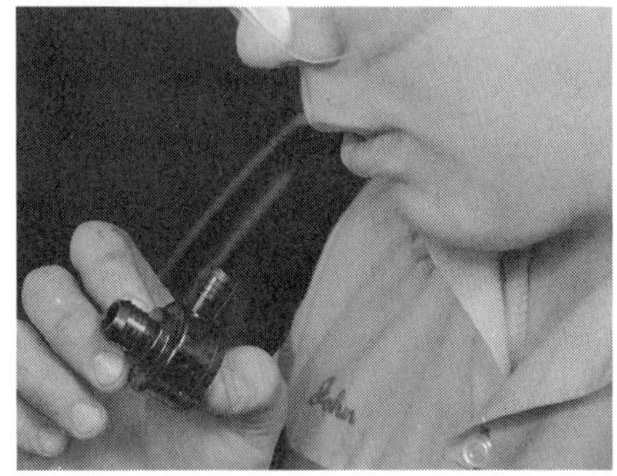

P6-6 Test the check valve by blowing into the intake manifold end of the valve. There should be a complete blockage of airflow.

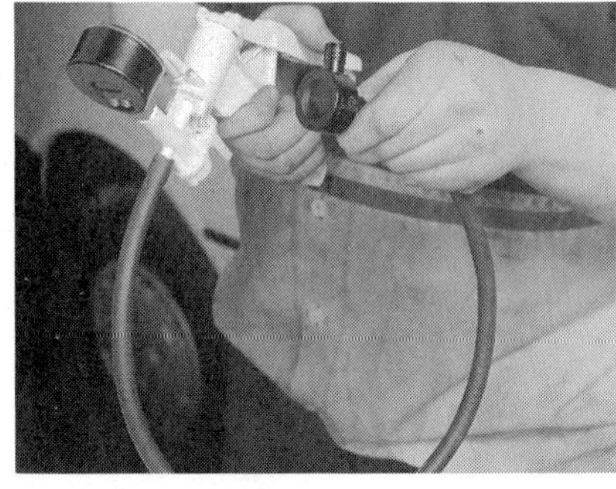

P6-7 Apply vacuum to the booster unit end of the valve. Vacuum should be blocked. If you do not get the stated results in steps 6 and 7, replace the check valve.

P6-8 Check the unit's air control valve by performing a brake drag test. With the wheels of the vehicle raised off the floor, pump the brake pedal to exhaust residual vacuum from the booster.

P6-9 Turn the front wheels by hand and note the amount of drag that is present.

Photo Sequence 6
Typical Procedure for Vacuum Booster Testing (continued)

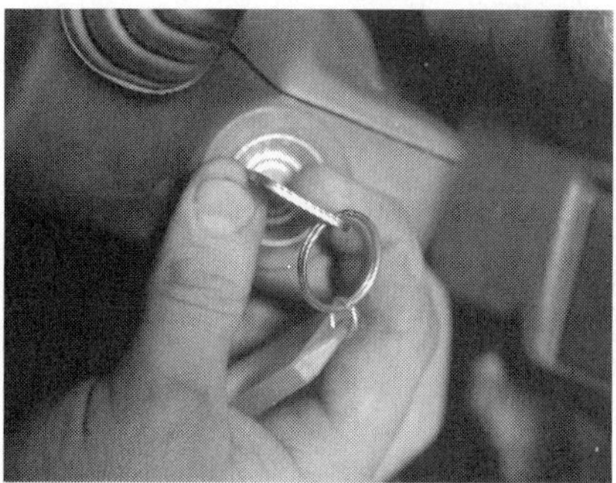

P6-10 Start the engine and allow it to run for 1 minute, then shut it off.

P6-11 Turn the front wheels by hand again. If drag has increased, this indicates that the booster control valve is faulty and is allowing air to enter the unit with the brakes unapplied. Replace or rebuild the booster.

Brake Pedal Checks

The brake pedal must be adjusted properly for dependable power-assist operation. In addition to the mechanical checks and brake travel check outlined in Chapter 3: Master Cylinder Service, you should also check pedal-free travel and pedal height settings. Excessive play or low pedal height settings may limit the amount of power-assist generated by the vacuum unit.

Pedal-Free Travel or Pedal Play

Free-pedal is the first easy movement of the brake pedal before the braking action begins to engage. Pedal-free travel or play should be very slight (1/16 in. – 3/16 in. in most cases). To determine the amount of free-pedal present, place your hand or foot on the pedal and gently push it down until an increase in pushing effort is felt (Figure 4-4).

Figure 4-4 Checking pedal-free travel with your hand rather than your foot will give a more accurate indication.

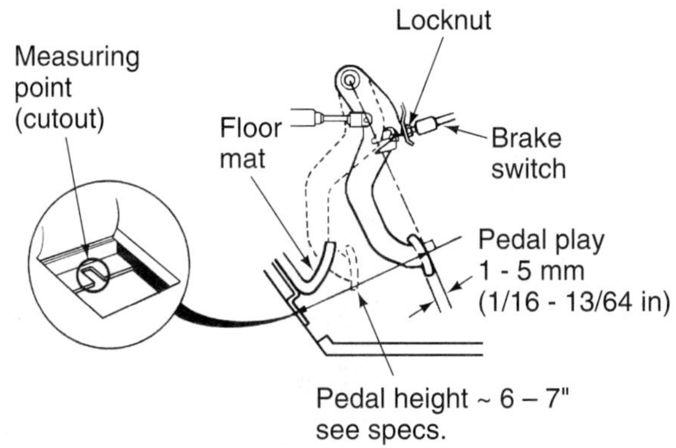

Figure 4-5 Taking the pedal height adjustment (Courtesy of Honda Motor Co., Ltd.)

Pedal Height Adjustment

Brake pedal height specifications are listed in the vehicle service manual. Typical heights range between 6 to 7 inches. Most pedal height measurements are taken from the floor mat to the base of the pedal, but in some cases the measurement is taken from a special point below the floor mat (Figure 4-5).

To adjust the pedal height, the pushrod is turned in or out. Before making this adjustment you should loosen the brake switch locknut and back off the brake switch until it no longer touches the brake pedal. Once this is done, use a pliers to screw the pushrod in or out as needed (Figure 4-6).

When the proper pedal height is set, screw in the brake switch until its plunger is fully depressed. Then back off the switch until the proper clearance between the switch and pad is obtained. Check the clearance with a feeler gauge and then tighten the locknut (Figure 4-7).

SERVICE TIP: After adjusting pedal height, check for proper brake stop light operation and, when equipped, proper cruise control operation.

Lower the pedal

Raise the pedal

Pushrod locknut

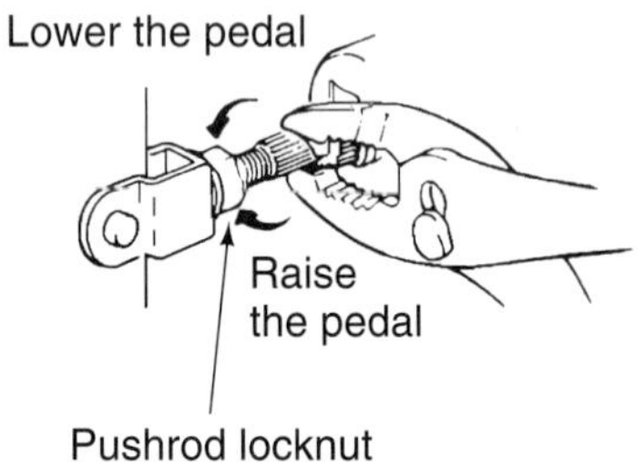

Figure 4-6 Adjusting brake pedal height (Courtesy of Honda Motor Co., Ltd.)

Caution: Check that the brake lights go off when the pedal is released.

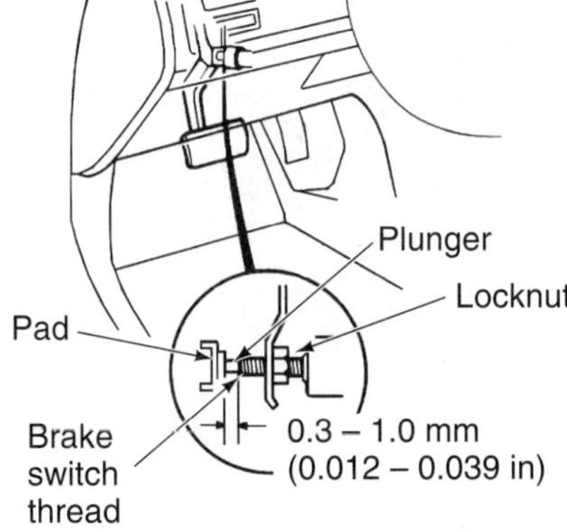

Note: After adjusting the pedal height, check for cruise control operation.

Figure 4-7 Adjusting the brake switch setting (Courtesy of Honda Motor Co., Ltd.)

Vacuum Booster Testing

It is important to remember that any condition that reduces the amount of vacuum the engine generates will affect power brake performance. These conditions include vacuum leaks, faulty valves, and improper valve timing. An engine rebuilt with a high-performance camshaft may also generate lower vacuum. When investigating poor brake performance, check engine vacuum before inspecting the vacuum booster system. Vacuum is measured on a gauge in inches of mercury (Hg). In most systems, a minimum of 14 in. Hg of vacuum is required to provide sufficient power brake operation.

Also check the vacuum tube or hose that runs from the engine intake manifold to the booster unit. It should be securely connected, not collapsed, punctured, or torn.

Some vehicles such as diesels or newer FWD models use an engine-driven or electric vacuum pump to supply the power unit with vacuum. Test it as the manufacturer recommends.

Vacuum Brake Servicing

SERVICE TIP: When fasteners are removed, always reinstall them in the same location. If a fastener is damaged, replace it with the same part number fastener. If an exact part number match is not available, install a fastener of the same size and strength. Stronger fasteners of the same size may also be used as replacements. Never use a fastener weaker than specified. Refer to the shop manual for proper torque values and if thread locking compound should be used.

Removal and Installation

To remove a typical vacuum booster unit (Figure 4-8) begin by setting the parking brake and disconnecting the battery ground cable. Disconnect the brake lines at the outlet ports to the master cylinder and plug the ports to prevent fluid loss. Next, disconnect the vacuum hose at the booster check valve. You can now remove all fasteners securing the master cylinder to the brake booster and carefully lift the master cylinder out of the engine compartment.

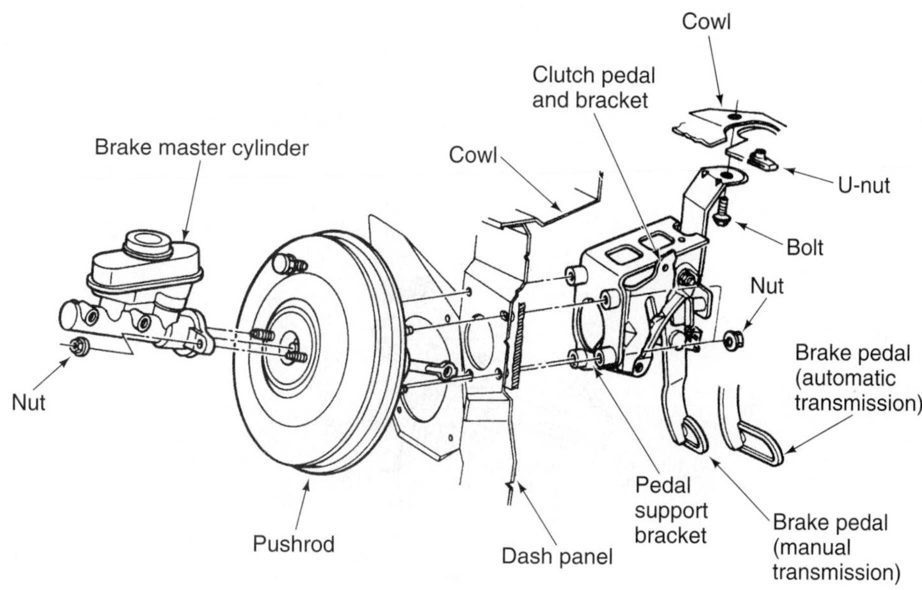

Figure 4-8 Exploded view of vacuum booster mounting arrangement (Reprinted with the permission of Ford Motor Company)

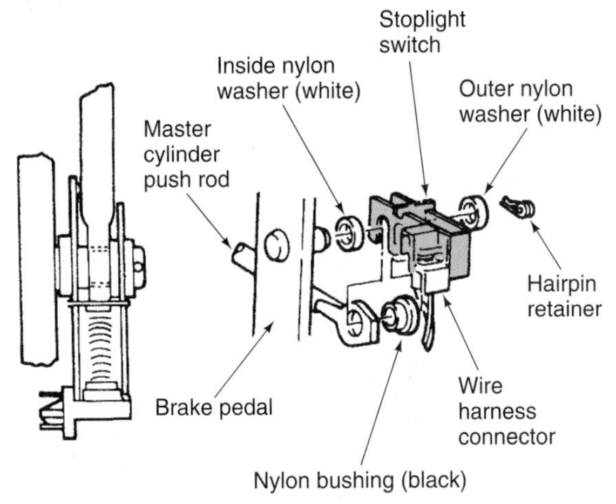

Figure 4-9 Removing stop light switch connector (Reprinted with the permission of Ford Motor Company)

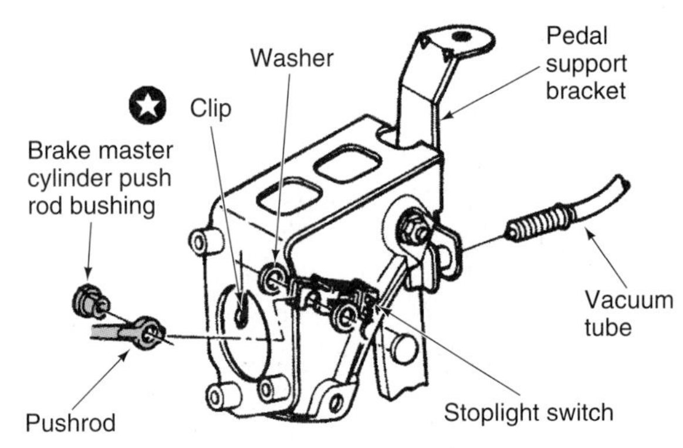

Figure 4-10 Removing pushrod retainer and brake master cylinder pushrod spacer from the pedal pin (Reprinted with the permission of Ford Motor Company)

With the master cylinder removed, move inside the vehicle and disconnect the stop lamp switch wiring connector from the stop lamp switch (Figure 4-9). Remove the switch from its mounting pin. Remove the nuts fastening the vacuum booster to the dash panel and slide the booster pushrod and bushing off the brake pedal pin (Figure 4-10).

The remainder of the disassembly is performed from the engine compartment. Clear the area around the booster. This may involve removing a vacuum outlet manifold (Figure 4-11), moving a wiring harness, or removing the transmission shift cable and bracket assembly. When the area is cleared of obstacles, move the power brake booster forward so that the booster studs clear the dash panel. The vacuum booster can then be lifted out of the engine compartment.

To install the vacuum booster, line up the brake support brake from inside the vehicle. Have a helper set the vacuum booster in position on the dash panel and hand thread the fastening nuts onto the studs. Reinstall the pushrod onto the pedal pin using a new pushrod bushing. You can now tighten the booster to dash panel retaining nuts to the specified torque. Next, reinstall the stop lamp switch on the pedal pin in the exact reverse of its disassembly. Reconnect the wiring connector to the switch body.

Move out to the engine compartment and reposition and install the wiring harness, transmission shift cable, and vacuum manifold. Connect the manifold vacuum hose to the check valve of the booster. Reinstall the master cylinder, connect all brake lines, and bleed the system of all air.

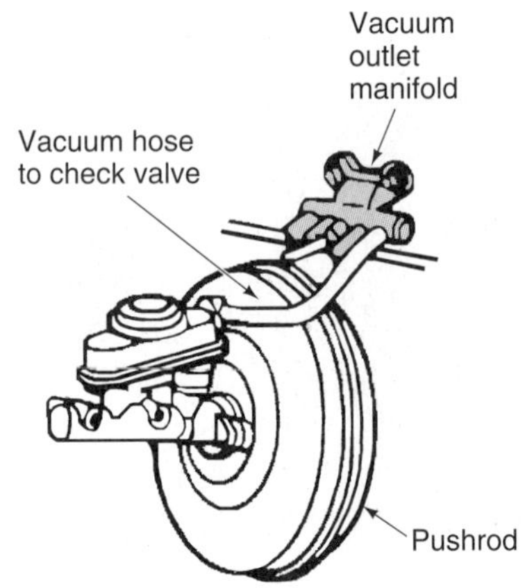

Figure 4-11 Remove the vacuum outlet manifold from the dash panel (Reprinted with the permission of Ford Motor Company)

On vehicles with manual transmissions and/or cruise control, adjust the manual shift linkage and cruise control dump valve. Reconnect the battery and test drive the vehicle to be sure the booster is operating properly.

SERVICE TIP: Some vacuum boosters cannot be rebuilt. They are designed to be serviced only as a complete assembly. Do not attempt to disassemble these units. See the vehicle shop manual for booster rebuild/replace instructions.

Overhauling a Tandem-Diaphragm Booster

Classroom Manual
Chapter 4, page 72

Classroom Manual Chapter 4, page 72

Special Tools

Vacuum power booster disassembly/assembly fixture

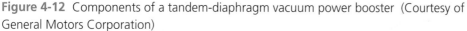

With a proper overhaul kit, a typical tandem-diaphragm booster (Figure 4-12) can be disassembled as follows:

1. Remove the boot and silencer from the rear of the housing.
2. Remove the check valve, grommet, and front housing seal from the front housing.
3. Scribe matching marks between the front and rear housings to aid in reassembly.

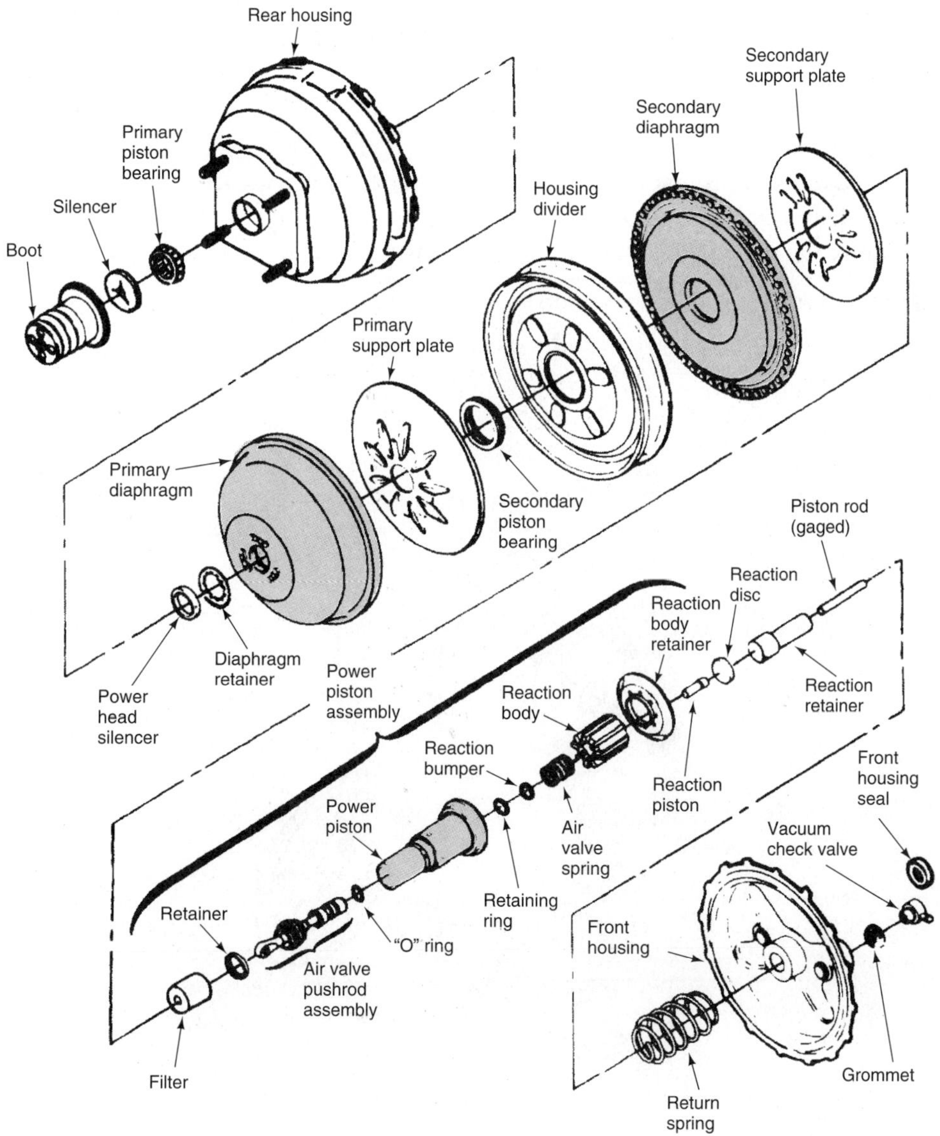

Figure 4-12 Components of a tandem-diaphragm vacuum power booster (Courtesy of General Motors Corporation)

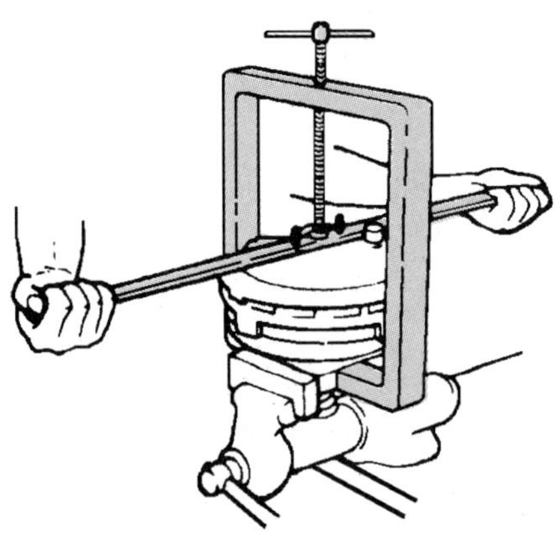

Figure 4-13 Unlocking the booster assembly (Courtesy of General Motors Corporation)

4. Position the booster in the disassembly fixture with the rear of the booster facing down. Tighten the hand screw down onto the booster. Allow just enough slack so the booster can turn.
5. To unlock the two halves of the booster use the lever to apply force in a counterclockwise direction.
6. Once the booster is unlocked, slowly loosen the fixture hand screw so the upper and lower halves separate (Figure 4-13).
7. Remove the diaphragm return spring and power piston group (Figure 4-14).

CAUTION: When separating the booster housing, be careful not to allow sudden release of spring tension, as damage or personal injury might result.

8. Press out the power piston bearing from the rear housing.

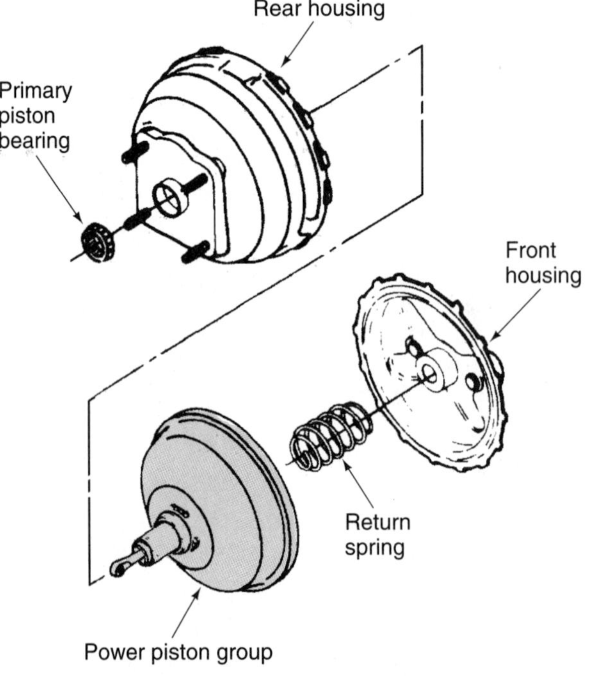

Figure 4-14 Booster inner components (Courtesy of General Motors Corporation)

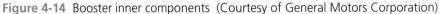

9. Remove the master cylinder piston rod, reaction body retainer, and the reaction retainer from the power piston (Figure 4-15).
10. Remove the power head silencer.
11. To remove the power piston assembly and pushrod, grasp the assembly at the outer edges of the housing divider and diaphragms. Position the assembly with the pushrod down against a hard surface and using a slight force or impact, dislodge the diaphragm retainer.
12. Remove the primary diaphragm and primary support plate from the housing divider, and separate the primary diaphragm from the plate.
13. Remove the secondary diaphragm and secondary support plate from the housing divider.
14. Remove the reaction body from the power piston. Then remove the reaction disc and reaction piston from the reaction body.
15. Separate the air valve spring and reaction bumper from the end of the air valve pushrod.

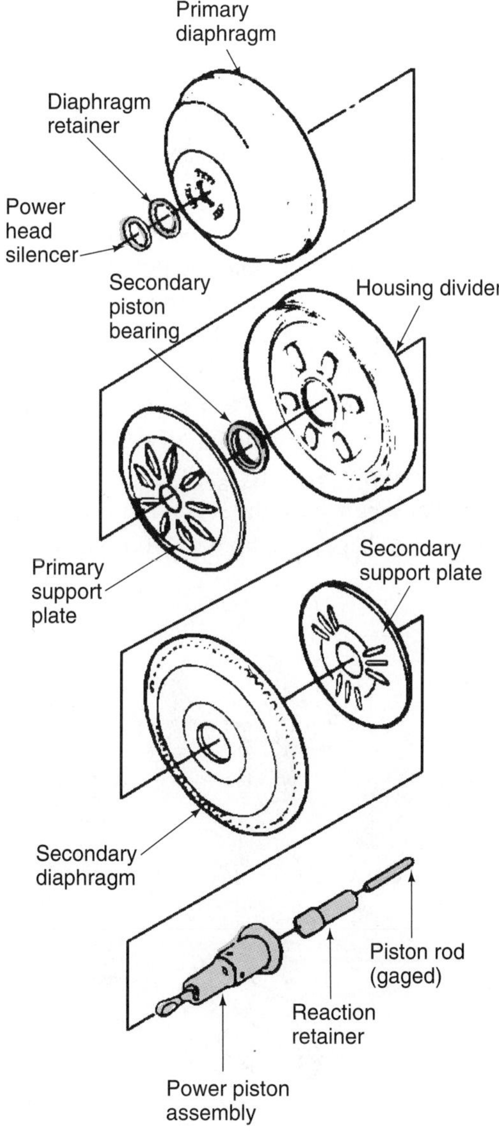

Figure 4-15 The vacuum booster power piston group (Courtesy of General Motors Corporation)

16. Remove the retaining ring from the air valve pushrod assembly. Insert a screwdriver or small pry tool through the pushrod eyelet and pull the air valve assembly straight out. Be careful as considerable force is required to accomplish this.
17. Complete the disassembly by removing the filter, retainer, and "O" ring from the air valve pushrod assembly.

Inspection and Cleaning

Inspect all parts for corrosion, nicks, cracks, cuts, scoring, bends or distortion, or other signs of wear. Using a crocus cloth, polish away any minor corrosion from the housings or diaphragm support.

Clean all components in clean, denatured alcohol. Do not use solvent cleaners. Never dip the power piston assembly in alcohol. Simply wipe it clean with a shop cloth dampened with alcohol. Use dry, unlubricated compressed air to dry all parts.

Reassembly

Always use all the parts furnished in the repair kit. Discard all old rubber parts.

WARNING: The power piston and pushrod assembly should not be disassembled. If replacement is required, these components are serviced as an assembly.

Special Tools

Vacuum power booster disassembly/assembly fixture

Air valve pushrod retainer installer

Power piston seal protector

To assemble the typical tandem vacuum power booster:

1. Lubricate the "O" ring with a thin layer of silicone lubricant or other recommended lubricant. Then install it on the air valve pushrod assembly.
2. Slide the air valve pushrod assembly into the power piston.
3. Using the special retainer install tool, install the new retainer and seat it (Figure 4-16).
4. Slip the filter over the pushrod eyelet and into the power piston, then install the retaining ring onto the air valve pushrod assembly.
5. Install the reaction bumper and air valve spring onto the end of the air valve pushrod. Next install the reaction piston and reaction disc into the reaction body. Then install the reaction body into the power piston.
6. Place the secondary diaphragm onto the secondary support plate and lubricate the inside diameter or the secondary diaphragm lip with silicone lubricant. Slide the secondary

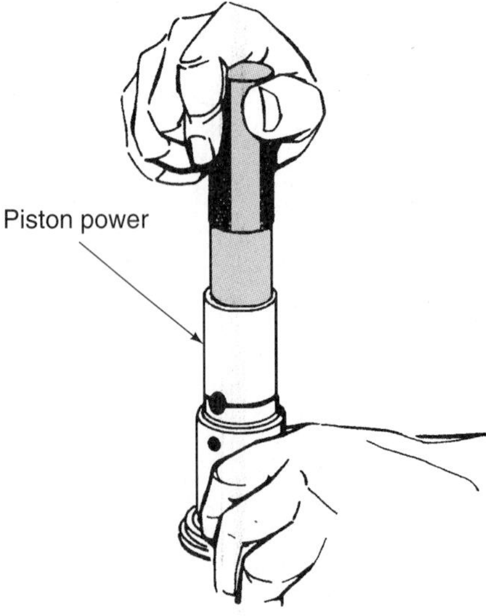

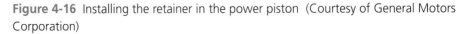

Figure 4-16 Installing the retainer in the power piston (Courtesy of General Motors Corporation)

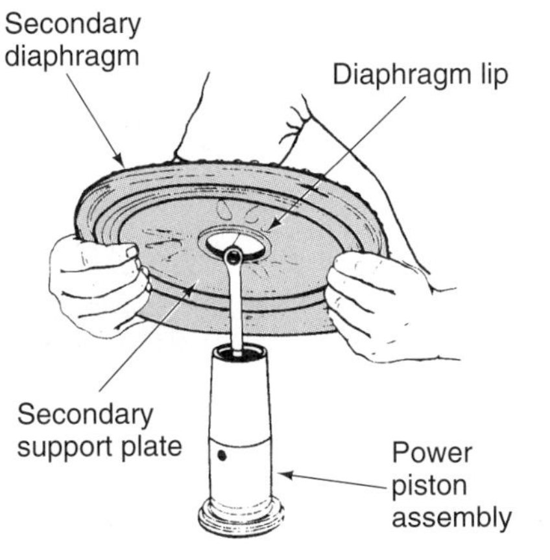

Figure 4-17 Assembling the secondary diaphragm and support (Courtesy of General Motors Corporation)

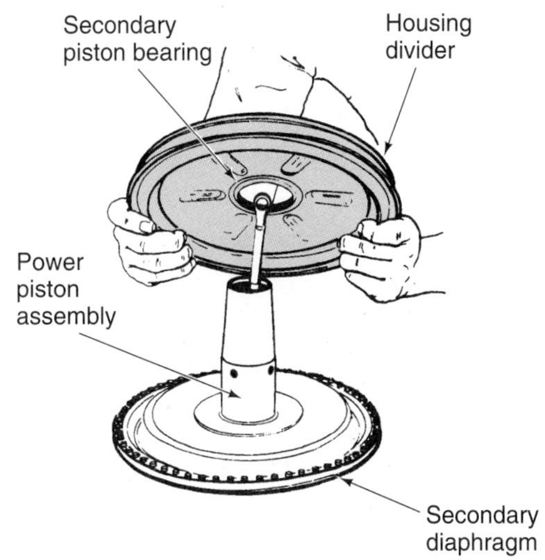

Figure 4-18 Assembling the booster housing divider (Courtesy of General Motors Corporation)

diaphragm and support plate over the power piston assembly and pushrod using the power piston seal protector to protect the power piston (Figure 4-17).

7. Install the secondary piston bearing into the housing divider. The flat surface of the bearing should be on the same side as the six raised lugs on the divider.
8. Spread a thin layer of silicone lubricant on the inside diameter of the secondary piston bearing. Now using the power piston seal protector as a guide, carefully place the secondary piston bearing and the housing divider over the power piston assembly (Figure 4-18).
9. Spread a thin layer of silicone lubricant on the inside diameter of the primary diaphragm lip. Place the primary diaphragm into the primary support plate and fold the primary diaphragm up and away from the primary support plate.
10. Install the primary diaphragm and support plate over the power piston assembly (Figure 4-19). Unfold the primary diaphragm back into position and pull the outside edge of the diaphragm over the formed flange of the housing divider. The beads on the secondary diaphragm should be seated evenly around the complete circumference.

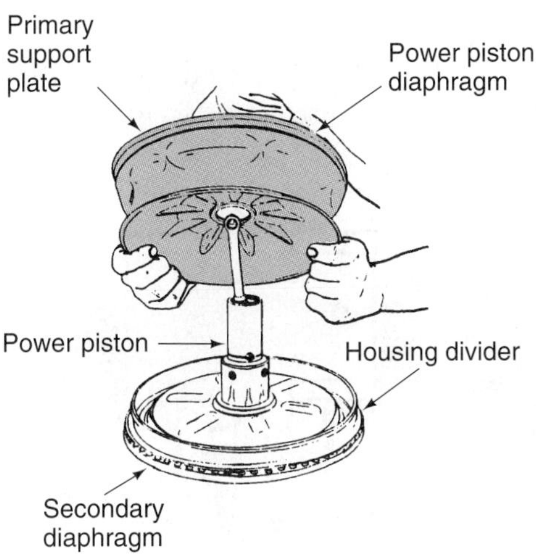

Figure 4-19 Installing the primary diaphragm (Courtesy of General Motors Corporation)

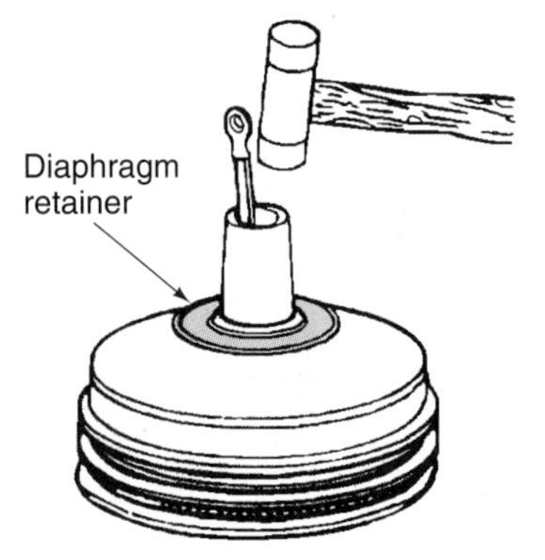

Figure 4-20 Installing the diaphragm retainer (Courtesy of General Motors Corporation)

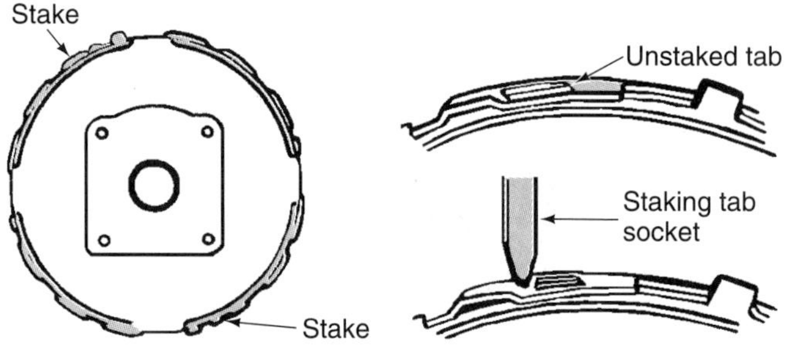

Figure 4-21 Staking the tabs to secure the booster together (Courtesy of General Motors Corporation)

11. Discard the original diaphragm retainer. Use a rubber mallet and the power piston seal protector to seat the new retainer (Figure 4-20).

12. Install the silencer, reaction retainer, reaction body retainer, and piston rod.

13. Lubricate the inside and outside diameters of the primary piston bearing with silicone lubricant and install the primary piston bearing into the rear housing.

14. Install the power piston group into the rear housing followed by the return spring.

15. Use the booster assembly fixture to lock the front and rear housing together. Set the assembled rear housing group into the fixture and align the scribe marks on the housings. Tighten the fixture hand screw just enough to allow the housing to rotate. Lock the front and rear housings together by applying force in a clockwise direction. Carefully back off the fixture hand screw and stake the housing at the two tabs located 180 degrees apart (Figure 4-21).

WARNING: Do not stake a tab that has been previously staked. It will weaken and fail at a later time, causing complete system failure.

SERVICE TIP: Assembly of the booster is easier if you connect a vacuum source to the booster.

16. Spread a thin layer of silicone lubricant on the inside and outside diameters of the grommet and front housing seal. Complete the reassembly by installing the grommet, vacuum check valve, front housing seal, silencer, and boot.

Check Piston Rod Length

Using the recommended gauging tool, check the length of the piston rod (Figure 4-22). If the piston rod is not within limits, obtain a service adjustable piston rod and adjust the rod to the correct measurement.

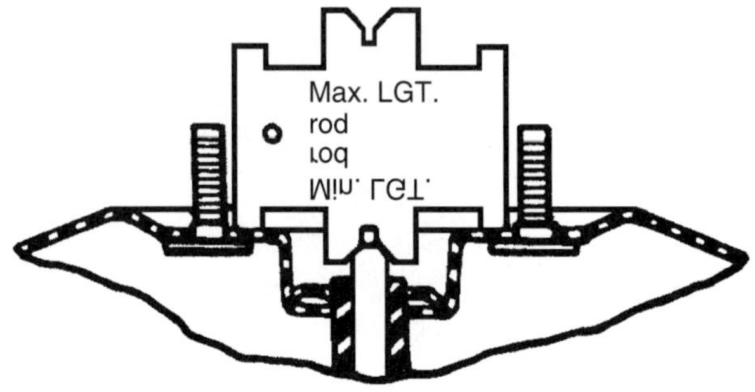

Figure 4-22 Gauging the piston rod length (Courtesy of General Motors Corporation)

Single-Diaphragm Vacuum Booster Overhaul

With a proper overhaul kit, a typical single-diaphragm booster (Figure 4-23) can be disassembled as follows:

1. Remove the boot and silencer from the rear of the housing.
2. Remove the check valve, grommet, and front housing seal from the front housing.
3. After scribing matching marks between the front and rear housings to make reassembly easier, use the booster disassembly fixture to separate the two halves. The procedure is the same as steps 4 through 6 in tandem booster disassembly.

CAUTION: When separating the booster housing, be careful not to allow sudden release of spring tension, as damage or personal injury might result.

4. After separating the housing, remove the diaphragm return spring and power piston group.
5. Press out the power piston bearing from the rear housing.
6. Remove the reaction body retainer, the piston rod, and the reaction retainer.
7. Using an awl or similar tool, pry out the filter.
8. To remove the power piston assembly and pushrod, grasp the assembly at the outer edges of the housing divider and diaphragms. Position the assembly with the pushrod down against a hard surface. Using a slight force or impact, dislodge the diaphragm retainer.
9. Separate the diaphragm from the diaphragm support.
10. Inspect, clean, and dry all components as described for the tandem booster overhaul on page 85.

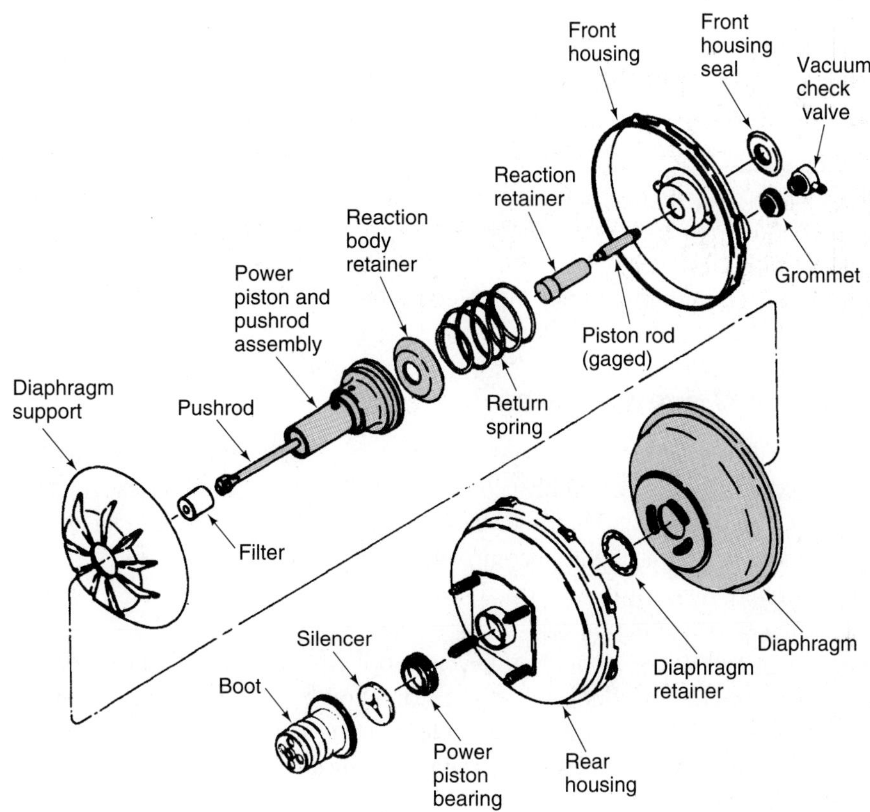

Figure 4-23 Components of a single-diaphragm vacuum booster (Courtesy of General Motors Corporation)

WARNING: Do not disassemble the power piston and pushrod assembly. If replacement is required, these components are serviced as an assembly.

To assemble the typical single-diaphragm vacuum power booster:

1. Spread a thin layer of silicone lubricant on the inside diameter of the diaphragm and fit the diaphragm into the diaphragm support plate.
2. Position the power piston on the workbench with the pedal pushrod facing up, then mount the diaphragm and support plate assembly over the piston.
3. Install the new diaphragm retainer, then seat the retainer and diaphragm assembly using the recommended driver and a mallet.
4. Seat a new filter into the rear of the power piston assembly, then install the reaction retainer and piston rod.
5. Install the reaction body retainer into the power piston and pushrod assembly.
6. Install the power piston bearing and piston assembly in the rear housing.
7. Apply a thin layer of silicone lubricant on the inside and outside of the bearing.
8. Install the power piston group into the rear housing, then install the return spring.
9. Position the rear housing assembly over the front housing, ensuring that matching marks are aligned. Mount the front housing in the holding fixture. Be sure that all components are properly aligned. Apply pressure to the rear housing with the holding fixture forcing screw, then rotate the rear housing clockwise to the lock position. Vacuum can be applied to the port in the front housing to help with housing assembly.
10. Stake housing tabs in two locations, 180 degrees apart to secure the assembly. Do not reuse tabs that have already been staked.
11. Lubricate and install the front housing grommet, then press the check valve and vacuum switch, if equipped, into the grommets.
12. Install the front housing seal, silencer, and boot.

Auxiliary Vacuum Pumps

Classroom Manual
Chapter 4, page 76

For some vehicles, the engine alone cannot produce sufficient vacuum to operate the vacuum booster. An auxiliary pump is used to provide the extra vacuum needed. Vacuum pumps used on cars and light-duty trucks are usually the diaphragm or vane type. Auxiliary vacuum pumps available for passenger cars are driven by an electric motor, a belt, or a gear/cam lever from the engine.

In some cases, the vacuum pump is mounted to the rear of the alternator and driven by the alternator rotor shaft.

Electric Motor Driven Pumps

A typical electric auxiliary vacuum pump consists of a small electric pump motor and an electronic controller combined into a single assembly (Figure 4-24). The controller houses an integral on-off switch, timer/relay, and a piston and umbrella valve assembly. Vacuum hoses connect the vacuum pump inlet hose to the vacuum-operated components. The vacuum pump outlet hose is usually connected to the intake manifold. A charcoal filter located in the outlet line prevents harmful vapors from entering the vacuum pump.

The system is wired so the auxiliary pump is supplied with current whenever the ignition switch is in the RUN position. Under low-vacuum conditions, the internal vacuum switch closes, completing the timer/relay circuit. This activates the vacuum pump motor, which generates auxiliary vacuum to help maintain 14.0 in. Hg of vacuum the vehicle requires. When 14.0 in. Hg of vacuum is reached, the internal vacuum switch opens. The timer/relay continues to supply current to the pump motor for several seconds before cutting off current to the pump motor. A low-vacuum warning switch located in the inlet hose line activates a dash warning lamp when a low-vacuum condition exists.

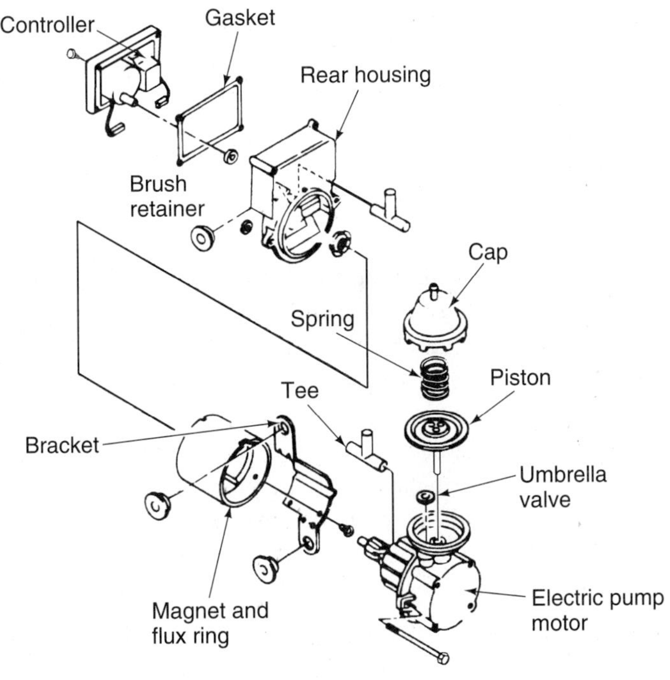

Figure 4-24 Auxiliary electric vacuum pump (Courtesy of Raybestos)

SERVICE TIP: In addition to providing increased vacuum power for both the brake system and the engine, an auxiliary vacuum pump provides extra vacuum for the following areas: the electronic ignition system, heat and air conditioning system, EGR system, and cruise control.

Troubleshooting

Before turning your attention to the auxiliary vacuum pump, check the hydraulic system, engine vacuum system, and brake booster. If these systems appear to be in order, check the condition of the vacuum line hoses and the connections at the vacuum pump. Also check for loose mounting bolts, an improperly tensioned drive belt, or improper engine idle speed on mechanically operated vacuum pumps. On inoperative electrically operated vacuum pumps, check for current at the pump motor electrical connector to ensure the problem is in the pump motor, not the circuit to it.

A hand-operated vacuum pump is used for most diagnostic procedures. The pump is connected to various ports on the auxiliary vacuum pump. These ports typically include an inlet port, a vacuum switch port, a pump housing port, and a pump outlet port (Figure 4-25). Reading vacuum at these connections can pinpoint problems in the auxiliary pump.

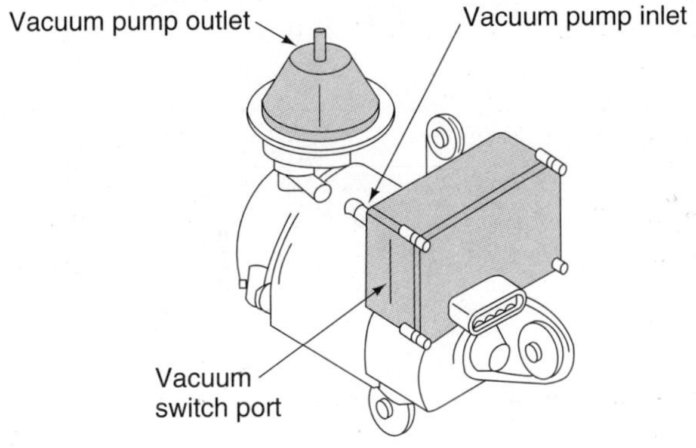

Figure 4-25 Vacuum ports on the auxiliary pump that are used for testing

WARNING: During testing never block the outlet nozzle of the vacuum pump. Permanent damage to the diaphragm can result from a blocked outlet nozzle.

SERVICE TIP: Refer to the service manual electrical wiring diagrams to locate the proper power and ground terminals on the electric vacuum pump connector.

If excessive brake pedal effort is needed, the brake warning light is on or the vacuum pump is inoperative. Troubleshoot as follows:

1. Turn on the ignition. Disconnect the electrical connector from the vacuum pump. Use a digital voltmeter to check for battery voltage across the appropriate connector pins. Also check for proper ground. If battery voltage is present at the connector, proceed to step 2. If there is no power, check the fusible link, fuse, and wiring circuits and repair as needed.
2. Connect a 12-volt battery directly to the appropriate connector terminals and ground the circuit via the correct terminal pin. If the pump is still inoperative, proceed to step 3. If the pump now operates, go on to the next diagnostic procedure.
3. Remove the vacuum pump. Disassemble and check for sticking or shorted brushes. Check the controller for damaged wires and repair/replace as needed.

If the vacuum pump operates but the brake warning light still lights and excessive brake pedal effort is needed, troubleshoot as follows:

1. Remove the pump's inlet hose and connect a hand-operated vacuum pump to its fitting. Turn on the ignition (Figure 4-26).
2. If the pump draws a vacuum of 10 to 15 in. Hg and runs for 5 to 10 seconds, check for vacuum leaks in other parts of the system. But if the pump runs intermittently and fails to maintain steady vacuum, proceed to step 3.
3. Remove the vacuum pump from the vehicle. Remove the controller and vacuum connector from the pump. Connect the hand-operated vacuum pump to the vacuum switch inlet port and apply approximately 20 in. Hg. If the switch leaks more than 2 in. Hg per minute, replace the controller. If the switch holds a vacuum, go to step 4.
4. Connect the hand-operated vacuum pump to the pump housing inlet. Apply 20 in. Hg of vacuum. If the pump leaks more than 2 in. of Hg per minute, the pump's umbrella valve is leaking and must be replaced. If the pump holds a vacuum, proceed to step 5.
5. Connect the hand-operated vacuum pump to the pump outlet. Plug the pump housing inlet and apply 20 in. Hg of vacuum. If the pump leaks more than 2 in. Hg per minute, check for a loose bonnet. Repeat the test. If the pump continues to leak, replace the piston. If the pump holds vacuum, the system is operating properly.

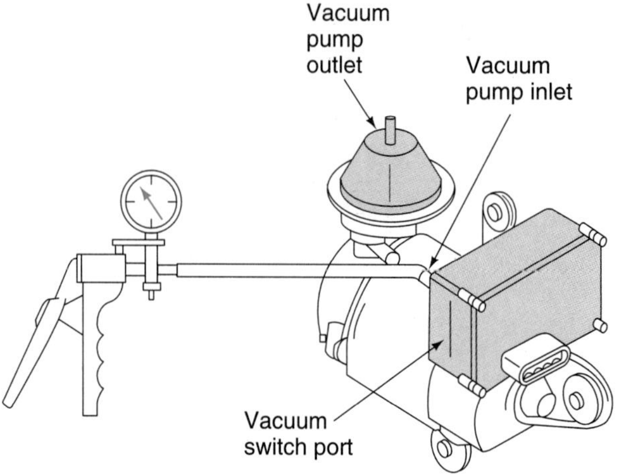

Figure 4-26 Vacuum gauge connected to the inlet port of the auxiliary vacuum pump

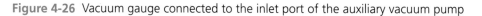

If the vacuum pump runs continuously with normal or excessive brake pedal effort and the brake warning light on or off, troubleshoot as follows:

1. Check all vacuum hoses for leaks and repair or replace as needed. If no leaks are found, go to step 2.
2. With the ignition off, disconnect the inlet hose from the vacuum connection. Attach the hand-operated vacuum pump and turn on the ignition. Replace it. If the pump runs then stops when the vacuum reaches 10 to 15 in. Hg, the system is normal. Bleed off the vacuum to below 10 in. Hg. At levels below 10 in. Hg, the pump should start, run until the vacuum exceeds 10 to 15 in. Hg, and automatically shut off. If the pump runs continously and draws a vacuum of 10 in. Hg or greater, the controller is defective and should be replaced.

Servicing System Components

The replacement and installation steps for various system components follow:

Vacuum Pump. To remove the auxiliary vacuum pump, raise the vehicle on the lift or safety stands. Remove any protective shields to access the pump. Carefully remove the locking-type electrical connectors and vacuum hoses from the pump. Remove all retaining nuts and carefully dismount the pump. Reverse the procedure to install.

Controller. To remove the controller (Figure 4-27), first remove the vacuum pump. Next remove all protective shields from around the controller and remove the fasteners that hold the rear housing and pump housing together. Carefully separate the rear housing from the pump housing and remove the fasteners securing the controller to the rear housing. Remove the brushes from the brush holders, then remove the controller, gasket, and washer.

Install the controller using a new gasket and washer. Place both brushes in the rear brush holder cavity. Preload the brush springs by locking them in the slots provided above the spring access slots. Position the brushes in channels, checking to be sure the shunts are positioned properly. Install the brush retainer in the proper position and return springs to the load position. Install the rear housing on the pump housing, replace all controller shields, and reinstall the vacuum pump.

Piston Assembly. To remove and replace the pump piston assembly, remove the vacuum pump as described. Remove the pump upper and lower shields. Carefully release the tabs of the cap, then remove the cap and spring (Figure 4-27).

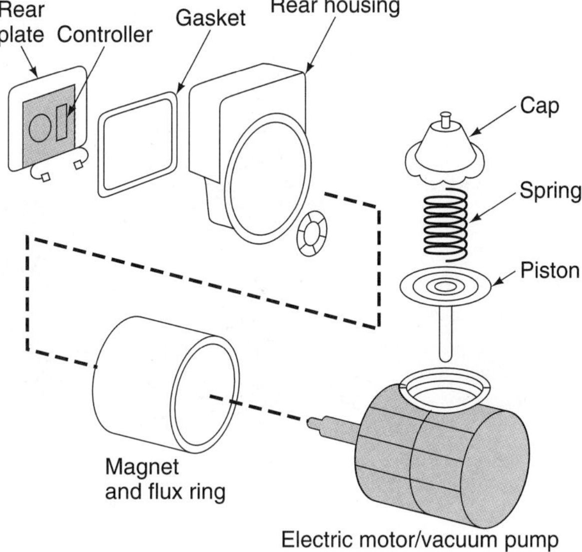

Figure 4-27 Exploded view of electric vacuum pump components

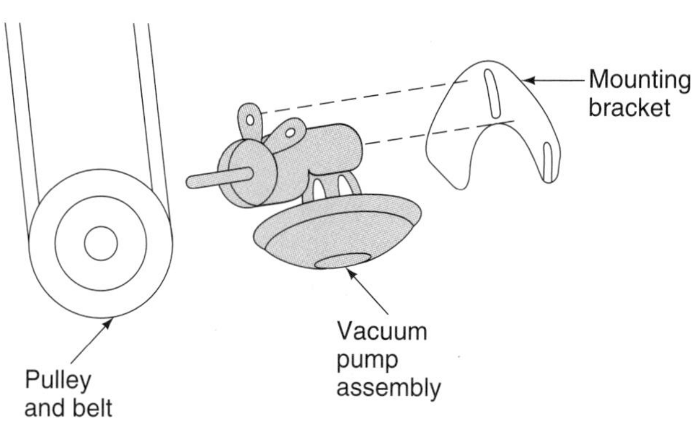

Figure 4-28 Typical belt-drive auxiliary vacuum pump mounting

CAUTION: The bonnet is spring loaded. The bonnet and pump housing must be held together during disassembly.

Remove the piston from the housing and the umbrella valve. Replace the umbrella valve, piston assembly, and spring. Place the bonnet over the spring and compress the spring and bonnet onto the pump housing. Bend the bonnet tabs around the pump housing, checking to be sure that the bonnet is secure. Reinstall the shields and install the pump.

Belt-Driven Vacuum Pump Systems

Belt-driven vacuum pumps are driven off the engine. The belt turns a pulley mounted on the pump's input shaft (Figure 4-28). The shaft runs a cam that operates a spring-loaded diaphragm to produce vacuum. If you suspect the pump is performing poorly, check for worn or misadjusted belts or pulleys. Connect a vacuum gauge to the pump inlet and plug the outlet port. The gauge should read approximately 15 in. Hg after running the engine at idle for about 30 seconds.

Gear-Driven Vacuum Pump Systems

Gear-driven vacuum pumps are usually driven off the camshaft gear of the engine (Figure 4-29). They are diaphragm pumps that require no maintenance. The entire pump should be replaced if a vacuum reading of approximately 20 in. Hg cannot be attained at the pump with the engine running at idle.

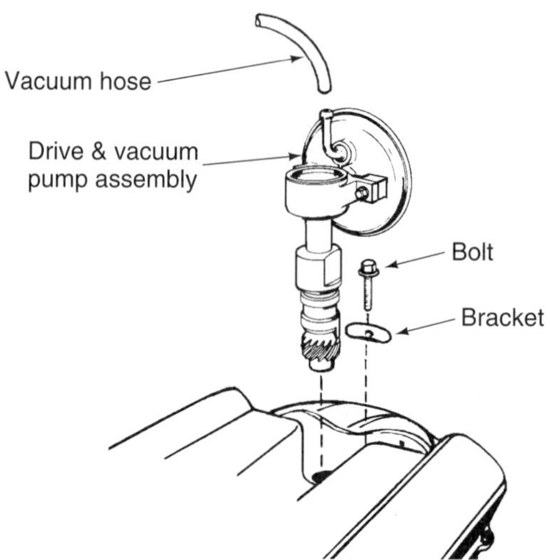

Figure 4-29 Typical gear-driven auxiliary vacuum pump (Courtesy of Oldsmobile Division of General Motors Corporation)

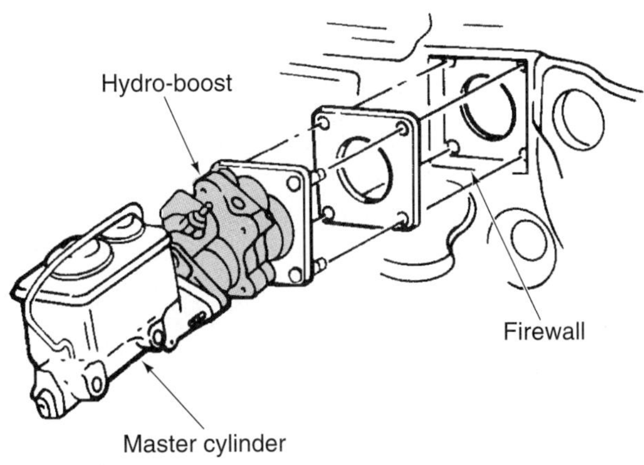

Figure 4-30 Hydro-boost power brake booster (Courtesy of Raybestos)

Hydro-Boost Power Brakes

Classroom Manual
Chapter 4, page 77

A hydro-boost power-assist brake system consists of the booster assembly, the accumulator, the power steering fluid circuit, including the gear pump and reservoir, the master cylinder, hydraulic lines, and disc and drum brake assemblies.

The hydro-boost unit is mounted in the engine compartment in much the same way as a vacuum booster (Figure 4-30). The pedal rod is connected at the booster input rod end. The master cylinder is bolted to the hydro-boost and is operated by a pushrod projecting from the booster cylinder bore.

Pressurized (100 to 150 psi) power steering fluid is delivered to the booster from the vehicle's power steering pump (Figure 4-31). The flow of fluid inside the booster is controlled by a hollow center spool valve. Pressing the brake pedal moves the booster pedal rod forward. This motion is transferred to the power piston and spool valve by mechanical linkage. As the spool valve is moved off center, ports open, allowing pressurized fluid into the power cavity of the booster. This fluid presses against the rear of the power piston, increasing the application force. When the brake pedal is released, pressure in the power cavity is released. A return spring pushes the power piston back to the unapplied position.

Do not reuse power steering fluid. Discard all used fluid. Traces of contaminants in the fluid may damage seals and rubber parts in the booster system. Contaminated fluid may lead to hydraulic brake system failure.

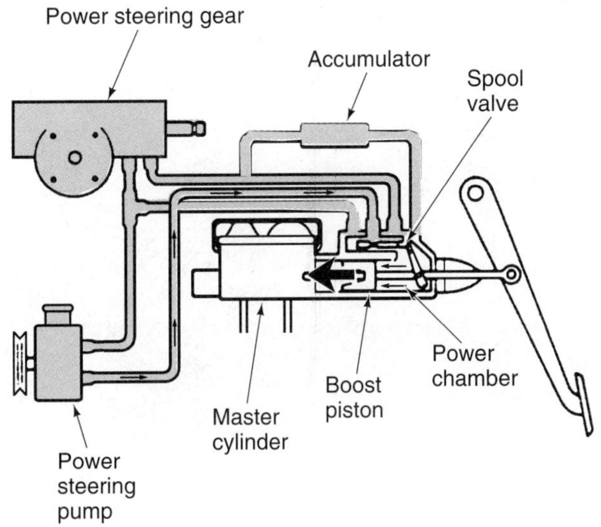

Figure 4-31 Hydro-boost power brake booster operation (Courtesy of Raybestos)

The booster's accumulator holds a charge of pressurized steering fluid to use as a reserve if fluid flow from the steering pump is lost or diminished. Pressurized fluid from the steering pump enters the accumulator through a check valve assembly. Fluid only enters the accumulator if its pressure is greater than that inside the accumulator. Once inside the accumulator, the fluid compresses an internal piston and spring or a gas charge.

Power steering fluid and brake fluid must not be mixed. If brake seals contact power steering fluid or steering seals contact brake fluid, seal damage will result.

If steering fluid pressure is lost, the force of the compressed spring or gas charge forces the pressurized steering fluid out of the accumulator so it can assist in applying the brakes. Depending on the system, one to three power applications are possible once power steering fluid pressure is lost. The spring-type accumulator is equipped with a pressure relief port. It allows pressurizing fluid to return to the steering pump reservoir when the piston is compressed to a certain point.

CAUTION: On models with hydro-boost assist systems, do not disassemble the accumulator. The accumulator may contain compressed gas. Always use the proper tools and follow the shop manual procedures to avoid personal injury. If the accumulator is inoperative, damaged, or diagnosed as defective, it must be replaced as an assembly. Do not apply heat to the accumulator or attempt to repair it. Dispose of an inoperative accumulator by drilling a 1/16 in. diameter hole through the end of the accumulator can opposite the "O" ring.

SERVICE TIP: On hydro-boost systems, use the recommended power steering fluid in the boost system circuit. The proper fluid is essential to system operation. Do not mix brake fluid with power steering or other hydraulic fluids. The use of improper fluids will damage the various seals and valves. Take precautions to keep dirt out of the hydro-boost system.

Testing the Hydro-Boost System

The hydro-boost unit will not work properly unless it receives a continuous supply of clean, bubble-free power steering fluid at the proper pressure. The power steering pump performs this function. The pump is belt driven from the engine crankshaft pulley. It is mounted to the engine with a bracket (Figure 4-32).

SERVICE TIP: The hydraulic pump drive belt may also be used to drive the coolant pump or other components. If this is so, the belt should not be disconnected. When disconnected, the hydraulic pump becomes an idler pulley. If it is allowed to run dry, it will burn the bearings and cause the coolant pump to stop and possibly damage the engine.

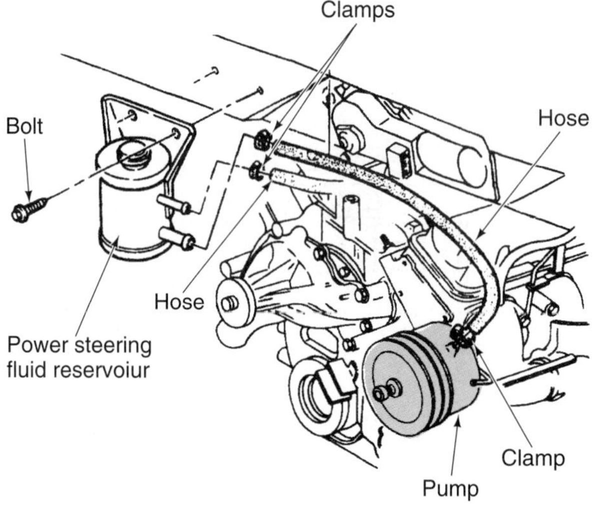

Figure 4-32 Power steering pump (Courtesy of General Motors Corporation)

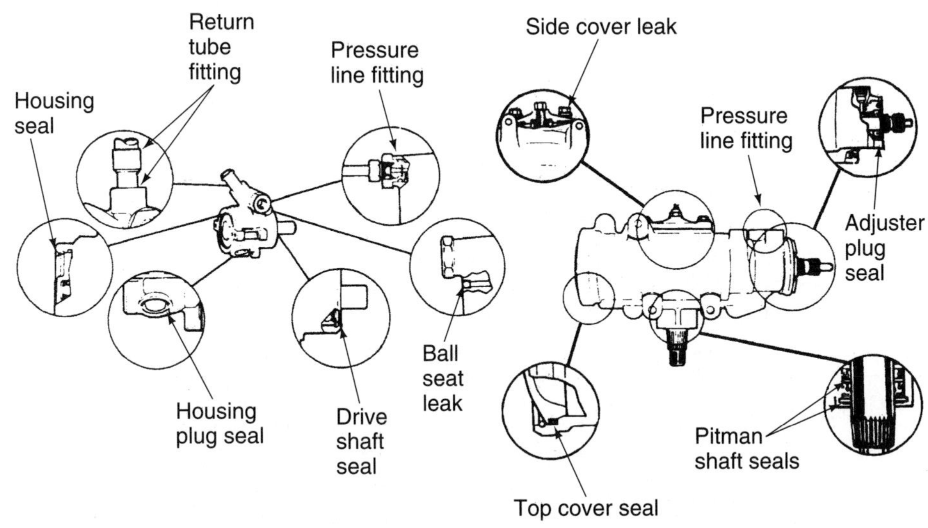

Figure 4-33 Possible sources of leakage in a power steering pump (Courtesy of General Motors Corporation)

The fluid reservoir for the pump is remote from the pump location. Check the fluid level in the pump reservoir with the engine warm. Add the recommended power steering fluid when required. If dirt, sludge, or water is found in the fluid, flush the system. Contaminated fluid will swell or deteriorate rubber parts in the system and lead to loss of braking power.

Do not operate the hydraulic pump without fluid in the pump reservoir. The hydraulic pump bearings and seals may become damaged.

If the hydraulic system fails due to a lose of a fluid, make the needed repairs. Fill and bleed the hydraulic booster system prior to starting the vehicle engine. If the failed component is not the hydraulic pump, run the pressure pipe directly back into the fluid reservoir. Whenever the hydraulic pump is replaced, clean and flush the entire hydraulic booster system. A worn pump can generate metal shavings that can contaminate the entire system.

Disassemble the booster head and clean any metal particles from all parts. Replace the seals and "O" rings. Remove the pipes and hoses and blow them clean of all metal shavings.

Hydro-Boost Fluid Leakage

If leakage is evident around the pump, clean and tighten all fittings and bolts. If the leak continues, the pump should be serviced or replaced. Figure 4-33 illustrates common leak locations in a power steering pump.

Check the operation of the power steering pump and check for improperly adjusted drive belts, a defective pressure relief valve (spring-loaded accumulators only), leaking hoses and fittings, or a defective accumulator.

Inspect the hydro-boost unit for leaks at the following locations (Figure 4-34).

Input Seal. Fluid leakage from the housing cover end of the booster near the reaction bore requires booster replacement.

Piston Seal. Fluid leakage from the vent at the front of the unit near the master cylinder indicates a leaking piston seal. Replace the seal.

Housing. Fluid leakage between the housing and housing cover indicates a defective seal. Replace the seal.

Spool Valve Seal. Fluid leakage near the plug area indicates the need to replace the spool valve seal.

Accumulator Cap Seal. Fluid leakage near the plug area indicates the need to replace the accumulator cap seal.

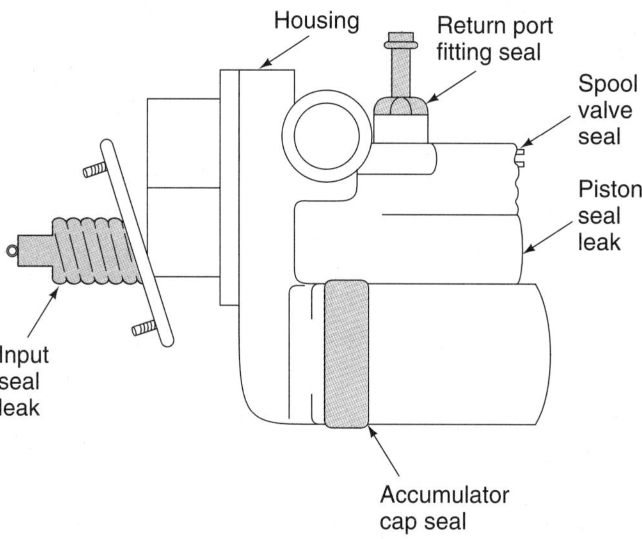

Figure 4-34 Possible sources of leakage in hydro-boost units

Return Port Fitting Seal. Fluid leakage at the return port indicates the need to replace this seal.

Basic Operational Test

Perform the hydro-boost basic operational test as follows:

1. With the engine off, pump the brake pedal repeatedly to bleed off the residual hydraulic pressure that is stored in the accumulator.
2. Hold firm pressure on the brake pedal and start the engine. The brake pedal should move downward, then push up against the foot.

Accumulator Test

To be certain that the accumulator is performing properly, proceed as follows:

1. With the engine running, rotate the steering wheel until it stops and hold it in that position for no more than 5 seconds.
2. Return the steering wheel to the center position and shut off the engine.
3. Pump the brake pedal. You should feel two to three power-assisted strokes.
4. Now repeat steps 1 and 2. This will pressurize the accumulator.
5. Wait 1 hour, then pump the brake pedal. There should be one or two power-assisted strokes.

If the accumulator has lost its charge, you should be able to rotate or wobble it with respect to the housing. Replace the accumulator assembly.

Bad valves cause the most common accumulator problem. If the valves are leaking, the accumulator may only hold a charge for a short period or may fail to hold a charge at all. In either case the booster must be disassembled and the valves replaced.

Noise Troubleshooting

It's easy to confuse hydraulic pump noise with noise from the transmission, ABS module, rear axle, or generator.

Air in the fluid causes noise.

Hydro-boost differs from vacuum brake boosters not only in the source of power (hydraulic versus vacuum) but also because hydro-boost is part of another major system in the vehicle—the power steering. Therefore, problems or malfunctions in the steering system can affect the booster, and a problem in the booster can affect the steering.

The following noises often occur with the hydro-boost system and may be cause for customer complaint. Some noises are normal and usually occur for a short time. Other noises are a sign of wear in the system or the presence of air in either the hydro-boost unit or the steering system. See Table 4-3 at the end of the chapter.

Servicing the Hydro-Boost

On models with hydro-boost brakes, check for contamination of the fluid, which results in restrictions within the spool valve, a sticking assist piston within the power unit, or a defective accumulator. The power unit will have to be disassembled for this check. If the spring is in good condition, continue going through the power unit, checking the fluid, air, and vacuum valves.

Before removing a hydro-boost unit from a vehicle, turn the engine off and pump the brake pedal several times to deplete accumulator pressure. Disconnect the master cylinder from the hydro-boost, leaving hydraulic lines connected. Carefully lay the master cylinder aside, being careful not to kink or bend the steel tubing. Next, disconnect the hydraulic hoses from the booster parts. Plug all tubes and the booster ports to prevent fluid loss and system contamination. Detach the pedal rod from the brake pedal. Remove the nuts and bolts from the booster support bracket and remove the hydraulic booster from the vehicle.

Hydro-Boost Disassembly

To disassemble and overhaul a hydro-boost unit (Figure 4-35) follow this procedure.

Special Tools

Accumulator piston compressor

Seal protector

C-clamp

1. Mount the housing and accumulator in a vise. Place the accumulator piston compressor over the end of the accumulator and install a nut onto the stud.
2. Using a C-clamp, compress the accumulator.
3. Insert a punch into the hole on the unit housing and dislodge the C-ring retainer.
4. Using a small screwdriver, remove the C-ring retainer.
5. Release the C-clamp, remove the nut and piston compressor, and remove the accumulator and its "O" ring.
6. Remove the retainer that holds the plug, "O" ring, and spring in the housing bore. Remove the plug "O" ring, and spring.
7. Remove the retainer that secures the output pushrod and baffle, and remove these items.

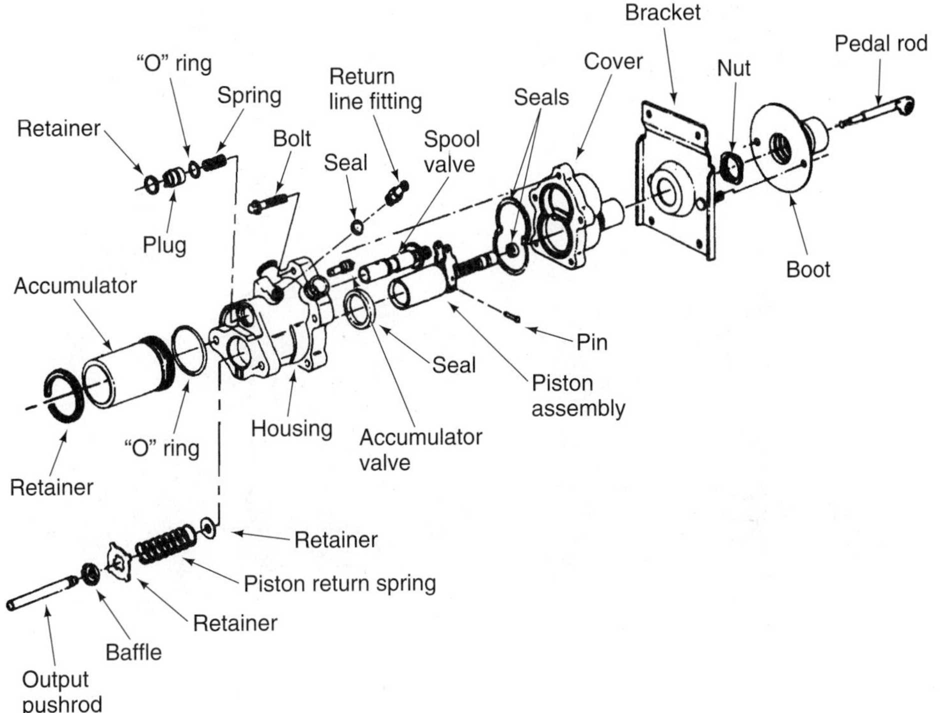

Figure 4-35 Hydro-boost components (Courtesy of General Motors Corporation)

8. Remove the piston return spring and retainer from the bore.
9. Saw off the eyelet of the pedal rod.
10. Remove the boot, nut, and bracket.
11. Remove the bolts holding the cover to the housing and separate the two.
12. Remove the seals from the cover.
13. Lift out the piston assembly, seal, and spool valve.
14. Using a wire hook, remove the accumulator valve.
15. Remove the return line fitting and seal.

Cleaning and Inspection

Once the unit is disassembled, clean all parts in denatured alcohol or clean power steering fluid. (Do not use transmission fluid.) If the parts are left exposed for 8 hours or more, they must be rewashed at the time of assembly.

Inspect the spool valve and valve bore for scratches or signs of wear. If scratches can be detected by rubbing a finger over the surface, the complete hydro-boost unit must be replaced. Due to critical clearance tolerances, the spool valve or housing cannot be substituted.

Reassembly

Special Tools

Accumulator piston compressor

Seal protectors

Discoloration of the spool or bore is normal and is not a sign of wear damage.

Use all parts included in the rebuild kit. If the unit has tube seats, remove and replace those that are damaged (Figure 4-36). Use a spare tube nut to reseat the new tube seat. Do not tighten the tube nut because the seat might be damaged. Remove the tube nut and inspect the seat for metal flakes or foreign material. Plug the ports until it is time to connect the lines. If the unit does not use tube seats, replace the "O" rings on the hose lines.

1. Lubricate all parts and bores with clean power steering fluid.
2. Install the seal and return line fitting.
3. Install the accumulator valve and spool valve.
4. Lubricate the seal protector with clean power steering fluid and use it to install the piston assembly and its seal into the housing bore (Figure 4-37).
5. Install the replacement seals onto the piston assembly and the housing and bolt the cover to the housing. Torque the bolts to specifications.

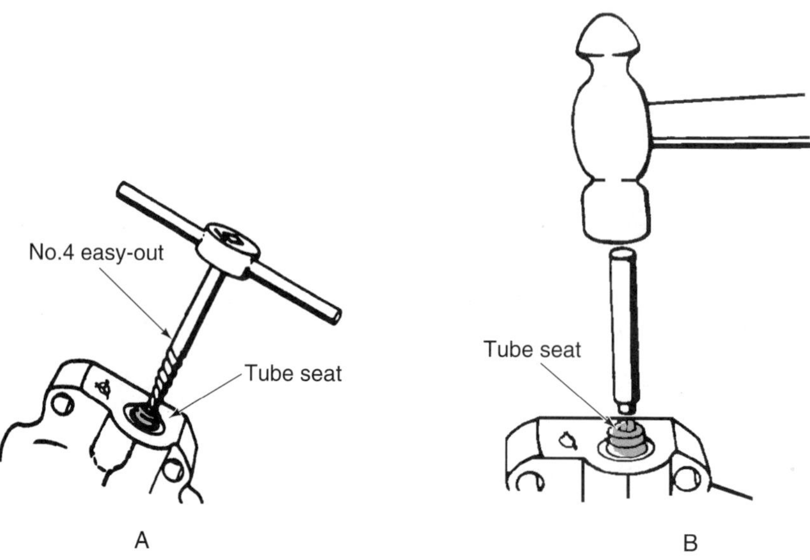

Figure 4-36 (A) Removing a damaged tube seat, and (B) installing a tube seat (Courtesy of General Motors Corporation)

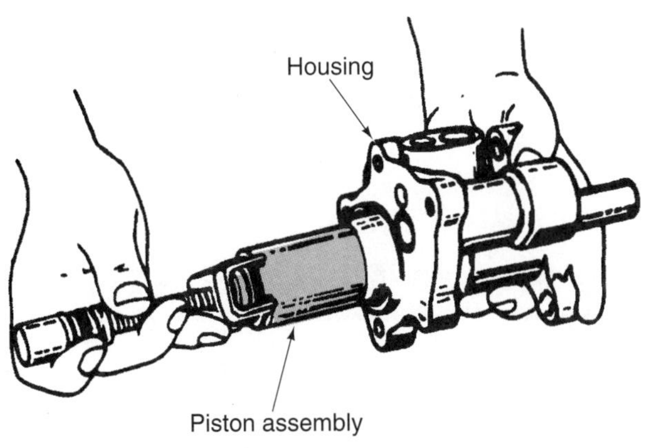

Figure 4-37 Installing the piston assembly using a seal protector (Courtesy of General Motors Corporation)

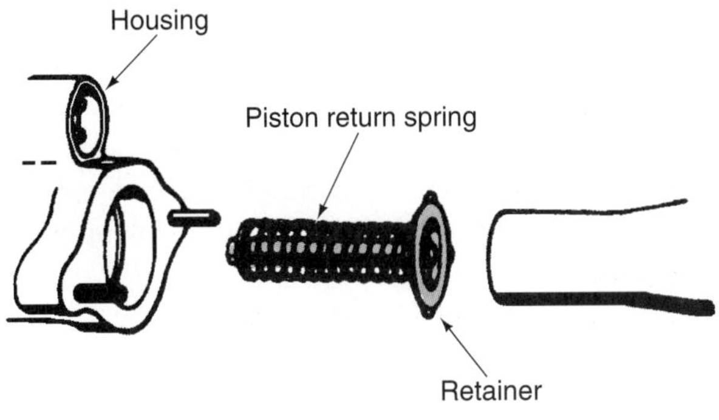

Figure 4-38 Installing the output rod retainer (Courtesy of General Motors Corporation)

6. Install the bracket, nut, and boot.
7. Using the proper seal protector, install the output pushrod assembly, consisting of the retainer, piston return spring, baffle, and output pushrod (Figure 4-38). Install the retainer to secure the assembly.
8. Slip the "O" ring onto the plug and install the spring, "O" ring, and plug into the housing (Figure 4-39). Secure with the retainer.
9. Slip the "O" ring onto the accumulator and install the assembly using the accumulator piston compressor and the C-clamp as in dissassembly. Depress the accumulator and install the retainer. Release the C-clamp and remove the compressor.

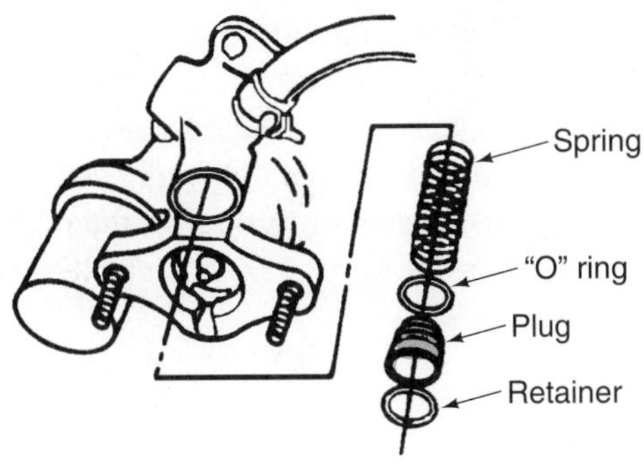

Figure 4-39 Installing the plug and spring assembly into the upper housing bore (Courtesy of General Motors Corporation)

10. Install the jam nut from the repair kit onto the pedal rod and screw the eyelet back onto the pedal rod.
11. Mount the hydro-boost on the vehicle.
12. Connect the hydraulic hoses. Torque to 20 foot-pounds.
13. Install the master cylinder.

Flushing the Hydro-Boost System

If the hydraulic fluid in the hydro-boost system becomes contaminated, flush the entire system with clean, recommended power steering fluid.

1. Completely drain the system. Disconnect the fluid return pipe from the hydraulic pump inlet port and plug the inlet port to keep contaminates from entering the system. Let the return pipe drain into a large container.
2. Have a helper pour fresh power steering fluid into the reservoir while you run the engine at idle and slowly pump the brake pedal. Allow fluid to drain from the system as fresh fluid is pumped through the system.
3. Continue flushing until no dirt or contamination is seen in the draining fluid. If foreign material still shows in the fluid, disassemble and clean the booster head, hydraulic pump, and power steering gear. Replace all seals and hoses according to the manufacturer's recommendations. Once all new parts are installed, repeat the flushing process.
4. Reconnect the fluid return pipe at the hydraulic pump inlet port and fill the system with the recommended fluid.
5. Bleed the system as outlined in the following section.

Bleeding Procedure

Whenever the hydro-booster is removed and reinstalled, the steering system must be bled of all air.

1. Fill the power steering pump reservoir to the full mark and allow it to set undisturbed for several minutes.
2. Start the engine and run it for approximately one minute.
3. Stop the engine and recheck the fluid level. Repeat steps 1 and 2 until the level stabilizes at the full mark.
4. Raise the front of the vehicle on safety stands.
5. Turn the wheels from stop to stop. Check and add fluid if needed.
6. Lower the vehicle and start the engine.
7. Apply the brake pedal several times while turning the steering wheel from top to top.
8. Switch off the engine and pump the brake pedal five or six times.
9. Recheck the fluid level. If the steering fluid is extremely foamy, allow the vehicle to stand for about 5 minutes with the engine off. Then repeat steps 7 through 9.
10. Check the steering fluid for signs of air. Aerated fluid looks milky. The level in the steering fluid reservoir will also rise when the engine is turned off. If the fluid has air in it that cannot be purged using steps 1 through 9, the problem may lie in the steering system pump. Refer to the proper section in the vehicle shop manual for further troubleshooting instructions.

CASE STUDY

A customer complains of difficulty applying the brakes on a vehicle. A test drive confirms the hard pedal condition. The vehicle is equipped with a vacuum booster, front disc brakes, and rear drum brakes. Inspection of the front and rear brake assemblies does not reveal any major problems. The shoes and pads are not glazed due to overheating or otherwise contaminated with brake fluid or oil. The amount of wear on parts seems normal based on mileage, and the brakes are properly adjusted. While there could be a problem with the master cylinder, the technician believes the problem is more likely in the vacuum-boost system. The manifold vacuum tests good at 16 inches at idle. The vacuum lines and fittings are not kinked or leaking. The technician removes the master cylinder from the booster and disconnects the booster pushrod from the brake pedal. A close inspection of the exterior components of the booster reveals the problem. The rubber grommet sealing the check valve to the booster is badly deteriorated. The technician removes the check valve and tests its operation. It is worn and sticks in the open position. The technician replaces these components as well as the front housing seal, boot, and silencer. The booster functions properly and the hard pedal condition disappears.

Terms to Know

Accumulator
Air control valve
Check valve
Diaphragm
Hard pedal
Hydro-boost
In. Hg
Pedal-free travel
Power piston
Tandem vacuum boost unit
Vacuum boost
Vacuum suspended diaphragm

ASE Style Review Questions

1. Diagnosing vacuum power booster problems is being discussed:
Technician A considers the conventional brake system and power booster to be two separate systems. She inspects conventional brake components before turning her attention to the power unit.
Technician B says hard pedal is often the first step in complete vacuum booster failure.
Who is correct?
A. A only
B. B only
C. Both A and B
D. Neither A nor B

2. Tandem vacuum power boosters are being discussed:
Technician A says they contain two diaphragms instead of the usual one.
Technician B says tandem units are smaller in diameter than standard vacuum units.
Who is correct?
A. A only
B. B only
C. Both A and B
D. Neither A nor B

3. The check valve installed in the vacuum booster system is being installed.
Technician A says it is used to prevent atmospheric air pressure from leaking from the booster chamber once the brakes are applied.
Technician B says it is used to prevent vacuum from being lost from the unit once the engine is turned off.
Who is correct?
A. A only **C.** Both A and B
B. B only **D.** Neither A nor B

4. Vacuum booster testing is being discussed:
Technician A says any condition that reduces engine vacuum will also reduce braking performance.
Technician B says a blocked passage in the power piston, a sticking air valve, or a broken piston return spring are all causes of hard pedal.
Who is correct?
A. A only **C.** Both A and B
B. B only **D.** Neither A nor B

5. Testing the vacuum boost system is being discussed:
Technician A says that with the engine off and residual pressure, pressing and holding the brake pedal should result in a slight pedal drop after a short time.
Technician B says there should never be even the slightest drop in pedal when this test is performed.
Who is correct?
A. A only **C.** Both A and B
B. B only **D.** Neither A nor B

6. Hydro-boost servicing is being discussed:
Technician A says that whenever the hydro-boost unit is serviced the power steering system must be bled.
Technician B says that power steering fluid and brake fluid are interchangeable in small amounts.
Who is correct?
A. A only **C.** Both A and B
B. B only **D.** Neither A nor B

7. Hydro-boost service is being discussed:
Technician A cleans hydro-boost parts in clean transmission fluid or brake fluid.
Technician B cleans the parts in a solvent cleaner.
Who is correct?
A. A only **C.** Both A and B
B. B only **D.** Neither A nor B

8. A customer complains of a "running faucet" noise during quick stops in his hydro-boost equipped car.
Technician A says there is a low fluid condition in the system.
Technician B says the condition is normal.
Who is correct?
A. A only **C.** Both A and B
B. B only **D.** Neither A nor B

9. Vacuum booster overhaul procedures are being discussed:
Technician A disassembles and cleans the power piston and pushrod assembly.
Technician B services the power piston and pushrod assembly as a unit and does not perform disassembly.
Who is correct?
A. A only **C.** Both A and B
B. B only **D.** Neither A nor B

10. The effects of contaminated fluid in a hydro-boost unit is being discussed:
Technician A says there could be restrictions within the spool valve, a sticking assist piston within the power unit, or a defective accumulator.
Technician B says the unit will have to be disassembled and checked.
Who is correct?
A. A only **C.** Both A and B
B. B only **D.** Neither A nor B

Table 4-1 ASE TASK

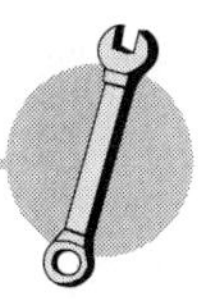

Diagnose vacuum power assist unit problems.

Problem Area	Symptoms	Possible Causes	Classroom Manual	Shop Manual
BRAKES GRAB	When applied, brakes modulate between on and off.	1. Low fluid in master cylinder	47	55
		2. Bad master cylinder seals	48	60
		3. Cracked master cylinder casting or reservoir	49	58
		4. Leaks at calipers or wheel cylinders or in lines or at connections	99	114
		5. Air in brake system	47	124
HARD PEDAL	When applied, the brake pedal gives hard resistance and is difficult to depress.	1. Broken or damaged hydraulic lines	96	114
		2. Faulty vacuum check valve or grommet	76	78
		3. Collapsed or damaged vacuum hose	75	79
		4. Plugged or loose vacuum fitting	75	79
		5. Faulty air valve seal or support plate	69	78
		6. Damaged floating control valve	69	78
		7. Bad stud welds on front or rear housing or power head	69	85
		8. Restricted air filter element	74	85
		9. Worn or distorted reaction plate or levers	71	87
		10. Cracked or broken power pistons or retainer	71	87
		11. Brake pedal binding on shaft	44	59
		12. Brake pedal interference in linkage	44	59
		13. Insufficient clearance between master cylinder pushrod and piston that allows pressure to build in system	44	68
BRAKES FAIL TO RELEASE	Braking action continues after driver removes foot from pedal.	1. Blocked passage in power piston	69	87
		2. Air valve sticking shut	69	78
		3. Broken piston return spring or air valve spring	69	86
		4. Tight brake pedal linkage	44	57

Table 4-2 ASE TASK

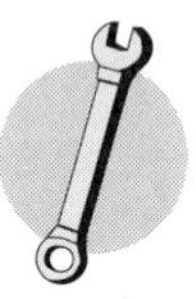

Diagnose hydro-boost power assist unit problems.

Problem Area	Symptoms	Possible Causes	Classroom Manual	Shop Manual
BRAKE PEDAL SLOW TO RETURN		1. Excessive seal friction in booster	82	99
		2. Faulty spool operation	79	97
		3. Restriction in return line from booster to pump reservoir	81	99
		4. Damaged input rod end	81	99
BOOSTER CHATTERS, BRAKES GRAB, OR PEDAL VIBRATES	When applied, brakes modulate between on and off.	1. Faulty spool action due to contamination in system	82	99
		2. Slipping power steering pump belt	83	98
		3. Low fluid level in power steering pump reservoir	78	99
ACCUMULATOR LEAK-DOWN; SYSTEM WILL NOT HOLD CHARGE		1. Contamination in hydro-boost system	84	99
		2. Internal leakage in accumulator system, especially accumulator valves	83	99
BRAKES SELF-APPLY WHEN STEERING WHEEL IS TURNED		1. Contamination in hydro-boost system	84	99
		2. Restriction in return line from booster to pump reservoir	81	99

Table 4-3 ASE TASK

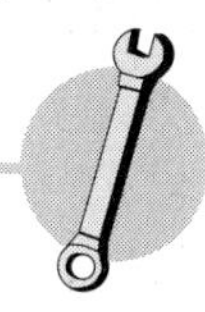

Diagnose hydro-boost power assist unit problems (Noise Troubleshooting).

Noise	Possible Causes	Solution	Classroom Manual	Shop Manual
MOAN OR LOW HUM ACCOMPANIED BY VIBRATION IN PEDAL AND/OR STEERING COLUMN. HEARD DURING PARKING OR OTHER VERY LOW-SPEED MANEUVERS.	Low fluid level in the power steering pump, or air in the power steering fluid caused by holding the pump at relief pressure (steering wheel held all the way in one direction) for an excessive amount of time (more than 5 seconds).	First check the fluid level and add fluid if needed. To eliminate the air, the system should sit for 1 hour with the cap removed. If the condition persists, it may be a sign of excessive pump wear. Check the pump according to the manufacturer's specifications.	78	99
"RUNNING FAUCET" NOISE WITH BRAKE PEDAL NEAR FULLY DEPRESSED.	Normal condition heard in emergency braking conditions.	None needed	78	100
SLIGHT HISSING NOISE.	Hydraulic fluid escaping through the accumulator valve is normal.	None needed	84	100
"GULPING" NOISE HEARD DURING BRAKE APPLICATION AFTER BLEEDING SYSTEM.	Normal condition, which should disappear under normal driving conditions.	None needed	78	100
AFTER ACCUMULATOR HAS BEEN EMPTIED AND ENGINE IS STARTED AGAIN, A HISSING SOUND OCCURS THE FIRST TIME BRAKES ARE APPLIED OR A STEERING MANEUVER IS MADE.	Caused by fluid rushing through accumulator charging orifice. Occurs one time after accumulator is emptied. If the sound continues with no additional accumulator pressure assist, it may indicate the accumulator is not holding pressure.	Check using Accumulator Test Procedures discussed in this chapter.	84	100

Hydraulic Line and Valve Service

Upon completion and review of this chapter, you should be able to:

- ❑ Diagnose poor stopping and brake pull or grab conditions caused by problems in the brake fluid, brake lines, or brake hoses and perform needed repairs.
- ❑ Diagnose poor stopping and brake pull or grab conditions caused by problems in the hydraulic system valves and perform the needed repairs.
- ❑ Inspect brake lines and fittings for leaks, dents, kinks, rust, cracks, or wear. Tighten loose fittings and supports.
- ❑ Inspect flexible brake hose for leaks, kinks, cracks, or bulging or wear. Tighten loose fittings and supports.
- ❑ Remove and replace double flare and ISO-type brake lines, hoses, fittings, and supports.
- ❑ Inspect, test, and replace the following types of brake system hydraulic valves: metering (hold-off), proportioning (balance), pressure differential, and combination.
- ❑ Inspect, test, adjust, and replace a load-sensing or height-sensing proportioning valve.
- ❑ Reset a brake pressure differential valve.
- ❑ Properly select, handle, install, and store standard and silicone brake fluids.
- ❑ Bleed and/or flush the hydraulic brake system.
- ❑ Pressure test the hydraulic brake system.

Brake Fluid

Today, brake fluid must meet many strict standards in order to provide safe, dependable stopping power and minimum system maintenance. The brake fluid must have a low freezing point, a high boiling point, and the ability to absorb water. It must possess good lubricating characteristics, and be compatible with rubber and metals used in hydraulic system components.

Basic Tools

Basic mechanic's tool set

Bleeder wrench

Flare nut wrench

The Department of Transportation (DOT) sets standards for all brake fluids sold in the United States. The following three grades are currently used in motor vehicles: DOT 3, DOT 4, and DOT 5 (silicone) (Figure 5-1).

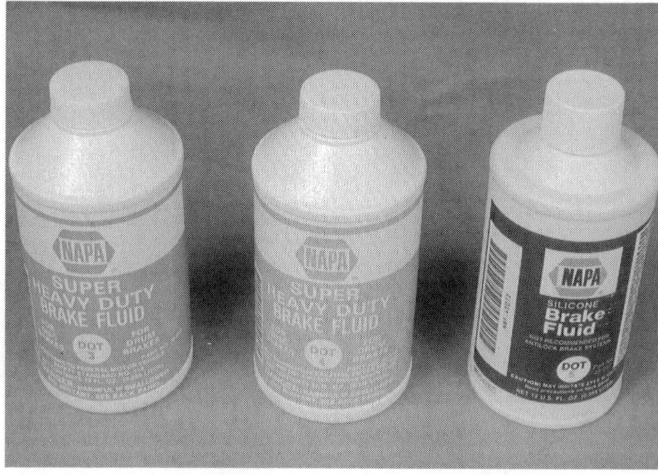

Figure 5-1 Buy and store brake fluid in the original containers. Keep the containers tightly capped.

DOT 4 absorbs less moisture and has a higher boiling point than DOT 3. Brake fluid with a high boiling point is important to systems that generate a great amount of heat. If the brake fluid boils, the hydraulic stopping power of the system will be reduced. The pedal will have to be pumped repeatedly to compress the vaporized fluid and build up pressure at the brakes to stop the vehicle.

Because both DOT 3 and DOT 4 fluids absorb moisture from the air, always keep containers tightly capped. Reseal brake fluid containers immediately after use. It is also wise to buy smaller containers of brake fluid and keep them sealed until needed. Taking these steps to minimize water in the brake fluid helps reduce corrosion of metal parts and deterioration of rubber components. It also keeps the brake fluid boiling point high to minimize the chance of vapor lock.

CAUTION: Brake fluid may irritate your skin or eyes. If you get fluid in your eyes, rinse them thoroughly with water. If fluid gets on your skin, wash the area thoroughly with soap and water.

Silicone Brake Fluid

DOT 5 silicone brake fluid has many desirable characteristics, but it is not recommended for all applications. This purple-colored fluid has an extremely high boiling point, does not readily absorb moisture, is noncorrosive to hydraulic system components, and does not damage paint like ordinary fluid does. There are some problems when using DOT 5 with seal wear and accumulation of water in the system. Silicone is currently much more expensive than regular brake fluid.

WARNING: DOT 5 silicone brake fluid should never be used in an antilock brake system. Excessive wear of brake components and other conditions can occur from using the wrong fluid.

Contaminated Fluid Problems

Always use the brake fluid type recommended in the vehicle's shop manual or owner's manual. Using the wrong type of fluid or using fluid that is contaminated with water, mineral oil, power steering fluid, or other liquids can cause the brake fluid to boil or rubber parts in the hydraulic system to deteriorate.

Swollen master cylinder piston seals are the best indicator of water or other contamination concerning the brake fluid. Deterioration may also be indicated by swollen wheel cylinder boots, caliper boots, or a damaged master cylinder diaphragm.

If you determine that water or other contaminants have entered the brake system and the master cylinder piston seals have been damaged, replace all rubber parts in the brake system, including the brake hoses. Also check for brake fluid on the brake linings. If fluid is found, replace the linings.

If there is water or contaminants in the system, but the master cylinder seals appear undamaged, check for leakage throughout the system or signs of heat damage to hoses or components. Replace all damaged components found. After repairs are made, or if no leaks or heat damage are found, drain the brake fluid from the system, flush the system with new brake fluid, refill, and bleed the system.

Flushing the Hydraulic System

Since DOT 3 and DOT 4 fluids do absorb moisture, it is good maintenance practice to flush the hydraulic system with fresh fluid on a regular basis (some automakers recommend flushing every two years). Flushing removes the accumulated water and any dirt or rust particles that may be present.

It is also important to remember that oil- or petroleum-based solvents can be very destructive to any rubber components such as seals, boots, and hoses found in the brake hydraulic system. Even very small amounts of oil or solvents may cause the rubber parts to swell, soften, and disintegrate. If any of these liquids have been improperly put into the hydraulic system, the entire system should be flushed and all rubber parts replaced. To help minimize potential problems, make sure your hands are clean before installing new rubber parts such as wheel cylinders, caliper seals, or boots. During cylinder or caliper overhaul, do not use ordinary solvents to clean the bores. Use only denatured alcohol or brake fluid.

Flushing is done at each bleeder valve in the same manner as bleeding. Open the bleeder valve approximately 1 1/2 turns and force fluid through the system until the fluid emerges clear and uncontaminated. Do this at each bleeder valve throughout the system. After all lines have been flushed, bleed the system.

Brake Lines, Fittings, and Hoses

Hydraulic system lines are made of steel tubing and rubber hosing. They transmit fluid under pressure between the master cylinder and the drum or disc brake units (Figure 5-2). Rigid tubing is used to direct fluid between stationary brake parts. Most hydraulic tubing is made of doublewall, welded steel tubes that are coated to resist rust. The tube ends are double flared or have a chamfer-type flare to guard against leakage. Thread fittings are used to connect the tube to brake parts.

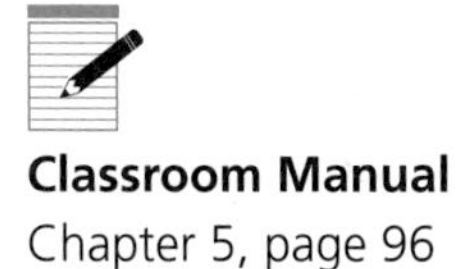

Classroom Manual
Chapter 5, page 96

The master cylinder and brake lines are mounted to the vehicle frame and body. But as the vehicle is operated, the disc and drum brake assemblies move up and down with the vehicle's suspension system. Flexible hoses are needed to connect the rigid brake lines to brake components that move. Brake hose must be free to flex and move as the wheel moves up and down or turns through its entire turning radius. Hose must also be able to withstand the high pressures within the system. Brake hose is available in varying lengths and end fittings (Figure 5-3).

Exposure to the elements, road salts in winter, salt air, water, and contaminants in the system, all contribute to rusting and corrosion of brake fittings, lines, and hardware.

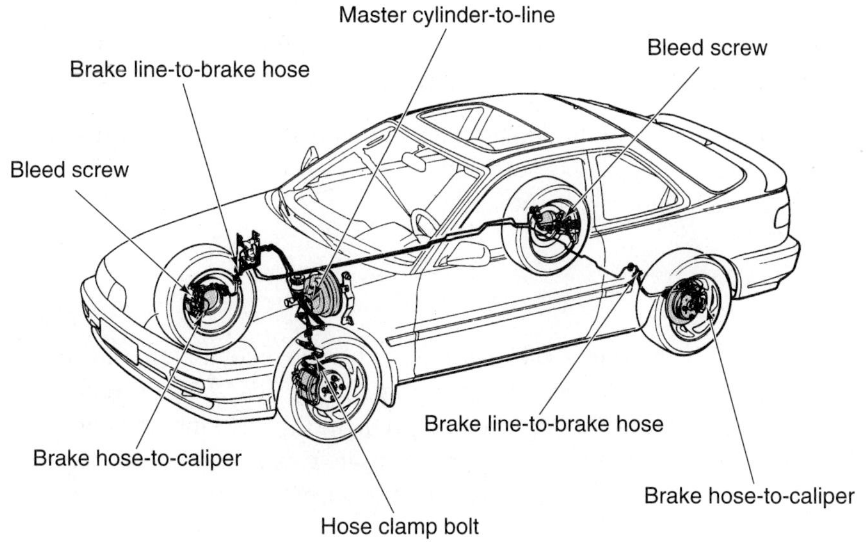

Figure 5-2 Brake lines and hose connections (Courtesy of Honda Motor Co., Ltd.)

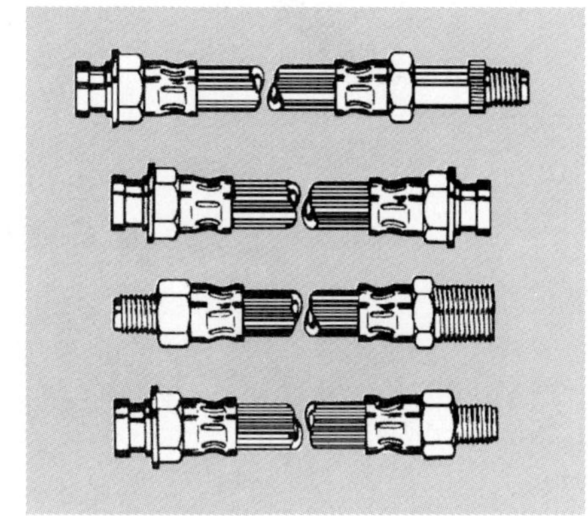

Figure 5-3 Typical brake hose end fittings (Courtesy of Chevrolet Motor Division)

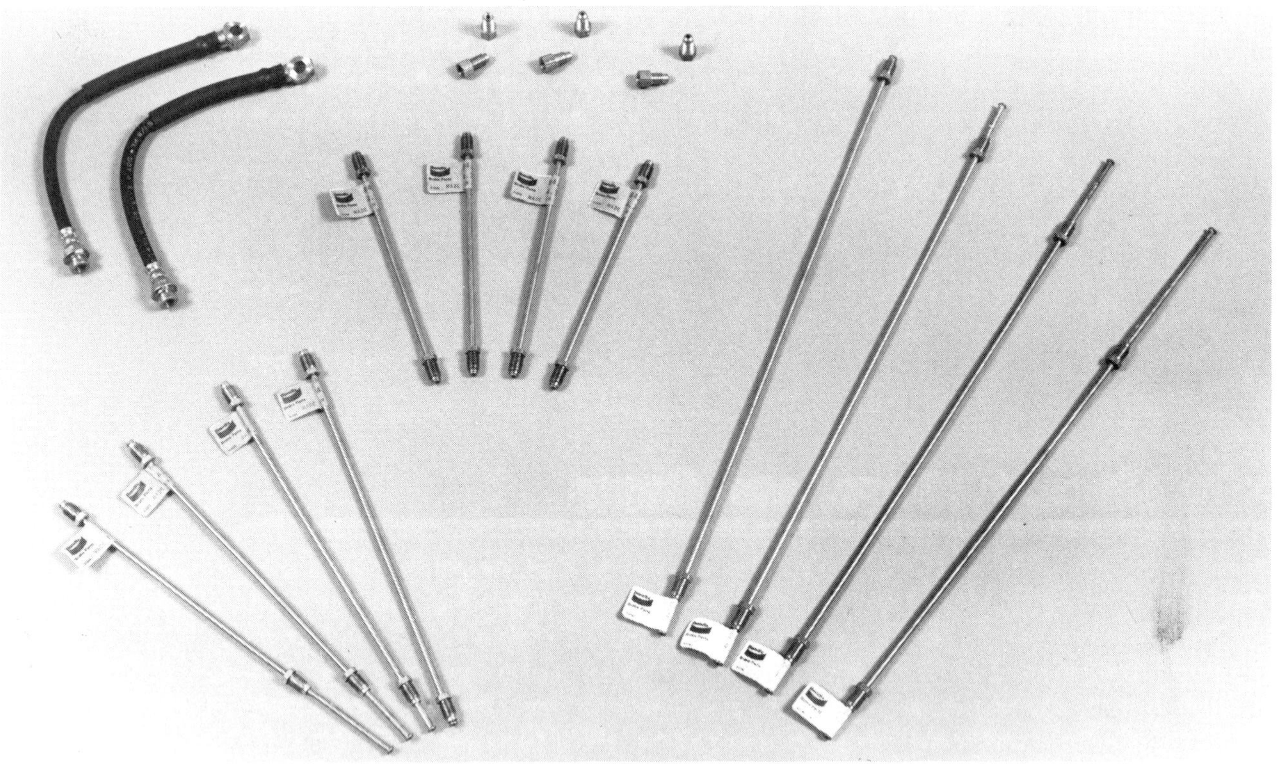

Figure 5-4 Stock brake lines and hoses with preinstalled flare nuts

Brake Lines

Brake lines or tubes are normally 3/16″ double wall steel tubing. Because of the many bends and lengths used, the lines are supplied in various dimensions and lengths. When replacing a brake line, always use steel piping designed to withstand high pressure and resist corrosion.

Stock brake line tubing is available in various lengths with ends preflared and flare nuts installed (Figure 5-4). Always use prefabricated tubing whenever possible.

 **WARNING:** Copper tubing should not be used in a hydraulic brake system. It is subject to fatigue cracking and corrosion. Both can result in brake system failures.

Brake line tubing should be inspected for kinks, cracks (ruptures), dents, and corrosion or rust (Figure 5-5). It is important that corroded or rusted lines be replaced immediately. They can burst under a panic stop situation or under high pressures.

Brake lines and flare nuts are available in both metric and conventional sizes. Care should be exercised not to confuse or mix the two.

Flare Connectors

Brake line tubing is connected to a component, brake hose, or other length of tubing using flared connections (Figure 5-6). The flare nut is slid onto the tubing with its threads facing the end of the tube to be flared. The tube is then double flared. The nut can then be turned into the body of a matching fastener. All brake tubing should be flared properly to ensure good, leakproof connections.

When connecting a tube to a hose, a tube connector, or a brake cylinder, use a lb.-in. torque wrench to tighten the tube fitting nut to specifications.

Four commonly used types of brake connections (Figure 5-7) are:

- ❑ SAE double 45-degree flare
- ❑ ISO metric flare

A

B

Figure 5-5 Inspecting brake line tubing running to a drum brake assembly

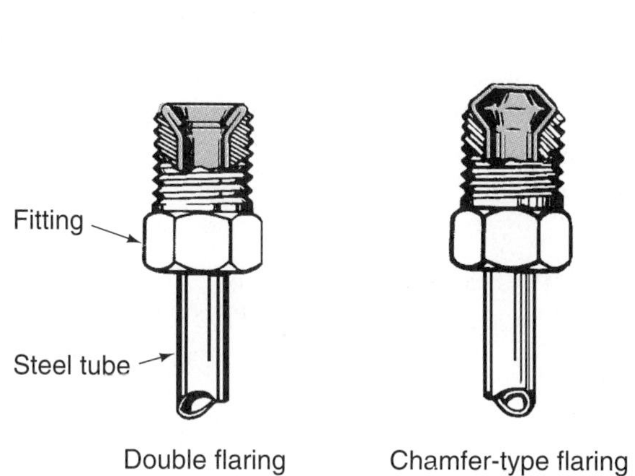

Figure 5-6 Brake lines are constructed of doublewall stainless steel tubing with flared-type fittings. (Courtesy of Chevrolet Motor Division)

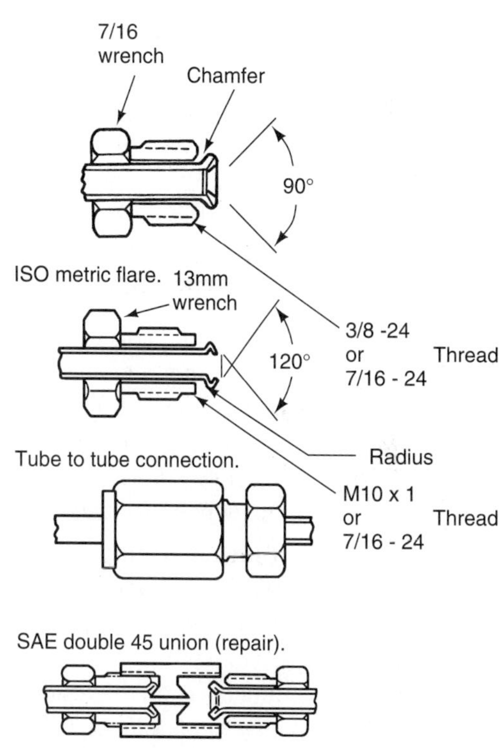

Figure 5-7 Commonly used types of brake line flared connections (Reprinted with the permission of Ford Motor Company)

❑ Tube-to-tube connection
❑ SAE double 45 union (repair)

Flare nuts do not usually corrode or rust, but the tubing that passes through them may. If the line corrodes and freezes to the nut, then the line will twist if any attempt is made to back out the nut with a flare-nut wrench.

To free a flare nut frozen to the line, apply penetrating oil to the connection. The connection can also be heated with a torch provided all plastic and rubber components are removed from the immediate area. Using a pencil-thin flame, apply heat to all sides of the flare nut, never to the line.

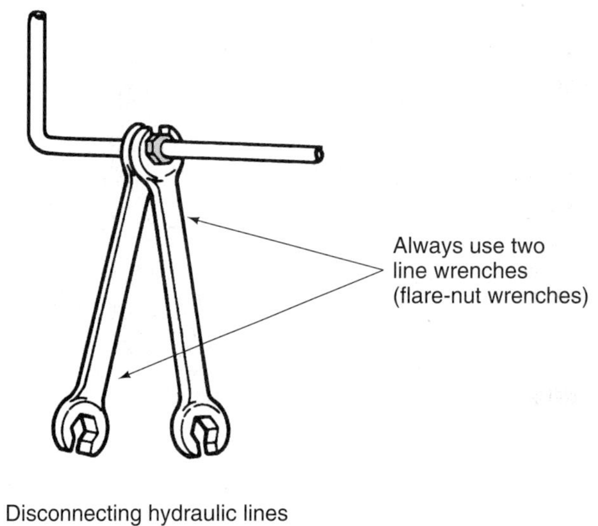

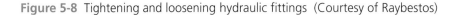

Figure 5-8 Tightening and loosening hydraulic fittings (Courtesy of Raybestos)

When the steel nut begins to glow from the heat, attempt to back out the nut with the flare wrench. If the nut cannot be freed, the line must be replaced.

Always use the correct type and size wrench when tightening or loosening hydraulic fittings (Figure 5-8). To avoid distorting a female fitting, screw a spare tube nut into the fitting to reinforce it. On fittings requiring gaskets, always install new copper gaskets. Used gaskets have taken a set and will not seal properly if reinstalled.

Fabricating Brake Lines

Many brake tube bundles are designed to be serviced as a unit. This reduces the amount of cutting, bending, and flaring steps needed to service the system.

Brake line replacement is one of the most common types of brake system servicing. To fabricate a brake line, select the correct tube and fitting dimension. Choose a standard length that is slightly longer than the old line or cut a length of tubing to size. With the old line as a reference, use a tubing bender to form the same bends in the new line as existed in the old. Each end of the new line must be flared to provide the seal and strength required by high hydraulic pressures. See Photo Sequence 7 for details.

Cutting Tubing. A tubing cutter should always be used to cut tubing to length. When properly used, the cutter ensures the ends of the tubing will be perfectly square, and the tubing itself will be perfectly round. The tubing cutter does this by supporting the tubing with a pair of rollers while a sharp cutting wheel is rolled around the edge of the tubing. After the tubing has been cut to size, any burrs on the inside edge of the tubing must be removed. A reamer is used for this job. Most tube cutters are equipped with a built-in reamer on one end.

Bending Tubing. When replacing brake lines, it is often necessary to bend the new line to duplicate the shape of the one being replaced. Steel tubing can be bent by hand to form gentle curves. If you attempt to bend tubing into a tight curve by hand, you will usually kink the tubing. Because a kink in a brake line weakens the line at that point, you should never use a kinked line. To avoid kinking, always use a bending tool. There are several types of tube bending tools available.

On small diameter tubing, a bending coil spring can be used. This coil spring is slipped over the outside of the tubing and prevents the tubing from kinking as it is slowly bent by hand (Figure 5-9). Bend the tubing slightly further than required and back off to the desired angle. This releases spring tension in the bender so it can be easily removed.

On larger diameter tubing, or where more precise bends are needed, a lever-type or gear-type bender should be used. Slip the bender over the tubing at the exact point the bend is required. Once the tubing is bent into the proper shape, assemble the flare nuts on the tubing before flaring the tube ends. Once the ends are flared, the flare nuts will not fit over the end of the tubing.

Photo Sequence 7
Typical Procedure for Fabricating and Replacing a Brake Line

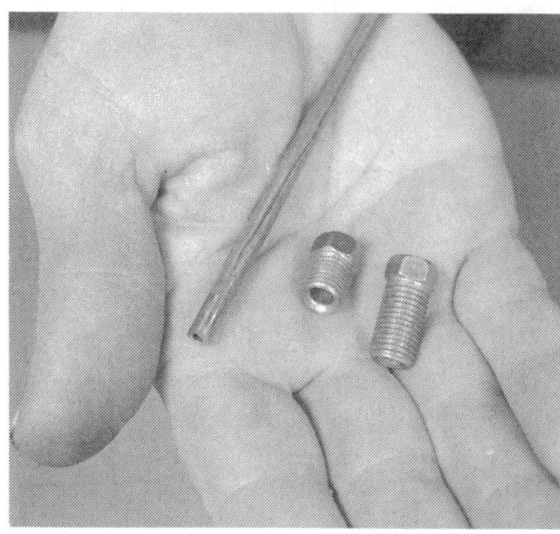

P7-1 Be sure to use the recommended bulk 3/16-inch doublewall steel brake tubing and the correct size and type tube nuts.

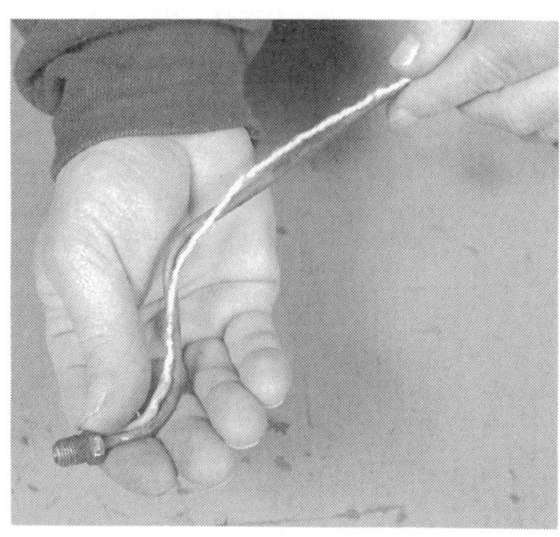

P7-2 To determine the correct length, measure the removed tube with a string and add about 1/8 in. for each flare.

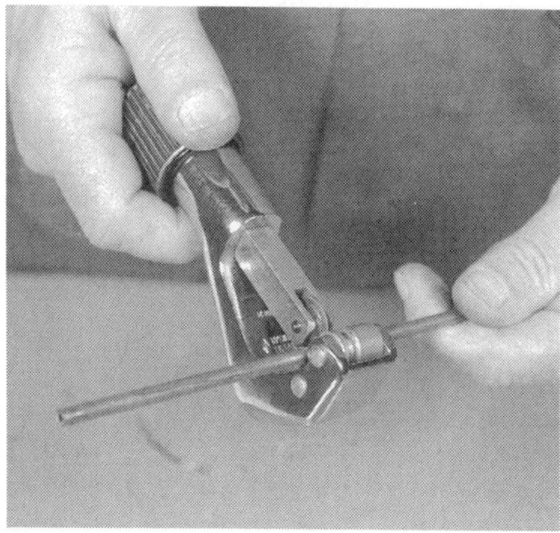

P7-3 Using a tube cutter, cut the tubing to the required length.

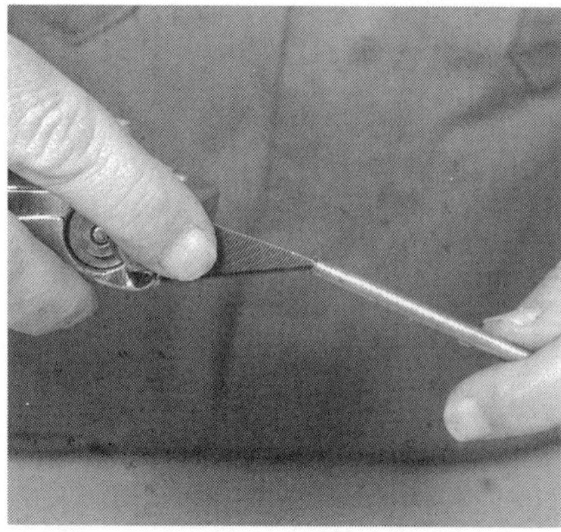

P7-4 Clean any burrs after cutting.

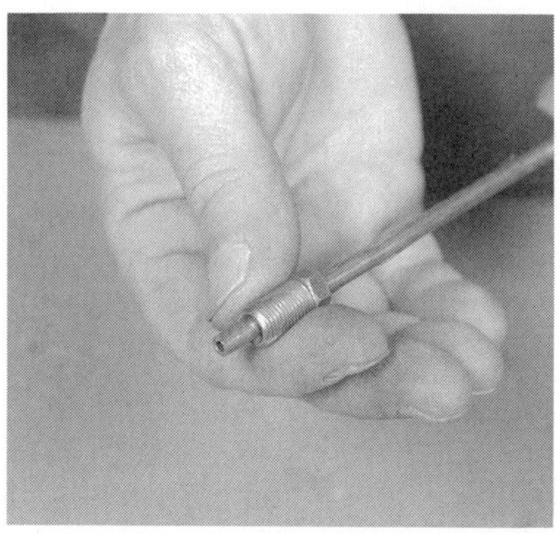

P7-5 Place the tube nut in the correct direction.

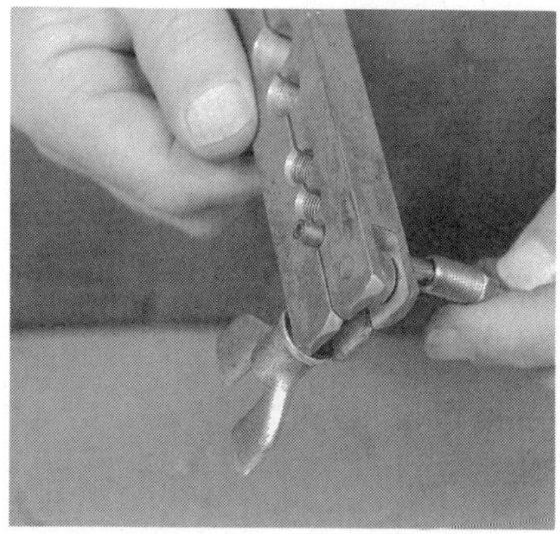

P7-6 Place the tubing in the flaring bar with the end protruding slightly above the face of the bar.

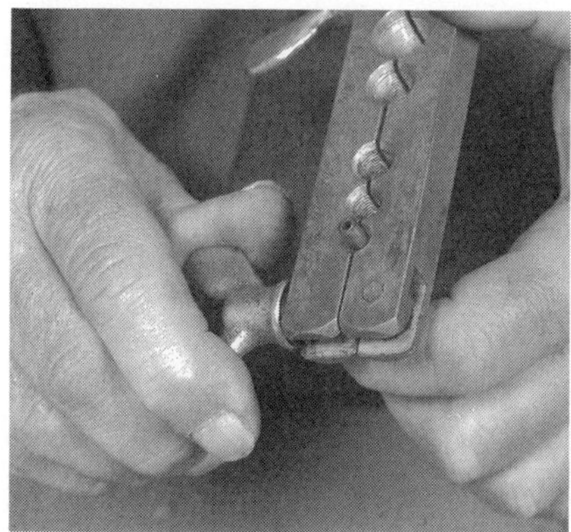

P7-7 Firmly clamp the tube in the bar so the force exerted during flaring does not push the tubing down through the bar.

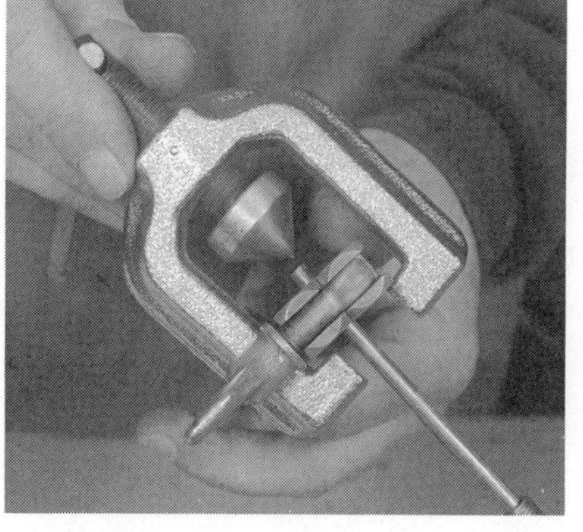

P7-8 To fold over the end of the tubing, place the adapter or anvil in place over the tube opening.

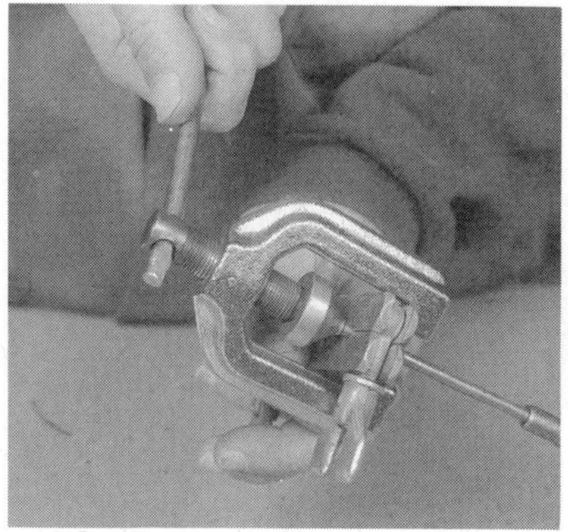

P7-9 Tighten down the flaring clamp.

Photo Sequence 7
Typical Procedure for Fabricating and Replacing a Brake Line (continued)

P7-10 Loosen the flaring clamp and check to see that the end of the tubing is properly belled.

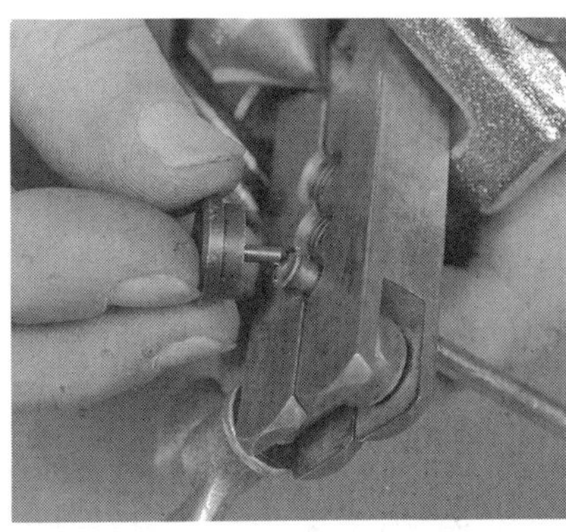

P7-11 Install the cone onto the tube opening and retighten the flaring clamp.

P7-12 The cone completes the double flare by folding the tubing back on itself. This doubles its thickness and creates two sealing surfaces.

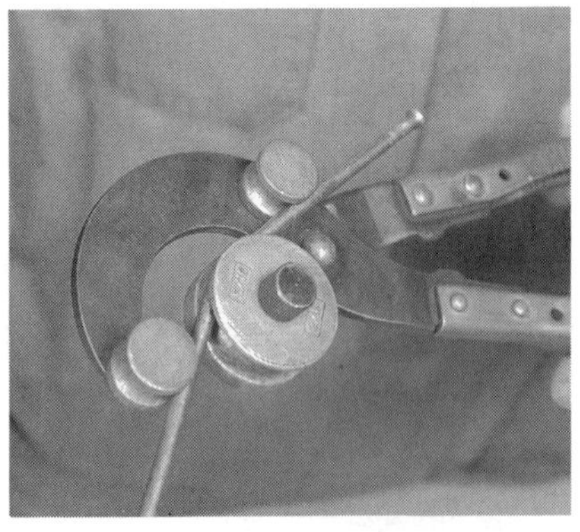

P7-13 Bend the replacement tube to match the original tube using a tubing bender.

P7-14 Clean the brake tubing by flushing it with clean brake fluid.

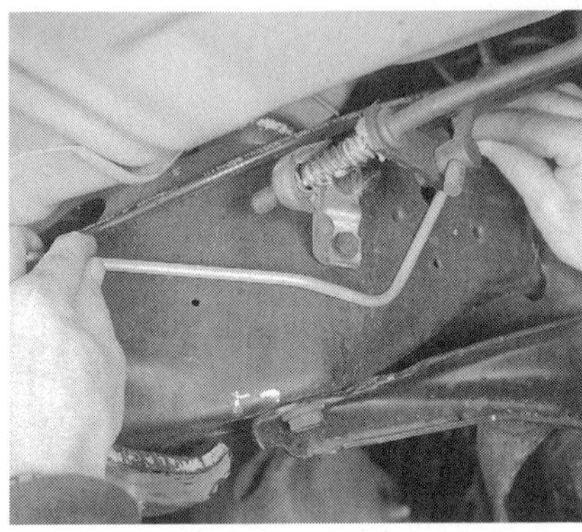

P7-15 Install the replacement brake tube, maintaining adequate clearance to metal edges and moving or vibrating parts.

P7-16 Install the brake tube and tighten the tube nuts to shop manual specifications with a lb.-in. torque wrench.

P7-17 Bleed the serviced primary or secondary circuit of the hydraulic system.

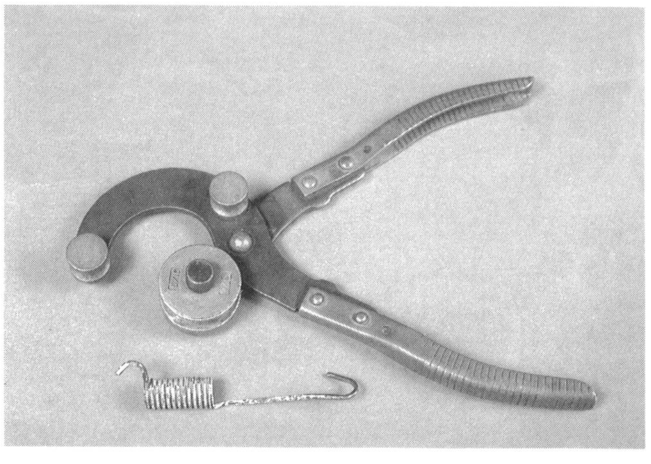

Figure 5-9 Tube bending tool used to shape brake tubing without kinking

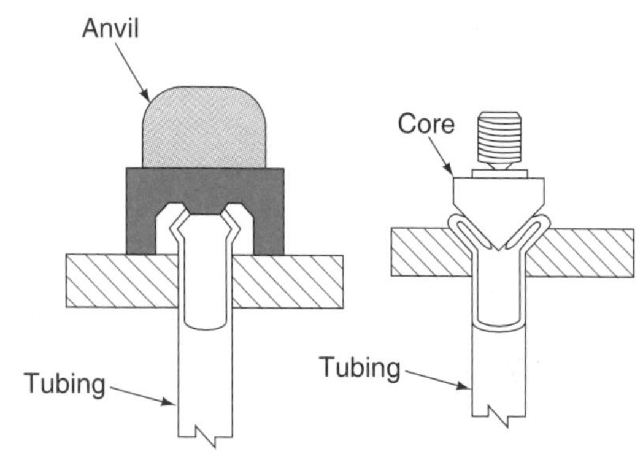

Figure 5-10 In the first step of forming a double flare, the anvil is used to fold over the tubing. In the second step, the cone creates a double thickness of tubing.

Flaring Tubing

There are two types of flares used in brake line fabrication: the *double flare* and the *ISO flare*. Single flares are never used in brake line work.

The double flare and the ISO flare each use a particular style of fitting, and they are not interchangeable.

Forming a Double Flare. A double flare is made in two stages using a special flaring tool. A typical flaring tool consists of a flaring bar and screw feed flaring yoke. The double flaring process is shown in Figure 5-10 and demonstrated in Photo Sequence 7.

Figure 5-11 shows the importance of the correct flare. The angle of the flare and the nut is 45 degrees, while the angle of the seat is 42 degrees. When the nut is tightened into the fitting, the difference in angles, called an *interference angle*, causes both the seat and the flared end of the tubing to wedge together. When correctly assembled, brake lines connected with flare fittings provide joints capable of withstanding high hydraulic pressure.

Forming an ISO Flare. The ISO flare or bubble flare meets the requirements of the International Standards Organization (ISO). It is used on many import vehicles and some domestic front wheel drive vehicles. The ISO flare offers a number of advantages:

ISO stands for International Standards Organization.

- ❑ When tightened, the shoulder of the nut bottoms out in the body of the part, creating a uniform pressure on the tube flare.
- ❑ The design is not subject to overtightening. Simply tighten the nut firmly on the seat to produce the correct sealing pressure.

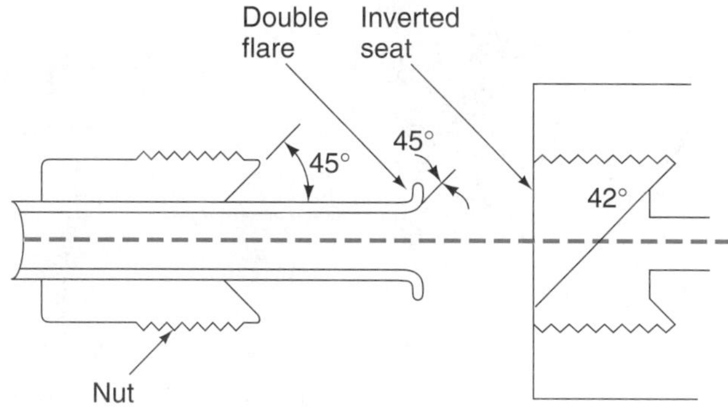

Figure 5-11 Cross-sectional view of properly flared tubing and fitting angles

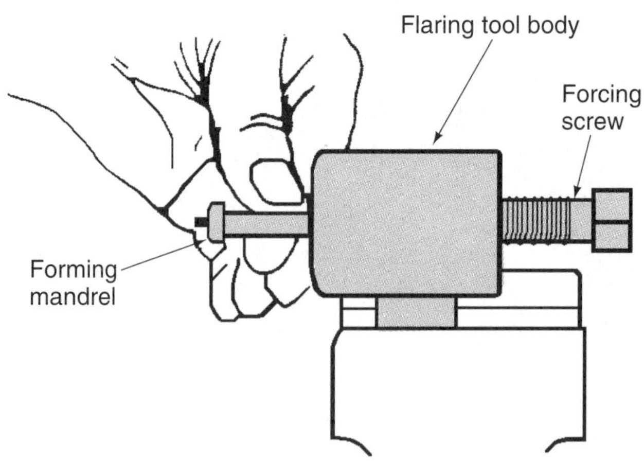

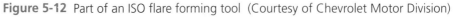

Figure 5-12 Part of an ISO flare forming tool (Courtesy of Chevrolet Motor Division)

A special tool is also required to form the ISO flare. To form an ISO flare:

1. Cut the tubing to length and install the fittings before forming the flare.
2. Chamfer the inside and outside diameter of the tube with the deburring tool and remove all traces of lubricant from the tubing and flaring tool.
3. Clamp the ISO flaring tool in a bench vise. Select the proper size collet and forming mandrel for the diameter of steel tubing being used. Insert the mandrel into the body of the flaring tool. Hold the mandrel in place with your finger and thread in the forcing screw until it contacts and begins moving the mandrel (Figure 5-12). After contact is felt, turn the forcing screw back one full turn.
4. Slide the clamping nut over the tubing and insert the tubing into the correct collet. Leave about 3/4 in. of tubing extending out of the collet (Figure 5-13).
5. Insert the assembly into the tool body so that the end of the tubing contacts the forming mandrel. Tighten the clamping nut into the tool body very tightly to prevent the tubing from being pushed out during the forming process.
6. Using a wrench, turn in the forcing screw until it bottoms out. Do not overtighten the screw or the flare may be oversized.

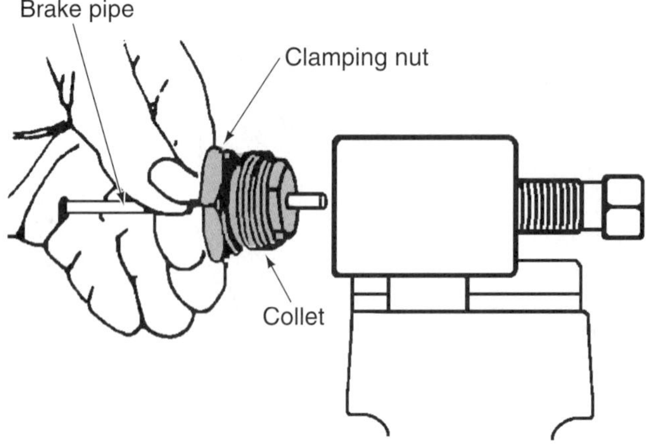

Figure 5-13 Clamping nut and collet used to secure brake tubing into the flaring tool (Courtesy of Chevrolet Motor Division)

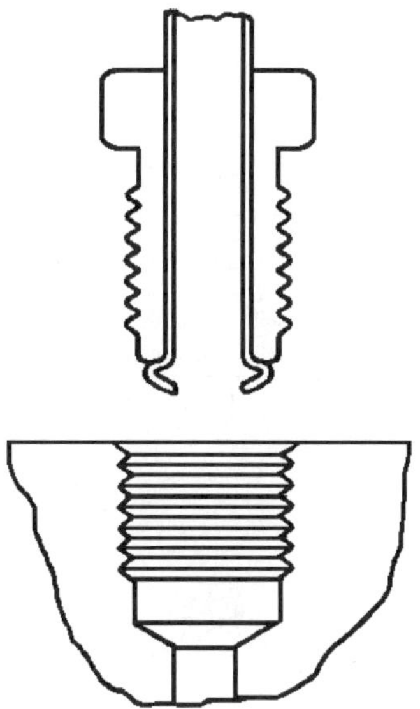

Figure 5-14 The completed ISO or bubble flare (Courtesy of Chevrolet Motor Division)

Whether making a double flare or an ISO flare, remember to follow the directions for the flaring tool you are using. After flaring and before installing the tubing, apply air pressure to blow out any metal chips or particles of dirt from inside the tubing. The tubing can also be cleaned by flushing it with fresh brake fluid.

7. Back the clamping nut out of the flaring tool body and disassemble the clamping nut and collet assembly.
8. The ISO flare is now ready for assembly (Figure 5-14).

WARNING: Using any other attachment methods, such as compression fittings, single flares, brazing, or soldering, will jeopardize the integrity and safety of the entire braking system.

Replacing Brake Hoses

Inspect all brake hoses for cracks, cuts, any indication of chafing, twists, and loose supports. Have a helper apply pressure to the brake pedal (using a pumping action) and watch or feel for bulging or expansion that indicates a weak hose (Figure 5-15).

When connecting a rigid brake line to a hose, tube connector, or brake cylinder, always tighten the tube fitting nut to proper torque specifications.

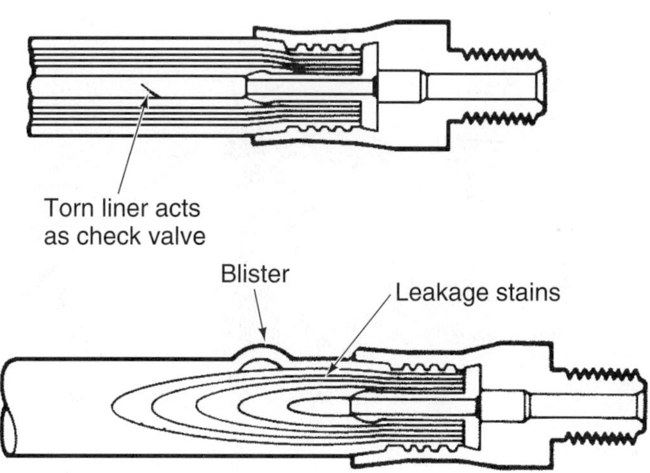

Figure 5-15 Possible internal defects in brake hose (Courtesy of Raybestos)

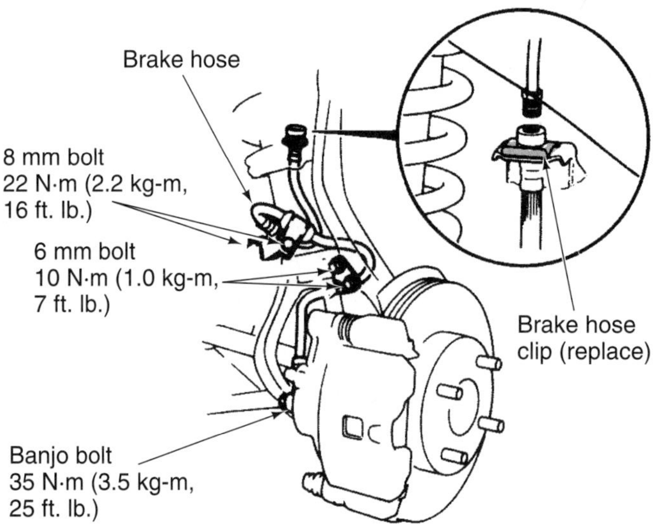

Figure 5-16 Brake hose routing and clip connection at a disc brake assembly (Courtesy of Honda Motor Co., Ltd.)

Be sure to use a replacement hose that meets OEM safety standards. It is also vital that the replacement hose be the same size as the original one. A hose that is too long may rub on the chassis. One that is too short may break when the movable component reaches the limits of its travel.

Classroom Manual
Chapter 5, page 98

Occasionally, although not too often, some caliper components and hoses can be reverse threaded; that is, they turn left to tighten instead of to the right. Sometimes the fasteners are noted with a slash through the flat surfaces of the nut or bolt. Before removing the old hose, clean all dirt from the ends of the hose to reduce the risk of dirt entering the system during assembly. Most hoses are attached to the chassis and secured with a hose lock clip. The clip keeps the hose from rubbing on the frame or suspension members (Figure 5-16).

To replace a typical brake hose, use a flare nut wrench to undo the connection at the hose and steel tubing junction (Figure 5-17). Any metal clips that are used to secure the hose in place

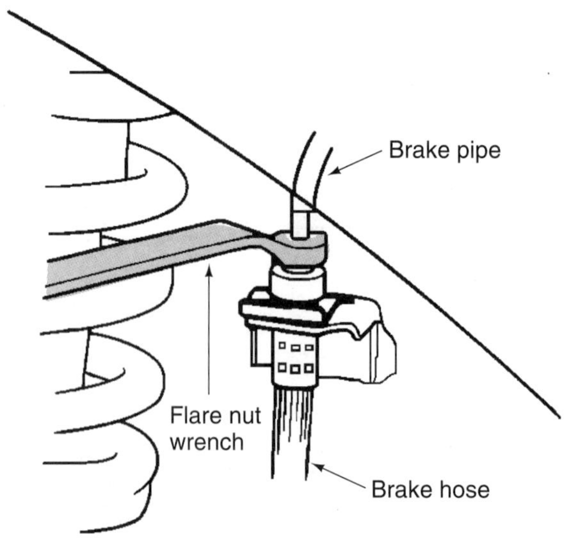

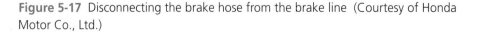

Figure 5-17 Disconnecting the brake hose from the brake line (Courtesy of Honda Motor Co., Ltd.)

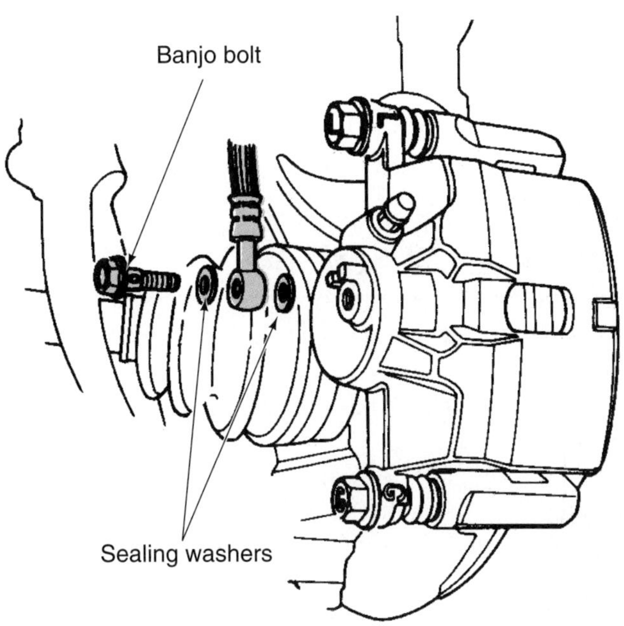

Figure 5-18 Brake hose banjo bolt connection details at wheel caliper (Courtesy of Honda Motor Co., Ltd.)

have weakened over time and must also be removed and replaced. To disconnect the brake hose at the caliper, simply remove the banjo bolt and pull the hose free of the assembly (Figure 5-18).

To install the replacement hose, place a new brake hose clip into the mounting bracket (Figure 5-19). Connect the hose to the brake line tubing. Refasten the hose to the caliper using a banjo bolt and new sealing washers.

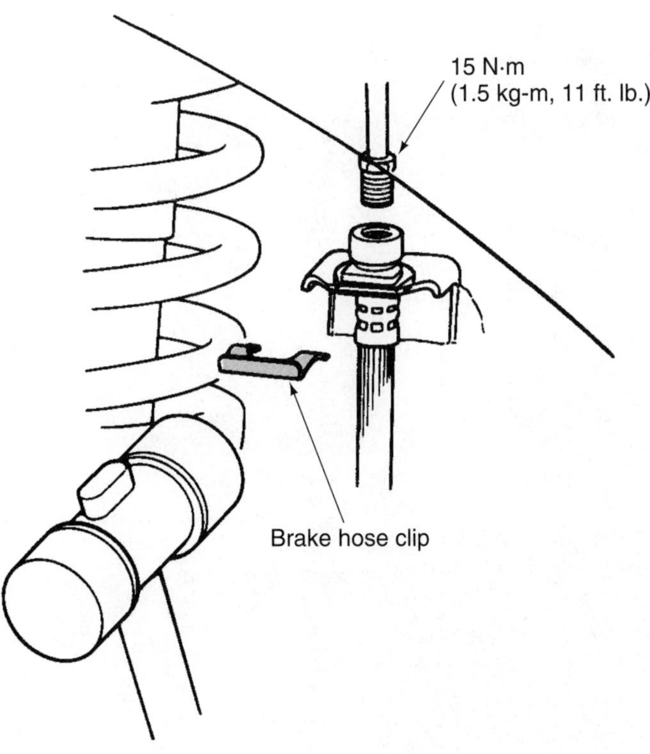

Figure 5-19 Installing new brake hose clip (Courtesy of Honda Motor Co., Ltd.)

After installing the new brake hose, check the hose and line connections for leaks, and tighten if needed. Check for clearance during suspension rebound and while turning the wheels. If any contact occurs, reposition the hose, adjusting only the female end. Tampering the male end once the female end has been tightened could result in twisting the hose. The treads and ends of the new hose must also be identical to those used originally to ensure correct fitting.

Hydraulic System Bleeding

Always bleed the applicable primary or secondary brake line system after brake line or hose replacement.

The hydraulic brake system must be free of air to work properly. Air can enter the system when hydraulic parts are disconnected or if the system is operated with a low fluid level in the master cylinder. Air in the system is compressed when the brake pedal is applied. This results in greater pedal travel and a spongy brake pedal.

Bleeding is a service procedure that feeds fresh brake fluid into the system and forces out air bubbles and the old aerated fluid through special bleeding ports. These ports are sealed with bleeder screws that can be partially backed open to allow fluid and air to escape the system. Bleeder screws are located strategically at high points throughout the brake system (Figure 5-20). A bleeder screw is normally installed in each drum brake wheel cylinder, in each disc brake caliper, adjacent to the outlet port of some master cylinders, and on some combination valves on disc brake systems. Some bleeder screws have a threaded passage and a protective dust cap screw that must be removed before a drain hose can be installed on the bleeder screw.

When opening or removing a bleeder screw, use a special six-sided bleeder screw wrench of the correct size. The shoulders or a bleed screw are easily rounded if the wrong tool is used. Hold the wrench in the correct position. Do not exert too much force on the screw. It is possible to break off the small screw in the housing. Fixing this problem requires drilling and tapping, so avoid it by working carefully. If drilling and tapping are not possible, the entire cylinder or caliper will have to be replaced.

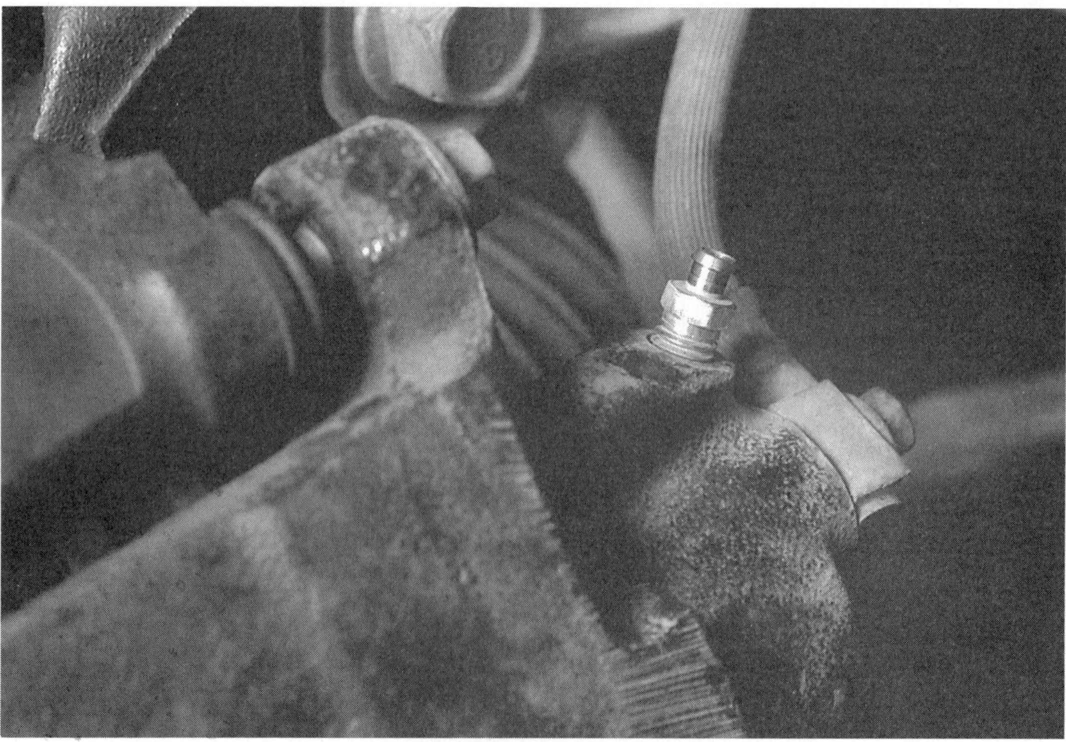

Figure 5-20 Bleeder screw on a disc brake caliper

Freeing a Frozen Bleeder Screw

Before performing cylinder or caliper service, always check the bleeder screw to see if it can be loosened. If the caliper or cylinder is serviceable, but the bleeder screw is frozen fast, use one of the following methods to free it.

Use a special wrench that allows you to exert pressure on the screw by striking it with a hammer. The shocks to the housing from the hammer blows and the tension against it might free the screw.

Heat from a welding torch is used to bring the housing around the screw to a dull red. When hot, apply light pressure with the correct size wrench. When using this method, however, be sure to remove the rubber and plastic components in the caliper or piston to prevent damage from the heat.

Bleeding Locations

Depending on the location at which the hydraulic system was opened to air, bleeding may be needed at all four wheels and the master cylinder. Or it might be needed at only at two wheels. If both tubes are disconnected from a dual-reservoir master cylinder, or if air was introduced into the system through low fluid level in both master cylinder reservoirs, bleeding is required at the master cylinder and all four wheels.

A bleeder screw is often located adjacent to the outlet port of master cylinders mounted on an angle.

If the tube serving one set of wheels in a dual system is disconnected at the master cylinder, or if air was introduced through a low fluid level in only the reservoir serving that set of wheels, only those wheels and lines affected need be bled. But if there is any doubt as to air in the system, the entire system should be bled.

SERVICE TIP: On some vehicles, it is possible to install a right-hand caliper on the left side wheel and vice versa. If you inadvertently do this, the bleeder screw will be located at the bottom of the caliper's cylinder. This low position makes it impossible to bleed all air out of the system.

Bleeding Fundamentals

Any time the hydraulic system is opened for service, it is absolutely necessary to bleed out any air that has entered it. It is also good practice to bleed the brakes when disc pads or shoes are replaced.

The job involves opening bleeder screws at individual wheels, one at a time. Pressure is then applied to the system either by stroking the pedal (Figure 5-21) or through the use of a pressure

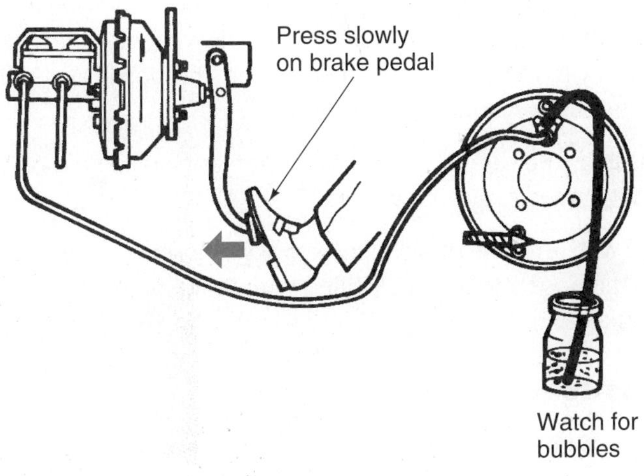

Figure 5-21 Setup for manual bleeding (Courtesy of Raybestos)

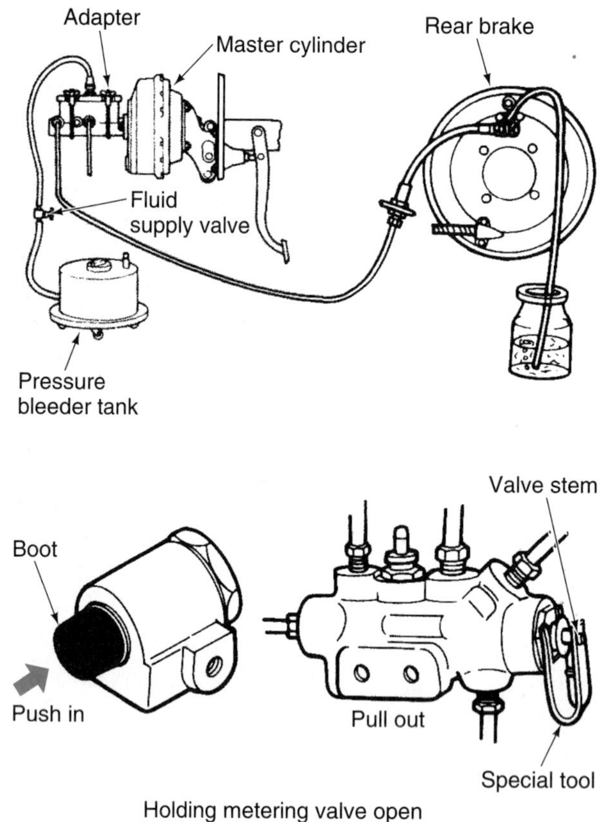

Figure 5-22 Setup for pressure bleeding (Courtesy of Raybestos)

bleeder (Figure 5-22). Either method forces a solid column of fluid through the brake lines, wheel cylinders, and calipers, expelling any air bubbles that might be present. Removing air from the fluid is important because air, unlike fluid, is compressible. Any air in the system will cause a spongy pedal and require that the pedal be pumped several times to generate sufficient stopping pressure.

SERVICE TIP: If the rear calipers are difficult to bleed, lightly tap them with a hammer to regroup the air bubbles so they can be expelled.

The bleeding procedure is fairly simple. The first step is cleaning the dirt from the filler cap before removing it from the master cylinder. This should be done thoroughly. Anytime the fluid in the reservoir falls below the level of the intake port(s), air will enter the system. If this has happened, all you will usually have to do is refill the reservoir and slowly pump the pedal a few times to purge the air. If this does not work, bleed the system.

WARNING: Be careful not to spill or drip any brake fluid on the car finish. If you do, flush it immediately with water.

General practice with any type of bleeding is to slip a hose over the end of the bleeder screw. Place the free end of the hose into a jar half-filled with brake fluid. Always keep the end of the hose submerged in brake fluid.

Bleeding Sequence

It is important to always check with the vehicle's service manual to be sure that the proper sequence for that vehicle is followed. The following bleeding sequence is typical for most hydraulic systems:

1. Bleed at the master cylinder bleeder screws, if so equipped. If the master cylinder does not have bleeder screws, crack the outlet tube nuts at the master cylinder and have a helper apply medium pressure to the brake pedal to force the air out.
2. Bleed at the bleeder screw on the combination valve, if so equipped.
3. Bleed at each wheel cylinder and caliper starting at the wheel cylinder or caliper farthest from the cylinder and working back to the master cylinder. This procedure prevents trapping air in brake lines between the two units. If there are two bleeder screws per wheel, bleed the lower one first. On vehicles with diagonally split systems, do one-half of the system at a time.

With a GM quick take-up master cylinder, for example, it is necessary to bleed all wheel cylinders and calipers in the following sequence:

1. Right rear wheel cylinder
2. Left front caliper
3. Left rear wheel cylinder
4. Right front caliper

If you bleed this system using the conventional sequence, you might chase air throughout the system.

Manual Bleeding

Manual bleeding uses the brake pedal and master cylinder as a hydraulic pump to expel air and brake fluid from the system when a bleeder screw is opened. Manual bleeding is normally a two-person operation—one to pump the brake pedal and another to operate the bleeder screws.

Relieve the vacuum or hydraulic reserve by applying the brake pedal several times with the ignition switch in the OFF position.

Fill the master cylinder to the proper level with clean brake fluid before starting and after completing the bleeding. Also check the master cylinder fluid level frequently during bleeding and keep the reservoirs at least half-full. Manual bleeding requires that the fluid level in the master cylinder be checked and replenished periodically. If the level drops below the intake port, air will enter the system. The bleeder screw should be open only when the pedal is depressed and closed before the pedal is released. Residual pressure valves in the system will not hinder bleeding. The metering valve (if so equipped) should be deactivated to properly bleed front disc brakes.

Open (during pedal application) and close (prior to pedal release) the bleeder screws, observing the emission of air bubbles. At the same time, have a helper pump the brake pedal slowly in order to move the fluid through the system and expel trapped air. Slip a piece of the hose over the bleeder screw and insert the free end into a jar partially filled with brake fluid. All air has been purged from the system when bubble-free fluid flows from the hose. Air in the system is visible when bubbles emerge from the end of the hose inserted in the fluid as the pedal is depressed. Keeping the end of the hose inserted in fluid will prevent the accidental admission of air into the system when the bleeder screw is open. Pedal pumping action must be done slowly to prevent any undue buildup of pressure and increase in turbulence.

If the master cylinder is suspected of having air in the bore, it must be bled before any wheel cylinder or caliper.

1. Remove the forward brake tubing connection at the master cylinder and allow brake fluid to flow from the port.
2. Reconnect but do not tighten the brake tube.
3. Slowly apply the brake pedal and hold it down. Observe for aerated or bubbled fluid escaping from the loose fitting.
4. Tighten the tube fitting before your helper releases the pedal.
5. Have your helper release the pedal. Wait 15 seconds.
6. Repeat this sequence, including the 15-second wait, until all air is bled from the bore.

To ensure that no air can be sucked back into the system, the bleeder valve must be closed at the end of each stroke before the brake pedal is released.

It may be necessary to repeat the bleeding operation ten or more times per wheel to expel all the air from the wheel cylinders and calipers.

7. Remove the rear brake tubing connection at the master cylinder. Repeat the bleeding process in steps 2 through 6.

CAUTION: On ABS vehicles, if the antilock unit is suspected of having air trapped in it, the unit must be bled. Refer to Chapter 8: Antilock Brake Service for details.

Once all air is removed from the master cylinder, the brake assemblies at each wheel can be bled. Photo Sequence 8 shows the proper method of manually bleeding a disc brake caliper. The steps shown apply to one bleeder screw and should be repeated at all other bleed points.

Manual Bleeding with Check Valve Bleeder Hose. Bleeding hoses are available with a one-way check valve that only allows fluid to flow out of the bleeder screw, not in the reverse direction. This bleeder hose eliminates the possibility of air being drawn back into the system when the brake pedal is released.

Pressure Bleeding

Pressure bleeding has two advantages over manual bleeding. It is faster because the master cylinder does not have to be refilled several times. Also, the job can be done by one person. The hose from the pressure bleeding tank is connected to the master cylinder by means of an adapter fitting that fits over the reservoir, taking the place of the reservoir cap (Figure 5-23). Pressure bleeding adapters exist in different configurations for the different types of reservoirs.

Figure 5-23 Bleeder adapter for a pressure bleeding system (Courtesy of General Motors Corp.)

WARNING: To prevent air in the pressure tank from getting into the lines, do not shake the tank while air is being added to the tank or after it has been pressurized. Set the tank in the required location, bring the air hose to the tank, and do not move it during the bleeding operation. The tank should be kept at least one-third full.

The hydraulic pressure generated by manual bleeding is sufficient to open the metering or combination valve and allow fluid to flow to the front wheels. When pressure bleeding, it is necessary to hold the valve stem open manually.

To hold the metering or combination valve open, either push the valve stem in or pull it out, depending on the valve type. Do not use more than 25 pounds of pressure to push the stem in or

Photo Sequence 8
Typical Procedure for Manually Bleeding a Disc Brake Caliper

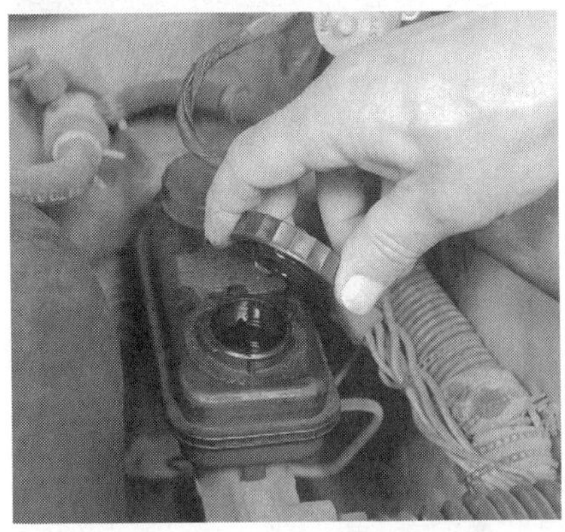

P8-1 Be sure the master cylinder reservoir is filled with clean brake fluid. Recheck it often to replace fluid lost during the bleeding process.

P8-2 Attach a bleeder hose to the bleeder screw.

P8-3 Place the other end of the hose in a glass jar partially filled with brake fluid. Be sure that the free end of the hose is submerged in brake fluid. This helps to show air bubbles as they come out of the system and prevents air from being accidentally sucked into the system through the bleeder screw.

P8-4 Have a helper apply moderate (40 to 50 pounds), steady pressure on the brake pedal and hold it down.

P8-5 Open the bleeder screw.

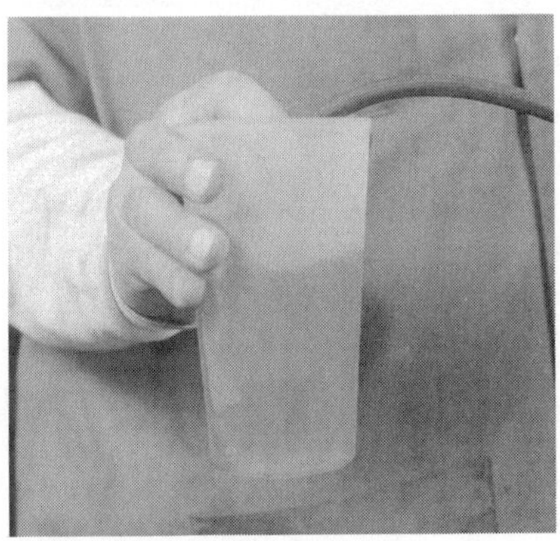

P8-6 Observe the fluid coming from the submerged end of the hose. At the start, air bubbles should be seen.

Photo Sequence 8
Typical Procedure for Manually Bleeding a Disc Brake Caliper (continued)

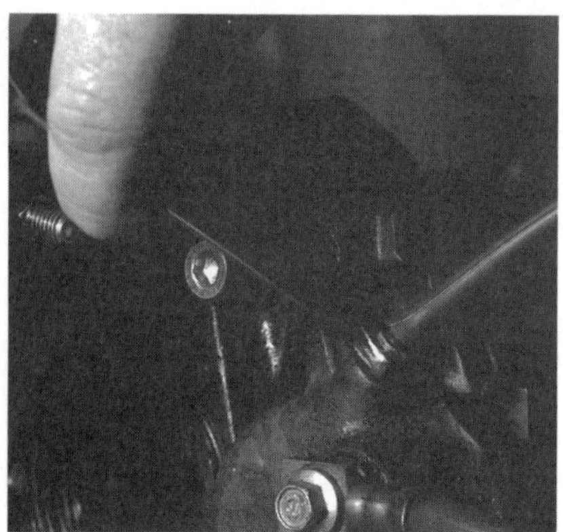

P8-7 When the fluid is clear and free of air bubbles, close the bleeder screw.

P8-8 Have your helper release the brake pedal. Wait 15 seconds and repeat steps 2 and 3 until no bubbles are seen when the bleeder screw is opened. Close the bleeder screw at that wheel and move to the next wheel in the bleeding sequence. Bleed all four wheels in the same manner.

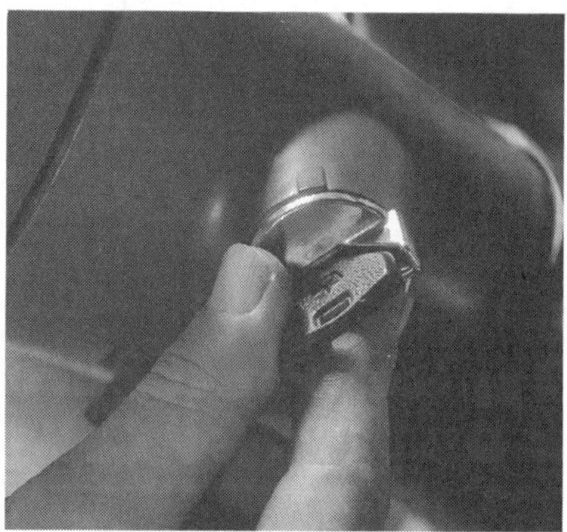

P8-9 When the entire system has been bled, turn the ignition switch to the ON position.

P8-10 Check the pedal for sponginess.

P8-11 Check the brake warning lamp(s) for an indication of unbalanced pressure. Repeat the bleeding procedure to correct either of these problems.

P8-12 Top off the master cylinder reservoir to the proper fill level.

pull it out. Otherwise, the valve can be damaged. To hold the metering valve open, use the special tool available for this task (Figure 5-24). Never use a screw clamp, wedge, or block that can put excessive pressure on the valve stem. Also, be sure to remove the special tool when the bleeding operation is completed.

The procedure for pressure bleeding follows:

1. Bring the unit to a working pressure of 15 to 20 pounds per square inch (psi). Be sure that there is enough brake fluid in the pressure bleeder to complete the bleeding operation.

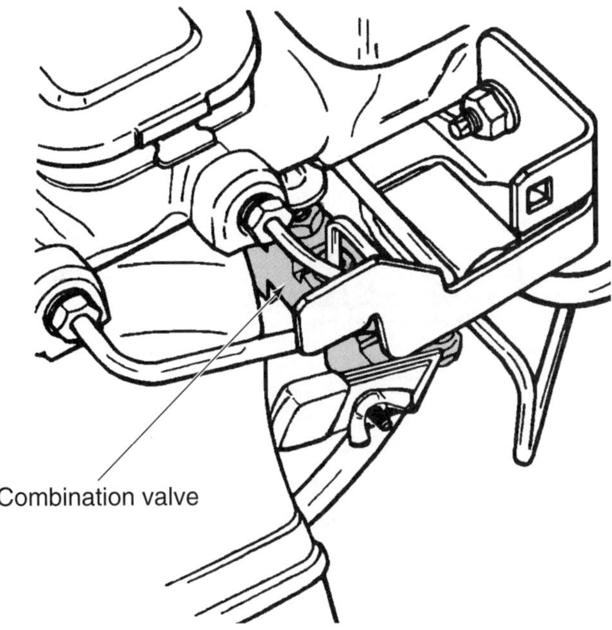

Figure 5-24 Combination valve pressure bleeding tool (Courtesy of General Motors Corp.)

2. Clean the master cylinder cover and remove it. Remove the reservoir diaphragm gasket if there is one. Clean the gasket seat and fill the reservoir.
3. Attach the adapter to the reservoir.
4. Connect the hose to the adapter fitting, making sure that the coupling sleeve is fully engaged.
5. Open the fluid supply valve.
6. Bleed the system in the normal sequence, beginning with the master cylinder.
7. Remove the combination valve pressure bleeding tool.
8. Disconnect the pressure bleeder and close the supply valve. Then, with the coupling wrapped in a rag to protect the car finish from drips, undo the coupler sleeve above the master cylinder. Make sure that the reservoir fluid level is within 1/4 inch of the top, then replace the gasket and cap.

Draining the System

To completely drain the system, open the bleeder screws and pump the pedal slowly until only air is expelled. If the fluid reservoir has a diaphragm-type seal, first remove it. If the vehicle is equipped with a pressure metering valve, hold it open. To catch the fluid, slip a hose over each bleeder screw and drain into containers.

Servicing Hydraulic System Valves

Metering valves, proportioning valves, and combination valves are used by auto manufacturers to regulate pressures within the system. Not all types of valves are found on all vehicles, so check the vehicle's shop manual for the exact type and location of hydraulic valves. All valves in the hydraulic system should be inspected whenever brake work is performed or a problem exists in the system. Brake grab or uneven braking may be an indication of a faulty metering, proportioning, or combination valve.

Classroom Manual
Chapter 5, page 101

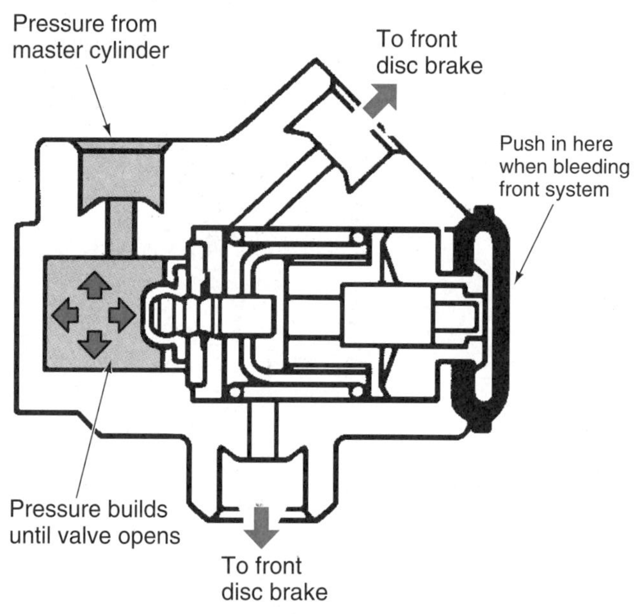

Figure 5-25 Typical metering valve. Check for leaks at all ports. (Courtesy of Raybestos)

Classroom Manual
Chapter 5, page 103

Metering Valves

Inspect the metering valve (Figure 5-25) whenever the brakes are serviced. Fluid leakage inside the boot on the end of the valve means the valve is defective and should be replaced. A small amount of moisture inside the boot does not necessarily indicate a bad valve.

Metering valves are nonadjustable and nonrepairable. If a valve is defective, replace it as an assembly. Always be sure to mount the new valve in the same position as the old valve.

Classroom Manual
Chapter 5, page 105

Proportioning Valves

If the vehicle's rear brakes lock up during moderate-to-hard braking and all other possible causes of lockup have been eliminated, the proportioning valve is the likely problem. If the valve is leaking, replace or service it. In many cases, the valve cannot be disassembled and serviced; it must be replaced as a unit. To change a bad proportioning valve, disconnect the brake line from the valve body and plug the lines to prevent dirt from entering the lines. Screw out the proportioning valve body from the master cylinder body. Remove and discard the old "O" ring (Figure 5-26).

To install the new valve, lubricate the new "O" ring and valve threads with clean brake fluid. Install the "O" ring into the cylinder body and turn in the valve body. Torque the valve to the specification listed in the shop manual. Finally, reconnect the brake lines. On some vehicles, the proportioning valve does not screw into the master cylinder, but is installed in the brake line near the cylinder body (Figure 5-27).

On some vehicles, proportioning valves are integrated into the master cylinder body (Figure 5-28). These valves often can be serviced with reconditioning kits. A typical service procedure follows:

1. Remove the master cylinder reservoir, the proportioning valve caps, and the cap "O" rings. Discard the "O" rings.
2. Using a needle nose pliers, remove the proportioning valve piston springs and valve pistons. Take care not to damage or scratch the piston stems. Remove the valve seals from the pistons.
3. Wash all parts with clean denatured alcohol and dry them with unlubricated compressed air. Inspect the pistons for corrosion and damage. Replace them as needed.

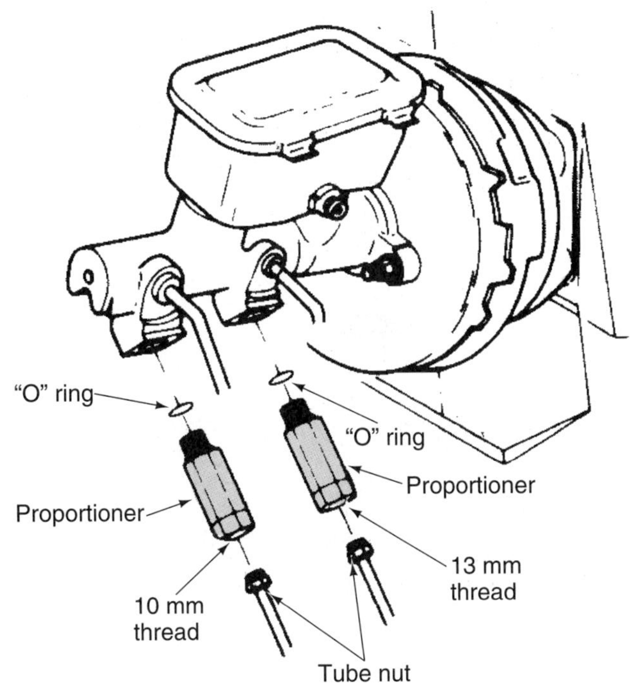

Figure 5-26 Removing a proportioning valve from a master cylinder (Courtesy of General Motors Corporation)

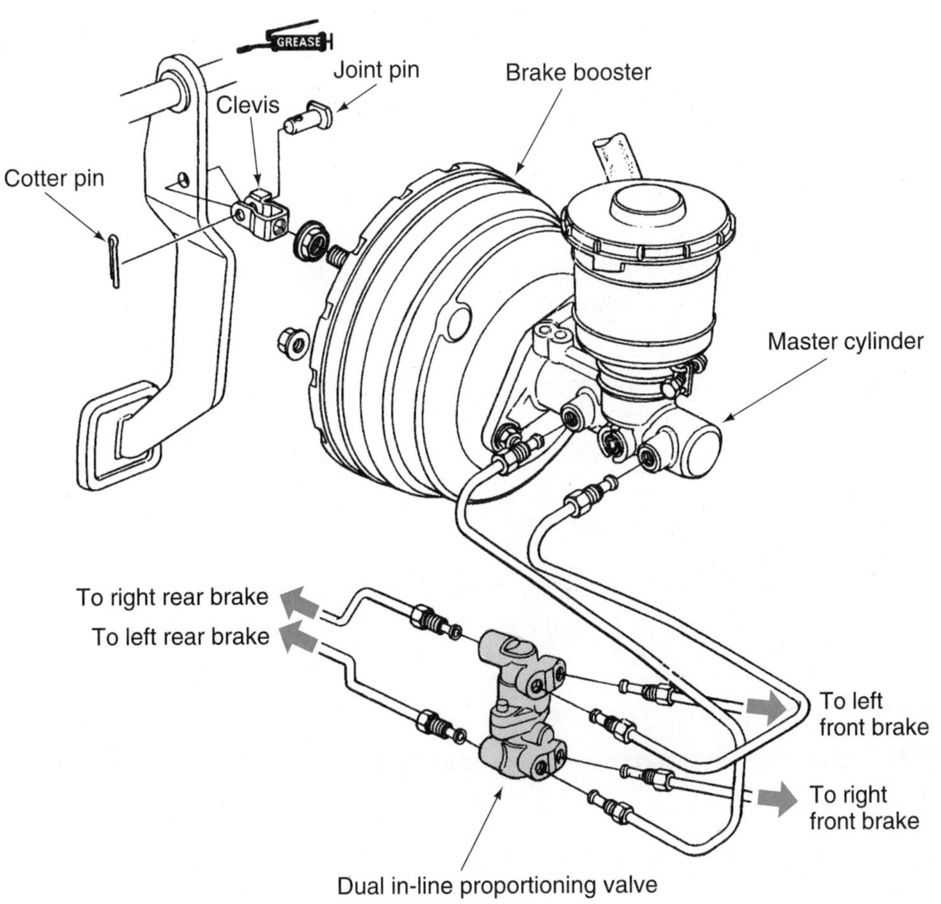

Figure 5-27 Location of dual in-line proportioning valve (Courtesy of Honda Motor Co., Ltd.)

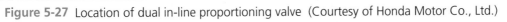

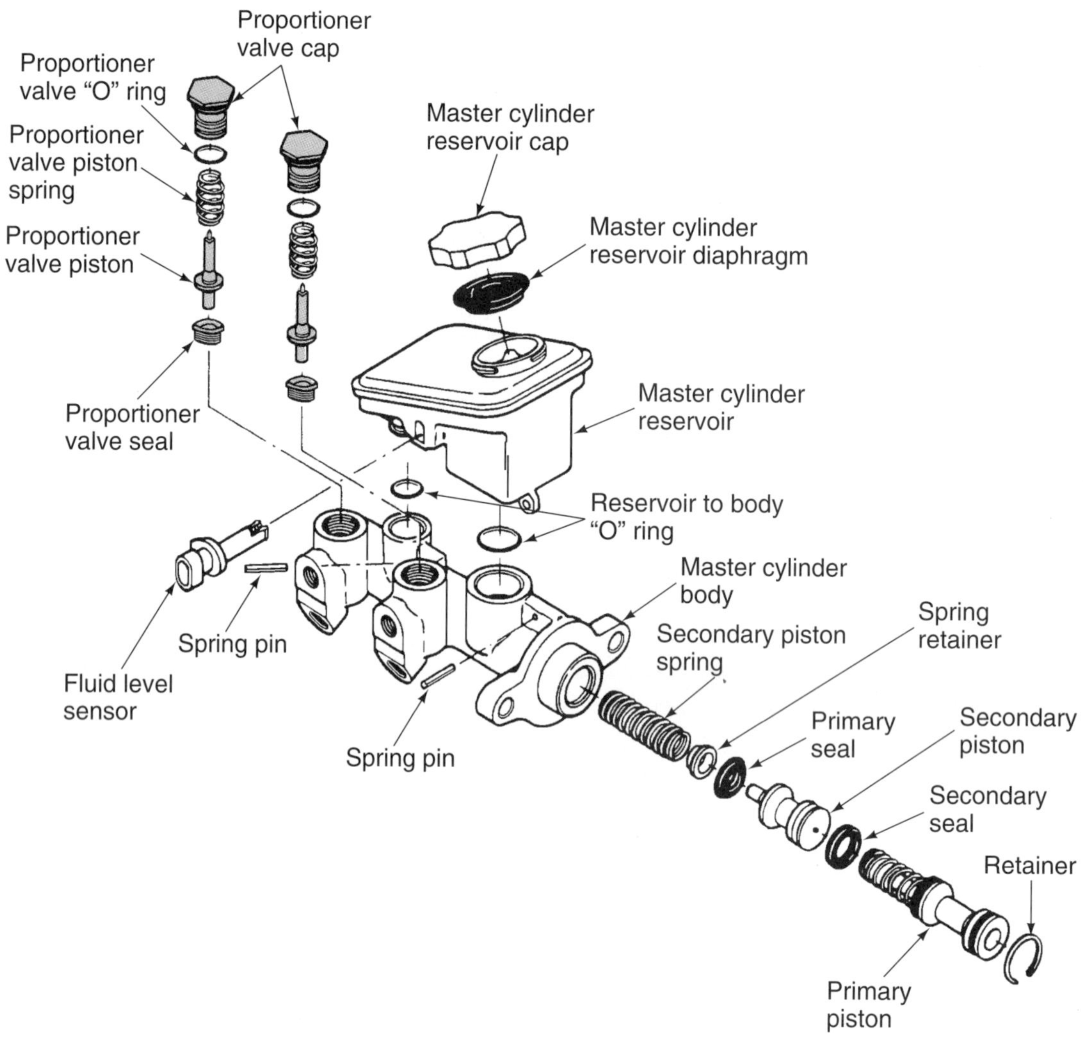

Figure 5-28 Components of proportioning valves integral to the master cylinder body (Courtesy of Chevrolet Motor Division)

4. Lightly lubricate the new proportioning valve cap "O" rings, valve seals, and piston stems with the silicone grease supplied in the repair kit.
5. Install the valve seals on the valve pistons so the seal lips are facing upward toward the caps.
6. Install the pistons and seals into the master cylinder body, followed by the valve springs.
7. Place the new cap "O" rings into the grooves in the proportioning valve caps and install the valve caps, torquing them to specifications.
8. Reinstall the master cylinder reservoir.

Height-Sensing Proportioning Valves

A brake height-sensing proportioning valve regulates the amount of brake pressure according to vehicle load condition. More cargo weight allows more braking while less cargo weight restricts brake pressure.

There are several important points to remember about height-sensing proportion valves. The valve must be adjusted correctly if it is to balance rear braking force to the load. Load-proportioning

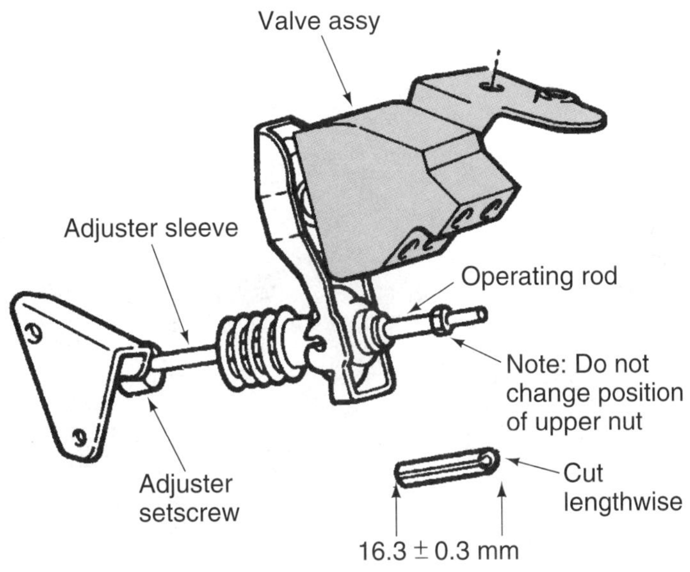

Figure 5-29 Adjusting the load-sensing proportioning valve (Reprinted with the permission of Ford Motor Company)

valves are also calibrated to work with the stock suspension. Any modification that improves load-carrying capability such as helper springs or air-assist shocks can adversely affect valve operation.

Modifications that make the suspension stiffer can prevent the suspension from deflecting the normal amount during hard braking or heavy load conditions. As a result, the proportioning valve may not increase rear brake effort sufficiently and stopping distance may increase dangerously. Due to this potential problem, modifications to the rear suspensions of these vehicles should be avoided.

Adjustment

Adjusting a load-sensing proportioning valve such as the one shown (Figure 5-29) involves setting the operating rod to a specified length. Because it is difficult to accurately measure the operating rod, a short length of 1/4 in. ID plastic tubing is cut to size and used as a gauge to accurately set operating rod length.

To begin the adjustment, position the vehicle on a lift or alignment machine so that its wheels are on a flat surface and the vehicle is at curb load level. Back off the valve adjuster setscrew, but do not change the position of the upper nut.

Cut the length of 1/4 in. tubing to the adjustment specification length (in this example 16.3 mm) and slit the tubing lengthwise so it can be installed onto the operating rod. Slip the tubing onto the operating rod to set the proper operating length between the valve body and the upper nut. With the adjuster sleeve resting on the lower mounting bracket, tighten the adjusting setscrew to lock in the setting. Remove the tubing from the operating rod. The operating rod adjustment is now set for operation under normal driving conditions.

Test drive the vehicle. If you find that rear braking pressures are too little or too great, slight adjustments can be made. With the suspension at curb height, loosen the adjuster setscrew and move the adjuster sleeve toward or away from the brake pressure control valve. Each 1 mm distance the adjuster sleeve is moved changes braking pressure by 60 psi. Move the adjuster sleeve down away from the valve body on the operating rod to increase braking pressure. Move the adjuster sleeve up toward the valve body to decrease braking pressure. Once the setting is adjusted, tighten the setscrew into the desired position.

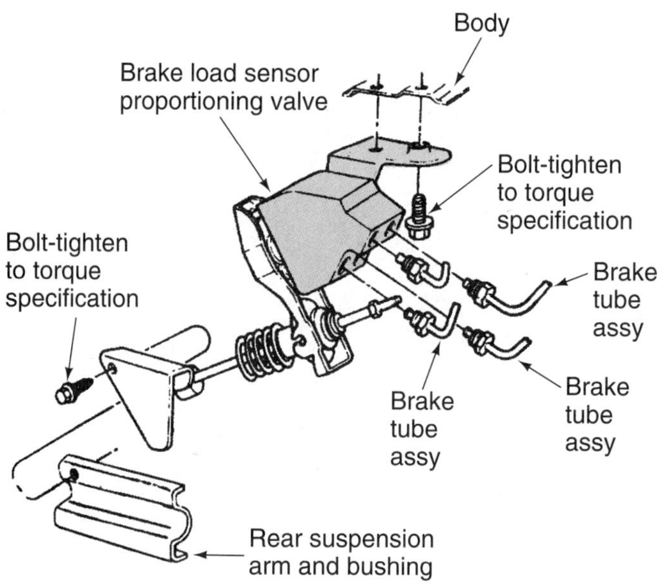

Figure 5-30 Typical load-sensing proportioning valve mounting (Reprinted with the permission of Ford Motor Company)

Removal and Replacement

To remove and replace a typical load-sensing proportioning valve assembly (Figure 5-30) raise the vehicle on the lift and disconnect the four brake fluid lines from the valve body. Tag the line positions so you will be certain they are reinstalled correctly.

Next, remove the fastener securing the load-sensing valve bracket to the rear suspension arm and bushing. You can now remove the two screws that secure the valve bracket to the underbody and remove the assembly.

SERVICE TIP: In some cases, the new replacement brake load-sensing proportioning valve will have a red plastic gauge clip on the brake pressure differential valve. Do not remove the clip until the valve is installed on the vehicle.

Before installing the new valve be certain the rear suspension is in full rebound. If the valve has a red plastic gauge clip, make sure it is in position on the proportioning valve and the operating rod lower adjustment screw is loose.

Position the valve and install the bolts that hold it to the underbody. Secure the valve's lower mounting bracket to the rear suspension arm and bushing using the retaining screw. Tighten all fasteners to specifications. Check to make certain the valve adjuster sleeve rests on the lower bracket and then tighten the lower adjuster setscrew.

Reconnect the brake lines to their original ports on the valve body and bleed the rear brakes. Remove the plastic gauge clip and lower the vehicle to the ground.

The new brake height-sensing proportioning valve automatically becomes operational when the vehicle's suspension is in the curb height condition. Remember that the addition of extra leaf springs to increase load capacity, spacers to raise the vehicle curb height, and air shocks to allow heavier loads without sagging should not be used on vehicles with height-sensing proportioning valves.

Combination Valves

Inspect and test the combination valve (Figure 5-31) whenever the brakes are serviced. If there is leakage around the large nut on the proportioning end, the valve is defective and must be

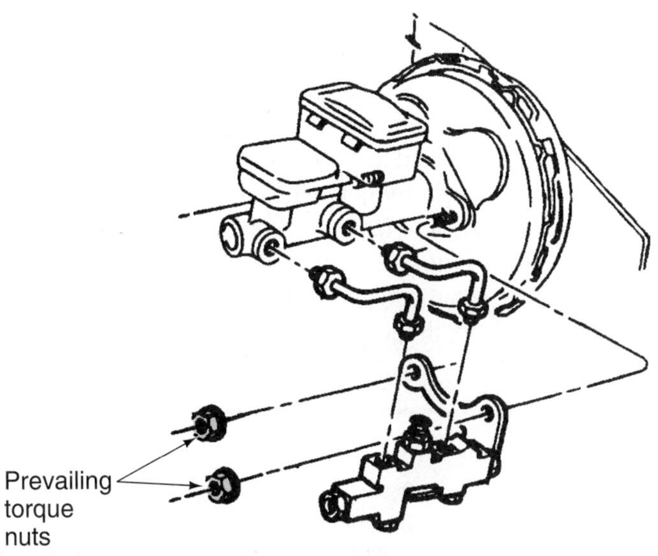

Figure 5-31 Typical combination valve mounting location (Courtesy of Chevrolet Motor Division)

replaced. A small amount of moisture inside the boot or a slight dampness around the large nut does not indicate a defective valve. Combination valves are nonadjustable and nonrepairable. If a valve is defective, it should be replaced as an assembly.

Combination Valve Warning Switch Test

To test the operation of the combination valve switch, it is necessary to bleed brake fluid from both the rear wheel and front wheel assemblies.

1. Raise the vehicle on the lift and attach a bleeder hose to a rear wheel bleeder screw connection. As in any bleeding operation, place the opposite end of the hose in a container filled with fresh brake fluid.
2. Have a helper open the bleeder screw. Turn the ignition switch to the ON position and apply the brake pedal using moderate to heavy pressure. Watch the brake system warning light. It should turn on.
3. Have your helper close the bleeder screw, then release the brake pedal. Apply the brake pedal using moderate to heavy pressure. The warning light should turn off.
4. Refill the master cylinder reservoir.
5. Repeat steps 1 through 3 at a front wheel bleeding location. You should see the same results.
6. Turn the ignition to OFF. If the warning light did not come on during the test, but does light when a jumper is connected to ground, the warning switch portion of the combination valve has malfunctioned.
7. Replace a defective valve; do not attempt to disassemble and service a bad combination valve.
8. Remember to top off the master cylinder reservoir after the test is complete.

CASE STUDY

A 4 × 4 pickup truck is experiencing low pedal and a lack of braking power. When the technician checks the master cylinder reservoir, it is low on fluid. There is no leakage around the master cylinder itself, so the vehicle is placed on the lift for a more complete inspection. The problem is easy to spot. The brake hose to the right rear wheel cylinder is badly scuffed in one spot and fluid is seeping out. Oddly, the hose itself seems quite new. It also seems to be somewhat longer than needed. The support clips are in place, but the hose still forms a wide loop between the clips. This is the damaged area. When the customer is questioned, it is learned that the hose is a replacement. The original was cut during some off-road driving up a mountain trail about two months ago. The technician removes the damaged hose and compares it to the recommended replacement hose pulled from inventory. The damaged hose is eight inches longer than the recommended replacement. The extra play in the hose caused it to rub against the underbody during the driver's off-road excursions. With the proper length hose correctly installed, extra play is not created and the hose will last its normal service life.

Terms to Know

Banjo nut
Bleeder hose
Bleeder screw
Bleeding
Bleeding sequence
DOT-3 brake fluid
DOT-4 brake fluid
DOT-5 silicone brake fluid
Double-flare fitting
Doublewall steel tubing
Flare nuts
Flushing
Height-sensing proportioning valve
Interference angle
ISO or bubble flare fitting
Manual bleeding
Pressure bleeding
Tube seat

ASE Style Review Questions

1. Brake fluid contamination is being discussed:
Technician A says swollen master cylinder seals are a sign water is in the system.
Technician B says that the system requires flushing if this condition exists.
Who is correct?
A. A only **C.** Both A and B
B. B only **D.** Neither A nor B

2. Brake fluids are being discussed:
Technician A says one of the major problems with silicone brake fluids is their tendency to absorb water.
Technician B says silicone fluids are not recommended for ABS systems due to the excessive wear they can cause within the system.
Who is correct?
A. A only **C.** Both A and B
B. B only **D.** Neither A nor B

3. Brake line tubing is being discussed:
Technician A says doublewall steel tubing with single flare fittings is acceptable.
Technician B says copper tubing with double flare or ISO flare fittings is acceptable.
Who is correct?
A. A only **C.** Both A and B
B. B only **D.** Neither A nor B

4. Replacing brake lines is being discussed:
Technician A uses prefabricated lines whenever possible.
Technician B always installs the flare fittings onto the tube before forming the flared ends.
Who is correct?
A. A only **C.** Both A and B
B. B only **D.** Neither A nor B

5. Air in the hydraulic system is being discussed:
Technician A says air can enter the system whenever a line or component is disconnected.
Technician B says air can enter the system when the fluid level in the master cylinder reservoir is low.
Who is correct?
A. A only **C.** Both A and B
B. B only **D.** Neither A nor B

6. Bleeding is being discussed:
Technician A says bleeder screws are located at low points in the hydraulic system.
Technician B says lightly tapping a caliper assembly with a hammer can regroup tiny air bubbles into larger ones that can be expelled through bleeding.
Who is correct?
A. A only **C.** Both A and B
B. B only **D.** Neither A nor B

7. Pressure bleeding is being discussed:
Technician A says that metering and combination valves must be held open using a special tool during the bleeding operation to ensure good results.
Technician B says that the pressure bleeder requires special adapters to connect it to the master cylinder reservoir.
Who is correct?
A. A only **C.** Both A and B
B. B only **D.** Neither A nor B

8. Hydraulic valves are being discussed:
Technician A says slight moisture inside the protective boot of a combination valve or metering valve is normal.
Technician B says proportioning valves are never serviced and are always replaced as a unit.
Who is correct?
A. A only **C.** Both A and B
B. B only **D.** Neither A nor B

9. Load- or height-sensing proportioning valves are being discussed:
Technican A says these valves are factory set and are nonadjustable.
Technician B says changing the suspension or load-carrying capacities of the vehicle can adversely affect valve operation.
Who is correct?
A. A only **C.** Both A and B
B. B only **D.** Neither A nor B

10. Bleeding the system is being discussed:
Technician A says the master cylinder should be bled before any individual wheel assembly.
Technician B says the bleeder screw should be closed before the brake pedal is released during manual bleeding of the system.
Who is correct?
A. A only **C.** Both A and B
B. B only **D.** Neither A nor B

CHAPTER 6

Disc Brake Service

Upon completion and review of this chapter, you should be able to:

- ❑ Diagnose disc brake system problems, including poor stopping, pulling, or dragging caused by problems in the disc brake caliper, hydraulic system, or disc brake mechanical assembly.
- ❑ Remove, inspect, and replace brake pads.
- ❑ Remove and replace a caliper assembly.
- ❑ Overhaul a caliper assembly, including disassembly, inspection, adjustment, replacement, and reassembly of all parts.
- ❑ Remove and replace brake rotors.
- ❑ Inspect and measure rotors for wear.
- ❑ Machine a rotor to correct dimensions and finish on a brake lathe.
- ❑ Service wheel bearings, including removal, cleaning, repacking, and installation.
- ❑ Reinstall wheels, torque lug nuts, and make final brake system checks and adjustments.

Diagnosing Disc Brake System Problems

Basic Tools

Basic mechanic's tool set

Rotor micrometer

Dial indicator

Poorly performing disc brakes are usually the result of worn parts or brake pads, poorly fitted or incorrectly assembled components, or rotor problems, such as grooving, distortion, or grease and dirt accumulation on the rotor surface. Worn pads increase the braking effort needed to stop the vehicle. But the same problem can be caused by a sticking or sluggish caliper piston. Installing the wrong type of brake pad can result in brake fade—a dangerous loss of braking power.

A vehicle that pulls to one side when the brakes are applied may have a bad caliper or loose caliper at that wheel. Grease or brake fluid may have contaminated the pads and rotor. Or the pad and lining may be bent or damaged. Grabbing brakes may also be caused by grease or brake fluid contamination or by a malfunctioning or loose caliper.

Worn rotors or pads may also result in roughness or pedal pulsation when the brakes are applied.

While most service manuals refer to disc brake linings as "pads" several manufacturers, including GM, refer to them as shoes.

If you suspect problems in the disc brake system, carefully road test the vehicle and compare your findings with those listed in Table 6-1, which is found at the end of this chapter. Steps on performing a safe, complete road test are given in Chapter 3. You should also inspect the wheel and brake assembly for obvious damage that could affect brake system performance. Check the:

1. Tires for excessive wear or improper infiltration
2. Wheels for bent or warped rims
3. Wheel bearings for looseness or wear
4. Suspension system for worn or broken components
5. Proper mounting of caliper and rotor
6. Brake fluid level for leaks at the master cylinder, in brake lines or hoses, at all connections, and at each wheel

SERVICE TIP: Most shop manuals provide detailed step-by-step instructions for a systematic diagnosis of the brake system. Always follow these procedures when they are provided by the manufacturer.

Service Precautions

When servicing disc brakes, never use an air hose or dry brush to clean disc brake assemblies. Use an OSHA-approved vacuum cleaner to avoid breathing brake dust. See Chapter 1 for details on working safely around airborne asbestos fibers.

Work on one wheel at a time to avoid "popping" pistons out of the other caliper and to allow the other caliper assembly to be used as a guide.

When removing wheels, be sure to avoid damage to the rotor, external lines, bleeder screws, and splash shield.

Do not spill brake fluid on the car as it may damage the paint. If brake fluid does contact the paint, wash it off immediately. To prevent splashing, cover the brake hose joints with rags or shop towels when disconnecting them. Do not mix different brands of brake fluid as they may not be compatible.

During servicing, keep grease, oil, brake fluid, or any other foreign material off the brake linings, caliper, surfaces of the rotor, and external surfaces of the hub. Handle the brake rotor and caliper carefully to avoid damaging the rotor or nicking or scratching brake linings.

Inspecting Brake Linings

Vehicles with disc brakes on the front wheels or on all four wheels should have the brake pad linings inspected on a regular basis. Most manufacturers recommend pad inspection whenever the wheels are removed for rotation (normally every 12,000 to 15 ,000 miles). Unfortunately, many car owners neglect tire rotation, so a good technician will inspect the pads whenever the wheels are removed, such as when mounting snow tires.

Do not attempt to remove the hub with the wheel and tire still mounted. The wheel and tire must be dismounted and the caliper removed before the hub and rotor can be dismounted.

Raise the vehicle on the hoist or safety stands. Be sure it is properly centered and secured on the stands or hoist. Mark the relationship of the wheel to the hub and bearing assembly to ensure proper wheel balance upon reassembly. Remove the wheel and tire assembly from the front disc brake rotor mounting face. Be careful to avoid damaging the brake caliper, the disc brake rotor shield (if equipped), and the front wheel knuckle. Reinstall two wheel nuts to hold the rotor on the hub and bearing assembly.

Pads can be inspected without removing the calipers. Check both ends of the outboard caliper by looking in at each end of the caliper. As shown in Figure 6-1, these are the areas where the highest rate of pad wear occurs. Also check the lining thickness on the inboard pad to be certain it has not worn prematurely. Replace the pads if they appear glazed (shiny and smooth), heat damaged, or contaminated with dirt or brake fluid. Look down through the inspection port to view the inboard pad and lining (Figure 6-2). Some import vehicles do not have inspection ports.

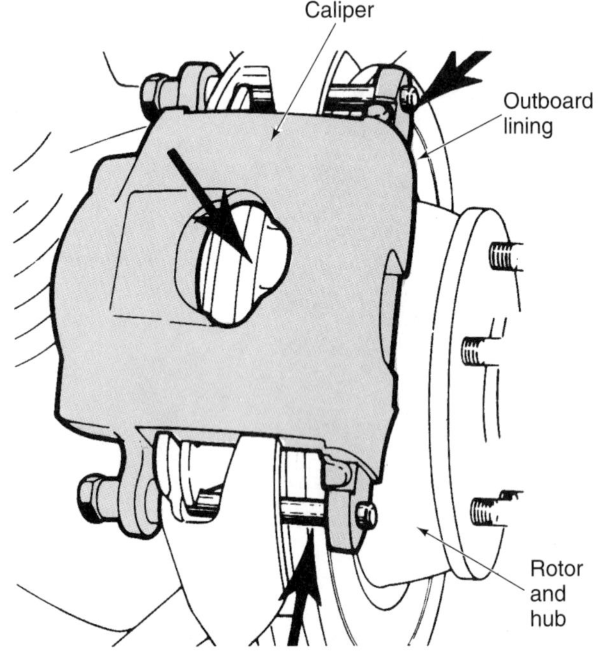

Figure 6-1 Inspecting the caliper and pad lining (Courtesy of Chevrolet Motor Division)

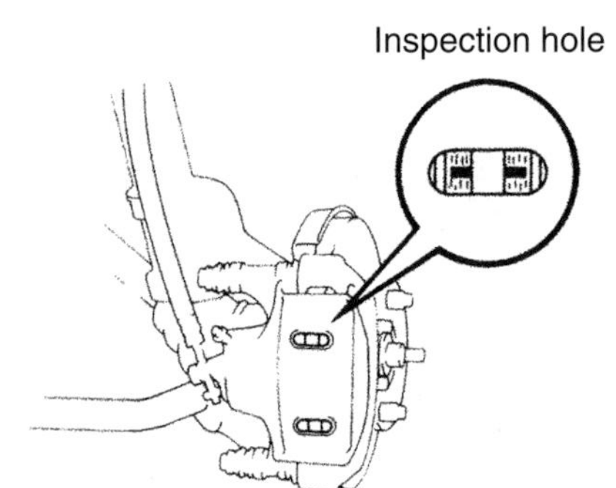

Figure 6-2 Some calipers are equipped with an inspection hole to check lining wear.

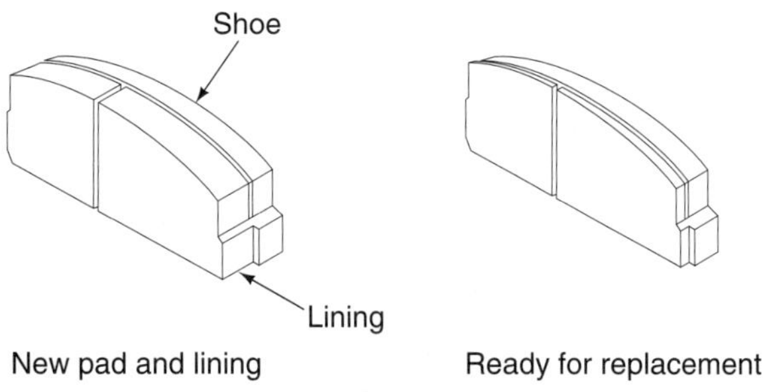

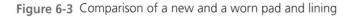

Figure 6-3 Comparison of a new and a worn pad and lining

Shop manuals all specify a minimum pad thickness, but the pad can only be measured if the unit is disassembled. When making a visual judgment concerning pad wear, consider that any pad worn to the thickness of the metal backing pad is in need of replacement (Figure 6-3).

On vehicles equipped with floating calipers, check for uneven wear on the inboard and outboard linings. If the inboard pad shows more wear than its outboard partner, the caliper should be overhauled. If the outboard pad shows more wear, the sliding components of the assembly may be sticking, bent, or damaged. In any case, uneven brake wear is a sign the pads and/or calipers need service.

Inspect the caliper for leaks. If leakage is present, the caliper must be overhauled.

Of course, if the customer complains that the brakes are making a high-pitched squeal, immediately suspect an audible brake sensor warning indicating that the system needs service. Electronic warning lights and tactile warning systems that produce a pulsating brake pedal are also indicators that the pads have worn past specifications.

Classroom Manual
Chapter 6, page 123

Preliminary Steps

If the inspection and road test pinpoint the disc brakes as the problem, the brake assembly will have to be disassembled to access the pads, caliper, and rotor. Proceed as follows:

1. Disconnect the battery ground cable.
2. Use a ball syphon to remove approximately two-thirds of the brake fluid from the front or disc brake reservoir on a front/rear split system. On a diagonally split system, remove fluid from both master cylinder reservoirs. If you forget to remove brake fluid, it may overflow the reservoir and spill when the piston(s) is forced back into the caliper bore. Replace the reservoir cover and safely discard the removed brake fluid. Another method of preventing reservoir overflow is to open the caliper bleeder screw and run a bleeder hose into a container to catch the fluid expelled when the piston is forced back into its bore. Opening the bleeder screw also makes it easier to move the piston. However, it is still good procedure to remove some fluid from the reservoir even when you plan to open the bleeder screw at the caliper.
3. If so equipped, turn the air suspension service switch off.
4. Remove the wheel and tire assembly as described under Inspecting Brake Linings earlier in the chapter.
5. Vacuum or wet-clean the brake assembly to remove all asbestos dust and fibers.
6. Inspect the bleeder screw ports. As explained in Chapter 5, a corroded or frozen bleeder screw can be freed by applying penetrating oil or by careful heating with a torch. Remove the caliper from the vehicle and perform the work at a workbench. If the bleeder screws cannot be loosened, they can be drilled out and the caliper body retapped for an insert. However, if the caliper body is also corroded or worn, it is best to replace the caliper with a new or rebuilt unit.

When the hydraulic hose is disconnected, plug it to prevent any foreign material from entering.

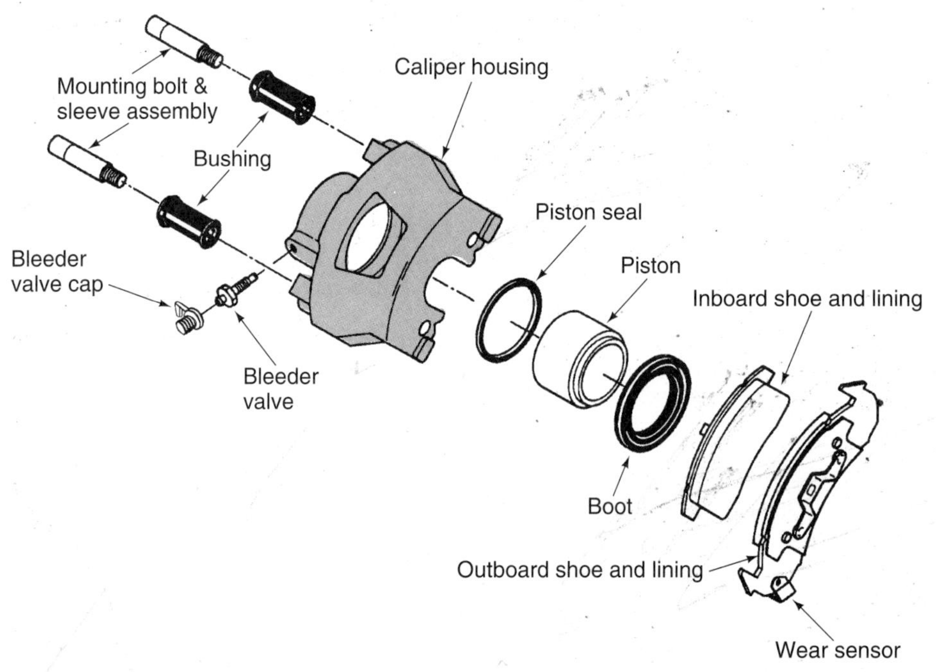

Figure 6-4 Exploded view of a sliding caliper disc brake assembly (Courtesy of Chevrolet Motor Division)

CAUTION: When using a propane torch to loosen a bleeder screw, exercise extreme care.

Brake Pad Replacement (Front Wheels)

Never permit the caliper assembly to hang with the weight on the brake hose. Support it on the suspension or hang it by a piece of wire.

Replace pads and linings in axle sets only. A typical replacement procedure for sliding caliper disc brake pads is shown in Photo Sequence 9 (see page 146). An exploded view of a typical sliding caliper disc brake assembly is illustrated in Figure 6-4. Once the vehicle is properly supported on the hoist or safety stands, remove the front wheels. On some calipers, it is necessary to push the piston into the caliper bore to provide clearance between the linings and the rotor. To do this, install a large C-clamp over the top of the caliper housing and against the back of the outboard pad (Figure 6-5). Slowly tighten the clamp to push the piston into the caliper bore far enough so that the caliper

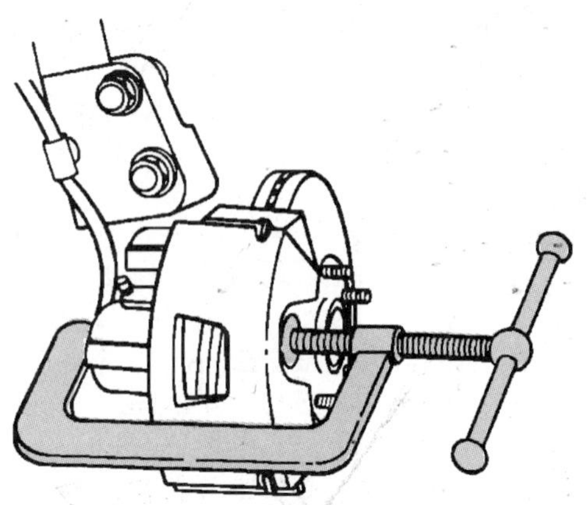

Figure 6-5 Compressing the piston in the caliper so the caliper can be removed from the rotor (Courtesy of Chevrolet Motor Division)

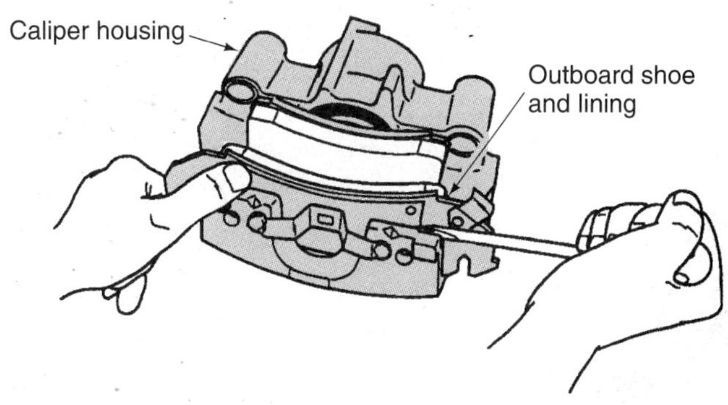

Figure 6-6 Removing the outer pad and lining (Courtesy of Chevrolet Motor Division)

assembly can be slid off the rotor. Remove the caliper bolts and sleeve assemblies and pivot the caliper up out of the way. There is no need to disconnect the brake hose from the caliper if the caliper is not going to be removed from the vehicle. Suspend the caliper from the vehicle underbody.

Use a screwdriver to disengage the buttons on the outer pad from the holes in the caliper housing (Figure 6-6). Next, remove the inner pad.

If the pads appear serviceable, use a vernier caliper to measure the thickness of each brake pad lining. Compare the pad thickness against service manual specifications. For example, the standard brake pad thickness may be 12.5 mm (0.50 in.). The service limit may be 1.6 mm (0.06 in.). Keep in mind that this measurement does not include the pad backing thickness. If the lining thickness is close to or less than the service limit, replace both brake pads on both front calipers as a set. If the pads are serviceable, always reinstall the pads in their original positions. Switching pad positions may reduce braking power. The pads must also be free of grease or brake fluid if they are to be reused. Replace contaminated brake pads and wipe any excess grease off the parts.

Inspect the mounting bolts and sleeve assemblies for corrosion and inspect all bushings for cuts or nicks. If damage to either part is found, install new parts when the caliper is reinstalled. Do not attempt to polish away corrosion.

Before installing the new linings, wipe the outside of the piston boot with denatured alcohol. Using the C-clamp, bottom the piston in the caliper bore, taking care not to damage the piston or boot. Once the piston is bottomed in the bore, lift the inner edge of the boot next to the piston and press out any trapped air. The boot must lay flat.

Install the inboard pad and lining by snapping the pad retaining spring into the piston's inside diameter (Figure 6-7). The pad retainer spring is already staked to the inboard pad. The pad must lay flat against the piston. After the pad is installed, check that the boot is not touching the pad. If it is, remove the pad and reposition the boot.

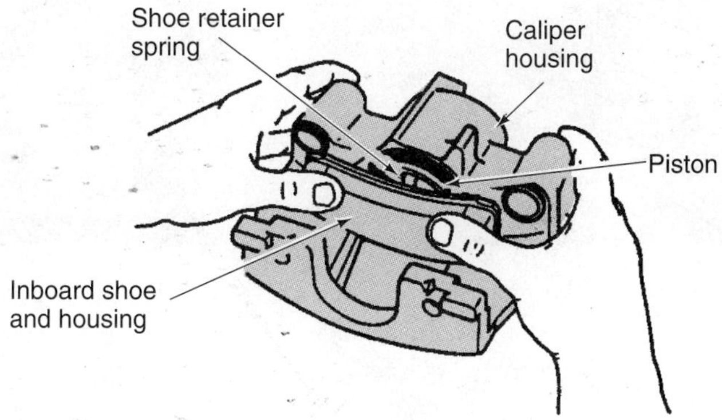

Figure 6-7 Installing the inboard pad and lining (Courtesy of Chevrolet Motor Division)

Install the outboard pad and lining with the wear sensor at the leading edge of the pad (Figure 6-8). During forward wheel rotation, the back of the pad must lie flat against the caliper.

When specified in the shop manual, apply recommended compounds such as "Molycote" M77 to shims and the shim-to-caliper contact surfaces during reassembly. On sliding calipers, liberally coat the inside diameter of the bushings with silicone grease before installing the mounting bolts and sleeves (Figure 6-9).

CAUTION: Before attempting to move the vehicle, depress the brake pedal several times to make sure the brakes work. Then road test the vehicle.

SERVICE TIP: The engagement of the brake may require a greater pedal stroke immediately after both brake pads have been replaced. Several applications of the brake pedal should restore the normal pedal stroke.

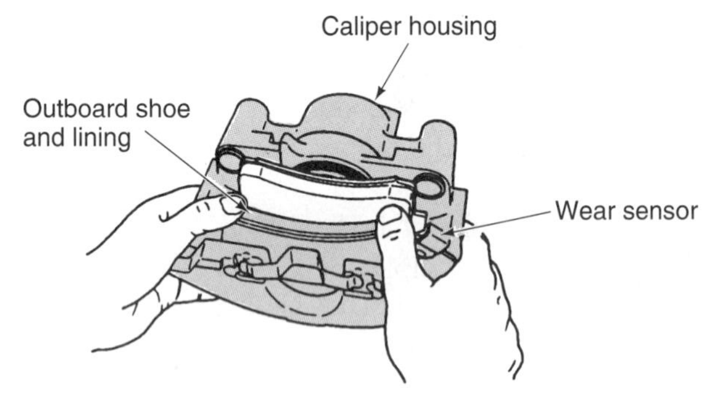

Figure 6-8 Installing the outer pad and lining (Courtesy of Chevrolet Motor Division)

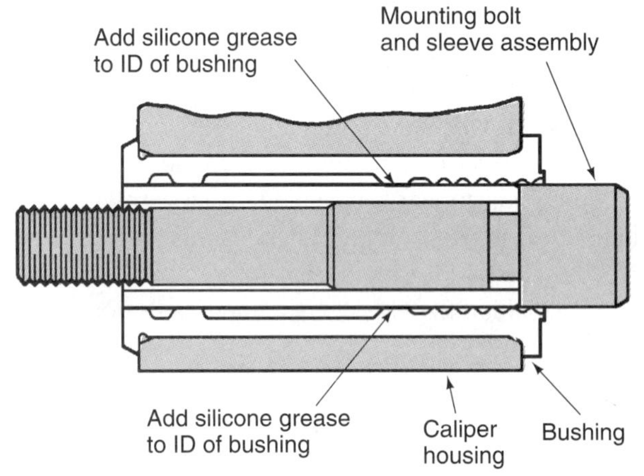

Figure 6-9 Lubricating the caliper cavity (Courtesy of Chevrolet Motor Division)

Photo Sequence 9
Typical Procedure for Replacing Brake Pads

P9-1 Begin front brake pad replacement by removing brake fluid from the master cylinder reservoir or reservoirs serving the disc brakes. Use a syphon to remove about two-thirds of the fluid.

P9-2 Raise the vehicle on the hoist, making certain it is positioned correctly. Remove the wheel assemblies.

P9-3 Inspect the brake assembly, including the caliper, brake lines and hoses, and rotors. Look for signs of fluid leaks, broken or cracked lines or hoses, and a damaged brake rotor. Correct any problems found before replacing the pads.

Photo Sequence 9
Typical Procedure for Replacing Brake Pads (continued)

P9-4 Loosen and remove the pad locator pins.

P9-5 Lift and rotate the caliper assembly up and off of the rotor.

P9-6 Remove the old pads from the caliper assembly.

P9-7 To reduce the chance of damaging the caliper, suspend the caliper assembly from the underbody using a strong piece of wire.

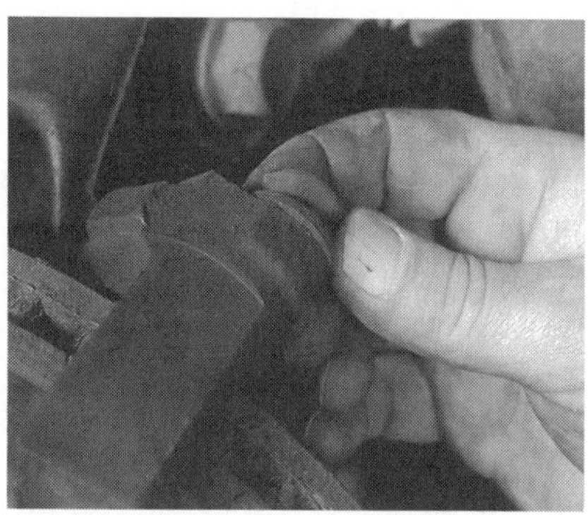

P9-8 Check the condition of the caliper locating pin insulators and sleeves.

P9-9 Place a block of wood over the caliper piston and install a C-clamp over the wood and caliper. Tighten the C-clamp to force the piston back into its bore.

P9-10 Remove the C-clamp and check the piston boot, then install new locating pin insulators and sleeves if needed.

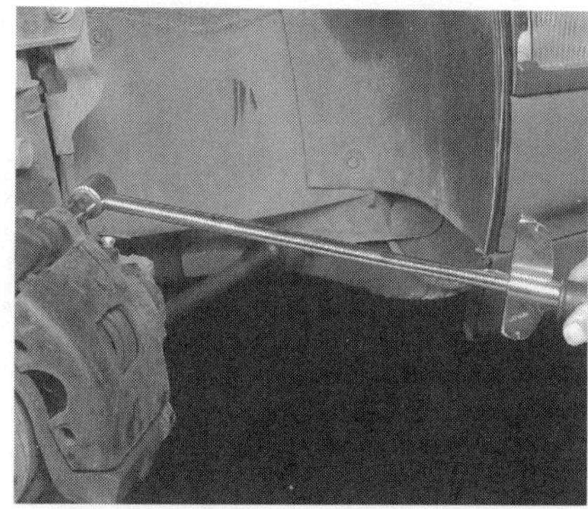

P9-11 Install the new pads in the caliper. Then set the caliper with its new pads onto the rotor and install the locating pins. Check the assembly for proper position. Torque the locator pins to proper specifications.

P9-12 Install the tire and wheel assembly and tighten to torque specifications. Then, press slowly on the brake pedal to set the brakes.

Special Tools

Wooden or plastic prying tool

C-clamp

Boot seating tool

Use crocus cloth to remove rust, corrosion, pitting, and scratches from the piston bore. If the bore cannot be cleaned with crocus cloth, light honing is permitted. Do not hone a plated bore.

Classroom Manual
Chapter 6, page 123

When using compressed air to remove a caliper piston(s), avoid high pressures. A safe pressure to use is 30 psi.

Before the brake pads are installed, apply a disc brake noise suppressor to the back of the pads to prevent brake squeal. For best results, follow the directions on the container.

Brake Caliper Overhaul

Use the following guidelines when servicing and overhauling caliper assemblies:

1. Clean the brake components in either denatured alcohol or clean DOT 3 or DOT 4 brake fluid. Do not use mineral-based cleaning solvents such as gasoline, kerosene, carbon tetrachloride, acetone, or paint thinner as they cause rubber parts to swell and soften in a very short time. For the same reason, do not use lubricated compressed air on brake parts.
2. Lubricate any moving member such as the caliper housing or mounting bracket to ensure a free-moving action. Use only recommended lubricant.
3. Lubricate rubber parts with clean brake fluid to make assembly easier.
4. Before reassembling, check that all parts are free of dust and other contaminants.
5. Replace parts when specified in the service manual.
6. Perform all work on a clean workbench free of grease and oil.

To remove and disassemble the brake caliper (Figure 6-10), follow the typical procedure shown in Photo Sequence 10.

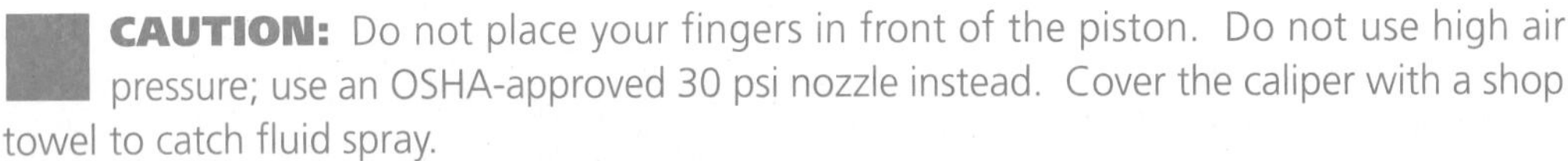

CAUTION: Do not place your fingers in front of the piston. Do not use high air pressure; use an OSHA-approved 30 psi nozzle instead. Cover the caliper with a shop towel to catch fluid spray.

WARNING: Always reinstall the brake pads in their original positions to prevent loss of braking efficiency.

Loaded Calipers

Loaded calipers are calipers that come with brake pads and mounting hardware fully installed. They eliminate the need to overhaul calipers and prevent many of the errors commonly committed when performing caliper service. These mistakes include forgetting to bend brake pad locating tabs that reduce vibration and noise, leaving off antirattle clips and pad insulators, or reusing worn or corroded mounting hardware that limits caliper movement and reduces pad life by as much as fifty percent.

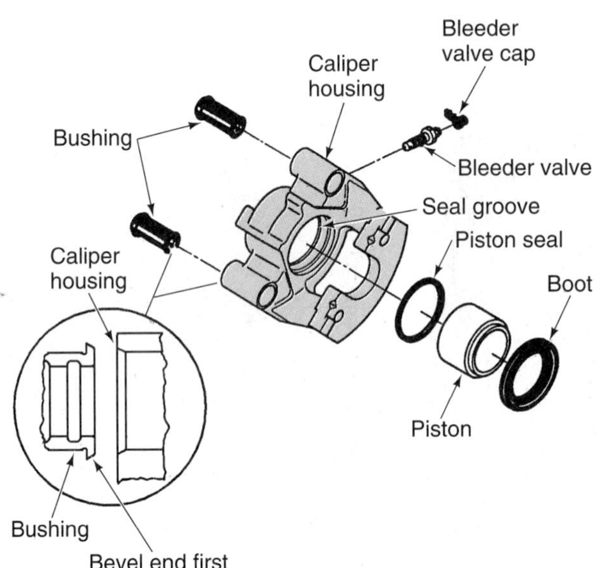

Figure 6-10 Brake caliper components (Courtesy of Chevrolet Motor Division)

Photo Sequence 10
Typical Procedure for Rebuilding a Disk Brake Caliper

P10-1 Disconnect the brake hose from the caliper. Remove the caliper bolts, then remove the caliper completely from the vehicle and move to the workbench.

P10-2 Inspect the bushings for cuts and nicks. Replace them if damage is found.

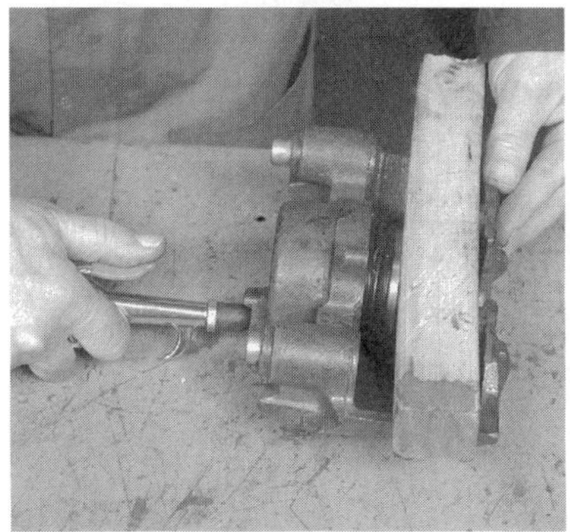

P10-3 Place a wooden block or shop rag in the caliper opposite the piston, then carefully remove the piston from the caliper by applying air pressure through the brake line hole. **WARNING:** Do not place your fingers in front of the piston. Do not use high air pressure; use an OSHA-approved 30 psi nozzle instead. Cover the caliper with a shop towel to catch fluid spray.

P10-4 Remove the piston boot and seal, taking care not to damage the cylinder bore. Use a small wooden tool or plastic tool to remove the seal.

P10-5 Inspect for wear, nicks, corrosion, or damage.

P10-6 Use a crocus cloth to polish out light corrosion. Replace the caliper if light polishing does not remove corrosion from around the seal groove.

P10-7 Clean the piston, caliper bore, and all parts with clean, denatured alcohol.

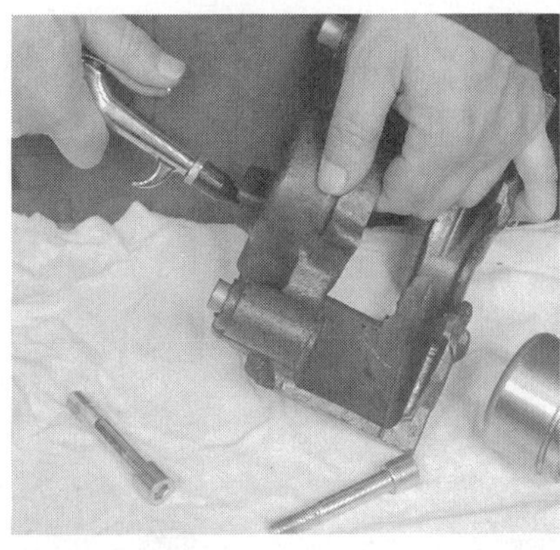

P10-8 Dry all parts with unlubricated compressed air. Blow out all passages in the caliper and the bleeder valve.

Photo Sequence 10
Typical Procedure for Rebuilding a Disk Brake Caliper (continued)

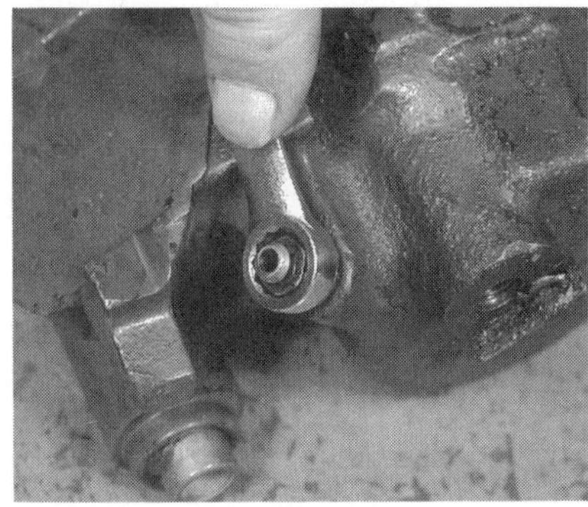

P10-9 Screw the bleeder valve and bleeder valve cap into the caliper housing. Tighten to specifications.

P10-10 Apply silicone grease to a new piston seal, then install the piston seal in the cylinder groove. Make sure the seal is not twisted.

P10-11 Apply rubber grease or brake fluid to a new piston boot, then install the boot onto the piston.

P10-12 Lubricate the caliper cylinder and piston with brake fluid. Install the piston into the caliper bore and push it to the bottom of the bore.

P10-13 Seat the boot in the caliper housing counter bore using the proper seating tool.

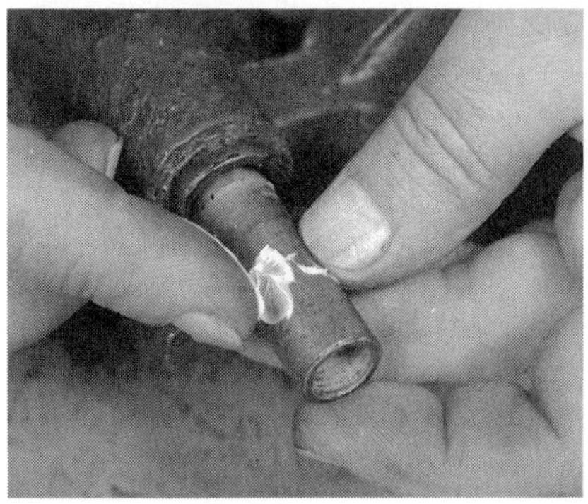

P10-14 Lubricate the beveled end of the bushings with silicone grease. Pinch the bushing and install it bevel end first into the caliper housing. Push the bushing through the housing bore.

P10-15 Reinstall the caliper in the reverse order of removal. **WARNING:** Always reinstall the brake pads in their original positions to prevent loss of braking efficiency. Fill the brake reservoir up and bleed the brake system.

Figure 6-11 Inspecting the rotor surface

Use of loaded calipers reduces labor time and ensures that all components that should be replaced are replaced. However, be sure the loaded calipers are equipped with quality pad materials. Calipers should also be matched side to side with the same type of friction materials. When one caliper is bad, both should be replaced with the same type of loaded caliper.

Rotor Service

Classroom Manual
Chapter 6, page 117

Brake rotors are typically designed so that tolerances in the braking surfaces for flatness, parallelism, and lateral runout must be maintained during the life of the rotor. Maintaining close tolerances on the shape of the braking surfaces helps minimize brake roughness and pedal pulsation.

The surface finish of the rotor also must be kept within a specific range of 60 Ra roughness or less. Control of the braking surface finish is needed to minimize problems of hard pedal apply, excessive brake fade, pulls, and erratic performance. Control of the surface finish can also improve lining life.

Inspect disc brake rotors whenever the pads or calipers are serviced, or when the wheels are rotated or removed for other types of work (Figure 6-11). Keep in mind that many problems that occur with disc brake rotors may not be apparent on casual visual inspection. Rotor thickness, parallelism, runout, flatness, and depth of scoring can only be measured with exacting gauges and micrometers. Accurate measuring tools and modern refinishing equipment are essential to good rotor service.

Thickness Variation Check

Check for thickness variation by measuring the rotor using a micrometer calibrated in ten-thousandths of an inch. Make all measurements at the same distance from the edge of the rotor.

A light scoring of the rotor surfaces that does not exceed 1.5 mm (0.06 in.) in depth is normal and is not detrimental to brake operation.

A rotor that varies in thickness by more than the shop manual specifications can cause pedal pulsation and/or front end vibration during brake applications. A rotor that does not meet specifications should be refinished to specifications or replaced.

To perform the thickness variation check, raise the vehicle on the lift and remove the front wheels. To hold the rotor in position while the measurements are being taken, install flat washers and the wheel nuts onto the studs and tighten them down against the rotor. The service manual will normally list a torque value for tightening the wheel nuts. Use the tightening sequence given

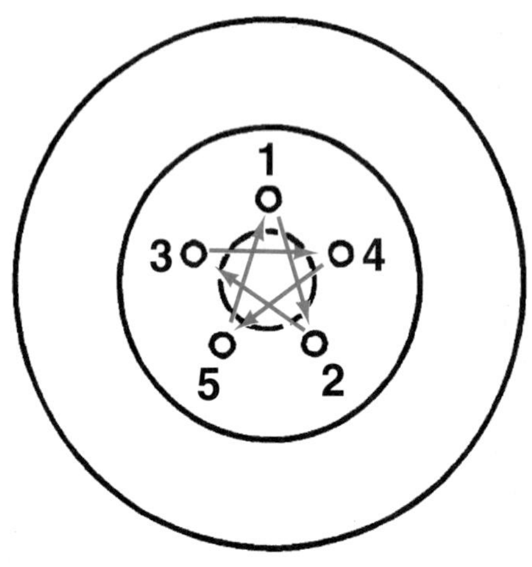

Figure 6-12 Star pattern tightening sequence for five lug nut wheel (Courtesy of Chervolet Motor Division)

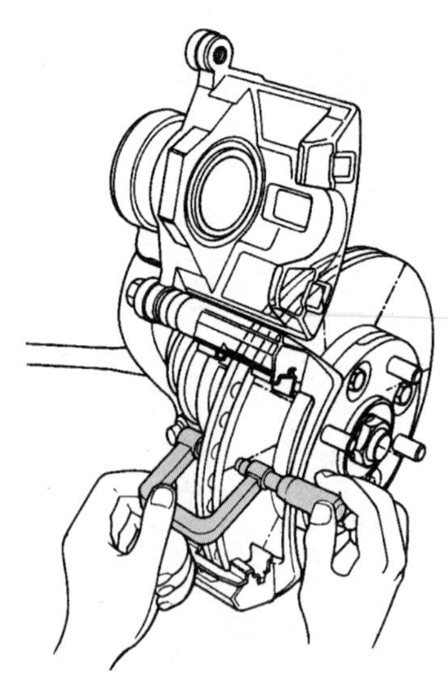

Figure 6-13 Measuring rotor thickness (Courtesy of Honda Motor Co., Ltd.)

in the manual. For a five lug nut configuration, a star pattern is used (Figure 6-12). Once the rotor is held securely, remove the caliper assembly from the rotor and hang it from the underbody so it will not drop or be otherwise damaged.

Using a rotor micrometer, measure the rotor thickness at points approximately 45° apart and about 1/2 in. in from the outer edge of the rotor (Figure 6-13). Check your measurements against the specification in the service manual. For example, the standard rotor thickness may be 0.95 in. with a maximum allowable difference between any two measurements of 0.0006 in. If you find a rotor that is beyond the stated limits for thickness or parallelism, the rotor must be refinished on a brake lathe.

Lateral Runout

Excessive lateral runout causes the rotor to wobble from side-to-side as it rotates. This wobble knocks the pads farther back than normal, which causes the pedal to pulse or vibrate as it is applied. You may also notice an increase in brake pedal travel because the pistons must move a greater distance to contact the rotor surface.

The most accurate method of checking for lateral runout is with the wheels still installed on the vehicle. This method provides a much more accurate reading of the total indicated runout (TIR) under real braking conditions but requires special equipment. If the equipment is not available to perform the check with the wheels installed, the reading can be made with the wheels removed but with the caliper still installed.

To perform a lateral runout check, raise the vehicle on the lift and remove the wheel. Install and torque the wheel nuts with flat washers as described in the thickness variation check procedure. Once the rotor is held securely, remove the caliper assembly from the rotor and hang it from the underbody so it will not drop or be otherwise damaged.

Inspect the rotor surface for grooving, cracks, or heat checks. Clean the rotor thoroughly to remove any rust or dirt buildup. Fasten the dial indicator to the steering knuckle so that the indicator button contacts the rotor surface about 1/2 in. in from the outer edge (Figure 6-14).

Set the dial indicator to zero and turn the wheel one complete revolution while carefully watching the indicator scale. The runout should not exceed the limit listed in the service manual, which is typically in the range of 0.003 to 0.005 in.

Indexing the rotor on the hub one or two bolt positions from the original position may improve the lateral runout readings. If the lateral runout is not brought into specifications by

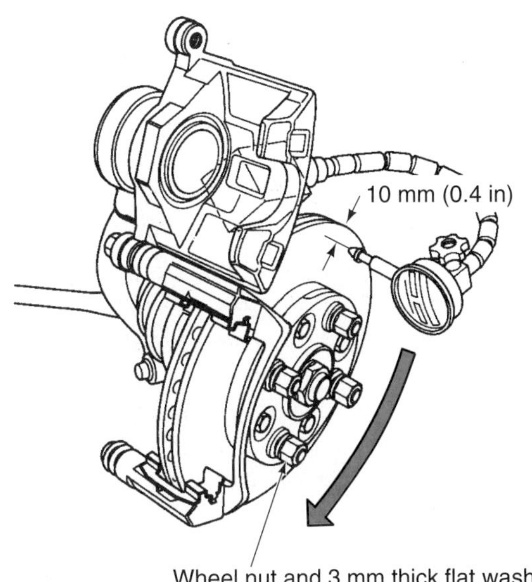

Figure 6-14 Measuring lateral runout (Courtesy of Honda Motor Co., Ltd.)

indexing the rotor, check the hub and bearing assembly for excessive lateral runout or looseness. If the hub and bearing assembly lateral runout exceeds specifications, then replace the hub and bearing assembly. If hub and bearing lateral runout is within specifications, the problem is with the rotor. Refinish or replace the rotor as necessary.

Rotor Inspection

Rotor scoring or excessive wear may be caused by linings that have worn through to the rivets on the backing pad (Figure 6-15). Low-quality friction materials or rust and dirt caught between the lining and rotor can also score the rotor surface. Any rotor having score marks deeper than 0.015 in. should be refinished or replaced.

SERVICE TIP: Some rotors found on GM vehicles have a single deep groove manufactured into their surface. This groove keeps the linings from moving radially outward and reduces operating noise.

The high heat generated when braking can also cause hard spots or heat checks to form in the rotor surface. The bluish hard spots can be removed by grinding, but will likely reappear once the rotor is heated again. Pulling, rapid lining wear, hard pedal, and braking noise are all possible signs that heat-related problems have occurred. Replace heat-damaged rotors.

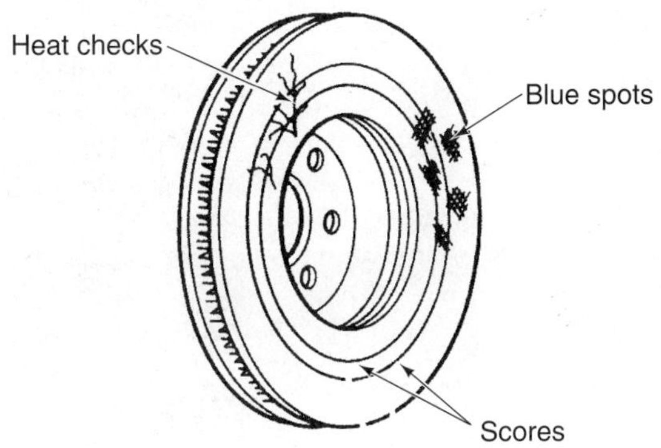

Figure 6-15 Typical rotor defects and problems (Courtesy of Raybestos Division, Brake Systems Inc.)

Removing the Rotor

To remove a rotor from the vehicle, raise the vehicle onto the hoist or safety stands and remove the wheel and tire assembly. Remove the caliper assembly from the rotor and suspend it from the vehicle underbody as described under pad replacement and caliper overhaul procedures given earlier in this chapter.

With the caliper safely out of the way, remove the rotor from the hub assembly by pulling it off the hub studs (Figure 6-16). If the rotor cannot be pulled off by hand, apply rust penetrate on the front and rear rotor/hub mating surfaces. Strike the rotor between the studs using a rubber or plastic hammer. If this does not free the rotor, attach a three-jaw puller to the rotor and pull it off.

Whenever you separate the brake rotor from the wheel bearing flange, clean any rust or foreign material from the mating surface of the wheel bearing flange and rotor (Figure 6-17). Neglecting to clean rust and dirt from the rotor and hub mounting faces prior to installing new or reused rotors will result in increased rotor lateral runout leading to premature brake pulsation and other problems.

Installing the Rotor

New rotors are shipped from the manufacturers with a protective oil coating on the rotor surface. To remove this coating, use carburetor cleaner or a service manual-recommended solvent.

If the original or refinished rotor is being reinstalled, sand the rotor surface with #80 grit sandpaper (asbestos pads) or #120 grit paper (semimetallic pads) to remove dirt or cutting dust. After sanding the refinished rotor, wash it down with denatured alcohol. Once the rotor is cleaned, do not touch the surface with your hands.

Make sure all mounting surfaces are also clean. Apply a small amount of silicone dielectric compound to the pilot diameter of the disc brake rotor before installing the rotor onto the wheel hub assembly. Reinstall the disc brake caliper assembly as detailed in the caliper rebuilding procedure. Install the wheel and tire assembly onto the rotor mounting face. Tighten the wheel hub bolt nuts with a torque wrench using the recommended tightening pattern. Failure to tighten in the correct pattern may result in increased lateral runout, brake roughness, or pulsation.

CAUTION: After lowering the vehicle on the lift, pump the brake pedal several times before moving the vehicle. This positions the brake linings against the rotor. If so equipped, turn the air suspension service switch back on. Finally, road test the vehicle.

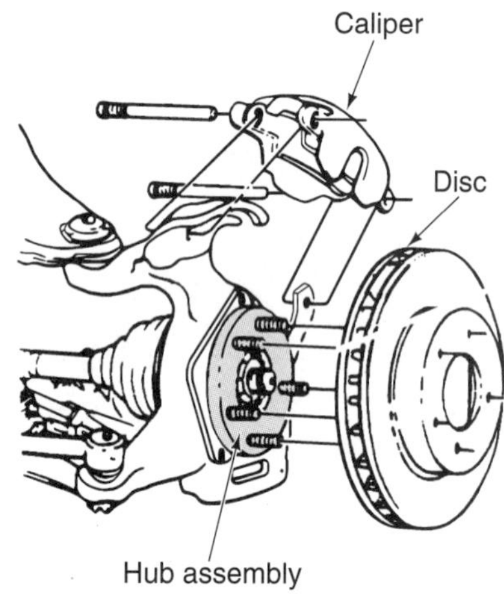

Figure 6-16 Removing the rotor from the hub (Courtesy of Raybestos Division, Brake Systems Inc.)

Figure 6-17 Cleaning the mating surface of the wheel bearing flange and rotor.

Refinishing Brake Rotors

Take cut from both sides.

Do not automatically refinish brake rotors when performing routine brake maintenance such as replacing worn disc brake pads. Refinish the rotor only under the following circumstances:

❏ If it fails lateral runout or thickness variation checks
❏ If there is noticeable brake pulsation
❏ If there are heat spots or excessive scoring

All brake rotors have a minimum thickness dimension cast into them. This dimension is the minimum wear dimension not a minimum refinishing dimension. Do not use a brake rotor that does not meet these specifications. Generally a refinished rotor must be thicker than its minimum thickness dimension. Refer to the service manual for exact specifications. A rotor that has been refinished too thin will not have proper heat transfer capabilities and should be replaced with a new brake rotor.

Never turn the rotor on one side of the vehicle without turning the rotor on the other side.

When installing new brake rotors, most shop manuals caution against refinishing the surfaces as these parts are already at the correct level of surface finish. Making a light cut on a new rotor may produce excessive lateral runout and result in brake shutter after only a few thousand miles of service. Clean any oil film off of the rotor with solvent, and allow the rotor to air dry before installing it on the vehicle.

NOTE: Some manufacturers do recommend that a new rotor be refinished if its runout is greater than a certain specification, such as 0.10 mm (0.004 in.).

CAUTION: When operating a lathe, be sure to wear the proper eye protection.

Brake Lathes

Accurate control for the rotor tolerances is necessary for proper performance of the disc brakes. Machining of rotors (and drums) should be done only with precision equipment. Both bench-type off-vehicle lathes (Figure 6-18) and on-vehicle lathes are used to refinish rotor and drum surfaces. Regardless of the type of lathe used, the equipment should be serviced on a regular basis following the manufacturer's recommended maintenance procedures.

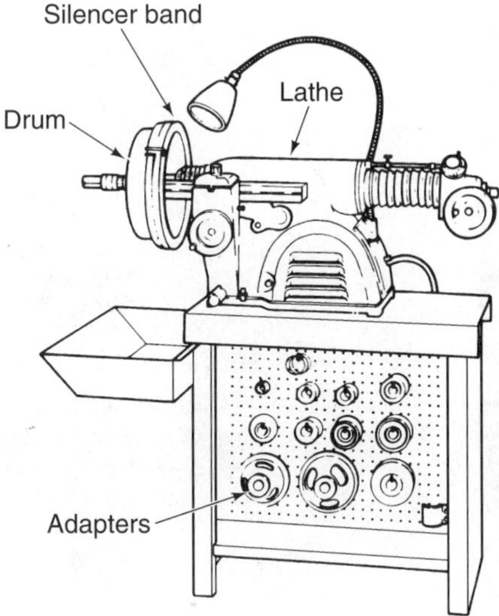

Figure 6-18 Typical bench-type brake lathe used for refinishing rotors and brake drums (Courtesy of Raybestos Division, Brake Systems Inc.)

With a bench-type lathe, the rotor (or drum) is mounted onto the lathe arbor and spun. A cutting bit is fed into the surface of the spinning rotor or drum to remove several thousandths of an inch of surface metal. The lathe has the ability to move the rotor or drum perpendicular to the bits so that the entire rotor or drum surface is refinished. Most lathes can operate at slow, medium, and fast speeds via a series of driving belts. Initial cleaning cuts can be made at higher speeds, but the final finishing cut should be at a slow speed.

A different cutter assembly is used for rotor and drums. Most rotor assemblies have two cutting bits. The rotor mounts between the bits and is pinched between them. As the cut is made, the exact same amount of surface material is cut from both sides of the rotor. Some lathes use only one bit for rotor refinishing and separate cuts of equal amounts must be taken from each side. The attaching adapters, tool holders, vibration dampers, and tool bits must be in good condition. Make sure mounting adapters are clean and free of nicks. Always use sharp cutting tools or bits and use only replacement cutting bits recommended by the equipment manufacturer. Dull or worn bits leave a poor surface finish, which will affect initial braking performance.

SERVICE TIP: When using a lathe with only one cutting bit, never take the rotor off the lathe until both sides are cut. Even a 0.001-inch to 0.002-inch variation in positioning the rotor on the lathe can produce a thickness variation of the rotor. This will cause pedal pulsation and loss of braking efficiency. Always cut both sides of the rotor without removing it from the arbor. This ensures that both sides of the rotor will be parallel after refinishing. Check the accuracy of the cuts by using a dial indicator and an outside micrometer.

Bench-Type Lathe Setup and Operation

The mounting procedure for rotors depends on if the rotor has wheel bearings mounted in its hubs. For rotors with bearing assemblies, knock out the inner bearing and grease seal before mounting the rotor on the arbor. The procedure is described under Wheel Bearing Service later in this chapter. Remove all grease and dirt from the bearing races before mounting the rotor. It is sometimes necessary to steam clean the grease out of the hubs. Index the rotor on the wheel bearing races to ensure that the machining job is true, and use the appropriate cones and spacers to lock the rotor firmly to the shaft (Figure 6-19).

For rotors without bearing assemblies, simply clean all rust and corrosion from the hub area using a crocus cloth or #120 grit paper (Figure 6-20). As with bearing-type rotors, use the proper cones and spacers to mount the rotor to the arbor shaft.

Figure 6-19 Mounting the rotor onto the lathe arbor using spacers and cones.

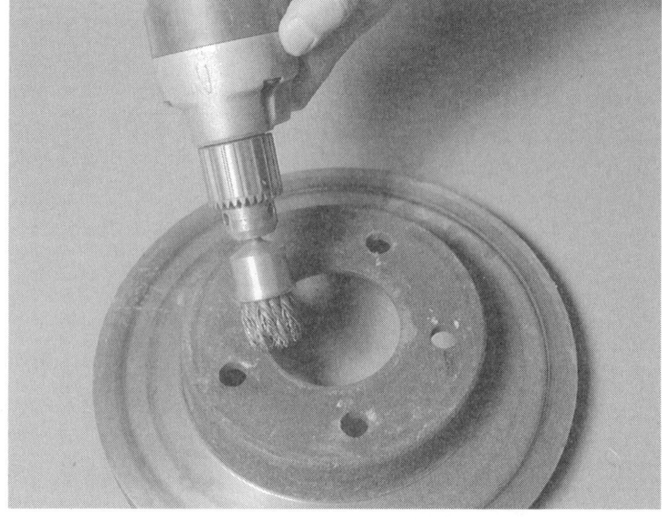

Figure 6-20 Cleaning the hub area of a nonbearing rotor

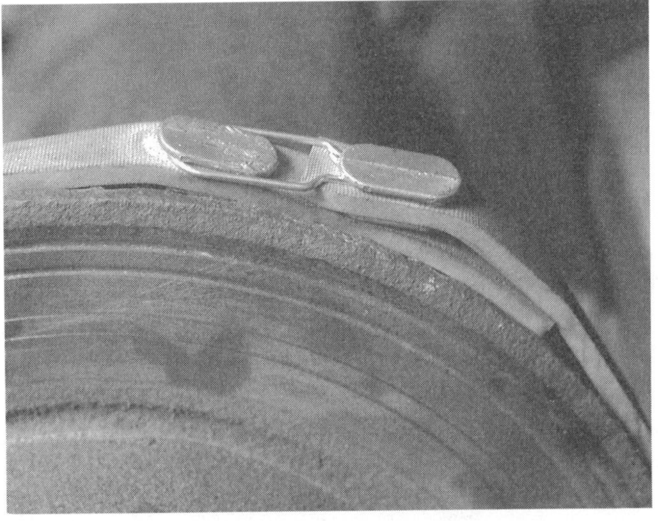

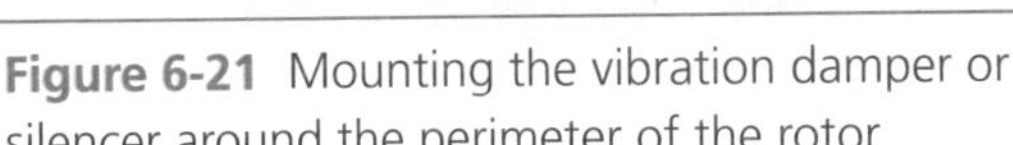

Figure 6-21 Mounting the vibration damper or silencer around the perimeter of the rotor

Figure 6-22 A properly formed test scratch

Once the rotor is on the lathe, a rubber or spring-type vibration dampener is installed on the outer diameter of the rotor to prevent the cutting bits from chattering during refinishing, which results in a smoother finished surface. The damper also helps reduce unwanted noise (Figure 6-21).

To determine the approximate amount of metal to be removed, turn on the lathe and bring the cutting bit up against the rotating disc until signs of a slight scratch are visible (Figure 6-22). Turn off the lathe and reset the depth-of-cut dial indicator to zero. Find the deepest groove on the face of the rotor and move the cutting bit to that point without changing its depth-of-cut position. Now use the depth-of-cut dial to bottom out the tip of the cutter in the deepest groove. The reading on the dial now equals or is slightly less than the amount of material needed to be removed to eliminate all grooves in the rotor surface (Figure 6-23). For example, if the deepest groove is 0.019-in. deep, the total amount to be removed may be 0.020 in. For best results with cuts that have a total depth greater than 0.015, take two or more shallow cuts rather than one very deep cut.

To make the series of refinishing cuts, reset the cutter position so the cutting bits once again just touch the ungrooved surface of either side of the rotor. Zero the cutting depth dial indicators on the lathe.

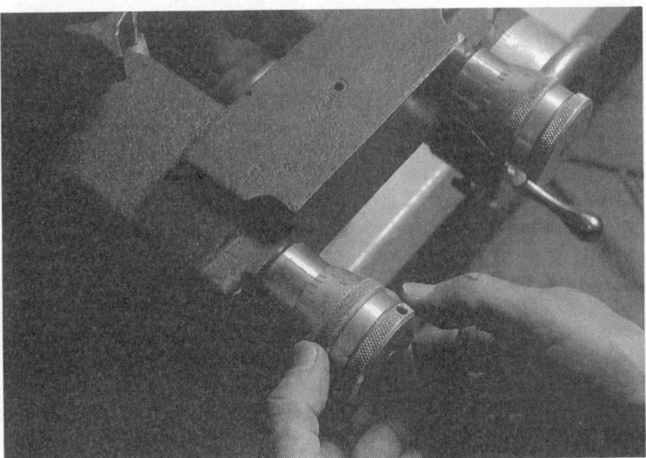

Figure 6-23 The depth-of-cut dial is calibrated in thousandths of an inch and is used to set the cutting depth.

Turn on the lathe and allow the arbor to reach full running speed. Turn the depth-of-cut setting dials for both bits to set the first pass cut. Turning these dials moves the bits inward. The dial is calibrated in thousandths of an inch increments. The first cut should only be a portion of the total anticipated depth of cut. For example, if the total depth of cut is expected to be 0.019 in., make a fast roughing cut of 0.010-in., followed by a medium speed 0.007-in. cut, and a slow final finish cut of 0.002 in. When the cutting depth is set for the first cut, activate the lathe motion to move the cutting bits along the surface of the rotor. After the first cut has been completed, turn off the lathe and examine the rotor surface. Areas that have not yet been touched by the bits will be darker than those that have been touched (Figure 6-24). If there are large patches of unfinished surface, make another cleaning cut of the same depth. When the majority of the surface has been refinished, make a shallow finishing cut at low arbor speed. Repeat the slow finishing cut until the entire rotor surface has been refinished. Make sure you do not cut the rotor thickness to beyond its service limit. To ensure this, remeasure the refinished rotor with a micrometer to determine its minimum thickness and compare this measurement to the manufacturer's minimum refinished thickness specification.

It is very important that the rotor surface be made nondirectional. Dress the rotor surfaces with a sanding disc power tool equipped with 120-grit aluminum oxide sandpaper. Sand each rotor surface with moderate pressure for at least 60 seconds. You can also do the job with a sanding block with 150-grit aluminum oxide sandpaper. With the rotor turning approximately 150 rpm, sand each rotor finish for a minimum of 60 seconds using moderate pressure (Figure 6-25). After the rotor has been sanded, clean each surface with denatured alcohol.

Machining Out Excessive Runout

If the test scratch does not make a complete circle on the rotor but appears as a crescent, the rotor is wobbling on the arbor more than 0.004 in. Do not resurface the rotor because a wobble could be machined into the rotor, causing the brake pedal to pulsate when the brakes are applied. One of three conditions is the likely cause of the wobble: a bent lathe arbor, distorted mounting adapters, or excessive rotor runout.

To determine if the arbor is bent, mount a dial indicator on the lathe. Disconnect power to the lathe. Release the pulley belt tension by moving its controlling lever and rotate the arbor slowly by turning the drive pulleys. Observe the movement of the dial indicator needle. Movement of the needle in excess of one division or 0.001 in. would indicate a bent arbor. Contact the lathe manufacturer's representative to check this.

Figure 6-24 The first cut should remove most of the surface area. Dark spots that remain have yet to be refinished.

Figure 6-25 Applying a nondirectional finish to the rotor using a small disc sander

A distorted mounting adapter can be corrected by installing it on the lathe's arbor and machining it.

Rotors with excessive wobble or runout pose another problem. Such a rotor may have been machined incorrectly or the rotor may have become distorted by overtightening the wheel lug nuts. Such a rotor can be resurfaced provided the amount and position of runout is marked when the rotor is mounted on the vehicle and these same conditions can be reproduced on the bench lathe arbor. Proceed as follows:

1. Remove the rotor from the hub and clean the inside surface of the rotor and the axle hub face of rust and grease.
2. Reinstall the rotor on the hub and torque the wheel stud nuts backward on the studs.
3. Perform a lateral runout check as described earlier in the chapter. The maximum travel of the needle in a clockwise direction indicates the place where the rotor has its maximum outward deflection. Mark this spot on the rotor hub with a plus (+) mark.
4. Once the rotor is marked, remove it from the vehicle and mount it on the lathe with the proper mounting adapters. Install the dial indicator on the lathe and adjust the indicator's point to the outside rotor surface. Slowly turn the rotor clockwise and observe whether the runout amount and location are the same as they were on the car. If the readings are not the same as they were on the car, loosen the arbor nut and reposition the rotor and the adapters on the rotor. Repeat this procedure until the rotor runout matches the runout measured on the car as previously described.
5. Once the conditions on the vehicle have been recreated on the lathe, machine off an amount of material equal to the maximum lateral runout.

NOTE: If the rotor's runout on the lathe cannot be made to match the runout found on the car, the rotor hub is distorted and should be replaced. To determine the runout of the rotor hub, mount the dial indicator point against the flange. Rotate the axle and observe the movement of the gauge needle.

On-Vehicle Lathes

Excellent refinishing results can also be achieved using an on-vehicle brake lathe (Figure 6-26). The main advantage of an on-vehicle lathe is that the rotor does not have to be unmounted from

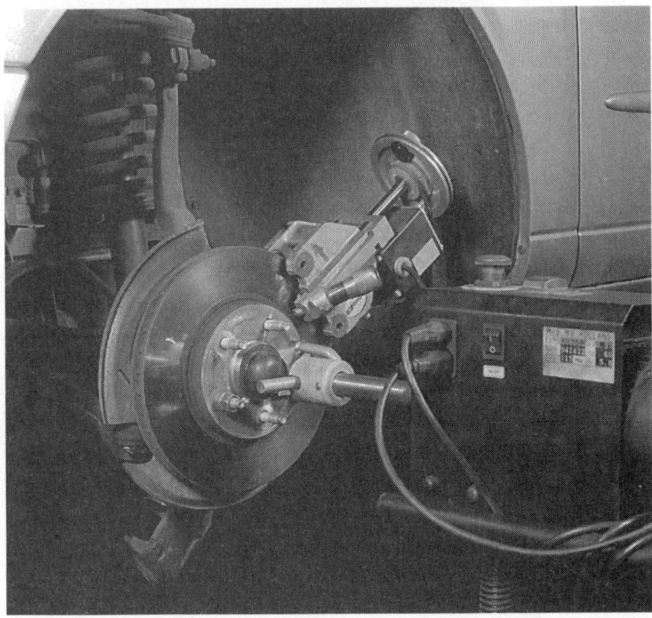

Figure 6-26 Typical on-vehicle lathe setup

the hub. On-vehicle lathes are also ideal for rotors with excessive runout problems. The time and trouble needed to reproduce the exact runout condition on the lathe arbor is eliminated. Simply refinish the rotor on the vehicle.

To install the lathe, remove the caliper from its mounting bracket and hang it out of the way with a wire hook. Then torque the wheel mounting nuts equally. Attach the lathe to the rotor using the hardware that comes with the lathe. Follow the manufacturer's mounting and operating instructions precisely. As with the bench-type lathe, the cutting bits of the on-vehicle lathe straddle the rotor and depth-of-cut settings are made using adjustment knobs. The dial marks determine the depth of cut each tool is taking from the rotor surface.

With an on-vehicle lathe, the engine is used to rotate the rotor, so the lathe can only be used on drive axle wheels. However, a problem exists because the differential gearing in the transaxle transmits the power to the opposite front axle, not to the rotor to be resurfaced. To prevent that half-axle from rotating, the opposite wheel can be lowered to the floor. However, this may reduce floor-to-lathe clearance to the point where it is difficult to run the lathe.

Another method of transferring drive power to the rotor is to lock on the brakes at the opposite wheel. To do this, first remove and tightly plug the brake hose to the caliper on the rotor you are refinishing. Apply the brakes so the caliper on the opposite wheel is applied. Hold the brakes on and have an assistant clamp the front brake hose to keep the caliper locked on. Release the brakes and inspect to see that the caliper on the opposite wheel remains applied. If it is, start the engine, shift the transmission into first gear, and idle the engine as slowly as possible. The rotor should rotate as slowly as possible to provide a smooth surface. Spinning the rotor too fast may cause the tool bits to overheat and wear out faster.

Burnishing Disc Pads and Brake Rotors

After replacing brake pads and/or servicing rotors the new braking surface must be broken in, or burnished. This is done by making 20 stops from 30 mph, using medium to firm pedal pressure. Wait a minimum of 15 seconds between stops to avoid overheating the brakes.

Rear Disc Brake Inspection and Replacement

Rear disc brakes are now common on many import and domestic vehicles. In most cases, the rear assemblies are identical to the front disc brakes with the exception of some type of parking brake mechanism (Figure 6-27). To service a typical rear disc brake assembly:

Special Tools

Locknut wrench with extension bar

Rear caliper guide

Brake spring compressor tool

1. Raise the vehicle on the lift and remove the rear wheels. The rear caliper is often protected by a plastic shield. Remove this shield and the bolts securing the caliper (Figure 6-28).
2. To remove the disc brake pads, remove any pad shims and retainers and pull the pads off the assembly (Figure 6-29).
3. Check the condition of the rotor as outlined earlier in this chapter. Thoroughly clean the rotor surface and inspect it for grooving, heat checks, and other signs of damage.
4. Perform rotor thickness checks and lateral runout measurements (Figure 6-30). Refinish or replace the rotor if thickness or runout is not within specifications.
5. Remove the bolts securing the caliper bracket and lift off the bracket (Figure 6-31). Thoroughly clean the caliper bracket.
6. If the rotor must be removed from the vehicle, remove the screws holding it to the wheel assembly (Figure 6-32) and pull off the rotor.

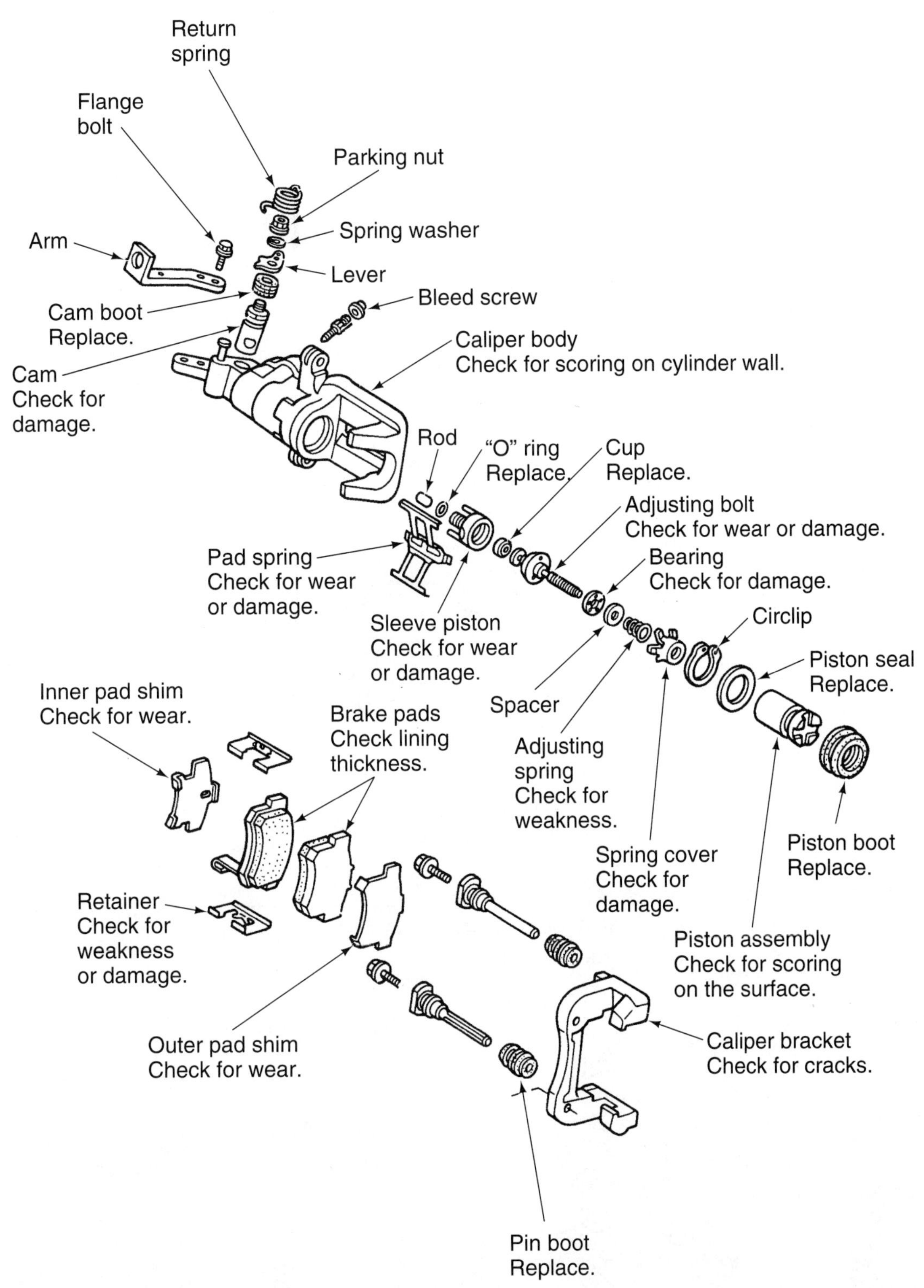

Figure 6-27 Rear disc brake assembly. The parking brake cable connects to the lever arm. (Courtesy of Honda Motor Co., Ltd.)

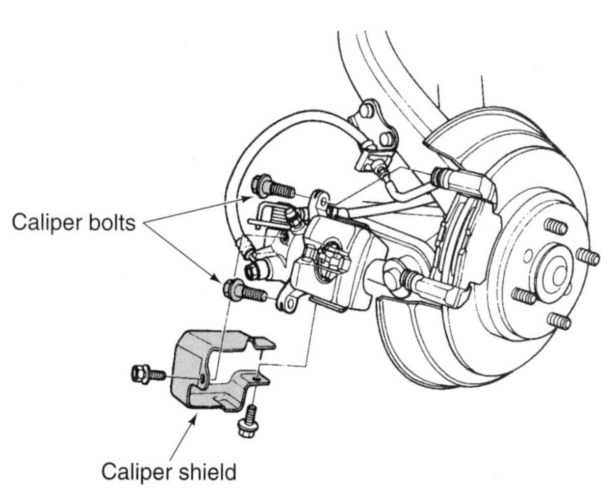

Figure 6-28 Removing the rear disc brake caliper retaining bolts and caliper shield (Courtesy of Honda Motor Co., Ltd.)

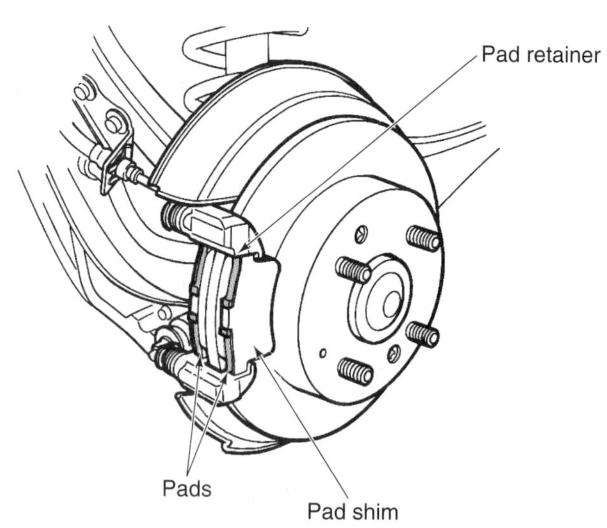

Figure 6-29 Removing the rear disc brake pad retainer, shims, and pads (Courtesy of Honda Motor Co., Ltd.)

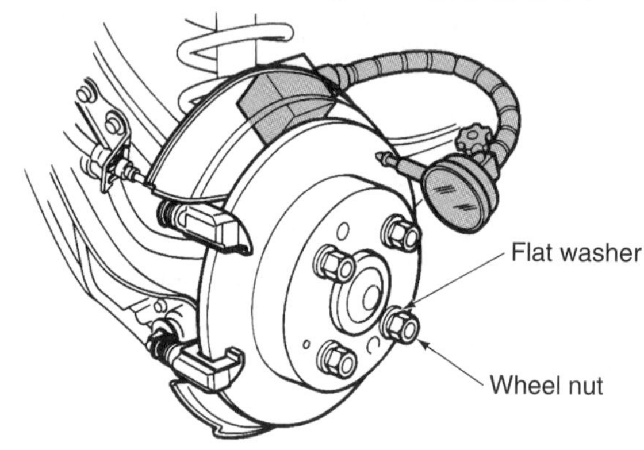

Figure 6-30 Measuring rear rotor runout (Courtesy of Honda Motor Co., Ltd.)

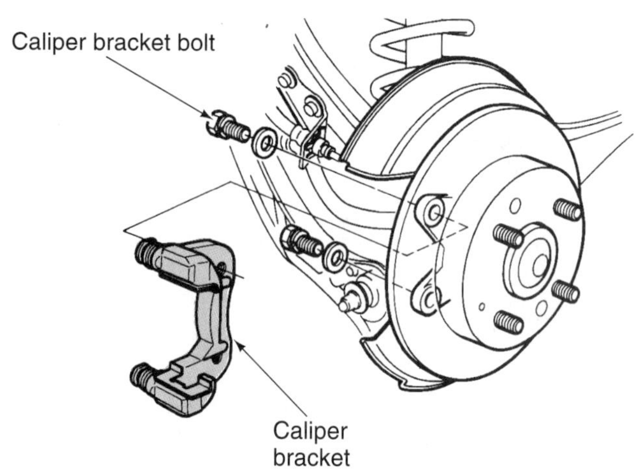

Figure 6-31 Removing the rear caliper bracket (Courtesy of Honda Motor Co., Ltd.)

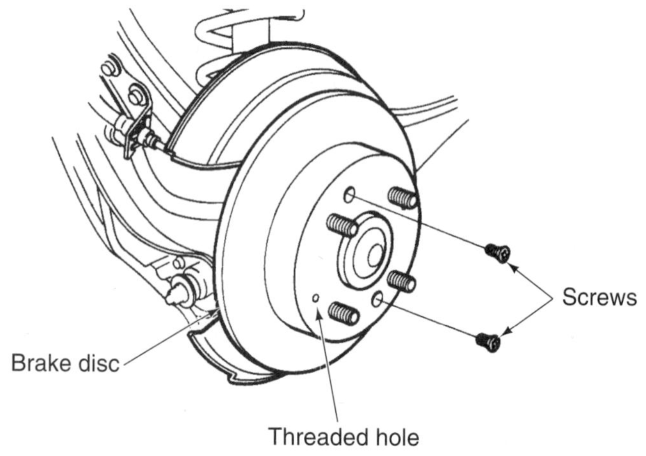

Figure 6-32 Removing the rotor retaining screws (Courtesy of Honda Motor Co., Ltd.)

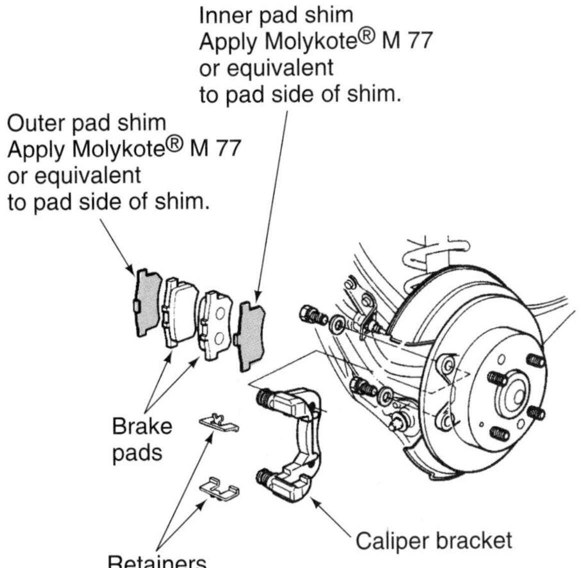

Figure 6-33 Installation of new pads and shims into the caliper bracket using retainer clips (Courtesy of Honda Motor Co., Ltd.)

7. Install the new or refinished rotor and the clean caliper bracket. Install the new brake pads and pad shims onto the caliper bracket as shown (Figure 6-33).
8. Lubricate the boot with a small amount of silicone grease before installing it to minimize the chances of twisting. On the rear disc brake assembly shown in this example, the caliper piston is rotated in a clockwise direction to install it in the cylinder. The cut-out in the piston is aligned with the tab on the inner brake pad by turning the piston back (Figure 6-34). The piston boot must sit properly without twisting. If the boot becomes twisted during installation, back it out until it is seated properly. Finally, reinstall the caliper assembly and caliper shield.

Servicing the Rear Caliper

The following is a typical service procedure for disassembling and rebuilding a rear disc brake caliper.

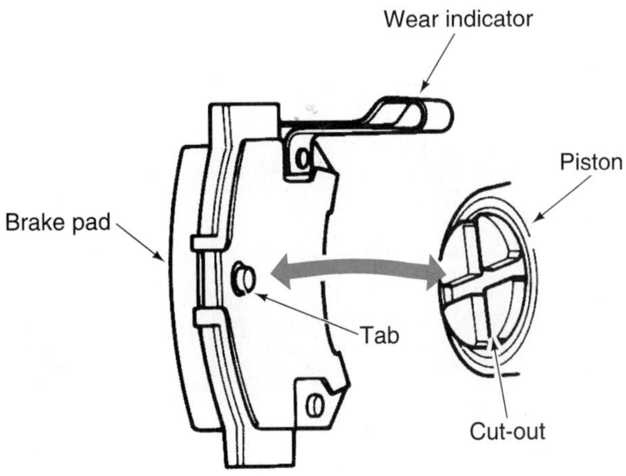

Figure 6-34 Some installations require the piston be aligned to a tab on the inner pad. (Courtesy of Honda Motor Co., Ltd.)

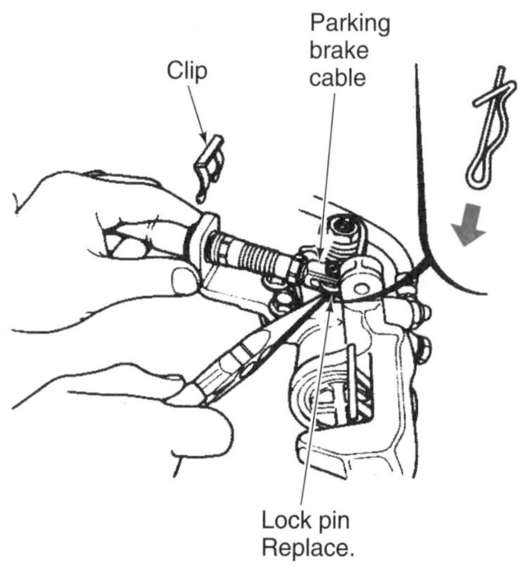

Figure 6-35 Disconnecting the parking brake cable from the rear caliper (Courtesy of Honda Motor Co., Ltd.)

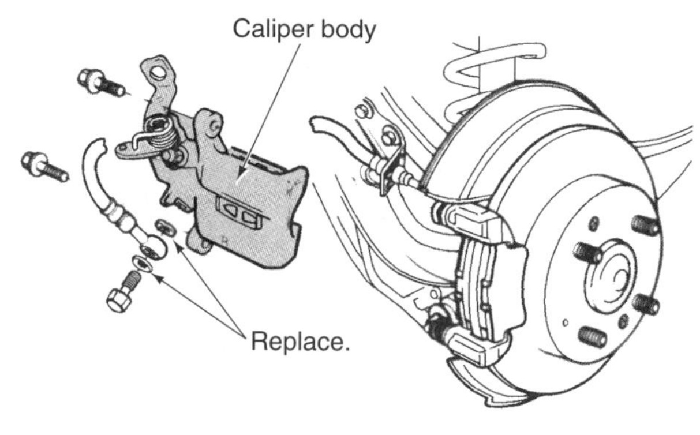

Figure 6-36 Disconnecting the brake hose at the rear caliper (Courtesy of Honda Motor Co., Ltd.)

1. Remove the caliper shield and disconnect the parking brake cable from the lever on the caliper by pulling out the lock pin (Figure 6-35).
2. Disconnect the flexible brake hose from the caliper body, remove the caliper mounting bolts, and lift the caliper off its mounting bracket (Figure 6-36). To prevent dirt from entering the caliper body, clean the outside of the caliper before beginning disassembly.
3. Lift out the pad spring from the caliper body.
4. To remove the caliper piston from the bore, use a properly-sized locknut wrench and extension bar. Turn counterclockwise to back the piston out of the bore (Figure 6-37). Once the piston is free, remove the piston boot.
5. Carefully inspect the piston body for signs of wear (Figure 6-38). Replace the piston if it is worn or damaged in any way.

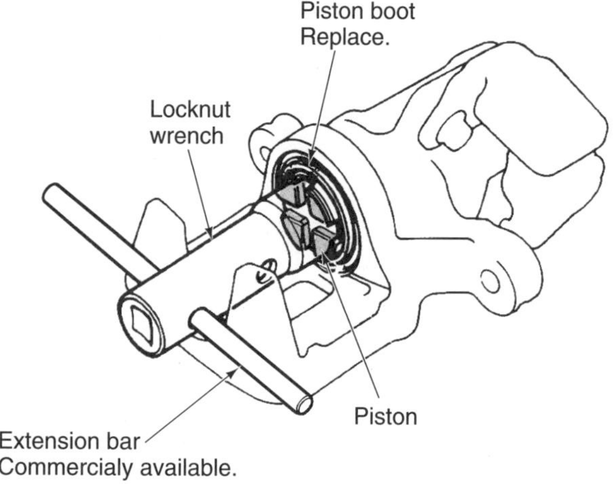

Figure 6-37 Removing the caliper piston using a locknut wrench (Courtesy of Honda Motor Co., Ltd.)

Figure 6-38 Rear caliper piston assembly (Courtesy of Honda Motor Co., Ltd.)

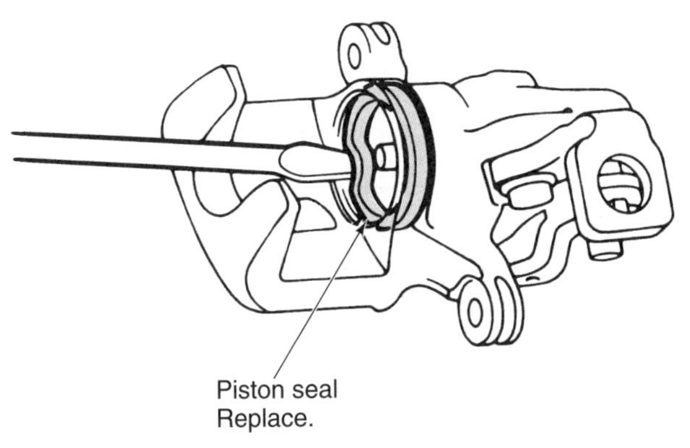

Figure 6-39 Removing the rear caliper piston seal (Courtesy of Honda Motor Co., Ltd.)

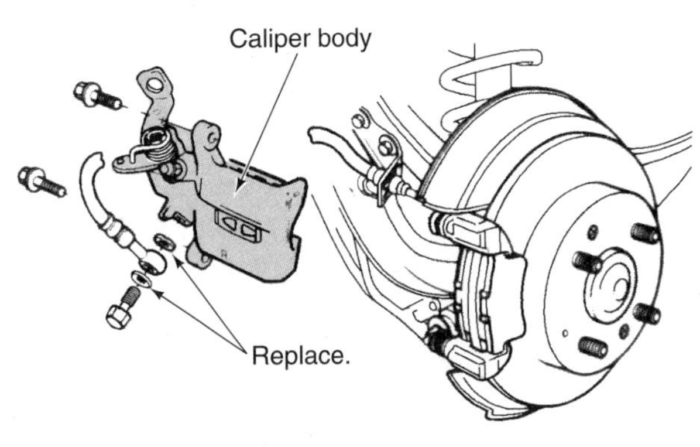

Figure 6-40 Installing the caliper guide needed for caliper disassembly on this type of rear disc brake caliper (Courtesy of Honda Motor Co., Ltd.)

6. Remove the piston seal from the caliper body using the tip of a screwdriver (Figure 6-39). Be careful not to scratch the bore.
7. Removing the brake spring and its related components from the caliper body requires the use of several special tools. A rear caliper guide is installed in the cylinder so the cutout on the guide aligns with a tab on the brake spring cover (Figure 6-40).
8. A brake spring compressor is then installed between the caliper body and the rear caliper guide. Turn the shaft of the compressor tool to compress the adjusting spring and using snap ring pliers, remove the circlip holding the spring in place (Figure 6-41).
9. Once the circlip is removed, remove the spring compressor tool from the caliper body. The following components can then be removed: spring cover, adjusting spring, spacer, bearing, adjusting bolt, and cup (Figure 6-42).

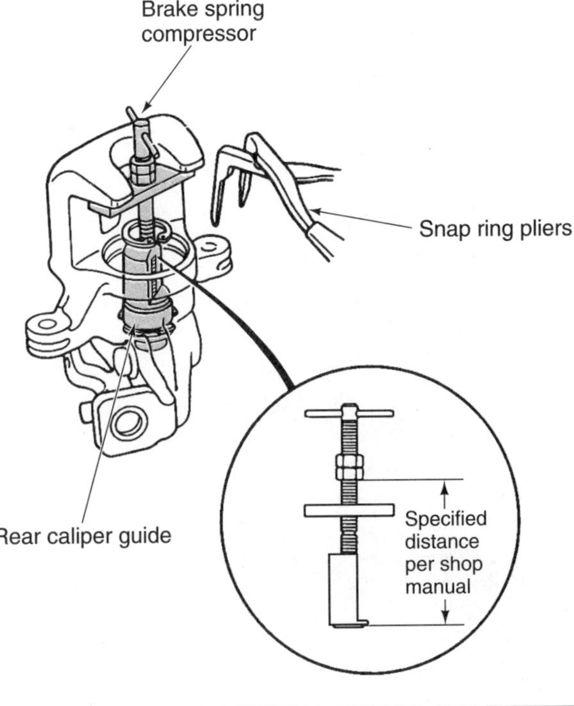

Figure 6-41 Disassembling the caliper using a brake spring compressor and snap ring pliers (Courtesy of Honda Motor Co., Ltd.)

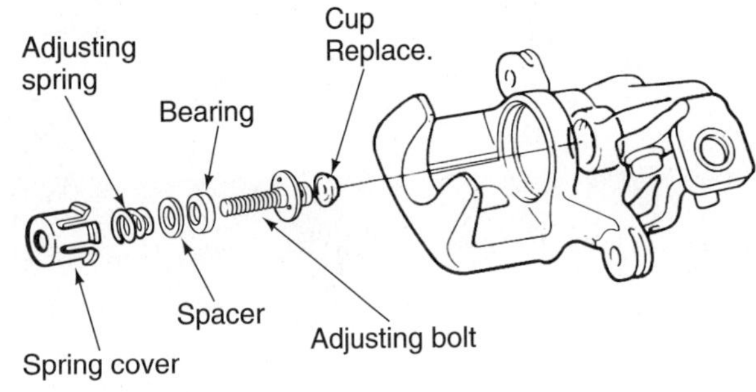

Figure 6-42 Inner components of the rear caliper assembly (Courtesy of Honda Motor Co., Ltd.)

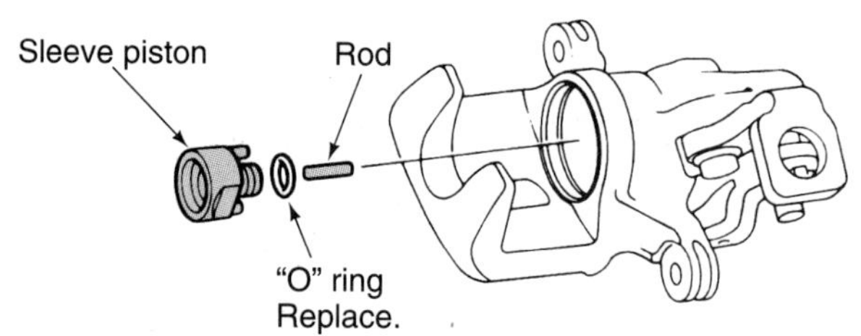

Figure 6-43 Disassembly of the sleeve piston and cam rod (Courtesy of Honda Motor Co., Ltd.)

10. Next remove the sleeve piston and "O" ring. Then remove the rod from the cam as shown in Figure 6-43.
11. Remove the return spring, the parking lever and cam assembly, and the cam boot from the caliper body (Figure 6-44).

WARNING: On the caliper shown in this example, do not loosen the parking nut on the parking lever/cam assembly with the cam installed in the caliper body. If the lever and shaft must be separated, secure the lever in a vise before loosening the parking nut.

12. Begin caliper reassembly by packing all cavities of the needle bearing with manufacturer's specified lubricant. Coat the new cam boot with assembly lubricant and install it in the caliper body.
13. Apply assembly lubricant to the area on the pin that contacts the cam. Install the cam and lever assembly into the caliper body.
14. Then install the return spring (see Figure 6-44).

WARNING: Since the cam and lever have been separated, they must be reassembled before installing the cam in the caliper body. Install the lever and spring washer, apply locking agent to the threads, and tighten the parking nut while holding the lever in a vise. When installing the cam, take care that the cam boot lips do not turn outside in.

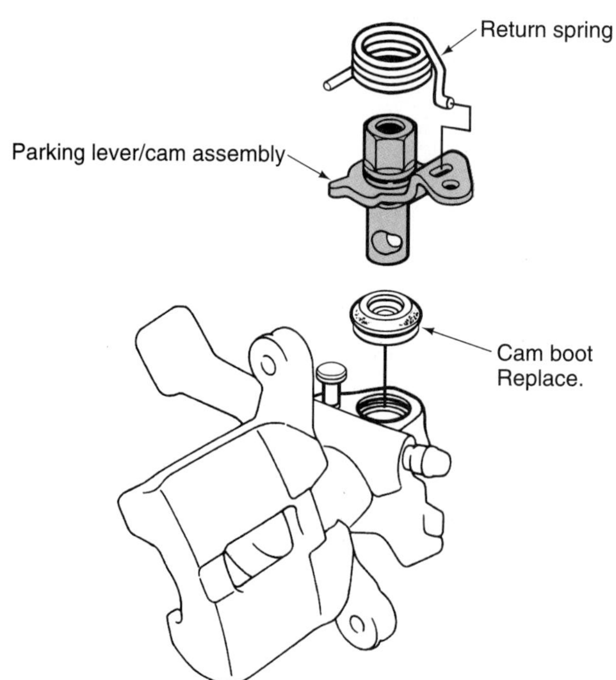

Figure 6-44 Disassembly of the parking brake lever subassembly (Courtesy of Honda Motor Co., Ltd.)

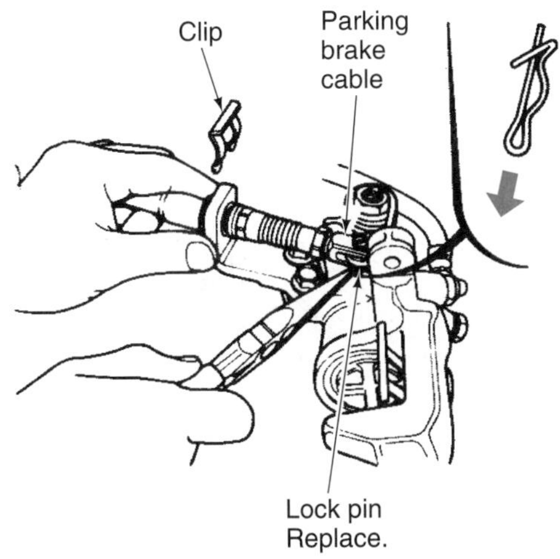

Figure 6-45 Installing the internal components of the caliper assembly (Courtesy of Honda Motor Co., Ltd.)

15. Install the rod in the cam, followed by a new "O" ring on the sleeve piston. Then install the sleeve piston so the hole in the bottom of the piston is aligned with the rod in the cam, and the two pins on the piston are aligned with the holes in the caliper (see Figure 6-43).

16. Install a new cup with its groove facing the bearing side of the adjusting bolt. Fit the bearing, spacer, adjusting spring, and spring cover on the adjusting bolt, then install it in the caliper cylinder (Figure 6-45).

17. As during disassembly, install the special caliper guide tool in the cylinder, aligning the cutout on the tool with the tab on the spring cover (see Figure 6-40). Adjust the special tool as shown in the detail (see Figure 6-45).

18. Install the brake spring compressor (Figure 6-46) and compress the spring until the tool bottoms out.

19. Make sure that the flared end of the spring cover is below the circlip groove. Then install the circlip (Figure 6-47), and remove the spring compressor. Check that the circlip is properly seated in the groove.

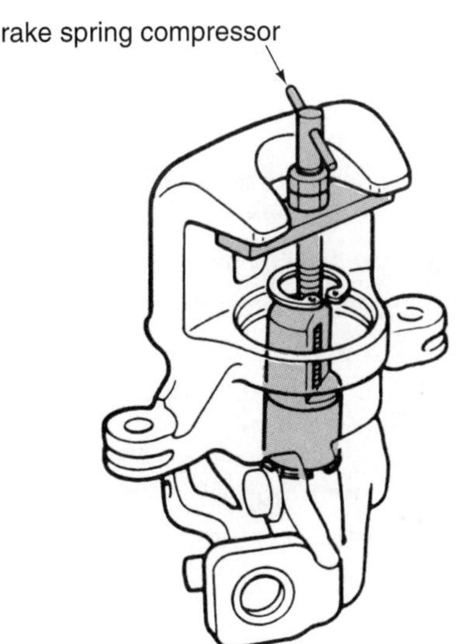

Figure 6-46 Installing the brake spring compressor tool in the caliper (Courtesy of Honda Motor Co., Ltd.)

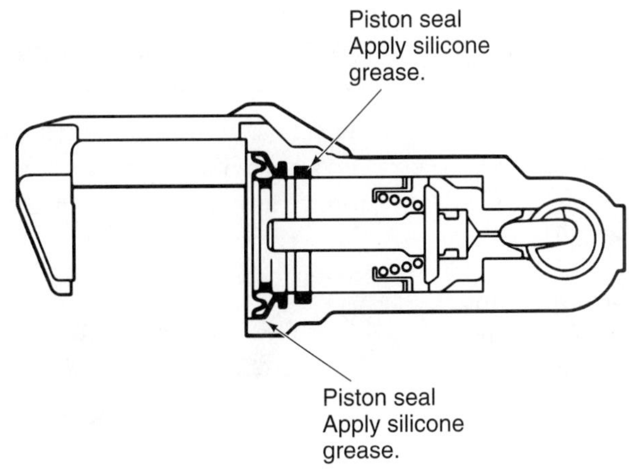

Figure 6-47 Installing the circlip (Courtesy of Honda Motor Co., Ltd.)

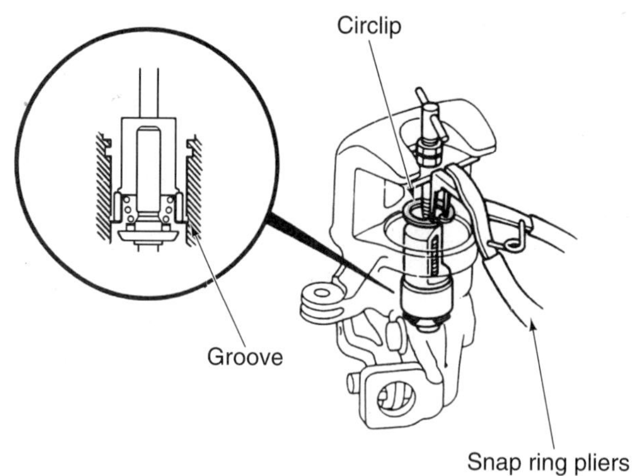

Figure 6-48 Lubrication points for the piston seal and boot (Courtesy of Honda Motor Co., Ltd.)

20. Coat the new piston seal and piston boot with silicone grease (Figure 6-48) and install them in the caliper.
21. Coat the outside of the piston with brake fluid and install it on the adjusting bolt while rotating it clockwise with the locknut wrench (see Figure 6-37).

WARNING: Avoid damaging the piston and piston boot.

22. Install the brake pad retainers, brake pads, and pad spring on the caliper.
23. Install the caliper on the caliper bracket and tighten the caliper bolts to torque specifications.
24. Reconnect the brake hose to the caliper with new sealing washers and tighten the banjo bolt to specifications.
25. Reconnect the parking brake cable to the arm on the caliper.
26. Top off the master cylinder reservoir and bleed the brake system. Depress the brake pedal several times, then adjust the brake pedal. Before making adjustments, be sure that the parking brake arm on the caliper touches the pin.
27. To complete the installation, reinstall the caliper shield.

Wheel Bearing Service (Nondriving Axles)

Special Tools

Dust cap pliers

Bearing driver

Classroom Manual

Chapter 6, page 117

Wheel bearings on nondriving wheels equipped with disc brakes must be cleaned, inspected, and repacked with the proper type of wheel bearing grease at regular intervals. The job is often performed as part of major brake service.

The service procedure differs between disc and drum brake equipped wheels. On disc brake systems, the brake caliper must be removed before the wheel hub assembly can be removed.

WARNING: Most vehicles are equipped with right-hand thread bearing retaining nuts. However, some older cars may use a left-hand thread on the left side of the car. You can be sure by checking the vehicle's shop manual.

Raise the vehicle on a hoist or jack with jack safety stands. Remove the hubcap or wheel cover. Remove the wheel lug nuts; then remove the wheel. Locate, loosen, and remove the brake caliper holding bolts and then remove the caliper. The brake line does not need to be disconnected, but the caliper must be supported while unbolted to avoid brake line damage.

Figure 6-49 Removing the grease cap using grease cap pliers

Using dust cap pliers, remove the grease cap (Figure 6-49). Pull out the cotter pin and remove the adjusting nut and thrust washer. Shake the rotor assembly to unseat the outer bearing cone and remove the cone. Remove the hub and rotor assembly, taking care not to drag the inner bearing against the spindle threads (Figure 6-50).

CAUTION: Wear full eye protection when driving bearing cones (races) in or out. The parts can shatter under the force of the blows.

Place the rotor bearing side down on a clean workbench. Position a clean shop towel under the hub area to cushion the inner bearing cone when it pops free. The cup (race) is press fitted into the rotor hub and can be removed from the inside out by striking it with a brass drift punch. Place two wooden blocks beneath the hub to provide enough clearance to remove the race. Place the brass drift punch through the opposite end of the hub until it contacts the edge of the cup (race), then strike the punch with a hammer to drive out the cup (race) as shown (Figure 6-51). Carefully note the direction the seal fits so you can properly install the new one.

The front wheel bearing should be adjusted to the manufacturer's specifications.

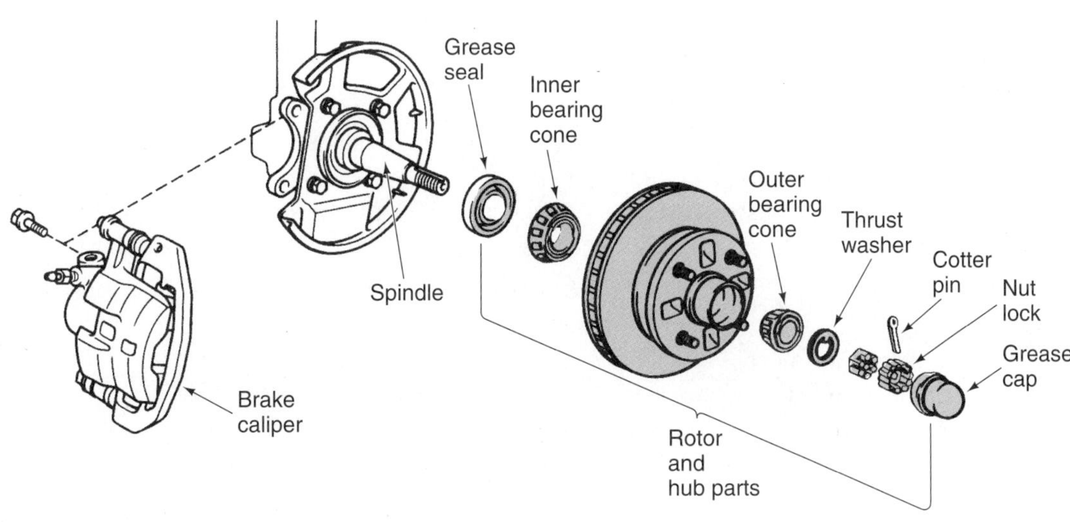

Figure 6-50 Exploded view of disc brake rotor hub parts on a nondriving axle

Figure 6-51 Removing bearings from a rotor hub

WARNING: Do not drive on a bearing cone (race) with a hard tool. A hard tool could slip and scratch the bearing surface.

Clean the hub and grease cap of all oil and old lubricant. Pay particular attention to the center or cavity of the hub, which is full of lubricant. Remove all the lubricant from the cavity. Wet clean all brake dust from the brake parts, taking care not to get any lubricant on brake parts. Wipe the spindle clean.

WARNING: Bearings must always be reinstalled into the same cone (race). Do not interchange left- and right-side bearings. A bearing installed into the wrong cone (race) will wear out quickly.

Wash the bearings, grease cap, thrust washer, and adjusting nut in cleaning solvent. Stubborn lubricant can be removed using a bristle brush. Dry the bearing with a clean lintfree rag or compressed air.

CAUTION: Never spin dry bearings using an air nozzle. The bearings may spin too fast and fly apart.

Carefully inspect the bearings after cleaning. Turn each roller completely around and check the surface. Carefully inspect each of the cups (races). A good bearing along with the most common bearing problems is illustrated (Figure 6-52). Driving on a bearing cage may bend the cage. Common roller surface damage conditions include:

1. Galling–indicated by metal transfer or smears on the ends of the rollers. It is caused by overloading, lubricant failure, or overheating. Overheating is usually the result of a too-tight adjustment.
2. Etching–a condition that results in a grayish-black bearing surface. It is caused by insufficient or incorrect lubricant.
3. Brinelling–occurs when the surface is broken down or indented. It is caused by impact loading, usually because the adjustment is too loose.
4. Abrasive wear–results in scratched rollers. It is caused by dirty or contaminated bearing lubricant.

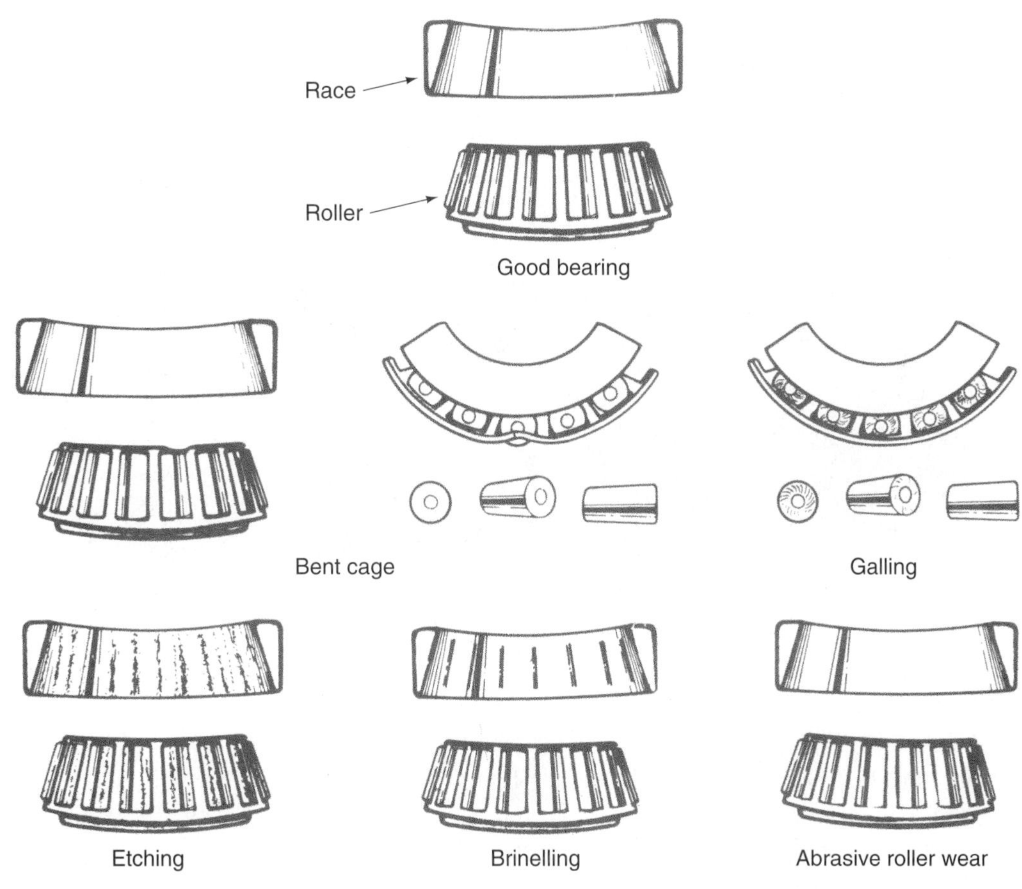

Figure 6-52 Types of wheel bearing damage (Reprinted with the permission of Ford Motor Company)

If the roller or cup shows signs of any of the preceding conditions, replace the bearing. A replacement bearing comes with both the roller and cup (race). Always compare the old parts with their replacement to be sure you have the correct items (Figure 6-53).

Figure 6-53 Comparison of new and worn bearing assemblies

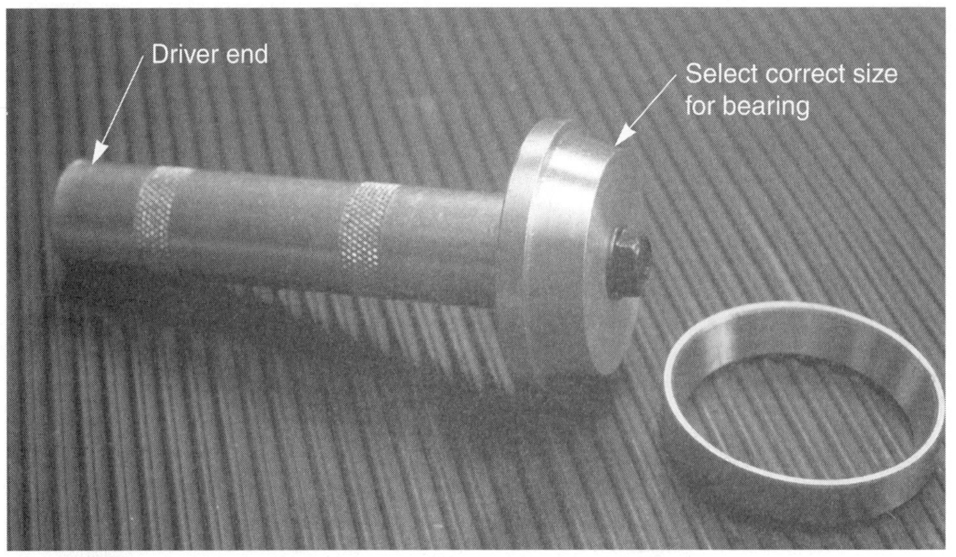

Figure 6-54 Bearing driver used to install the new cone (race)

Installation

A bearing driver is used to drive the new cone (race) back into the hub (Figure 6-54). Set the race in the hub, making certain it is squarely seated. Drive the cone (race) into the hub by hammering on the end of the driver. Keep the cone (race) going into the hub square (Figure 6-55). There is a step inside the hub, and the cone (race) must seat against it. You will notice a change in the sound of the driver when the cone (race) seats against the step. Inspect to ensure that it is seated all the way around.

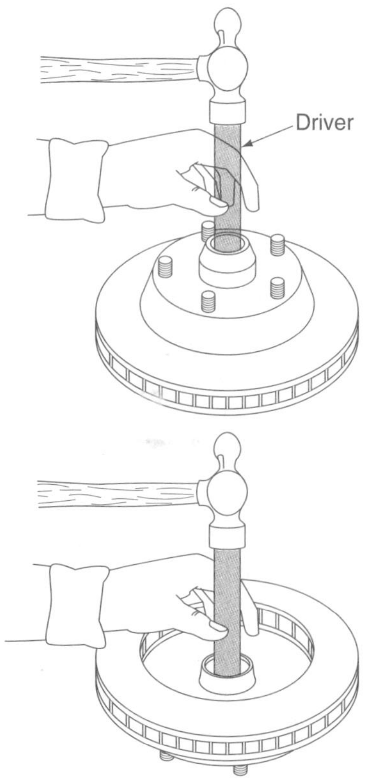

Figure 6-55 Installing the new cone (Courtesy of General Motors Corporation)

Figure 6-56 Wheel bearing packer tool

Once the cone is seated, pack new grease into the cleaned bearings and hub. Use the wheel bearing grease specified in the shop manual. High-temperature grease must be used in disc brake assemblies because of high braking temperatures generated by disc brakes. Never use grease that has been open to the air for any length of time because it attracts dirt and dust that can damage bearings.

You can pack the bearings by hand or with a tool called a bearing packer (Figure 6-56). A bearing packer has lubricant stored inside its housing. The bearing is placed on top of the packer and a special cone device is positioned on top of the bearing. Pushing on the cone handle forces the bearing down into the packer and lubricant in the tool is forced into the bearing.

To hand pack new grease between the rollers, take a small amount of grease in your palm and squeeze it between the rollers (Figure 6-57). Pack grease around each roller, then wipe off the excess grease using a clean, lintfree cloth.

Turn the rotor so its inner side is facing up. Fill the hub cavity with an approved wheel bearing grease to the level of the cup's smallest diameter. Coat the inner and outer cones (races) with a light coat of grease and carefully insert the inner bearing assembly into the hub. The small end always faces the center of the hub.

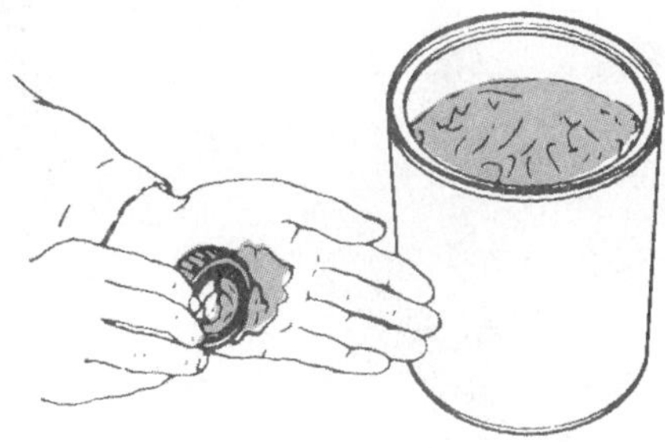

Figure 6-57 Packing bearings by hand (Courtesy of Nissan Motor Corporation in USA)

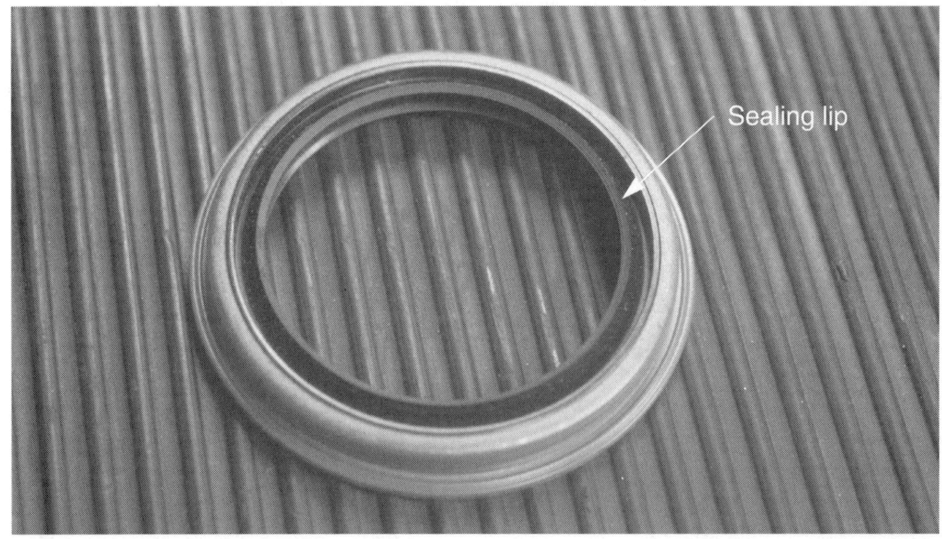

Figure 6-58 Grease seal used to keep grease in the hub

A new grease seal (Figure 6-58) must be installed whenever the bearings are repacked. This seal fits tightly in the hub with the seal tight around the spindle. The seal prevents grease from escaping the hub and contaminating the brakes. It also keeps dirt and water from finding its way into the hub. Compare the old and new seal to be sure you have the correct replacement. Lightly coat the sealing lip and outer edges of a new seal with grease, and then use a bearing driver to install the seal. Place the seal in position in the hub, making certain the seal is facing in the same direction as the one removed earlier. The sealing edge should face away from the tool and toward the bearing when installed. Hold the tool firmly in a straight line over the wheel hub and gently tap the tool until the seal is fully seated (Figure 6-59). Striking too hard will distort the seal. Listen for a change in the sound as the mallet strikes the driver. It indicates that the seal has bottomed.

Figure 6-59 Installing the new grease seal

WARNING: Never reuse an old seal. The old seal will have a worn sealing lip that will leak.

Reassembly and Adjustment

Carefully push the rotor assembly onto the spindle hub. Take care to center the wheel or rotor to prevent contact of the seal with the spindle threads. The wheel should be pushed on far enough so that the seal is in safe contact with its riding surface on the axle spindle. Install the outer bearing thrust washer and adjusting nut.

Replace the caliper and wheel. Rotate the wheel to ensure that brakes are not dragging. Dragging brakes will cause a false wheel bearing adjustment.

You must adjust the wheel bearing rollers and cups (races) correctly. Bearings that are too tight will overheat and fail; bearings that are too loose will gall or brinell. Always check the shop service manual for the correct adjustment procedure and specifications.

A typical adjustment procedure using a torque wrench is shown (Figure 6-60).

1. Tighten the spindle nut to specifications with a torque wrench while turning the wheel assembly forward by hand to fully seat the bearings. Turning the assembly will remove any grease or burrs that could cause excessive wheel bearing play later.
2. Back off the nut to its "just loose" position.
3. Hand tighten the spindle nut (Figure 6-61).
4. Loosen the spindle nut until either hole in the spindle lines up with a slot in the nut (not more than one-half flat of the nut).

Check the master cylinder fluid level and be sure the reservoirs are filled when the brake job is finished.

Be sure the brake pedal is firm after servicing the brakes and before moving the vehicle. Be sure to road test the vehicle.

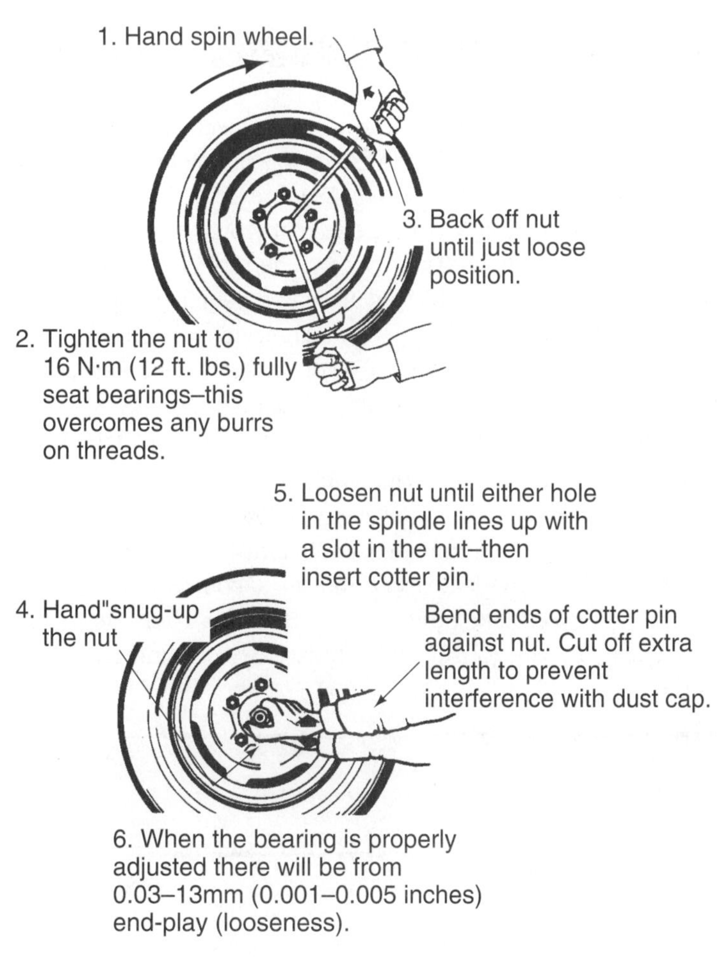

1. Hand spin wheel.
3. Back off nut until just loose position.
2. Tighten the nut to 16 N·m (12 ft. lbs.) fully seat bearings–this overcomes any burrs on threads.
5. Loosen nut until either hole in the spindle lines up with a slot in the nut–then insert cotter pin.
4. Hand"snug-up the nut
Bend ends of cotter pin against nut. Cut off extra length to prevent interference with dust cap.
6. When the bearing is properly adjusted there will be from 0.03–13mm (0.001–0.005 inches) end-play (looseness).

Figure 6-60 Typical bearing adjustment procedure (Courtesy of Chevrolet Motor Division)

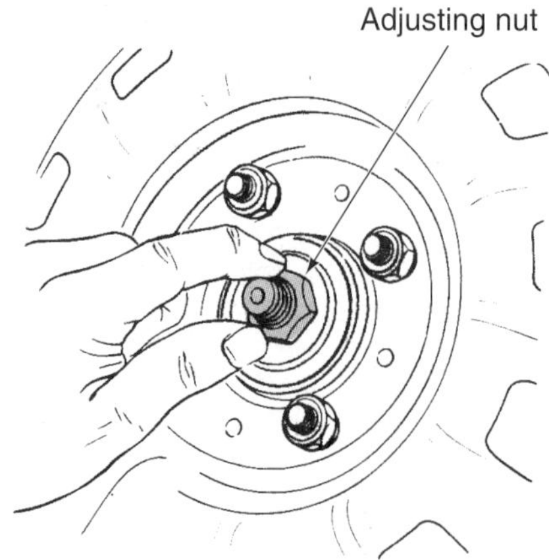

Figure 6-61 Tightening the spindle nut by hand (Courtesy of Chrysler Corporation)

5. Install a new cotter pin. Be sure to bend the ends of the cotter pin against the nut or cut off extra length to ensure that the ends will not interfere with the dust cap. Measure the looseness in the hub assembly. There will be from 0.001 to 0.005 in. (0.03 to 0.13 mm) endplay when properly adjusted.
6. Install the dust cap, replace the wheel cover or hubcap, and lower the car.

Always torque the lug nuts when installing a wheel on a vehicle with disc brakes. Never use an impact gun to tighten the lug nuts. Warpage of the rotor could result if an impact gun is used.

CASE STUDY

A customer returns his vehicle to the shop where he recently had the front disc brakes repaired. The brakes are now working, but they squeal on almost every application. He is upset and somewhat impatient. The shop manager reviews the work file in her records. The vehicle had extensive front brake service the week before. The rotors were turned and new pads and hardware were installed. The work was performed by a young technician whose work is ordinarily excellent. The vehicle is placed on the lift and the front wheels removed. Everything appears to be in order. The senior technician is called over. He inspects the brake assembly and runs his finger over the rotor surface. Closer inspection reveals the rotor does not have a nondirectional finish and antisqueal compound was not applied. Failure to perform two simple procedures has resulted in a come back and extra work.

Afterward, the shop manager questions the young technician who performed the service. The technician remembers both the vehicle and the owner. The technician had promised the work by five on a Friday and was running late. In a rush, he skipped an important step in finishing the rotors and also forgot to apply antisqueal compound. Fortunately, neither omission had compromised brake performance. But other oversights could have. Always take the time to do a complete job, even if you have to run a little late and take some heat from the customer.

Terms to Know

Discard dimension

Lateral runout

Loaded caliper

Nondirectional finish

Parallelism

Test scratch

ASE Style Review Questions

1. While servicing the front axle disc brakes on a front wheel drive vehicle:
   Technician A determines the right wheel pad is worn and replaces both the right and left wheel pads.
   Technician B determines the pads are not worn but rotates their position to ensure even pad wear.
   Who is correct?
   A. A only **C.** Both A and B
   B. B only **D.** Neither A nor B

2. After servicing the disc brakes on a vehicle:
   Technician A reinstalls the lug nuts using an impact wrench to ensure a tight fit.
   Technician B refills the master cylinder reservoirs to the proper level.
   Who is correct?
   A. A only **C.** Both A and B
   B. B only **D.** Neither A nor B

3. Loaded calipers are being discussed:
   Technician A says loaded calipers are replacement calipers that come with pads and hardware already installed.
   Technician B says loaded calipers should only be installed in axle sets.
   Who is correct?
   A. A only **C.** Both A and B
   B. B only **D.** Neither A nor B

4. Rotor refinishing is being discussed:
   Technician A says it is very important that the rotor surface be made nondirectional.
   Technician B says rotors should be refinished as part of routine disc brake service.
   Who is correct?
   A. A only **C.** Both A and B
   B. B only **D.** Neither A nor B

5. When performing disc brake work:
   Technician A works on one wheel at a time to avoid "popping" pistons out of the other caliper and to allow the other caliper assembly to be used as a guide.
   Technician B never permits the caliper assembly to hang with the weight on the brake hose.
   Who is correct?
   A. A only **C.** Both A and B
   B. B only **D.** Neither A nor B

6. When performing a rotor run-out inspection:
   Technician A uses a micrometer.
   Technician B uses a dial indicator.
   Who is correct?
   A. A only **C.** Both A and B
   B. B only **D.** Neither A nor B

7. Wheel bearing service is being discussed:
   Technician A says that newly installed bearings are self-adjusting and require no special adjustment.
   Technician B says bearing assemblies are interchangeable with one another.
   Who is correct?
   A. A only **C.** Both A and B
   B. B only **D.** Neither A nor B

8. Rotor serviceability is being discussed:
   Technician A says the minimum wear thickness of a rotor is the discard thickness of the rotor.
   Technician B says the minimum wear thickness is the minimum thickness after refinishing.
   Who is correct?
   A. A only **C.** Both A and B
   B. B only **D.** Neither A nor B

9. When refinishing a rotor on a lathe, the rotor wobbles excessively:
   Technician A says the lathe arbor may be bent.
   Technician B say the mounting adapters of the lathe may be distorted.
   Who is correct?
   A. A only **C.** Both A and B
   B. B only **D.** Neither A nor B

10. After servicing front disc brakes:
    Technician A reinstalls the wheel and tire assembly and tightens the nuts in the recommended tightening pattern.
    Technician B pumps the brake pedal several times to position the new pads before test driving the vehicle.
    Who is correct?
    A. A only **C.** Both A and B
    B. B only **D.** Neither A nor B

Table 6-1 ASE TASK

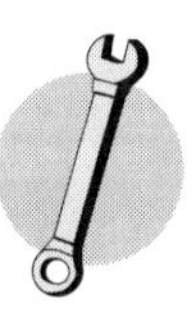

Diagnose disc brake system problems.

Problem Area	Symptoms	Possible Causes	Classroom Manual	Shop Manual
WEAK BRAKING ACTION	Depressing the brake pedal produces little or no braking action at the wheels equipped with disc brakes.	1. Reservoir fluid level low	47	55
		2. Excess air in hydraulic system	47	124
		3. Open bleeder screw	—	124
		4. The pistons may have been pushed back too far in the bores while servicing the calipers, resulting in improperly positioned pads	123	149
		5. Excessively tapered or warped pads	120	143
		6. One or more of the calipers may have sustained piston seal damage	125	149
		7. Leakage past the piston cups in the master cylinder	55	60
		8. Bad or excessively loose wheel bearings	119	168
		9. Excessive rotor runout or bent rotor	117	152
		10. Pad or plate knock back resulting from either loose wheel bearings or a faulty front suspension	119	168
		11. Leak in the hydraulic system	96	114
		12. Weak brake hoses have expanded	98	121
		13. Incorrectly adjusted manual master cylinder pushrod	44	68
		14. Defective master cylinder primary cup	55	60
		15. Improper seating or positioning of the brake pad assemblies	120	146
HARD PEDAL	Excessive brake pedal pressure is required to stop the vehicle.	1. Pads worn below the minimum thickness	120	143
		2. Master cylinder or power brakes not operating properly	55	60
		3. Glazed pads, faded/overheated condition, blued/heat checked rotors	121	142
		4. Linings have excess grease, oil, or brake fluid	120	142
		5. Pedal pushrod and linkage are binding	44	59
		6. Dented, kinked, collapsed, or blocked lines, hoses, or connections	96	114
		7. Lining worn below specifications	120	143
		8. Excessive lining friction or poor quality lining	122	141
		9. Frozen or sticking caliper piston	127	149

Table 6-1 ASE TASK (continued)

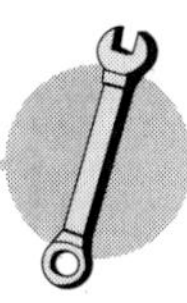

Diagnose disc brake system problems.

Problem Area	Symptoms	Possible Causes	Classroom Manual	Shop Manual
SPONGY PEDAL	Pedal has a soft, springy, or spongy feeling when pressed.	1. The brake fluid has a low boiling point	95	111
		2. Weak brake hoses expanding under pressure	98	121
		3. Air in the hydraulic system	47	124
PEDAL PULSATION Note: On some vehicles equipped with ABS, the brake pedal may pulsate during hard stops. This is normal and shows the ABS is doing its job.	Brakes chatter when applied. Vibration or chatter is felt in the pedal when applied. Rough or poor braking action.	1. The brake rotor demonstrates excessive lateral runout	117	152
		2. The brake rotor has excessive thickness variation	117	151
		3. Loose or worn steering/suspension parts	119	141
		4. Excessive front bearing clearance	119	168
		5. Worn or damaged front-wheel bearing	119	168
		6. Sharp cornering or rough road conditions have knocked the pistons back into the caliper bore	123	149
		7. Loose caliper mounting bolts	125	143
		8. Wheel bearing adjusted too loosely	119	175
		9. Improperly aligned front end caused by loose front end parts	—	141
		10. Lining contains brake fluid or grease	120	142
BRAKE GRAB	Severe and/or uneven braking when the brakes are applied.	1. Loosely mounted calipers	125	141
		2. Binding in the pedal linkage	44	59
		3. Incorrect tire pressure	—	54
		4. The front end has worn parts, broken springs, or is out-of-alignment	—	141
		5. A fault in the metering or proportioning valve	103	132
		6. Lining contaminated with grease, oil, or other fluids	120	142
		7. The lining is high-friction on one side, and low-friction on the other	121	144
		8. Partially or fully blocked lines, hoses, or other passageways	96	114
		9. Sluggish or frozen caliper pistons	123	149
		10. Soft or swollen seals on caliper cylinder bores	125	149
		11. The caliper cylinder bores feel rough or show signs of corrosion	123	149
		12. Pads fitted incorrectly or loose on the plate	121	145
		13. Master cylinder bore feels rough or shows signs of corrosion	48	60

Table 6-1 ASE TASK (continued)

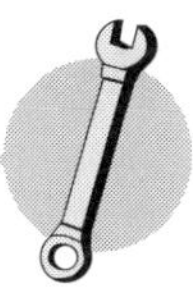

Diagnose disc brake system problems.

Problem Area	Symptoms	Possible Causes	Classroom Manual	Shop Manual
EXCESSIVE BRAKE PEDAL TRAVEL	Brake pedal closer than 1 to 2 inches from floor when fully applied.	1. Sharp cornering or rough road conditions have knocked the pistons back into the caliper bore	123	149
		2. Air in the hydraulic system	47	124
		3. Hydraulic system contains insufficient fluid	47	55
		4. Loose caliper attachment	125	141
		5. Loose wheel bearing attachment	119	168
		6. Incorrect power brake pushrod adjustment	44	68
		7. Excessive rotor lateral runout	117	152
		8. A leak in the hydraulic system	96	114
		9. Master cylinder not operating properly	55	60
		10. Damaged or worn caliper piston seals	125	149
		11. Piston knocked back into the bore by a loose or worn wheel bearing	119	168
		12. Piston knocked back into the bore on a rear disc brake due to excessive axle side play.	130	—
LEAKY CALIPER	Brake fluid leaks evident in caliper area during inspection.	1. Scoring or corrosion evident on the cylinder bore surface	123	149
		2. Damage or excessive wear on the caliper piston seal	125	149
		3. Damaged caliper piston	123	149
BRAKES FAIL TO RELEASE	Braking action does not stop when brakes are released, particularly when the system is hot.	1. Power brakes not operating properly	67	—
		2. Defective residual check valve in the disc brake circuit	103	131
		3. Sluggish or sticking pedal linkage	44	59
		4. Frozen or seized pistons	123	149
		5. No release of pressure in the master cylinder	55	60
		6. Incorrectly seated or positioned brake assemblies	125	147
		7. Sliding or floating caliper sleeves, pins, or anchors are frozen	127	149
BRAKE PULL	Vehicle pulls to one side when brakes are applied.	1. There is brake fluid, oil, or grease on the linings	120	142
		2. Distorted pads, unmatched linings, or uneven wear have damaged the linings	120	143

Table 6-1 ASE TASK (continued)

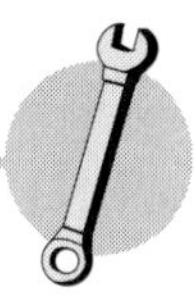

Diagnose disc brake system problems.

Problem Area	Symptoms	Possible Causes	Classroom Manual	Shop Manual
BRAKE PULL	Vehicle pulls to one side when brakes are applied.	3. One of the rotors has a rough surface	117	151
		4. Seized or frozen pistons	123	149
		5. Loose caliper mounting bolts or guide pins, or loose caliper	125	141
		6. Worn bushings or collapsed stabilizers	—	149
		7. Uneven tire pressure, tread wear, or a size difference from right to left	—	141
		8. Excessive rotor parallelism or runout	117	152
		9. Sticking caliper piston	123	149
		10. Misaligned front end	—	141
		11. Damaged or faulty suspension parts	—	141
BRAKE DRAG (ONE WHEEL)	Vehicle wanders to one side and wheel alignment is correct.	1. Sticking caliper piston	123	149
		2. Swelling in the caliper piston seal	125	149
		3. Loose or worn front wheel bearings	119	168
		4. Brake fluid is not returning properly due to a defective brake hose or hydraulic tube	95	114
		5. Caliper is sticking or frozen	127	149
BRAKE DRAG (ALL WHEELS)	Brakes get hot. Vehicle has poor gas mileage and performance.	1. Binding in the brake pedal	44	59
		2. Contaminated or incorrect brake fluid has damaged rubber parts	—	112
		3. Blocked master cylinder compensating ports	55	60
NOISE	Brakes chatter.	1. Excessive lateral rotor runout	117	152
		2. Rotor does not have parallelism	117	151
		3. Loose wheel bearings	119	168
		4. Loose front spindle	119	169
		5. Loose rear suspension	—	141
		6. Incorrectly torqued lug bolt	—	5
		7. A kink or bend in the rear axle	—	141
NOISE	Brakes scrape.	1. A rusty or muddy buildup on the edges of the caliper housing	123	141
		2. A worn pad, metal tabs, or the backing is exposed to the rotor	122	143
		3. The rotor is scraping against the housing due to incorrect caliper alignment	125	141
		4. Loose wheel bearings	119	168

Table 6-1 ASE TASK (continued)

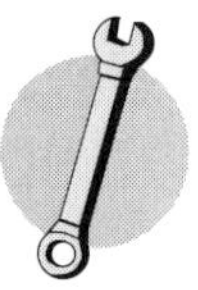

Diagnose disc brake system problems.

Problem Area	Symptoms	Possible Causes	Classroom Manual	Shop Manual
NOISE	Brakes rattle when applied.	1. Too much clearance between the pad and the caliper	124	144
		2. The antirattle spring in the pad is missing/not properly positioned	122	144
		3. Loose caliper mounting bolts	125	141
NOISE	Brakes squeal when applied.	1. Glazed pads	121	142
		2. The insulator shim located between the piston and pad plate is not in place	121	144
		3. Worn or weak antirattle springs/clips	122	144
		4. The pad wear indicators are contacting the rotor. (When the brakes are applied, the squeal disappears.)	131	143
		5. Corrosion buildup between the caliper and the pads	123	149
		6. Loose outboard pad tabs	121	146
		7. The linings are loose or in poor condition	122	143
NOISE	Brakes grind when applied.	1. The pads are worn down to the metal backing plate	122	143

CHAPTER 7

Drum Brakes

Upon completion and review of this chapter, you should be able to:

- ❏ Diagnose drum brake problems, including poor stopping, pulling, grabbing, and dragging caused by the drum brake wheel assembly. Determine and perform needed repairs.
- ❏ Diagnose poor stopping, pulling, grabbing, or dragging caused by the drum brake hydraulic assembly. Determine and perform needed repairs.
- ❏ Diagnose poor stopping, pulling, noise, dragging, or pedal pulsation caused by the drum brake mechanical assembly. Determine and perform needed repairs.
- ❏ Remove, clean, inspect, and service brake drums, including machining on a brake lathe.
- ❏ Remove, clean, and inspect brake shoes/lining and related hardware, including all springs, pins, clips, levers, adjusters, and self-adjust units. Determine and perform needed repairs.
- ❏ Clean and remove loose dirt, rust, and scale from the brake backing plates. Inspect and determine if plate replacement is needed.
- ❏ Disassemble, clean, inspect, and overhaul a wheel cylinder assembly. Replace all cups, boots, and damaged hardware.
- ❏ Properly lubricate brake shoe support pads on backing plate, adjusters, and other drum brake hardware.
- ❏ Determine correct brake shoe application.
- ❏ Install replacement brake shoes and related hardware.
- ❏ Adjust brake shoes and reinstall brake drums or drum hub assemblies and wheel bearings.
- ❏ Reinstall the serviced wheel, torque lug nuts, and make final adjustments.

Troubleshooting Drum Brakes

Basic Tools

Basic mechanic's tool set

Hydraulic floor jack

Although an increasing number of today's vehicles have disc brakes on all four wheels, many cars, minivans, and trucks use disc brakes on the front axle and drum brakes on the rear axle. Some older vehicles (pre-1980) have drum brakes at all four wheels.

You may need to test drive a car with drum brakes to locate the source of an owner's complaint. Follow the test drive guidelines outlined in Chapter 3. Following are common troubles usually associated with the brake shoes or brake shoe attaching parts:

Classroom Manual
Chapter 7, page 144

- ❏ Distorted or improperly adjusted brake shoe
- ❏ Loose, glazed, or worn lining
- ❏ Weak or bad return springs
- ❏ Loose backing plate
- ❏ Self-adjusters not operating
- ❏ Oil, grease, or brake fluid on lining

A distorted or improperly adjusted brake shoe may result in brake drag or grab, pulling to one side, wheel lockup, hard pedal, excessive pedal travel, or noisy operation. A loose lining may cause pull to one side or a persistent brake chatter.

Glazed or worn linings are the most common cause of drum brake problems. Problems include hard pedal, pulling to one side, wheel locking, brake chatter, excessive pedal travel, noisy brakes, grabbing brakes, or little or no braking power.

A broken or weakened retracting or return spring will cause the brake to drag or the vehicle to pull to one side. A loose backing plate can cause brake drag, brake chatter, or a locked wheel. If the self-adjuster operates incorrectly, the brake shoes will become improperly adjusted and the vehicle may then pull to one side, show excessive pedal travel, or develop a clicking sound in the shoes.

Oil, grease, or brake fluid on the linings can cause hard pedal, side pull, wheel lock, brake chatter, uneven braking, noisy brakes, grabbing brakes, or loss of braking power.

Damaged or distorted brake drums also adversely affect braking. Excessive pedal travel may be a sign of a cracked drum. But as discussed in Chapter 3, excessive pedal travel may also be caused by trouble in the hydraulic system. An out-of-round drum will cause the brake to drag and/or chatter. The vehicle will also pull to one side. A scored brake drum will cause uneven braking with noisy or grabbing brake action. Drums that have become bellmouthed or distorted cause only partial lining contact and a twisting of the brake shoe that results in excessive pedal travel and hard pedal pressure. Threaded drums result when a dull or poorly shaped cutting bit is used to resurface the drums. These threads cause the brake shoes to move toward or away from the backing plate according to the direction of the thread spiral. This action creates a snapping or clicking action, which is especially noticeable on light brake applications at low speeds. A faulty brake cylinder may be the cause of brake drag, wheel lock, side pull, or a noisy and grabbing brake.

If you find any of these conditions during your test drive, remove the drums and inspect the brakes. Any wear in the shoes, shoe holding and retracting hardware, drums, or wheel cylinder will make a complete brake system overhaul a necessity.

While disc brakes allow for a quick visual inspection every time the wheel assembly is removed, the components of a drum brake assembly remain covered by the drum when the wheel is removed. The drums must be removed to inspect the condition of the brake assembly. Fortunately, drum brakes wear about half as quickly as disc brakes, and the recommended inspection interval for many vehicles is now 30,000 miles or every two years.

Many state vehicle inspection laws specify that drums must be pulled at regular intervals and the system inspected. Always remove the drums for inspection whenever the customer voices a brake-related complaint or concern, or whenever you suspect a problem. Always service drum brakes and linings in axle sets.

CAUTION: When servicing wheel brake parts, do not create dust by cleaning with a dry brush or compressed air. Asbestos fibers can become airborne if dust is created during servicing. Breathing dust containing asbestos fibers can cause serious bodily harm.

See Chapter 2 for details on safely working around asbestos dust.

Drum Brake Removal (Rear Axle)

Mark the relationship of the wheel to the axle flange to ensure proper wheel balance upon assembly.

Typically, the rear brake drums are attached to the spindle by a wheel hub unit with push nuts. Wheel studs in the wheel hub are used for mounting the rear wheels (Figure 7-1). A typical procedure for drum brake removal is shown in Photo Sequence 11.

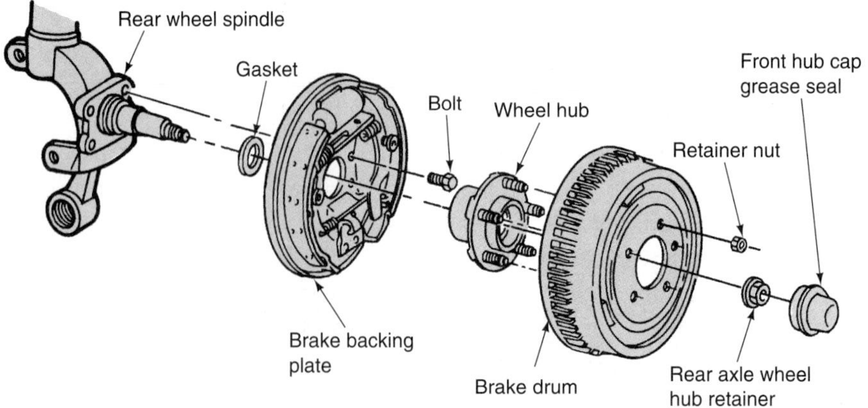

Figure 7-1 Rear wheel spindle, hub, and brake drum mounting (Reprinted with the permission of Ford Motor Company)

Photo Sequence 11
Typical Procedure for Removing a Drum Brake (Rear Axle)

P11-1 Release the parking brake and raise the car on a hoist. Mark the position of the wheel to the axle flange so the wheel can be reinstalled in the same relative position to the flange. This ensures proper wheel balance.

P11-2 Remove the nuts holding the wheel assembly to the hub and pull off the tire and wheel assembly.

P11-3 Mark the position of the brake drum to the axle flange so the drum can be reinstalled in the same relative position to the flange.

P11-4 Usually, the drum can then be slid off of the hub bolts. On some vehicles, two small screws threaded into the axle flange or a large nut on the end of the axle may hold the drum in place. Remove these and pull off the drum.

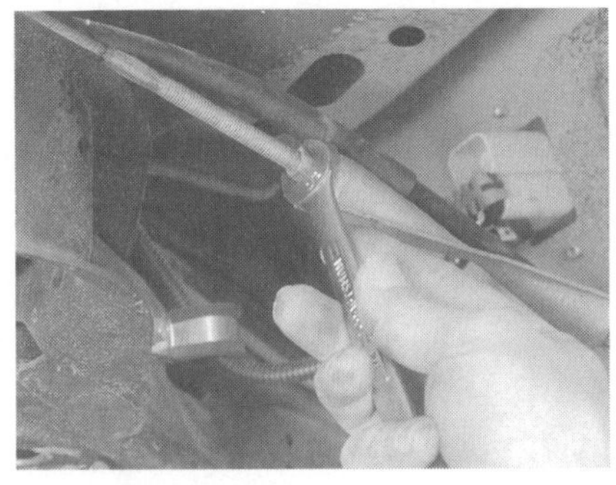

P11-5 If the brake drum does not slide off easily, double-check that the parking brake is released. Also, relieve all tension from the parking brake cables by loosening the adjusting nut at the equalizer.

P11-6 On self-adjusting drum brakes, disengage the brake shoe adjusting lever and back off the brake shoe adjustment. Begin by removing or drilling out the access plug or covering.

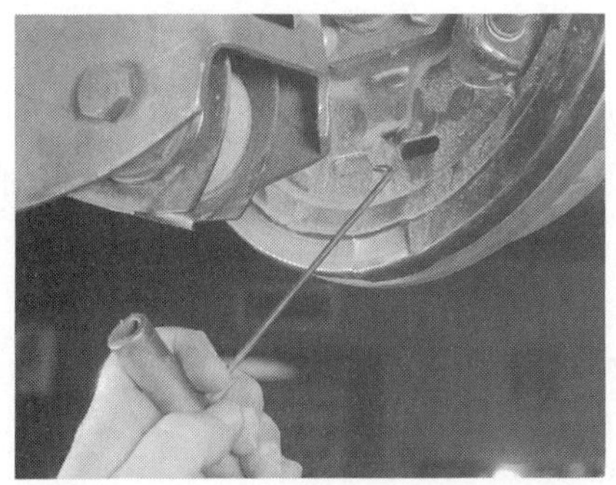

P11-7 Insert a small wire hook through the slot and pull the lever away from the star wheel no more than 1/16".

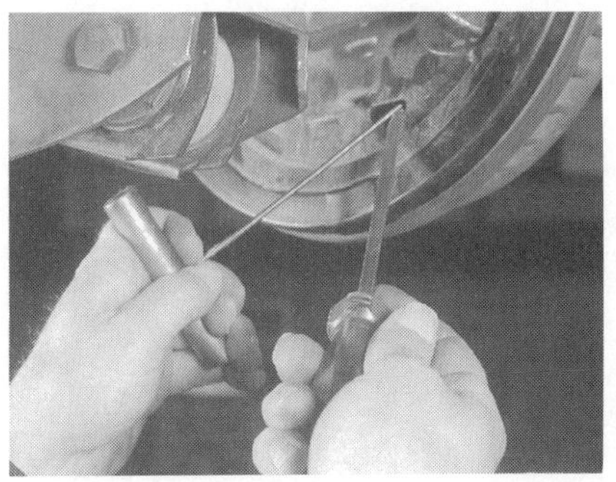

P11-8 Use a screwdriver to back off the adjusting screw star wheel.

P11-9 Once the adjuster is backed off, use a rubber mallet to tap gently on the outer rim and/or around the inner drum diameter. Take care not to use excessive force as this may damage the drum.

Photo Sequence 11
Typical Procedure for Removing a Drum Brake (Rear Axle) (continued)

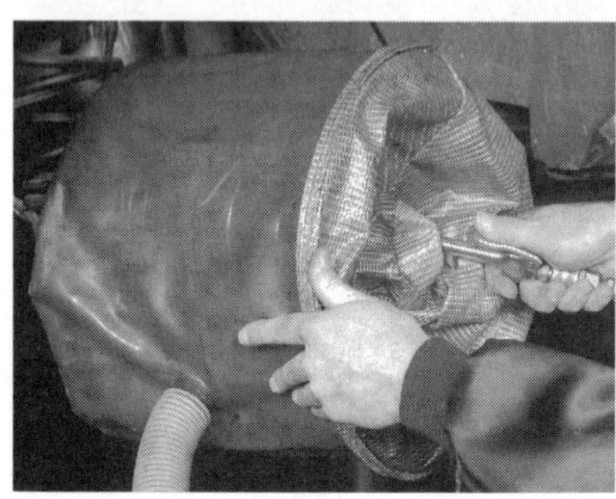

P11-10 Thoroughly clean the exposed brake assembly area and drum using a vacuum system equipped with a HEPA filter.

P11-11 Inspect the drum brake assembly for broken springs, signs of heat damage, leaking wheel cylinders, and other wear.

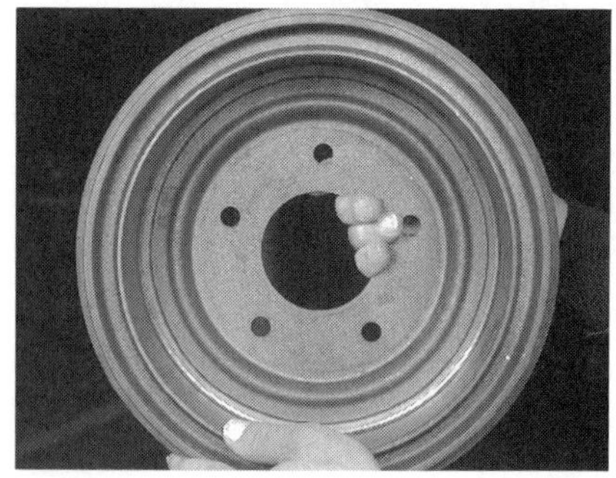

P11-12 Inspect the drum for signs of scoring or other damage.

To remove the rear axle drums and inspect drum brake components, release the parking brake and raise the car on a hoist or jack and jack safety stands. Remove the nuts holding the wheel assembly to the hub and pull off the tire and wheel assembly. Usually, the drum can then be slid off of the hub bolts. Some drums are retained with two small screws threaded into the axle flange. In some cases, a large nut on the end of the axle holds the drum in place.

Always inspect parking brake cables for internal drag or seizing.

If the brake drum assembly does not slide off easily, relieve all tension from the parking brake cables. On most vehicles, simply loosen or remove the adjusting nut at the equalizer (Figure 7-2). You may also need to back off the brake shoe adjustment before the drums can be removed. On some cars, this is done through a slot in the backing plate; on others, access is provided through a slot in the brake drum.

On older non-self-adjusting brakes, insert a brake adjusting tool through the hole and engage the teeth in the star wheel (Figure 7-3). Move the free end of the tool downward to turn the star wheel and increase the shoe-to-drum clearance. Turn the star wheel until you cannot hear any brake shoe-to-drum contact when you rotate the drum.

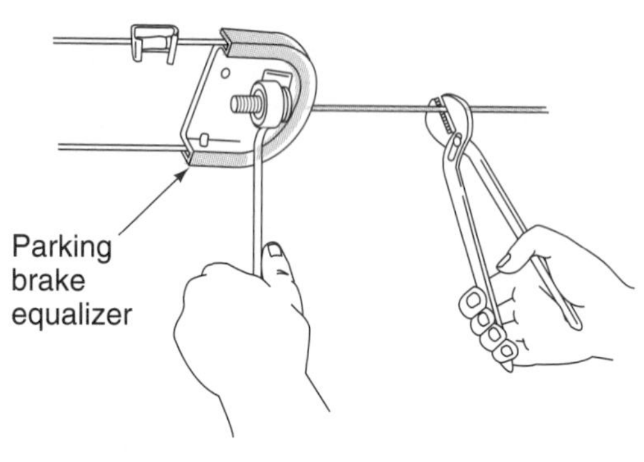

Figure 7-2 Adjusting an emergency brake system to remove a brake drum

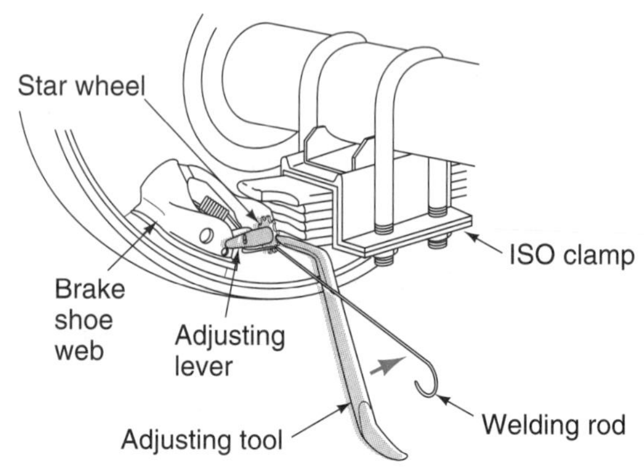

Figure 7-3 Backing off the brake adjustment to remove the brake drum (Courtesy of Chrysler Corporation)

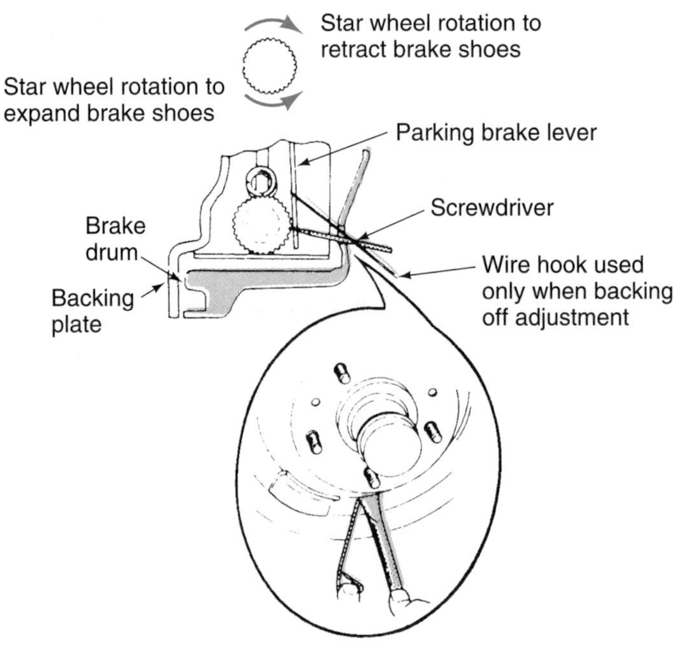

Figure 7-4 Backing off the self-adjusters to remove the brake drum (Courtesy of Pontiac Motor Division)

On self-adjusting drum brakes, you must also disengage the brake shoe adjusting lever before backing off the adjusting screw star wheel (Figure 7-4). This is done by carefully reaching in through the adjusting slot with a small screwdriver and pushing the lever away from the star wheel. Do this carefully; you should move the lever no more than 1/16 inch. It may be necessary to remove an access plug and the brake line-to-axle retention bracket to allow sufficient room to insert a screwdriver and the brake tools needed (Figure 7-5). On some General Motors vehicles, a knockout plate must be drilled out using a 7/16-in. (11-mm) drill. After adjustment, a rubber cover is fitted into the adjusting hole to prevent dirt or contamination from entering the drum brake assembly.

Classroom Manual
Chapter 7, page 148

Once the adjuster is backed off, use a rubber mallet to tap gently on the outer rim of the drum and/or around the inner drum diameter by the spindle. Take care not to use excessive force as this may damage the drum.

Mark the relationship of the brake drum to the axle flange to ensure proper wheel balance upon assembly.

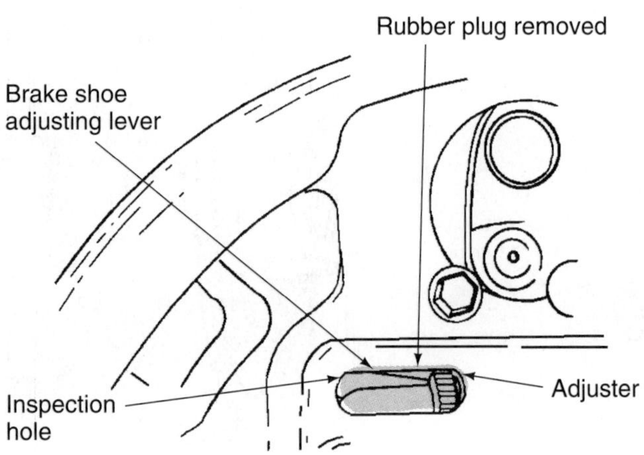

Figure 7-5 Rear drum brake inspection hole (plug removed) and adjuster (Reprinted with the permission of Ford Motor Company)

Rear Wheel Hub Bearing Unit Removal/Replacement

With the wheel assembly and drum brake removed, continue the disassembly by removing the front hub cap grease seal. Next, remove the axle wheel hub retainer and pull the wheel hub unit from the rear wheel spindle. Installation is the reverse of removal. The rear wheel hub is lubricated for life. If the bearings become worn, the entire hub bearing unit is replaced as a unit. The hub must also be removed to access the drum brake backing plate.

Special Tools

Cartridge-type respirator

Wet cleaning or vacuum cleaning system

Classroom Manual

Chapter 7, page 138

Drum Brake Removal (Front Axle)

On older vehicles equipped with front axle drum brakes, the drum and hub are usually an integral assembly so it is necessary to disassemble the front wheel bearings in order to remove the drum. Remove the dust or grease cap, cotter pin, castellated nut or plain nut and nut lock, thrust washer, and outer bearing (Figure 7-6). Pull the drum and hub assembly from the spindle as shown (Figure 7-7), protecting the bearings from dirt and being careful not to drop the drum.

As explained in Chapter 6, nondriving wheel bearings, whether on disc or drum brake assemblies, must be cleaned, inspected, and repacked with the correct grease at regular intervals. The inner bearing is held in the wheel hub by the grease seal. Lay the wheel on the floor with the brake drum side down. Place a clean rag under the hub area to catch the inner bearing. Using a mallet and a brass or wooden drift, tap out the seal and inner bearing as shown (Figure 7-8). Bearing inspection, repacking, and installation are similar to that covered in Chapter 6 for nondrive axle disc brake wheel hubs.

Some cars have demountable front drums. These may be removed without removing the hub or disturbing the front wheel bearing.

WARNING: Never step on the brake pedal while a brake drum is off or the wheel cylinder will pop apart.

Cleaning and Inspection

Always begin drum brake work by performing a thorough and safe cleaning and inspection procedure. To do this, use an industrial vacuum cleaner equipped with a HEPA filter to remove all dust

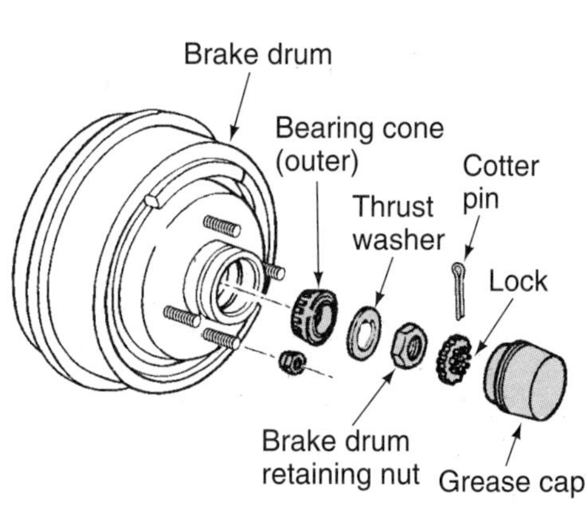

Figure 7-6 Parts disassembled to remove the hub on older front drum brake vehicles (Courtesy of Chrysler Corporation)

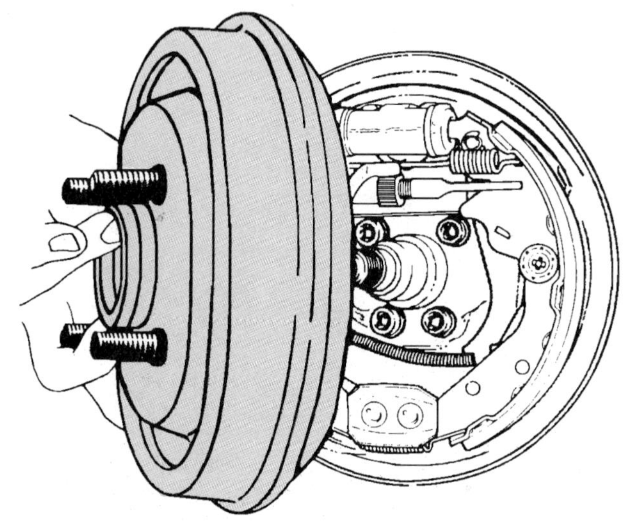

Figure 7-7 Removing the brake drum (Courtesy of Chrysler Corporation)

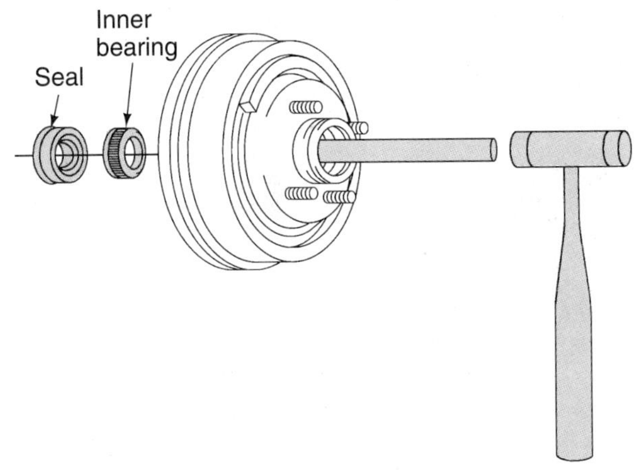

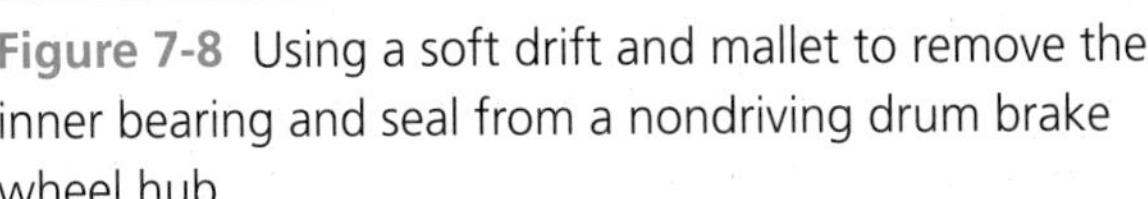
Figure 7-8 Using a soft drift and mallet to remove the inner bearing and seal from a nondriving drum brake wheel hub

Figure 7-9 Cleaning the drum brake assembly

from the rear brake backing plate and the interior of the brake drum. As an alternative, use a water-dampened cloth or water-based solution (Figure 7-9). As covered in Chapters 1 and 2, equipment is commercially available to perform washing functions on brake parts. Wet cleaning methods are preferred as they prevent asbestos fibers from becoming airborne.

Thoroughly clean the backing plates, struts, levers, and other metal parts to be reused. Examine the rear wheels for evidence of oil or grease leakage past the wheel bearing seals. Such leakage could cause brake failure and indicates the need for additional service work.

Servicing Brake Drums

Brake drums must be carefully inspected and pulled from service if damaged. The first step in assessing the drum is to visually inspect the brake shoes installed on the vehicle (Figure 7-10). Inspecting the brake shoes often reveals defects in the drums. For example, if the linings on one

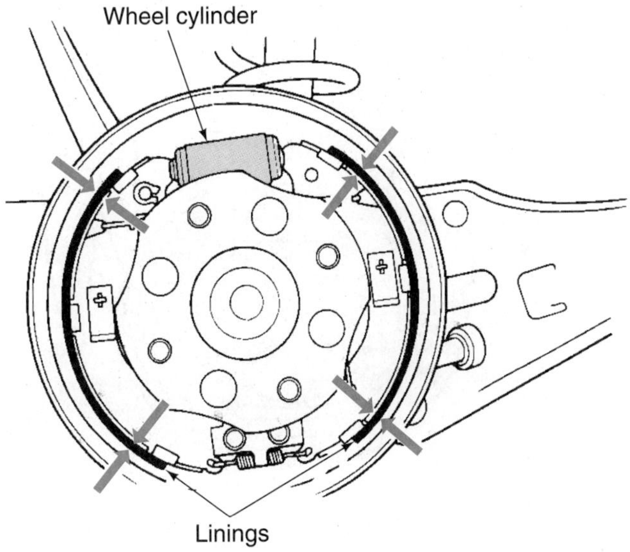

Figure 7-10 Inspecting linings with drum removed (Courtesy of Honda Motor Co., Ltd.)

Figure 7-11 Inspecting a drum for scoring

wheel are worn more than the other, that drum may be scored or rough. Uneven wear from side to side on any one set of shoes can be caused by a tapered drum. Linings worn badly at the toe or heel may indicate an out-of-round drum.

Thoroughly wet clean the drums. If the drums have been exposed to leaking oil or grease, thoroughly clean them with a non-oil-based solvent after washing to remove dust and dirt. Locate the source of the oil or grease leak and correct the problem before reinstalling the drums.

A brake drum surface that is highly polished can cause brake lockup or noise.

Visually inspect the drum braking surface for scoring by running your fingernail across the surface (Figure 7-11). Any large score marks mean that the drum will have to be resurfaced or replaced.

Drum Measurements

Special Tool

Brake drum micrometer

Every drum inspected must be measured using a drum micrometer or gauge to make sure that it is within the safe oversize limits. The brake drum micrometer is set to the drum diameter and measures the amount and type of wear. Take measurements at the open and closed edges of the friction surface and at right angles to each other (Figure 7-12). Measure the diameter at various points of 45° around the circumference and at the bottom of the deepest groove.

Drums with a taper or out-of-roundness exceeding 0.006 in. are unfit for service and must be turned or replaced. If the maximum diameter reading measured from the bottom of the deepest

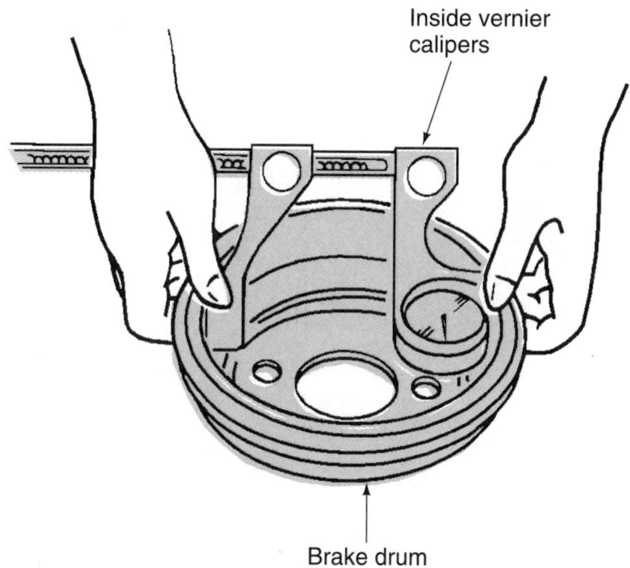

Figure 7-12 Measuring drum inside diameter (Courtesy of Honda Motor Co., Ltd.)

groove exceeds the new drum diameter by more than 0.060 in., the drum cannot be refinished and must be replaced. Drums that are smooth and true but exceed the new diameter by 0.090 in. or more must also be replaced.

If the drums are smooth and true and within safe limits, smooth up any slight scores by polishing with fine emery cloth. If scoring or light grooves cannot be removed by hand, the drum must be turned or replaced. Even slightly rough surfaces should be turned to ensure a true drum surface and to remove any possible contamination on the surface from previous brake linings, road dust, and so on.

Types of Drum Damage

Brake drums work like a heat sink. It is their job to absorb heat and dissipate it into the air. As drums wear from normal use or are thinned out by refinishing, the amount of metal available to absorb and release heat is reduced. As a result, the drums operate at increasingly higher temperatures. The drum's structural strength is also weakened by the loss of metal. Braking forces can distort the drum's shape or lead to cracking or other problems.

Scored Drum Surface. The most common cause of drum scoring (Figure 7-13A) is when road grit or brake dust becomes trapped between the brake lining and drum. Glazed brake linings that have been hardened by high heat or inferior linings that are very hard can also groove the drum surface. Excessive lining wear that exposes the rivet head or shoe steel will score the drum surface. If the scoring is not too deep, the drum can be refinished.

Bell-Mouthed Drum. Bell-mouth is shape distortion caused by extreme heat and braking pressure (Figure 7-13B). It is most common on wide drums that are weakly supported at the outside of the drum. Bell-mouth makes full drum-to-lining contact impossible, so braking power is reduced. Drums must be refinished.

Concave Drum. A concave wear pattern (Figure 7-13C) is the result of a distorted shoe that concentrates braking pressure on the center of the drum.

Convex Drum. A convex wear pattern is the result of excessive heat or an oversized drum, which allows the open end of the drum to distort.

Hard Spots. Hard spots or chisel spots in the cast iron surface result from a change in metallurgy caused by heat. They appear as small raised areas. Chatter, pulling, rapid wear, hard pedal, and noise may occur. Hard spots can be ground out of the drum, but since only the raised surfaces are removed, they can reappear when heat is reapplied. Drums with hard spots must be replaced.

Threading. An extremely sharp or chipped tool bit or a lathe that turns too fast can literally cut a thread into the drum surface. During brake application, the shoes ride outward on the thread, then

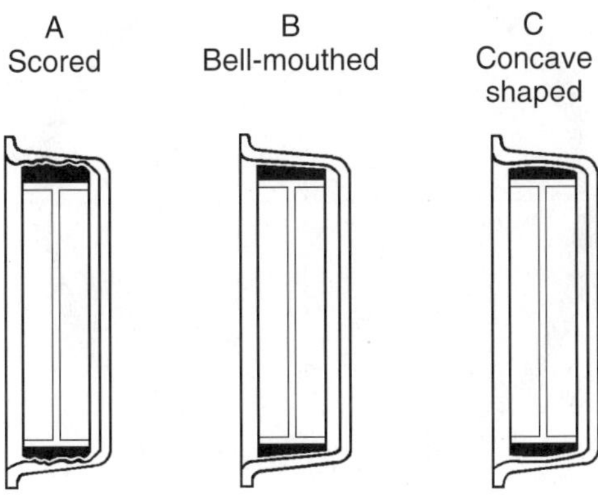

Figure 7-13 Typical brake drum defects (Courtesy of Raybestos Division)

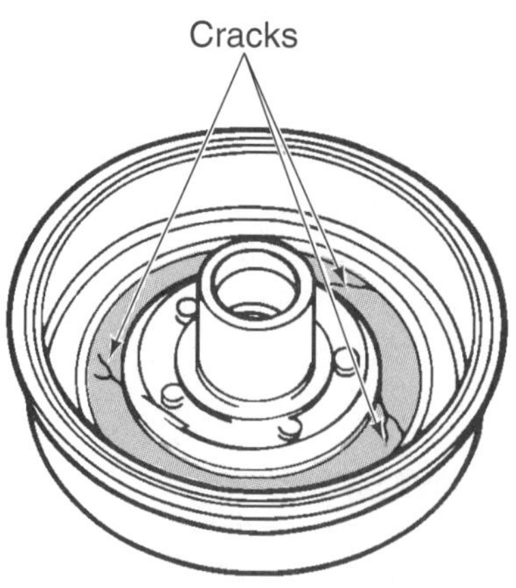

Figure 7-14 Typical location of drum brake cracks (Courtesy of Raybestos)

snap back with a loud crack. Threading can also cause faster lining wear and interfere with shoe alignment during braking action. To avoid threading the drum surface, use a properly shaped bit and a moderate-to-slow lathe speed.

Heat Checks. Unlike hard spots, heat checks are visible on the drum surface. They are caused by high operating temperatures. Heat-checked drums may have a bluish/gold tint, which is another sign of high operating temperatures. Hardened carbide lathe bits or special grinding attachments are available through lathe manufacturers to service these conditions. Excessive damage by heat checks or hard spots requires drum replacement.

Cracked Drum. Cracks in the cast iron drum are caused by excessive stress. They may appear anywhere, but they are most common in the vicinity of the bolt circle or at the outside of the flange (Figure 7-14). Fine cracks are often hard to see and often do not appear until after machining. Any crack, no matter how small, means the drum must be replaced.

Out-of-Round Drums. Slightly out-of-round drums usually appear good to the eye, but the problem causes pulling, grabbing, and pedal vibration or pulsation. An out-of-round or egg-shaped condition (Figure 7-15) is often caused by heating and cooling during normal brake operation. To test for an out-of-round drum before the drum is removed, adjust the brake to a light drag and feel the rotation of the drum by hand. Any areas of hard or no drag may indicate a problem. Remove

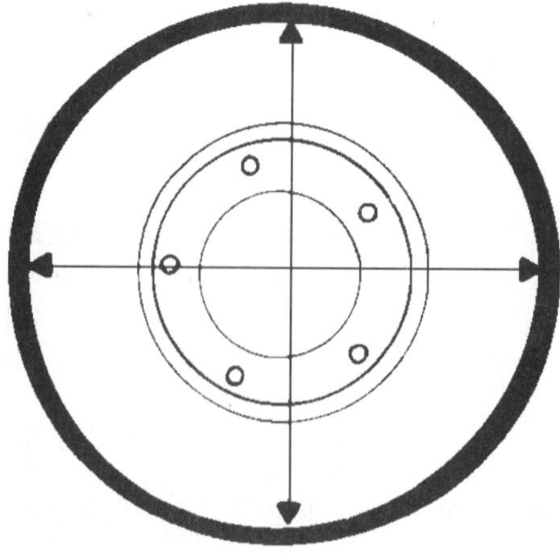

Figure 7-15 An out-of-round drum shows variations in drum diameter

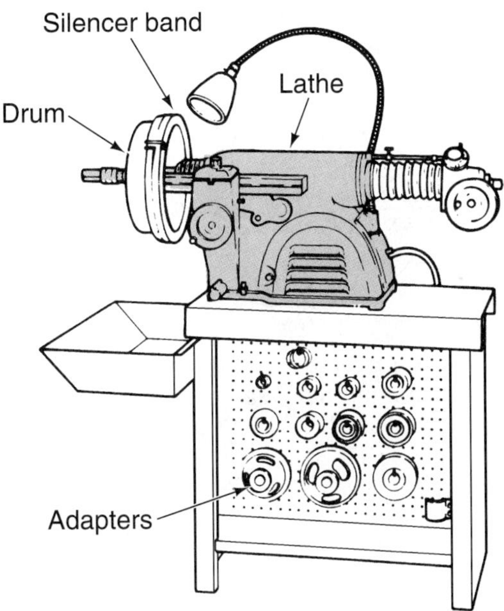

Figure 7-16 Drum mounted on brake lathe (Courtesy of Raybestos)

the drum and measure it at several points to determine the degree of eccentric distortion. A brake drum that is out of round enough to cause vehicle vibration or roughness when braking should be refinished. Remove only enough stock to return the brake drum to roundness.

Brake Drum Refinishing

Brake drums with moderate-to-severe scoring or other defects can be refinished either by turning (cutting) or grinding on a brake lathe (Figure 7-16). The best results are obtained by turning drums with a very fine feed. Only enough metal should be removed to obtain a true, smooth friction surface. If too much metal is removed from a drum, the following unsafe conditions can result:

- ❑ Brake fade caused by the thin drum being unable to absorb the heat produced during braking
- ❑ Poor and erratic braking due to distortion of the drum weakened
- ❑ Noise caused by a thin drum vibrating during operation
- ❑ Drums cracking or breaking during a very hard brake application

SERVICE TIP: Do not refinish the brake drum to remove minor score marks. Remove them with sandpaper.

If one drum must be machined to remove a major defect, the other drum on the same axle set must also be machined in the same manner and to the same diameter so that braking will be equal at both wheels.

Drum Discard Dimension

Brake drums manufactured after December 31, 1970 are stamped with a discard dimension, or maximum inside diameter (Figure 7-17). This is the allowable wear dimension and not the allowable machining dimension. There must be 0.030 in. left for wear after turning the drums. That is, the refinished drum diameter must be at least 0.030 in. (90.76 mm) less than the discard diameter. If this dimension is exceeded, the drum will wear beyond its maximum allowable diameter during normal operation.

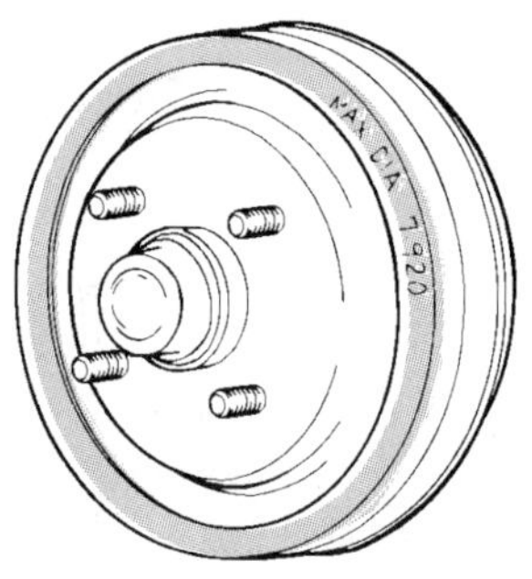

Figure 7-17 Brake drum maximum diameter stamped into drum (Courtesy of Raybestos)

Special Tools

Brake lathe

Mounting adapters

Always use sharp cutting tools or bits and install the vibration dampening device supplied with the lathe to help eliminate cutting chatter.

The ideal speed for refinishing drum surfaces is a spindle speed of 150 rpm. As with rotor finishing covered in Chapter 6, make a series of shallow cuts rather than one deep cut. Cross feed speeds for roughing cuts should range from 0.006 in.–0.010 in. (0.15 mm–0.20 mm) per revolution. The cross feed speed for the final finish cut should be no more than 0.002 in. (0.05 mm) per revolution.

Refinishing a brake drum increases the inside diameter of the drum and changes the lining-to-drum fit. When the brake drum maximum inside diameter is exceeded either through wear or refinishing, the brake drum must be replaced.

Rules for Drum Refinishing

When machining a drum, follow the equipment instructions for the specific tool you are using. However, the following general rules of drum machining should always be kept in mind:

Never pick up the drum with your fingers on the newly turned surface. Do not drop the drum as this can cause it to go out of round.

1. Be sure to use the correct lathe adapters and ensure the adapters are in good condition.
2. Never grind a lathe cutting bit to a sharp point; the tip of the tool must be rounded to prevent threading.
3. With a hub-mounted drum, torque all wheel nuts to the same tightness specification before machining. This applies whether the wheel is mounted or dismounted. Properly torqued nuts reinforce the drum and reduce the chances of distortion or chatter.
4. Some front drums can be removed without the hub. These drums must be turned with their individual hubs attached and not with a lathe adapter.
5. Refinish left-hand and right-hand drums in axle set pairs—do not machine one side only. Drum diameters must be within 0.005 in. of one another.
6. Different lathe models allow finishing the drum surface by either a fine cut, grinding, or honing to remove spiral tool marks.
7. Remove the ridge on the inner and outer edge of the drum friction surface before turning the entire drum. This prevents damage to the cutting tool.
8. If the cutting-bit tool chatters as it passes over hard spots in the drum surface, grind the surface or discard the drum.
9. After turning, keep the drum mounted on the lathe, and deburr it with no. 80 grit sandpaper (do not use emery cloth) to remove all minute rough and jagged surfaces.

SERVICE TIP: On drums with integral bearing hubs, check both the inner and outer bearings and races for damage. Replace the damaged bearing. If the bearings are in good condition, clean and repack them with the recommended grease. Always install new grease seals.

SERVICE TIP: Studs on the drums must be in good condition or should be replaced if damaged or missing.

10. If the drum does not clean up when turned to its maximum machining diameter, it must be replaced. The removal of more metal will affect the ability of the drum to dissipate heat and can cause drum distortion.

Some replacement brake drums are semifinished. A semifinished drum may require additional machining to obtain the proper dimensional specifications and surface finish.

Fully finished drums do not require additional machining unless it is needed to match the diameter of an old drum on the same axle set. New drums are also protected with a rustproofing coating that must be thoroughly cleaned off the friction surfaces. A volatile, nonoil base solvent, such as carburetor cleaner or lacquer thinner, should be used to remove all traces of the coating.

Cleaning Refinished Drums

The surface of a freshly refinished drum contains millions of tiny particles. These particles can remain free on the surface or also lodge into the open pores of the cast iron drum. If these metal particles are not cleaned away, they will become imbedded in the brake lining. Once the brake lining becomes contaminated with metal shavings, it becomes a fine grinding stone and soon scores the drum.

Remove metal particles from the drum surface by wet washing. **Do not** blow out the drum with air pressure. If the drum is part of the front hub assembly, as with older vehicles, the wheel bearings and the grease seals **must** be removed from the drum before cleaning. Be very careful to avoid contaminating the wheel bearing grease in the front hub with the cleaning solution. To avoid this potential problem, all old grease can be removed from the hub before cleaning the drum. After cleaning, the wheel bearings can be repacked with fresh grease.

Wash the brake drum thoroughly with hot water and wipe with a lintfree rag. Use compressed air to thoroughly dry the clean drum.

Wipe the inside of the brake drum (especially the newly finished surface) using a lintfree white cloth dipped in denatured alcohol or a commercial brake cleaning solvent. Use a cleaner that does not leave a residue. This operation should be repeated until dirt is no longer visible on the wiping cloth (Figure 7-18). Allow the drum to dry before reinstalling it on the vehicle.

Figure 7-18 Cleaning the drum after refinishing

Drum Installation

After completing your inspection and/or servicing of the drum brake assembly, reinstall the rear brake drums over the wheel studs. If the drum has an alignment tang or a hole for a locating screw, make sure it is lined up with the hole in the hub flange. If retaining clips or speed nuts are used to hold the drum in position, install these.

Torque values specified in shop manuals are usually for dry, unlubricated fasteners.

At a nondrive axle, install the inner bearing and grease seal in the hub, then install the hub and drum assembly, outer bearing, thrust washer, and nut on the spindle. Adjust the wheel bearing as described later in this chapter. Install the nut lock (if used) and cotter pin. Install the wheels and make service and parking brake adjustments as required.

Servicing Drum Brake Linings

With the drum removed, visually and manually inspect the brake shoes. Their condition often reveals defects in the drums. If the linings on one wheel are worn more than those on the other, it may indicate a rough drum. Uneven wear from side to side on any one set of shoes may be caused by a tapered drum. If some linings are worn badly at one end, an out-of-round drum may be the problem.

Replace shoes and linings in axle sets only.

Check the thickness of the brake shoe linings (Figure 7-19). The linings must be replaced when they are worn beyond the manufacturer's wear specifications at the thinnest point. Typically, this is to within 1/32 in. (0.79 mm) of the rivet heads. The shoes should be replaced when lining wear approaches these limits. Linings must also be replaced when covered with grease, oil, or brake fluid. Replace the brake shoe and lining in axle sets only. Never replace just one shoe of a brake assembly. Replace both leading and trailing shoes.

Before replacing the lining, check the drum diameter as described earlier in the chapter to determine that it is within specification. If the braking service diameter exceeds the specification, the drum must be replaced.

On single-anchor duo-servo drum brakes, the shorter lining always goes toward the front of the vehicle.

Inspect the brake shoes and linings for distortion, cracks, or looseness. If any of these conditions are found, discard the brake shoe and lining assembly. Never attempt to service a damaged brake shoe and lining.

Operating a vehicle for any length of time with worn brake shoes and linings will quickly result in a scored brake drum. Always replace parts that are near the end of their service life. Never replace

Figure 7-19 Checking the brake shoe lining using a depth gauge

linings on one brake assembly without replacing those on the opposite wheel. Service the linings in axle sets only and never change only one brake shoe and lining in a given brake assembly.

Brake Shoe and Lining Removal

Always review the shop manual illustrations and service procedures for the brake assembly you are servicing. While most drum brakes contain similar components that generally work in the same manner, there are slight differences between vehicles. Figure 7-20 shows the components of a typical nonservo, leading-trailing-shoe drum brake assembly. Figure 7-21 illustrates a duo-servo drum brake assembly.

Check the condition of the brake shoes and linings, brake shoe adjusting screw spring, brake shoe hold-down spring, and brake drum for signs of overheating. If brake shoes and linings have a slight blue coloring, this indicates overheating. In this case, the brake shoe adjusting screw springs and brake shoe hold-down springs should be replaced. Overheated springs lose their tension and could allow a new lining to drag and wear prematurely if not replaced. Inspect the backing plate and wheel cylinder for signs of damage and leakage.

Install all new components included in the repair kit used to service drum brakes and lubricate parts as specified. Do not use lubricated compressed air on brake parts because damage to

Special Tools

Brake spring remover and installer

Brake spring pliers

Plier-type spring remover and installer

Classroom Manual
Chapter 7, page 141

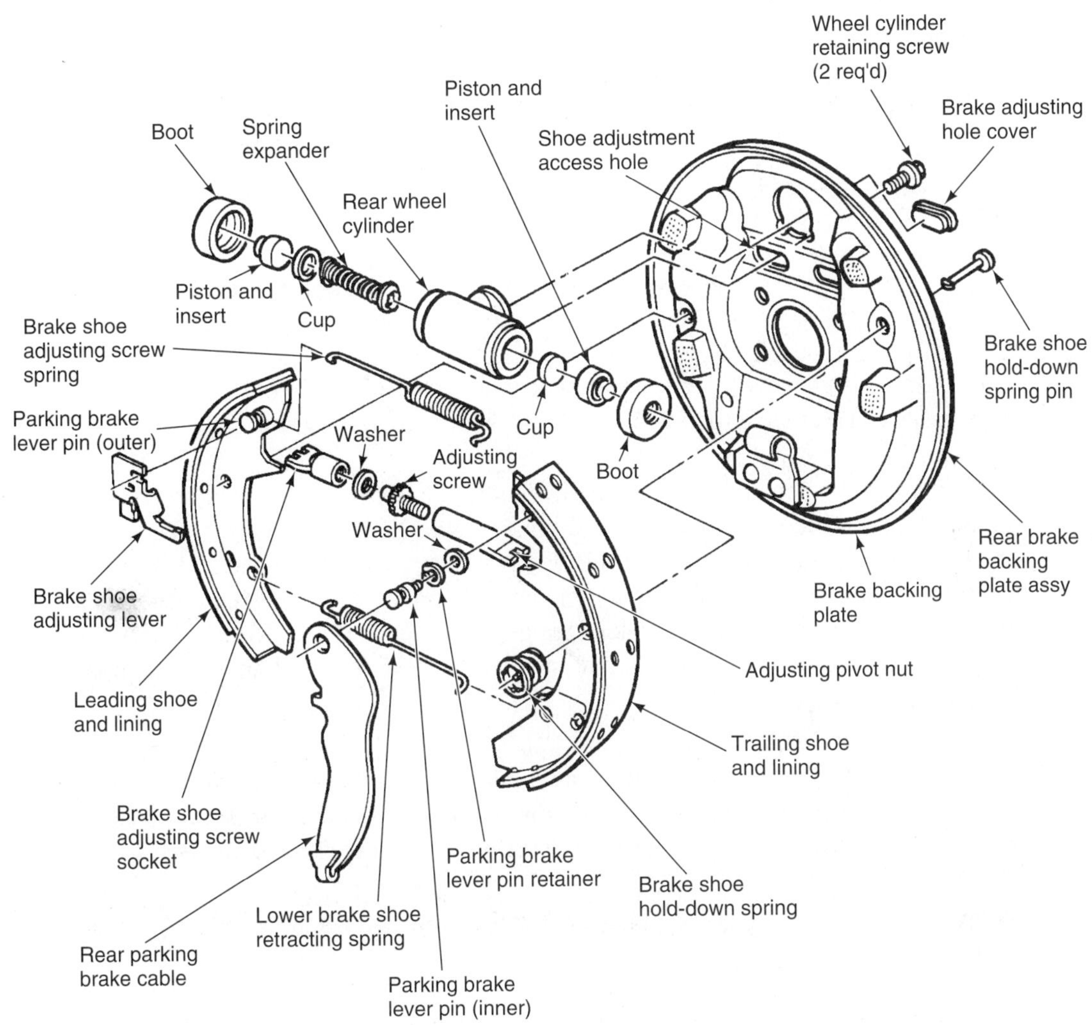

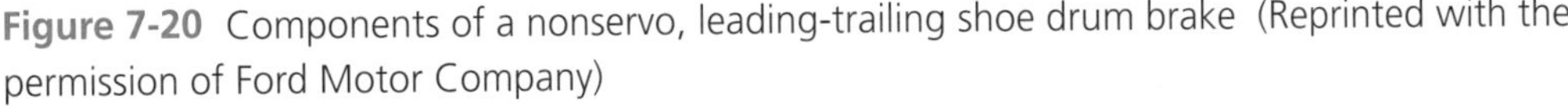
Figure 7-20 Components of a nonservo, leading-trailing shoe drum brake (Reprinted with the permission of Ford Motor Company)

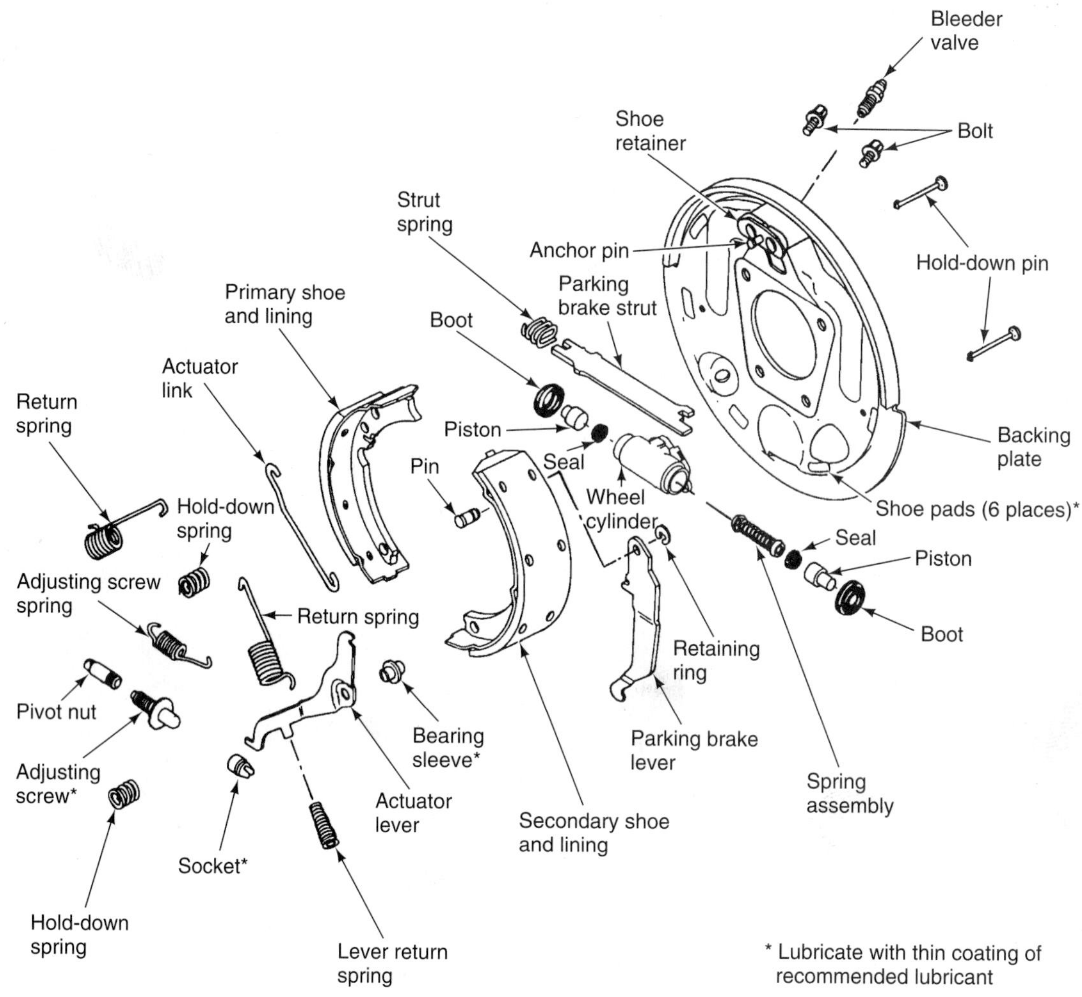

Figure 7-21 Components of a duo-servo, shoe drum brake (Courtesy of Chevrolet Motor Division, General Motors Corporation)

rubber parts may result. Remember that if any hydraulic component is removed or disconnected, it will be necessary to bleed all or part of the brake system. The following brake shoe lining replacement procedure is typical for many vehicles. During the lining removal and assembly, you must also perform any needed service on individual components and subassemblies, such as the backing plate, adjuster screw, wheel cylinder, springs, and other hardware. Service procedures for these individual components are covered in the sections that follow the relining procedure.

Whenever the rear brake shoe and lining are removed, the parking brake rear cable and conduit tension should be checked.

To remove brake shoes and linings:

1. Raise the vehicle on a hoist.
2. Remove the wheel, tire, wheel hub, and brake drum assembly.
3. Remove the rear parking brake cable from the parking brake lever (Figure 7-22).
4. Remove the two brake shoe hold-down springs and brake shoe hold-down spring pins (Figure 7-23).
5. Remove the upper return spring (adjusting screw spring) using the proper spring removal tool (Figure 7-24).
6. Lift the brake shoes and linings, brake shoe hold-down springs, and adjuster off the backing plate and rear wheel cylinder assembly (Figure 7-25) taking care not to bend the brake shoe adjusting lever.

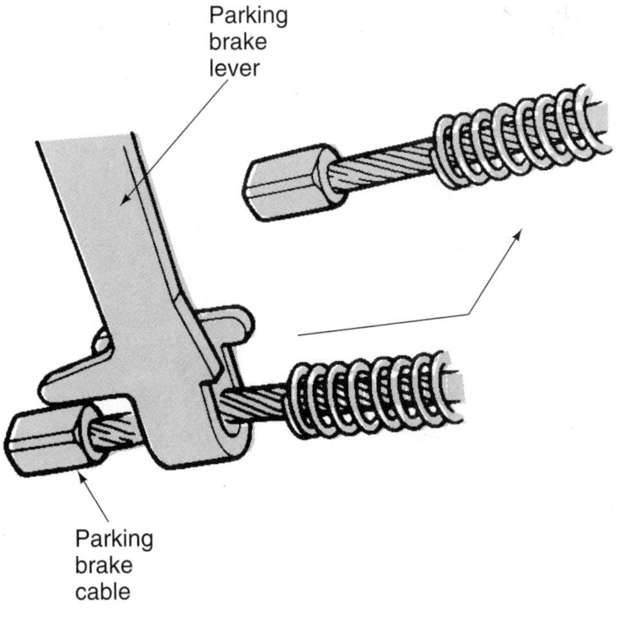

Figure 7-22 Disconnecting the parking brake cable from the parking brake lever (Courtesy of Honda Motor Co., Ltd.)

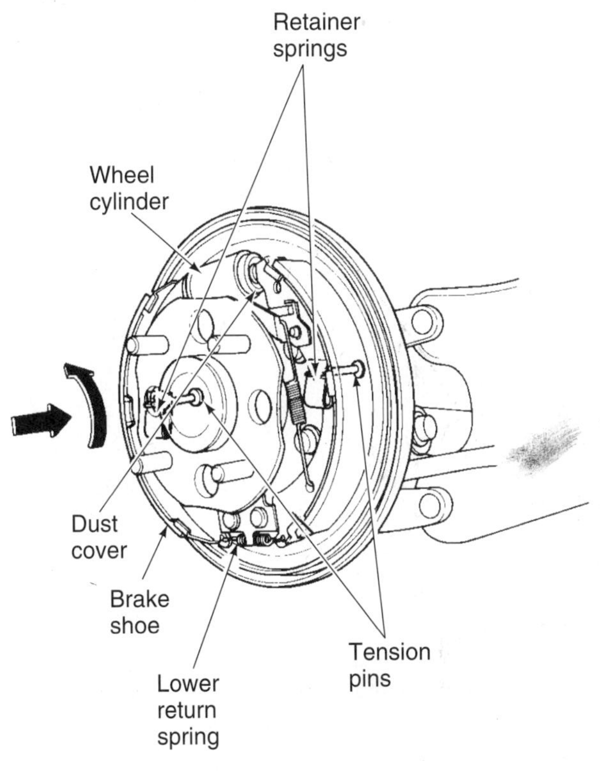

Figure 7-23 Removing tension pins and retainer springs (Courtesy of Honda Motor Co., Ltd.)

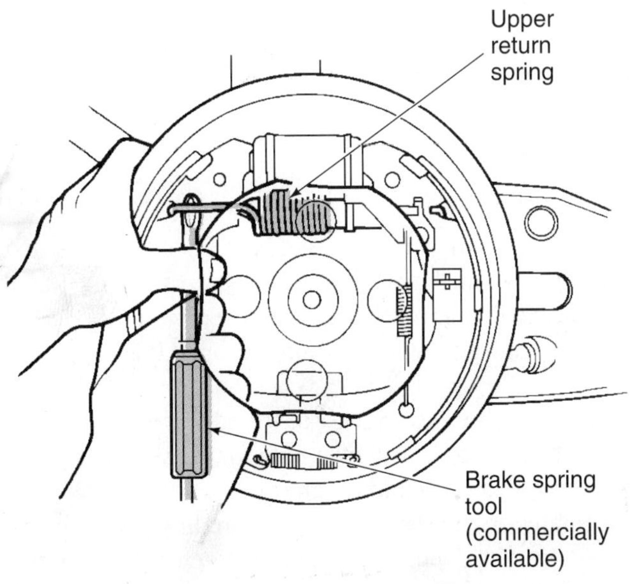

Figure 7-24 Removing upper return spring using a brake spring tool (Courtesy of Honda Motor Co., Ltd.)

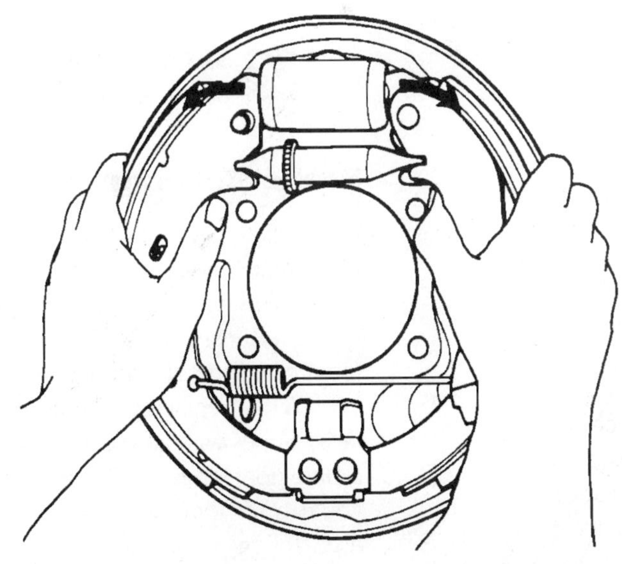

Figure 7-25 Removing brake shoes and linings off of rear backing plate (Reprinted with the permission of Ford Motor Company)

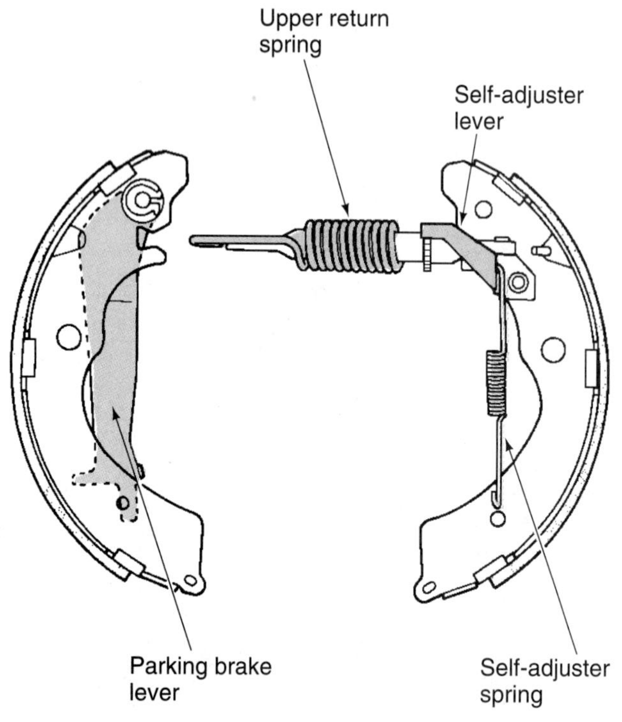

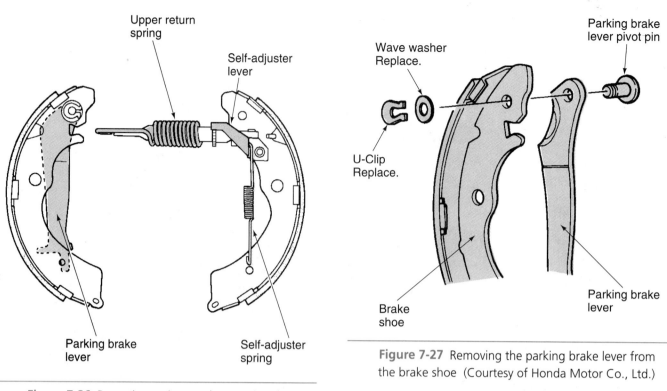

Figure 7-26 Removing springs and separating the brake shoes (Courtesy of Honda Motor Co., Ltd.)

Figure 7-27 Removing the parking brake lever from the brake shoe (Courtesy of Honda Motor Co., Ltd.)

7. Remove the brake shoe retracting springs from the lower brake shoe and lining attachments and upper shoe-to-adjusting lever attachment points. This will separate the brake shoes and linings and disengage the adjuster mechanism (Figure 7-26).
8. Remove the parking brake lever pin retainer and spring washer and remove from trailing shoe (Figure 7-27).

SERVICE TIP: Perform service operations on a clean bench, free from all mineral oil materials.

Brake Relining

Drum brake linings that are worn to within 1/32 in. of a rivet head or that have been contaminated with brake fluid, grease, or oil must be replaced. Two methods are used to secure the lining to the shoe: bonding and riveting (Figure 7-28). Bonding linings to the shoe surface requires the use of special adhesives and curing ovens. Riveting is more practical for most shops. It is faster and requires no special equipment.

Check wheel cylinders for fluid leakage whenever relining drums.

When replacing the linings, inspect the brake shoes for cracks, distorted surfaces, or looseness. Shoes with any of these conditions must be replaced. The rivet holes in the shoe should be round, not elongated. Clean the shoes and remove all burrs and rough spots. Discard the shoe if you can see that the new lining will not sit tightly on the shoe. If a specific relining kit is being used, follow its instructions and use all components provided. There may be differences in the size of primary and secondary shoe linings due to the fact that the secondary lining must provide more braking power. So do not accidentally switch linings or change the position in which they mount on the shoe. The secondary shoe lining is normally a full-length lining that is centered on the shoe. The primary lining is shorter and may be positioned at high, center, or low positions on the shoe.

Most friction linings are edge-marked or branded (Figure 7-29). The first marking usually identifies the lining manufacturer. The second marking identifies the material used to make the lin-

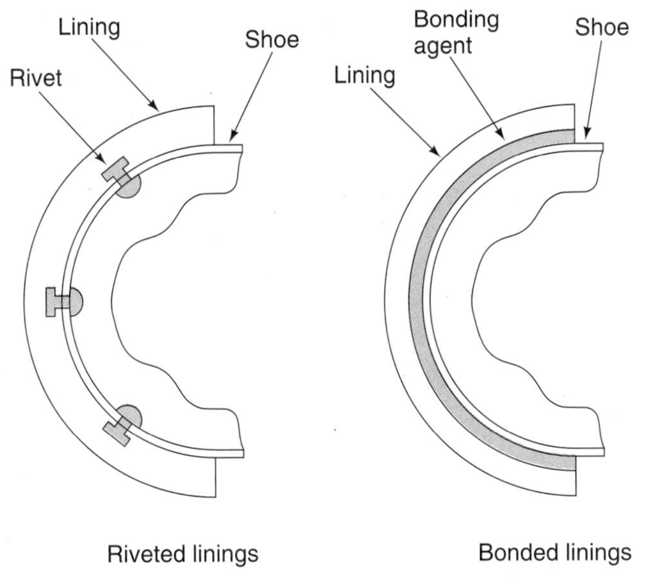

Figure 7-28 Linings are attached to the shoe using rivets or a bonding adhesive

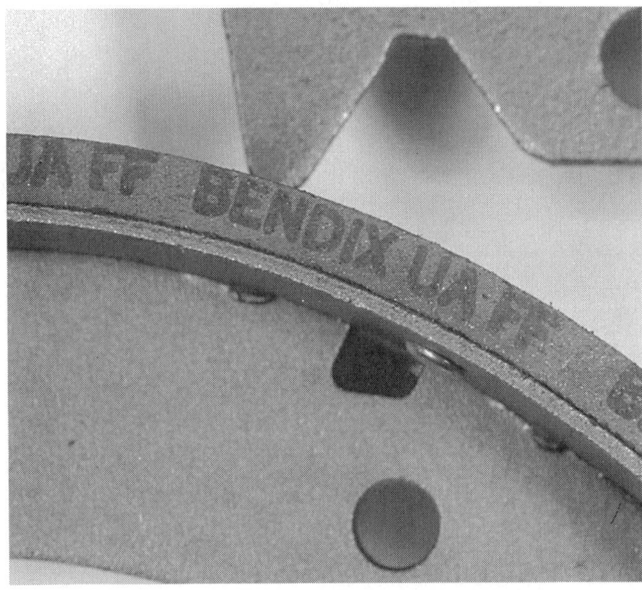

Figure 7-29 Edge-branding on the lining indicates the manufacturer, lining material, and coefficient of friction letter code.

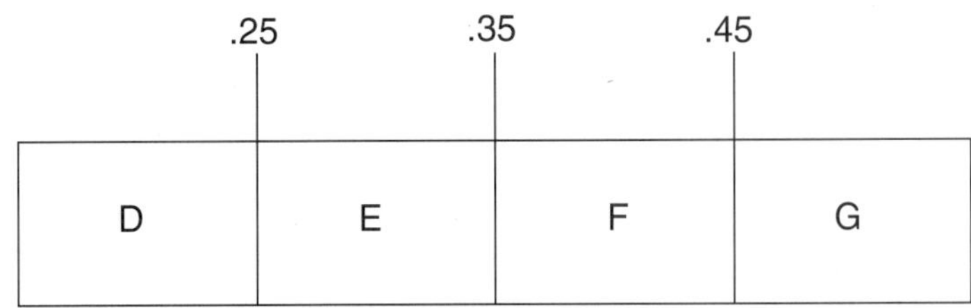

Figure 7-30 Coefficient of friction letter code scale

ing and the third marking identifies the friction code for that material. The letters "FE," "FF," and the like refer to the friction characteristics of the lining material (Figure 7-30). They do not reflect the quality of the lining. Each application has its own requirements and optimum coefficient of friction. Use the replacement lining specified for the vehicle by the vehicle manufacturer or by a quality aftermarket supplier. If linings with low-friction characteristics are used, excessive pressure on the brake pedal will be needed to stop the vehicle. If the coefficient of friction is too high, the brake will grab or pull.

Most brake linings sold today are preground so they are slightly thicker at the center than at the edges (Figure 7-31). Grinding in the field or arcing the lining to fit the shoe is no longer needed.

Inspect the replacement shoes to ensure that there are no burrs on the edges of the shoes where they will come in contact with the pads.

To install the replacement linings, begin by washing the rear brake shoes thoroughly with a clean solvent. Dry the shoe thoroughly. Remove all burrs or rough spots. Position a new lining on the cleaned brake shoe. Install and secure the fastening rivets beginning at the center of the new lining and working alternately outward to the ends. Because modern replacement linings are preground, no additional grinding is needed.

Once the new lining is installed, check the clearance between brake shoe and lining. The lining must seat tightly againt the shoe with no more than 0.008-in. (0.020-mm) clearance between any two rivets.

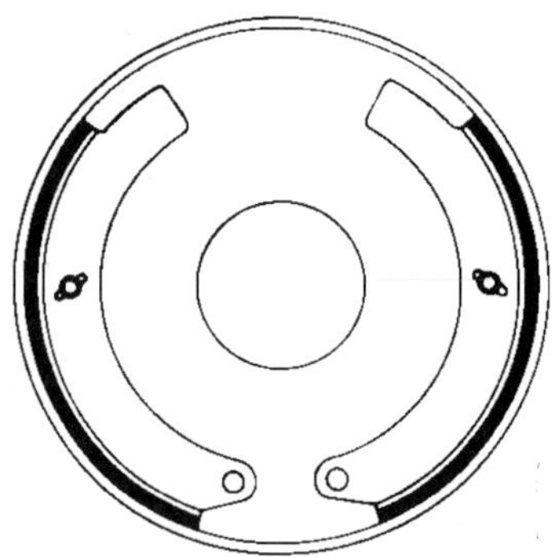

Concentric grind

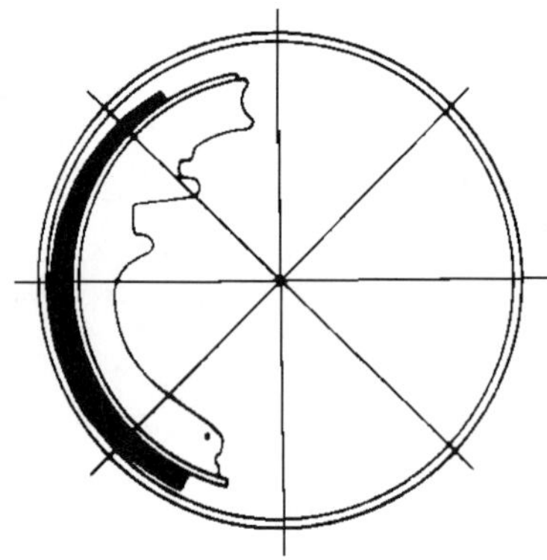

Typical OEM offset grind
(Furnished on new vehicles)

Figure 7-31 Concentric and eccentric grind linings

CUSTOMER CARE: Caution your customers to avoid unnecessary hard braking during the first 500 miles or so after relining. This is a critical break-in period for linings. A little care early on can dramatically increase the life of the linings.

Brake Shoe Installation

To install brake shoes and linings:

1. Lightly coat the surface of the parking brake lever pivot pin with brake cylinder grease and insert the pin into the brake shoe. Install the lever on the pivot pin using a new washer and clip (Figure 7-32).
2. Lightly coat the areas on the backing plate where the brake shoes contact the plate. Use the grease recommended in the service manual. Assemble the brake shoe adjuster as required. Some adjusters such as the one shown in Figure 7-20 require a stainless steel washer. Lightly coat the threaded areas of the brake shoe adjusting screw and any other points specified in the service manual with a good quality grease.
3. Turn the brake shoe socket all the way down on the adjusting screw, then back it off about one-half turn.
4. Place the trailing shoe on the rear brake backing plate and attach the rear parking brake cable. Next, position the leading shoe on the backing plate and attach lower brake shoe adjusting screw spring to the brake shoes.

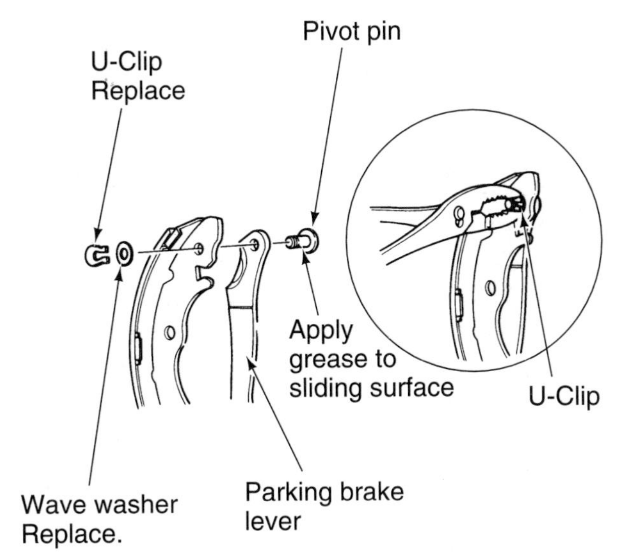

Figure 7-32 Installing the parking brake lever to the brake shoe (Courtesy of Honda Motor Co. Ltd.)

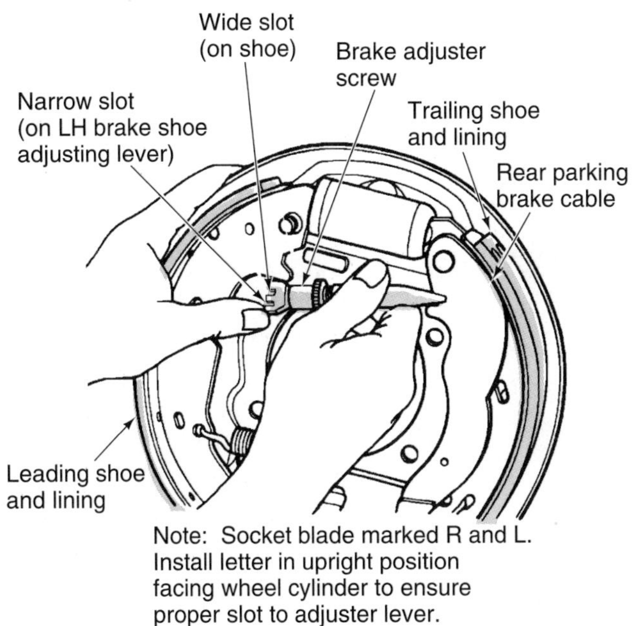

Figure 7-33 Installing adjuster assembly into slots in brake shoes and linings (Reprinted with the permission of Ford Motor Company)

5. Install the adjuster assembly. On some drum brakes, the adjuster assembly fits to the slots in the brake shoes and linings. The socket end fits into a slot in the leading shoe while the slot in the adjuster nut is fitted into slots in the trailing shoe and rear parking brake cable (Figure 7-33).
6. Fit the brake shoe adjusting lever onto the pin on the leading shoe and into the slot in the brake shoe adjusting screw socket (Figure 7-34).

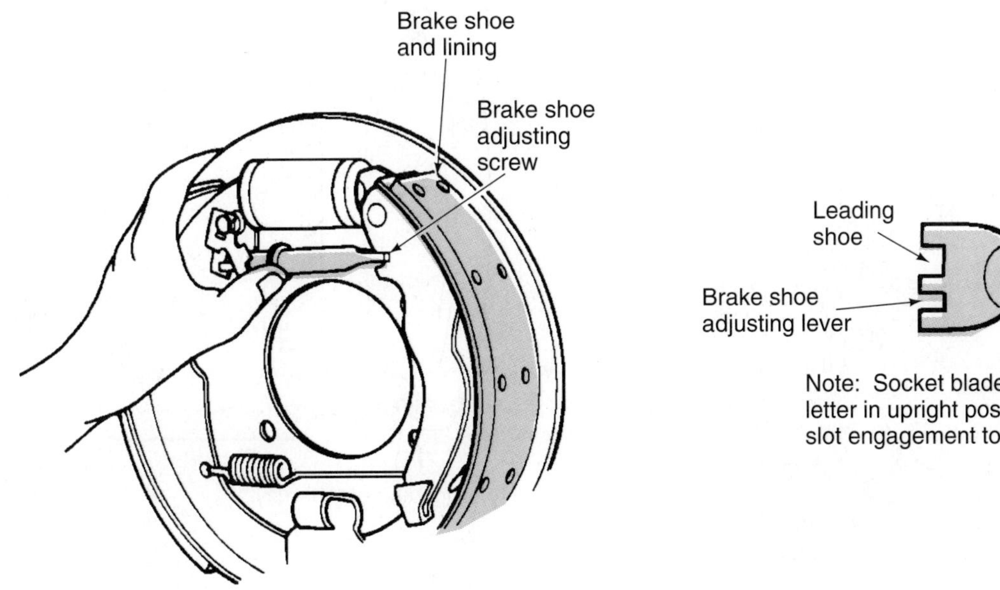

Figure 7-34 Properly installed brake shoe adjuster lever and adjusting screw (Reprinted with the permission of Ford Motor Company)

7. Using the spring installation tool, intall the upper brake shoe adjusting screw spring in a slot in the trailing shoe and a slot in the adjuster lever (Figure 7-35). Check to see that the brake shoe adjusting lever contacts the star and the adjuster assembly.
8. To complete the installation, install the brake shoe hold-down spring pins, brake shoe hold-down springs, and retainers (see Figure 7-23).

Special Tool

Brake drum/shoe clearance gauge

Brake Shoe and Lining Adjustment

After completing any brake service work, the vehicle must have a firm brake pedal before it can be safely moved. When properly adjusted, the brakes will not drag and the wheel will turn freely.

Begin the rear brake shoes and linings adjustment by checking the parking brake cable adjustment. The adjuster mechanism must operate freely with the brake shoes and linings centered on the rear brake backing plate. To set the shoe to lining adjustment, apply a small quantity of shop specified grease to the areas where the shoes contact the backing plate.

WARNING: Whenever you apply grease or lubricant to the brake assembly, do not get lubricant on the linings.

Use a brake adjustment gauge to measure the inside diameter of the drum braking surface. Adjust the brake shoe and lining diameter to fit the gauge (Figure 7-36). Turning the star wheel, adjust the shoe and lining diameter to be approximately 0.030 in. to 0.050 in. (0.76 mm to 1.5 mm) less than the inside drum diameter for each rear wheel. See the service manual for exact specifications.

Install the brake drum, then install the tire and wheel assembly. Apply the brake pedal several times, using moderate to firm foot pressure. Be sure there is a firm pedal. Finally, test the brake performance by making several stops at varying speeds.

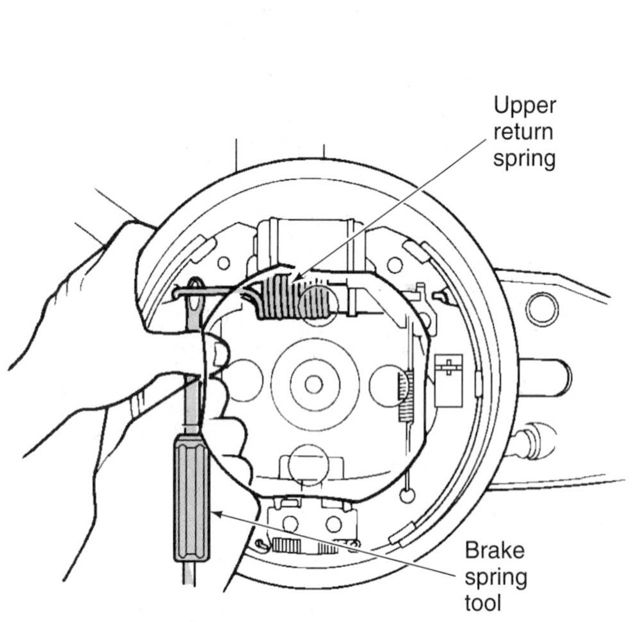

Figure 7-35 Typical installation position for the upper return spring (Courtesy of Honda Motor Co., Ltd.)

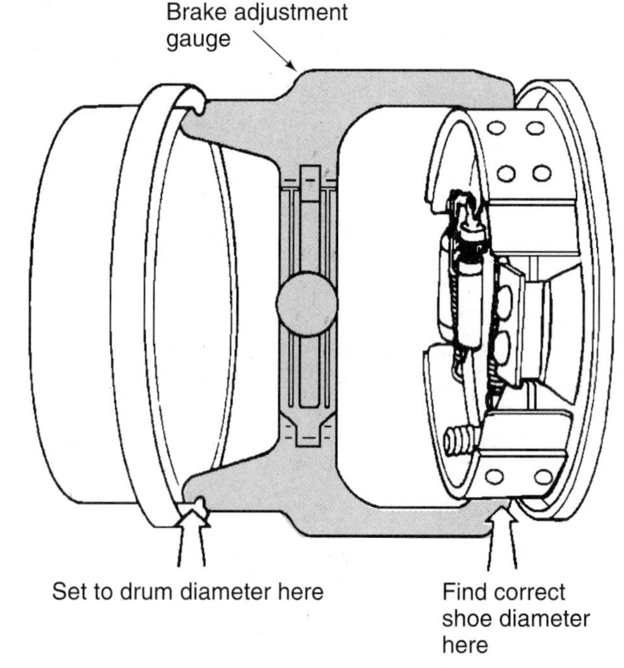

Figure 7-36 Making shoe and lining adjustment using an adjustment gauge (Reprinted with the permission of Ford Motor Company)

SERVICE TIP: A high brake pedal after lining replacement does not indicate a job well done. It may indicate tight clearances that can lead to seating problems with the linings. If this appears to be a problem, offer the customer a free lining-to-drum adjustment after 750 to 1,000 miles.

Servicing Individual Components

The following sections outline service for individual drum brake assembly components and subassemblies.

Springs

Check all return and hold-down springs for spread or collapsed coils and twisted, bent, or damaged shanks (Figure 7-37). As with most metal components, discoloration is a sign of brake overheating. Overheated springs lose some of their tension and should be replaced.

Some brake technicians check brake spring tension by the drop method. This method is not strictly scientific so the results are not always correct. However, it is often used as a first step in diagnosing brake springs. To perform the test, drop the spring on a clean concrete floor. If the spring bounces with a chunky sound, it is good. If, on the other hand, it bounces with a tinny sound, the spring is "tired" and should be replaced.

Because self-adjusters and parking brake linkages can be complicated, service one side at a time and use the other for reference.

Adjusting Screw Assembly

Check the operation of star-type automatic wheel adjusters by prying the shoe lightly away from its anchor or by pulling the cable to make sure the adjuster advances easily, one notch at a time. Adjuster cables tend to stretch, and star wheels and pawls may become blunted after a long period of use.

If problems are evident, disassemble the adjusting screw assembly (Figure 7-38) and clean the parts in denatured alcohol. Clean the threads using a wire brush. Make sure that the adjusting screw threads into the pivot nut over its complete length without sticking or binding. Check that none of the adjusting screw teeth are damaged. Lubricate the adjusting screw threads with brake lubricant, being careful not to get any on the adjusting teeth. Once lubricated, reassemble the unit. Apply brake lubricant to the adjusting screw threads, the inside diameter of the socket, and the socket face. To fully lubricate the adjusting screw, apply a continuous bead of lubricant to the

Do not use ordinary grease to lubricate drum brake parts, as it will not hold up under high temperatures.

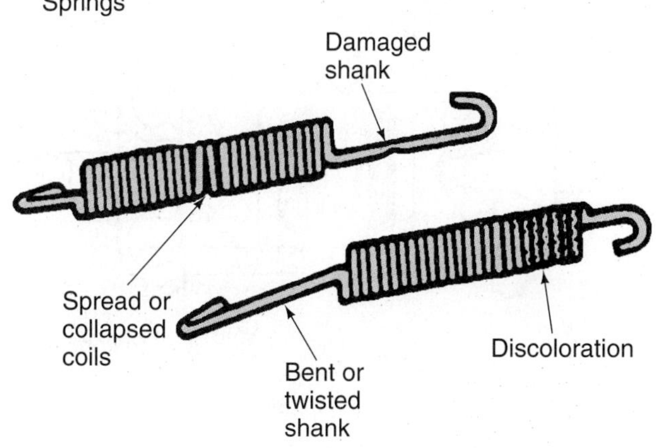

Figure 7-37 Types of drum brake spring damage (Reprinted with the permission of Ford Motor Company)

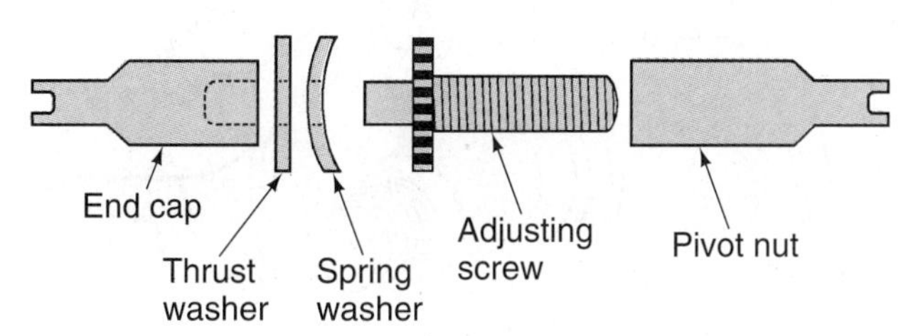

Figure 7-38 Components of a typical star wheel adjuster assembly (Courtesy of Raybestos Division, Brake Systems Inc.)

Always disassemble, clean, and inspect star wheel adjusters. Lubricate the threads with wheel bearing grease or antiseize compound.

open end of the pivot nut and socket (end cap) when the threads are fully engaged. Most adjusting screw assemblies use a thrust washer between the adjusting screw and socket (end cap). Some may also have an antinoise spring washer. Thread the adjusting screw in as far as it will go, tightening it to the recommended torque.

NOTE: When fasteners are removed, always reinstall them at the same location from which they were removed. If a fastener needs to be replaced, use the correct part number fastener for that application. If the correct part number fastener cannot be found, a fastener of equal size and strength (or stronger) may be used.

Wheel Cylinders

A drum brake wheel cylinder does not have to be disassembled and rebuilt unless it is leaking. To check for leaks, pull back each wheel cylinder dust boot (Figure 7-39). Normally, you will see a slight amount of fluid present. This is not a cause for alarm as it acts as a lubricant for the pistons. However, large amounts of fluid behind the boot indicate the fluid is leaking past the piston cups and an overhaul is in order. A leaking wheel cylinder must be replaced or overhauled to prevent brake fluid from contacting the drum and shoe lining surfaces. If the linings show signs of fluid on them, they must also be replaced.

Make sure both shoes properly engage the wheel cylinder.

Most modern drum brakes have the brake shoe fit directly against the wheel cylinder pistons. Or as shown in Figure 7-20, the piston may have an insert that presses outward onto the shoes. Other wheel cylinder designs have shoe links or pushrods between the piston and the shoes (Figure 7-40). On wheel cylinders with connecting links or pushrods, it is necessary to carefully remove one of the connecting links to check for leakage inside the boot.

Some wheel cylinders cannot be rebuilt. They are designed to be replaced as an assembly whenever there is leakage or signs of damage.

On some vehicles, it is not necessary to remove the rear wheel cylinder brake piston from the rear brake backing plate to disassemble, inspect, hone, and overhaul the rear wheel cylinder. Removal of the wheel cylinder from the backing plate is only necessary if the wheel cylinder must be replaced due to pitting or deep scratches in its surface. Some vehicles are equipped with wheel

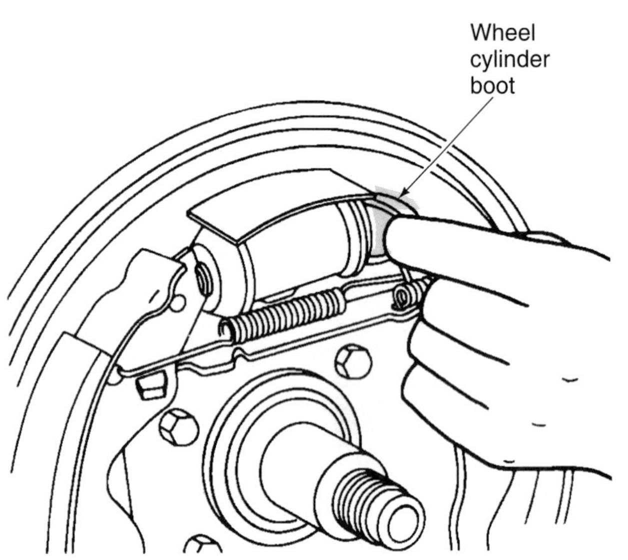

Figure 7-39 Inspecting wheel cylinder boot area for leakage (Reprinted with the permission of Ford Motor Company)

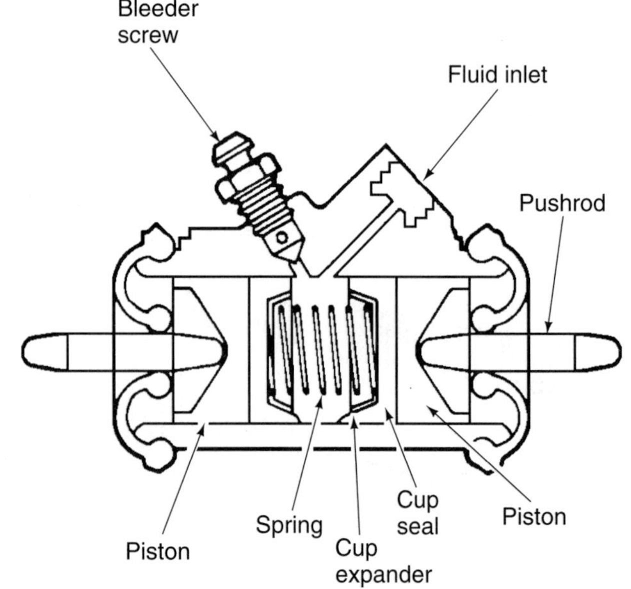

Figure 7-40 Wheel cylinder with brake shoe links or pushrods that transfer piston motion to the brake shoes (Courtesy of Raybestos)

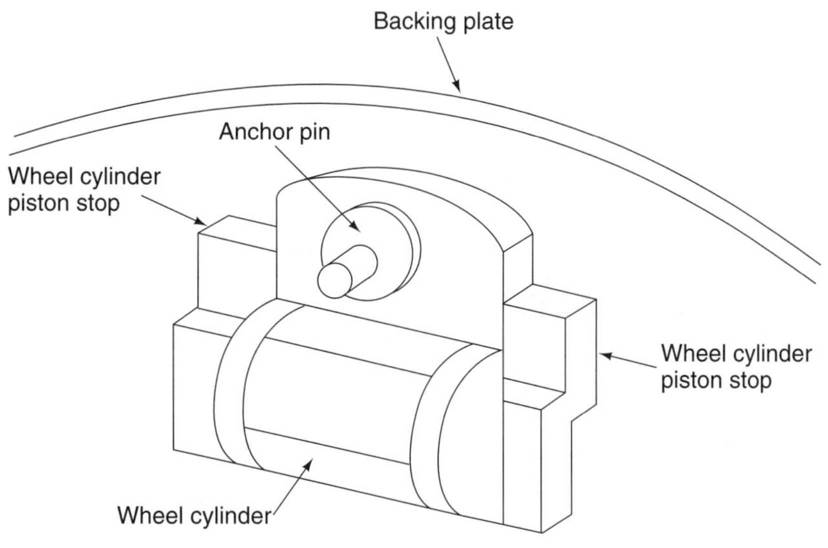

Figure 7-41 Backing plate with wheel cylinder piston stops

cylinder piston stops that protrude from the backing plate on either side to the wheel cylinder (Figure 7-41). The piston stops prevent the pistons from accidentally popping out of the bore during service. But the wheel cylinder must be partially removed to clear the stops before its internal components can be serviced.

When it is necessary to remove the wheel cylinder from the backing plate, begin by cleaning all dirt from around the wheel cylinder assembly. Disconnect the rear brake hose from the rear wheel cylinder. Disconnect the inlet tube and nut line. Plug an opening in the line to prevent fluid loss and contamination.

Remove wheel cylinder bolts. Loosen hub assembly bolts (if hub is not removed). Remove the wheel cylinder.

SERVICE TIP: While most wheel cylinders are bolted to the backing plate, some are held in position by a retaining ring (Figure 7-42) that can be removed using two small picks.

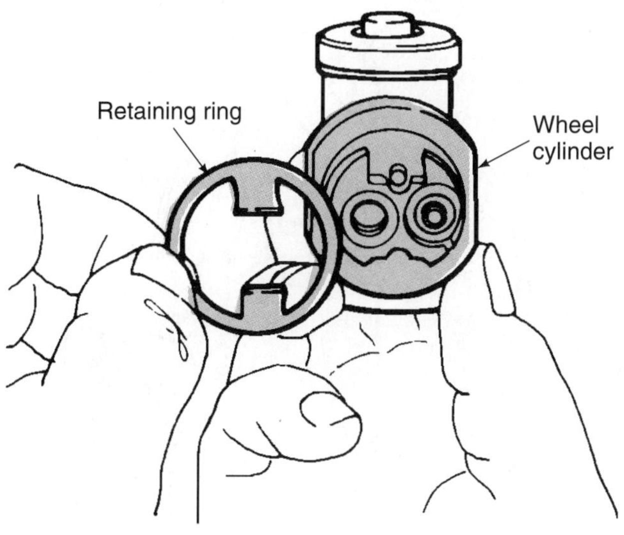

Figure 7-42 Removing wheel cylinder retaining ring (Courtesy of Raybestos)

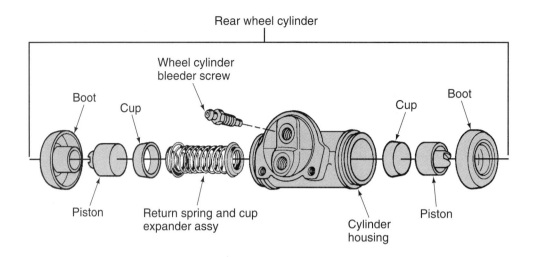

Rear wheel cylinder

Wheel cylinder bleeder screw

Boot

Cup

Cup

Boot

Piston

Return spring and cup expander assy

Cylinder housing

Piston

Figure 7-43 Wheel cylinder components (Reprinted with the permission of Ford Motor Company)

Wheel Cylinder Rebuild

With the brake shoe and linings removed as described earlier in the chapter, pop the wheel cylinder boots out of the retaining grooves in the rear wheel cylinder housing (Figure 7-43). Remove each boot and piston as an assembly. Next, remove the rubber cups and spring/expander assemblies from the cylinder bore. Rubber parts cannot be reused. Discard them to prevent accidental reuse.

Continue disassembly by removing the wheel cylinder bleeder screw from the cylinder. Wash all parts in clean, denatured alcohol and inspect the pistons for scratches, scoring, or other signs of damage. Check the cylinder bore for any scoring, nicks, scratch marks, or rust. Light scoring or corrosion in the cylinder bore can be cleaned with a crocus cloth, or if allowed by the manufacturer, a honing tool (Figure 7-44). Any honing performed must be light. The cylinder must not be honed more than 0.003 in. (0.08 mm) beyond its original diameter.

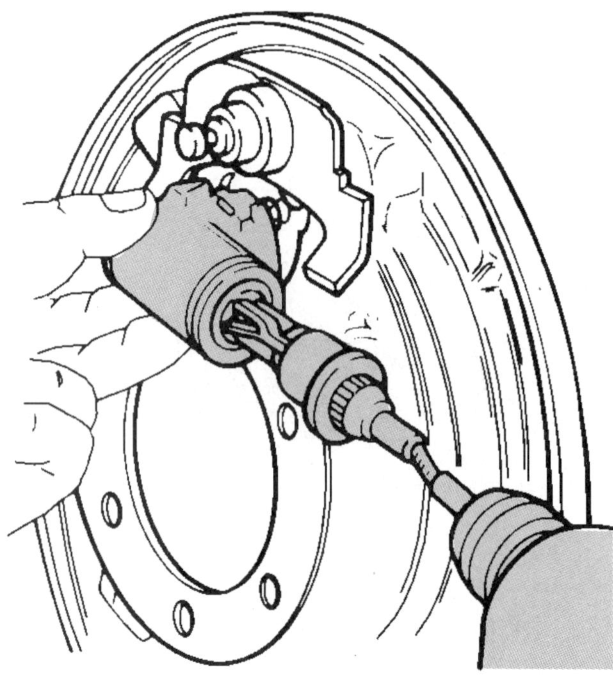

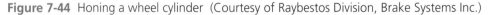

Figure 7-44 Honing a wheel cylinder (Courtesy of Raybestos Division, Brake Systems Inc.)

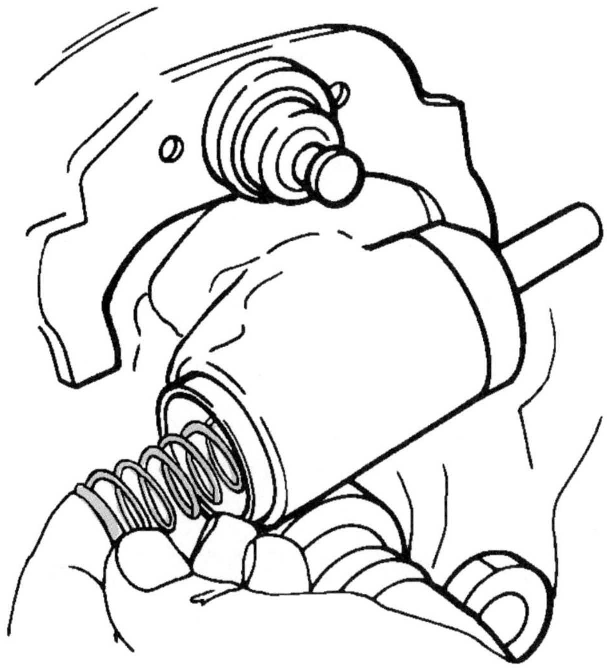

Figure 7-45 Installing the spring into the wheel cylinder (Courtesy of Raybestos)

Wash the rear wheel cylinder with clean, denatured alcohol after honing or polishing with the crocus cloth, then dry it with unlubricated compressed air. Be sure that the cylinder bleeder screw hole is open.

During reassembly be sure to install all of the parts in the wheel cylinder service kit. Replace the wheel cylinder assembly if the bore will not clean up with honing or polishing with a crocus cloth. Lubricate new seals and all internal parts with clean brake fluid. Replace if necessary. Always replace the rubber cups and dust boots. Thread the wheel cylinder bleeder screw into the rear wheel cylinder. Install the return spring (Figure 7-45) and cup expander assembly, and the pistons into their original positions in the cylinder bore. Place a boot over each end of the rear wheel cylinder. Install the brake shoe and lining assemblies as outlined. Install the brake drum assembly and the wheel assembly and bleed the brake system.

NOTE: Check to be sure that the brake line is installed in the lower brake drum hole and the wheel cylinder bleeder screw is in the upper hole. Always bleed the brakes before test driving the vehicle.

Brake Backing Plate

Carefully inspect the raised shoe pads on the backing plate. They must be free of corrosion or other surface defects that could prevent the shoes from sliding freely. Remove surface defects with a crocus cloth and then wet clean the area.

Examine the shoe anchor, support plate, and small parts for signs of looseness, wear, or damage that could cause faulty shoe alignment.

Check the backing plates for signs of cracking or bending. If there is any damage, they must be replaced. Make sure that the backing plate bolts and bolted-on anchor pins are torqued to specifications.

Examine the anchor pin rivets. The rivets must be tight and securely peened. You should also check the brake and shoe lining contact pads where the brake shoes and linings rest. Look for any deep grooves in the brake shoe and lining contact pads that could restrict brake shoe and lining movement. Hand sand any grooves in the braking surface. If grooves are still present after sanding, replace the rear brake backing plate.

NOTE: Any attempt to remove the grooves by grinding may result in improper brake shoe and lining-to-brake drum contact.

To remove a damaged backing plate, disconnect the brake line from the rear wheel cylinder. Remove the brake shoe and lining, the adjuster assemblies, the rear wheel cylinder and rear parking brake cable from the rear brake backing plate. Remove the bolts that secure the backing plate to the rear wheel spindle. Once removed, these bolts cannot be reused and should be discarded. Pull the backing plate from the spindle. If there is a foam gasket installed between the plate and spindle, pull it off and replace it with a new gasket.

Install the new backing plate with new retaining bolts. Reinstall the rear wheel cylinder and connect the brake line. Install the brake shoe and lining and adjuster assemblies. Slip the rear parking brake cable through the opening rear brake backing plate. The cable hold-down prongs must be securely locked in place. Connect the parking brake rear cable and conduit to the parking brake lever. Next, install the bearing and rear hub unit on the rear wheel spindle. Install the rear wheel axle hub retainer and tighten to torque specifications. Install a new front hub cap grease seal using a properly sized drive socket. To complete the installation, adjust the brake shoe and lining, install the brake drum and wheel, and bleed the brake system.

CASE STUDY

During a routine state-mandated brake inspection, a service technician finds a severely worn right rear drum brake assembly. The drum is blued and the shoes are burnt, both signs of extreme heat buildup. The vehicle is a four-year-old compact with an automatic transmission. It is driven by a retired woman. From the inspection records, the technician notes that the vehicle has only been driven 3,500 miles in the past year. The records also show that the vehicle also failed last year's inspection due to brake problems. The front pads were worn and the rotors were scored to the point where they had to be refinished. The vehicle is driven exclusively in city traffic and the woman describes herself as a very, very careful driver. A little too careful, the technician suspects. A quick look at the brake pedal confirms his suspicion. The left side of the long pedal is just as dirty and worn as the right side. The woman is a left-footed braker. By resting her left foot on the brake pedal, she is constantly generating slight braking action that quickly heats up and wears out brake parts. After some tactful questioning, the woman admits to the practice but is surprised to learn it is the cause of her problems. She only wants to be ready to stop, but promises to try to use the correct right foot braking method. Old habits are hard to break, though, and the technican anticipates that another brake job will be necessary at the same time next year.

Terms to Know

Adjusting screw assembly
Allowable machining dimension
Backing plate
Back-off
Bell-mouth
Discard dimension
Drum micrometer or gauge
Lining-to-drum fit
Out-of-round
Return spring
Self-adjusting brakes
Shoe links
Star wheel
Threaded drum
Wheel cylinder
Wheel hub
Wheel spindle

ASE Style Review Questions

1. Before attempting to remove a brake drum for inspection and servicing:
Technician A backs off the brake shoe adjuster.
Technician B takes up all slack in the parking brake cable.
Who is correct?
A. A only **C.** Both A and B
B. B only **D.** Neither A nor B

2. When inspecting a wheel cylinder:
Technician A finds a small amount of brake fluid behind the piston boot and rebuilds the wheel cylinder based on this fact.
Technician B does not rebuild the wheel cylinder and says the condition is normal.
Who is correct?
A. A only **C.** Both A and B
B. B only **D.** Neither A nor B

3. When adjusting the brake shoes on a vehicle equipped with a self-adjusting mechanism:
Technician A pushes the self-adjusting lever away from the star wheel a maximum of 1/16″.
Technician B pushes the self-adjusting lever away from the star wheel a minimum of 1/16″.
Who is correct?
A. A only **C.** Both A and B
B. B only **D.** Neither A nor B

4. Drum linings are badly worn at the toe and heel areas of the linings.
Technician A says the problem is an out-of-round drum.
Technician B says the problem is a tapered drum.
Who is correct?
A. A only **C.** Both A and B
B. B only **D.** Neither A nor B

5. The drum discard diameter is being discussed:
Technician A says this is the maximum diameter to which the drums can be refinished.
Technician B says the drum discard diameter is the maximum allowable wear dimension and not the allowable machining diameter.
Who is correct?
A. A only **C.** Both A and B
B. B only **D.** Neither A nor B

6. When machining a drum on a brake lathe:
Technician A uses a spindle speed of approximately 150 rpm.
Technician B makes a series of shallow cuts to obtain the final drum diameter.
Who is correct?
A. A only **C.** Both A and B
B. B only **D.** Neither A nor B

7. Drum servicing is being discussed:
Technician A says that as long as the drum-to-lining adjustment is correct, the diameters of the two drums on an axle set do not matter as long as they do not exceed the discard dimension.
Technician B says the drum diameters on given axles must be within 0.005″ of one another.
Who is correct?
A. A only **C.** Both A and B
B. B only **D.** Neither A nor B

8. Drum service is being discussed:
Technician A says new drums must be cleaned to remove the rustproofing compound from the drum surface.
Technician B says refinished drums must be cleaned to remove all metal particles from the drum surface.
Who is correct?
A. A only **C.** Both A and B
B. B only **D.** Neither A nor B

9. Brake linings are being discussed:
Technician A says the last code marking on the lining edge, such as FE or GG is a quality rating based on OSHA standards.
Technician B says it is a code indicating the lining's coefficient of friction.
Who is correct?
A. A only **C.** Both A and B
B. B only **D.** Neither A nor B

10. Drum brake problems are being discussed:
Technician A says weak or broken return springs can cause brake drag or pulling to one side.
Technician B says the same problems can be caused by a loose backing plate or an inoperative self-adjuster.
Who is correct?
A. A only **C.** Both A and B
B. B only **D.** Neither A nor B

Table 7-1 ASE TASK

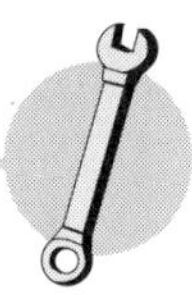

Diagnose drum brake system problems.

Problem Area	Symptoms	Possible Causes	Classroom Manual	Shop Manual
WEAK BRAKING ACTION OR BRAKE FADE UNDER HEAT	Depressing the brake pedal produces little or no braking action at the wheels equipped with disc brakes.	1. Improperly adjusted brake shoes	148	204
		2. Poor lining-to-drum contact	148	204
		3. Brakes adjusted too tightly	148	204
		4. Poorly lubricated backing plate assemblies	142	209
		5. Weak or damaged shoe retracting springs	142	205
		6. Thin or weak drum expanding under pressure	140	191
		7. Improper lining installed	141	200
HARD PEDAL	Excessive brake pedal pressure is required to stop the vehicle.	1. Frozen wheel cylinder pistons	138	209
		2. Mismatched lining types	141	196
		3. Dust of newly turned drum imbedded in brake linings	140	195
		4. Poorly lubricated backing plate and hardware	142	209
		5. Water on linings	141	196
		6. Grease, oil, or brake fluid on linings	141	196
SPONGY, LOW, OR SINKING PEDAL	Pedal has a soft, springy, or spongy feeling when pressed.	1. Excessively oversized drums that are expanding under heat and pressure	140	191
		2. Standard thickness lining with oversized drums	141	196
		3. Wheel cylinder pushrod pins not properly located in slot in the brake shoes	138	206
		4. Self-adjuster not working	148	205
PEDAL PULSATION NOTE: A rapid pulsation of the brake pedal is normal during hard stops on vehicles equipped with ABS.	Brakes chatter when applied. Vibration or chatter is felt in the pedal when applied. Rough or poor braking action.	1. Worn or loose wheel bearings	138	184
		2. Loose backing plates, anchors, or wheel cylinders	142	209
		3. Drum surface not smooth	140	190
		4. Bent axle	138	184
		5. Out-of-round drum or bent hub, drum, or wheel	140	192
		6. Drum loose on hub	138	184
EXCESSIVE BRAKE PEDAL TRAVEL	Brake pedal closer than 1 to 2 inches from floor when fully applied.	Worn/damaged brake shoes and/or linings	141	200
ERRATIC BRAKE ACTION	Braking action is inconsistent from application to application.	1. Worn or loose wheel bearings	138	184
		2. Frozen or sluggish wheel cylinder pistons	138	208
		3. Poorly lubricated backing plate assemblies	138	209
		4. Shoe retracting springs weak or damaged	142	205
		5. Lining not ground to fit drum	141	200
		6. Mismatched drum sizes on same axle	140	191

Table 7-1 ASE TASK (continued)

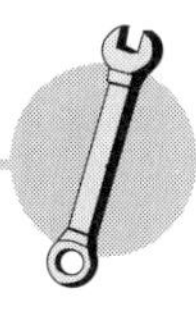

Diagnose drum brake system problems.

Problem Area	Symptoms	Possible Causes	Classroom Manual	Shop Manual
ERRATIC BRAKE ACTION (continued)	Braking action is inconsistent from application to application.	7. Loose backing plates, anchors, or wheel cylinders	142	209
		8. Bent backing plate	138	209
		9. Bent or distorted shoes	141	196
		10. Out-of-round drum	140	192
BRAKE GRAB OR PULL AT ONE BRAKE	Vehicle pulls to one side when brakes are applied.	1. Improper brake shoe adjustment	148	204
		2. Weak or broken springs	142	205
		3. Brake drum out of round or otherwise damaged	140	192
		4. Loose linings on shoes	141	200
		5. Loose backing plates, wheel cylinders, anchors, and so on	142	209
		6. Improperly matched linings on same axle	141	196
		7. Primary and secondary brake shoes and linings installed in wrong location on one side	141	200
		8. Leaking wheel cylinder or grease seals have contaminated linings	141	206
		9. Improper size wheel cylinder on one side	137	206
		10. Self-adjuster not working on one side	148	205
BRAKE DRAG OR LOCKING AT ONE OR MORE WHEELS	Brakes fail to release when brake pedal is released.	1. Improper lining-to-drum clearance	148	204
		2. Frozen or sluggish wheel cylinder pistons	138	206
		3. Swollen wheel cylinder cups	137	206
		4. Weak, broken, or wrong type brake shoe retracting spring installed	142	205
		5. Grease, oil, or brake fluid on lining	141	196
		6. Brake shoes not releasing due to poor lubrication on backing plate or worn backing plate	140	209
		7. Wheel cylinder-to-brake shoe pushrod pin not properly located	138	209
		8. Loose backing plate anchors and/or wheel cylinders	140	209
		9. Improper linings installed	141	200
		10. Worn wheel and axle bearings	138	184
NOISE	Brakes chatter.	1. Bad shoe adjustment or brake-to-drum contact	148	204
		2. Loose linings	141	200
		3. Loose backing plate, wheel cylinders, or anchors	138	209
		4. Rough drum surface	140	190
		5. Out-of-round drum	140	192

Table 7-1 ASE TASK (continued)

Diagnose drum brake system problems.

Problem Area	Symptoms	Possible Causes	Classroom Manual	Shop Manual
NOISE	Brakes clicking or snapping when brakes are applied.	1. Too much clearance between lining and drum	148	204
		2. Excessively ground backing plate bosses	138	209
		3. Poorly lubricated backing plates, anchors, and so on	138	209
		4. Cracked drums	140	192
		5. Bent brake shoe hold-down pins	140	196
		6. Threaded brake drum surface due to chipped or improperly ground bit	140	191

Antilock Brake Service

Upon completion and review of this chapter, you should be able to:

- ❑ Diagnose ABS problems, including poor stopping, wheel lockup, pedal feel, pulsation, and noise problems caused by the ABS system, and determine needed repairs.
- ❑ Diagnose ABS electronic control and component problems using the self-diagnostic capabilities of the system or the recommended test equipment, and determine needed repairs.
- ❑ Visually inspect the ABS brake fluid level and all electronic connectors.
- ❑ Perform common ABS self-diagnostic routines.
- ❑ Perform required electrical tests on a system.
- ❑ Troubleshoot using ABS warning lights.
- ❑ Service, test, and adjust antilock brake system speed sensors.
- ❑ Perform an ABS high-pressure fluid test and determine needed repairs.
- ❑ Remove and replace an ABS hydraulic assembly.
- ❑ Remove and replace an ABS accumulator.
- ❑ Remove and replace an ABS pump assembly.
- ❑ Bleed ABS systems.

Basic Service Skills Required

Basic Tools

Basic mechanic's tool set

Appropriate service manual

DMM

Antilock brakes (ABS) have become very popular on today's vehicles. The reason is simple. All ABS systems reduce the tendency of any wheel under their control to lock up, even during panic stops. To understand why an ABS system works, you must first remember that a disc or drum brake stops a vehicle by stopping a turning wheel. Once a wheel locks up and enters a skid condition, it stops turning. In a lockup condition, all the hydraulic pressure in the world becomes useless in stopping the vehicle. The pads or shoes can squeeze as tight as possible against the rotor or drum, but the only thing slowing down the vehicle is the friction between the tire and the road.

All ABS systems prevent the brakes from locking up in the same way. They rapidly release and apply hydraulic pressure to the disc or drum brake assemblies. The action is similar to a driver pumping the brake pedal to avoid a skid; but the ABS system can "pulse" the brakes on and off much faster—up to 15 times a second.

Most of the services done to the brakes of an antilock brake system are identical to those in a conventional brake system. There are, however, some important differences. One of these is the bleeding of the brake system. Always refer to the appropriate procedures in the service manual before attempting to service the brakes on an ABS-equipped vehicle.

CAUTION: Always check the service manual for special precautions before checking brake fluid level, bleeding the system, loosening a hydraulic line or hose, or replacing a part. It may be necessary to depressurize the system to safely service it.

Before servicing an antilock brake system, it is important that you understand the basics of electrical and electronic troubleshooting. Without this basic understanding, it will be difficult to follow the diagnostic procedures given in most service manuals.

Classroom Manual
Chapter 8, page 157

You must understand the basic theory of electricity and know the meaning of voltage, current, and resistance. You must be capable of measuring voltage in volts, current in amperes, and resistance in ohms using a DMM meter. You must also be able to perform other electrical tests involving jumper wires, test lights, and so on. You should be able to read and understand the electrical wiring diagrams (schematics) that are illustrated in vehicle service manuals. You must also know what happens in a circuit when a short or open condition occurs and the procedures for repairing wire and or connectors.

Chapter 2 describes the common equipment used in electrical and electronic troubleshooting. Chapter 10 contains basic information on the types of circuits used in automotive electronic systems and the problems that can occur in these circuits.

Although ABS is relatively new automotive technology, there are already many different ABS systems in use today. All systems use many of the same operating principles, but components vary slightly from manufacturer to manufacturer. For example, the system may or may not have an accumulator. The pump may be integral to the hydraulic module or it may be separate. The number and location of electrical relays and warning indicators may differ. Each system also has its own unique service and safety requirements. Become familiar with the warnings and cautions associated with the system you are working on. Read through the vehicle's service manual section on the ABS system before performing any service. While the ABS section of most service manuals is quite large, most of the information is troubleshooting flow charts for each possible digital trouble code generated by the system. Take the time to read over the system description and diagnostic sequence. As with computerized engine controls, a hit-and-miss method of troubleshooting ABS systems will waste both your time and your customer's money.

This chapter covers diagnostic and repair procedures for many common types of repairs made to ABS systems. It cannot possibly cover all systems found on the road today, and should only be used as a guide to understanding ABS systems in general.

System Self-Check

All ABS systems have some sort of self-test. This test is activated each time the ignition switch is turned on. You should begin all diagnosis with this simple test. To perform a typical ABS self-check sequence, place the ignition switch in the START position while observing both the red brake system light and the amber ABS indicator lights. Both lights should turn on. Start the vehicle. The red brake system light should turn off. With the ignition switch in the RUN position, the antilock brake control module performs a preliminary self-check on the antilock electrical system. The self-check takes three-to-six seconds, during which time the amber ANTILOCK indicator light in the instrument cluster remains on (Figure 8-1). Once the self-check is complete, the ABS indicator light should turn off. If any malfunction is detected during this test, the ABS warning light remains on and the ABS system is shut down. If the lights do not come on, you should proceed to diagnose that problem before moving on to others.

Diagnosis and Testing

Always follow the vehicle manufacturer's specified diagnostic and replacement procedures when servicing ABS systems. In general, ABS system diagnostics require three-to-five different types of

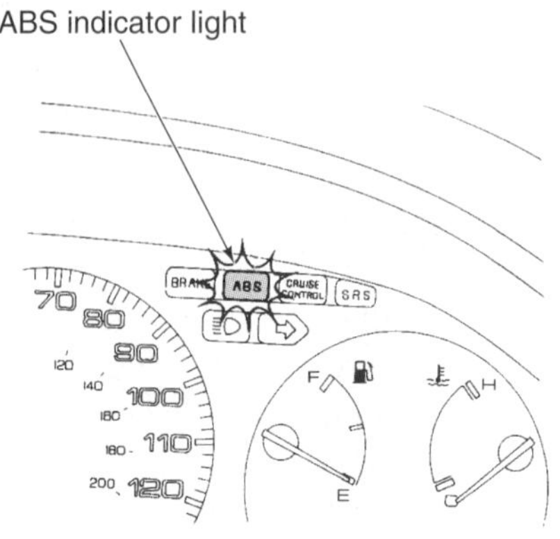

Figure 8-1 ABS indicator light (Courtesy of Honda Motor Co. Ltd.)

testing that must be performed in the order listed in the service manual. Types of testing may include the following:

1. Pretest checks or inspections
2. On-board ABS control module testing (trouble code reading)
3. Warning light symptom troubleshooting
4. Individual trouble code or component troubleshooting

WARNING: Following the wrong sequence or bypassing steps may lead to unnecessary replacement of parts or incorrect resolution of the symptom. The information and procedures given in this chapter are typical of the various antilock systems on the market. For specific instructions, consult the vehicle's service manual.

Special Tools

Scan tool

Jumper wire

The prediagnosis inspection consists of a quick visual check of specific system components that could create an apparent antilock system malfunction. Inspecting the system before diagnosing specific symptoms can result in the detection of a simple defect, which could be the cause of an inoperative system. This inspection should be the first step in diagnosing a complaint.

If a malfunction in the ABS system occurs, be sure that the control module connector is securely connected and that all ABS connectors are securely connected. Check for a blown fuse. If the malfunction is still present, refer to the procedure for reading trouble codes.

If pretest checks or inspections do not determine the problem, the ABS control module is then accessed for individual trouble codes. Some ABS systems have over 50 trouble codes, each corresponding to a different problem in the system. These codes are keyed to individual pinpoint tests and troubleshooting sequences listed in the service manual.

Classroom Manual
Chapter 8, page 162

To retrieve trouble codes, a scan tool is used. Often, the same scan tool recommended for diagnosing engine performance problems is used to retrieve ABS trouble codes. These scan tools are plugged into the vehicle's ABS diagnostic connector. The location of this connector can be found in the appropriate service manual. Some manufacturers recommend the use of antilock brake specific testers, like the one shown (Figure 8-2).

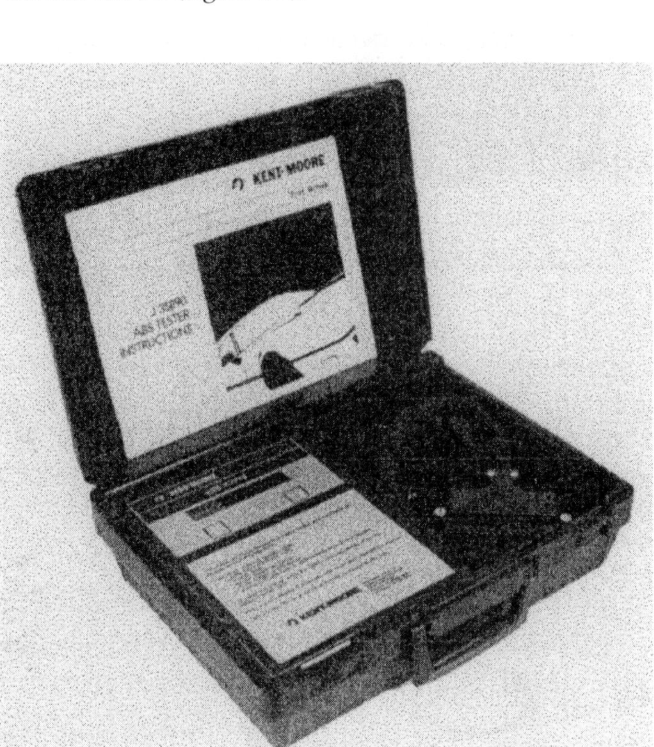

Figure 8-2 Antilock brake system tester (Courtesy of Kent-Moore Tools)

The brake and antilock warning lights are also designed to signal potential problems in the system. On/off combinations of the two lights are keyed to specific troubleshooting sequences listed in the service manual.

Many ABS components are simply remove-and-replace items. Normal brake repairs, such as replacing brake pads, caliper replacement, rotor machining or replacement, brake hose replacement, master cylinder or power booster replacement, or parking brake repair, can all be performed as usual. In other words, brake service on an ABS-equipped vehicle is similar to service on a conventional system with a few exceptions. However, before beginning any service, depressurize the accumulator to prevent personal injury from high-pressure fluid. This is accomplished by pumping the brake pedal (ignition off) until the pedal becomes hard.

Prediagnostic System Inspection and Test Drive

The diagnostic program of many systems can also inform the service technician whether or not the malfunction is intermittent.

The ABS system has an electronic control unit to control major functions. The control unit accepts input signals from sensors and instantly drives actuators. It is essential that both kinds of signals are proper and stable. At the same time, it is important that there are no conventional problems such as air leaks in the booster or lines, lack of brake fluid, or other problems with the brake system.

It is much more difficult to diagnose a problem that occurs intermittently rather than continuously. Most intermittent problems are caused by poor electrical connections or faulty wiring. In this case, careful checking of the suspected circuits may help prevent the replacement of good parts.

A visual check may not find the cause of the problems, so a road test should be performed. Before undertaking actual checks, take just a few minutes to talk with the customer who approaches with an ABS complaint. The customer is a very good source of information on such problems, especially intermittent ones. When talking with the customer, find out what symptoms are present and under what conditions they occur.

Start your diagnosis by looking for conventional problems first. This is one of the best ways to troubleshoot brake problems on an ABS-controlled vehicle.

Visual Inspection

Problems can often be spotted during this inspection, preventing time-consuming activities. Visually check all easily accessible components, including the following:

Using the compact spare tire supplied with the vehicle will not affect the performance of the ABS system. Tire size is important for proper performance of the ABS system. Replacement tires should be the same size load range and construction as the original tires.

1. Check the master cylinder fluid level.
2. Inspect all brake hoses, lines, and fittings for signs of damage, deterioration, and leakage. Inspect the hydraulic module for any leaks or wiring damage.
3. Inspect the brake components at all four wheels. Verify that no drag exists; also verify proper brake apply operation.
4. Inspect the wheel speed sensors and their wiring. Verify correct air gap range, solid sensor attachment, undamaged sensor toothed ring, and undamaged wiring, especially at vehicle attachment points.
5. Inspect all electrical connections for signs of corrosion, damage, fraying, and disconnection.
6. Inspect for worn or damaged wheel bearings that may allow a wheel to wobble.
7. Verify proper outer CV joint alignment and operation.
8. Verify that tires meet the legal tread depth requirements.

SERVICE TIP: Remember that normal braking components that fail may cause the ABS system to shut down. Do not be too quick to condemn the ABS system.

Test Drive

CAUTION: Test the feel of the brake pedal as well as the operation of the brake system before driving the vehicle. It is important for you to be confident that the brake system is working sufficiently to stop the vehicle during the test drive.

After the visual inspection is completed, test drive the vehicle to evaluate brake system performance. Accelerate to a speed of about 20 mph, then bring the vehicle to a stop using normal braking procedures. Look for any signs of swerving or improper operation. Next, return the vehicle speed to approximately 25 mph and apply hard braking pressure. You should feel the pedal pulsate if the ABS is working properly.

Classroom Manual
Chapter 8, page 168

During the test drive, both test lights should remain off. If any one of the warning lights turns on, take note of the condition that may have caused it. Both lights should remain off during the stopping procedure. Place the gear selector into PARK and observe the warning lights. They should both be off.

If the ECM detects a problem with the system, the amber ABS indicator lamp will either flash or light continuously (solid lamp) to alert the driver of the problem. In some systems, a flashing ABS indicator lamp alerts the drive to a problem that does not immediately hamper ABS operation. However, a flashing ABS indicator lamp is a signal that repairs must be made to the system as soon as possible.

Note the action of the brake pedal during the test drive. As outlined in Chapter 4, you can often determine the probable cause of a wide variety of brake system problems by brake pedal feel.

A solid ABS indicator lamp indicates that a problem that affects the operation of the ABS has been detected. No antilock braking will be available. Normal, non-antilock brake performance will remain. In order to regain ABS braking ability, the ABS system must be serviced.

The red BRAKE warning lamp will be illuminated when a low brake fluid level in the master cylinder is sensed, the parking brake switch is closed, or the bulb test switch section of the ignition switch is closed or under the control of the electronic control unit when certain ABS digital trouble codes are set. This lamp indicates that a base brake problem may exist.

Intermittent problems can often be noted during the test drive. If the scan tool is equipped with a "snapshot" feature, use it to capture system performance during normal acceleration, stopping, and turning maneuvers. If this does not reproduce the malfunction, perform an antilock stop on a low-coefficient surface such as gravel, from approximately 48–80 km/h (30–50 mph) while triggering the snapshot mode on the scan tool.

Malfunctions in the antilock brake system cause the antilock brake control module to shut off or inhibit the system. However, normal power-assisted braking remains. Loss of hydraulic fluid to the hydraulic module also disables the antilock system.

Self-Diagnostics

The electronic control system of most ABS systems includes sophisticated on-board diagnostics that, when accessed with the proper scan tool, can identify the source of many specific system malfunctions. For example, the General Motors' ABS-VI diagnostic program can record 57 different diagnostic trouble codes that pertain to possible problems in the ABS system into memory (Table 8-1). The service manual contains a detailed step-by-step troubleshooting chart for each of these trouble codes. The charts give specific instructions for confirming the suspected problem and performing the repair (Figure 8-3).

The ABS controller controls the ABS system. It is not the same control module used to control electronic engine control functions.

Testers and Scanning Tools

Different vehicle manufacturers provide ABS test and scan tools with varying capabilities. Some testers are used simply to access the digital trouble codes. Others may also provide functional test modes for checking wheel sensor circuits, pump operation, solenoid testing, and so on. For example,

Special Tools

Breakout box

DMM

Table 8-1 DIAGNOSTIC TROUBLE CODE AND SYMPTOM TABLE (Courtesy of Chevrolet Motor Division of General Motors Corporation)

DIAGNOSTIC TROUBLE CODE AND SYMPTOM TABLE		PAGE #
CHART	**SYMPTOM**	
A	ABS (Amber) Indicator Lamp "ON" Constantly, No DTCs Stored	5E1-24
B	ABS (Amber) Indicator Lamp "ON" Intermittently, No DTCs Stored	5E1-25
C	ABS (Amber) Indicator Lamp "OFF" Constantly, No DTCs Stored	5E1-26
D	Tech 1 Displays Undefined DTCs	5E1-27
DIAGNOSTIC TROUBLE CODE	**DESCRIPTION**	
A011	ABS Indicator Lamp Circuit Open or Shorted to Ground	5E1-28
A013	ABS Indicator Lamp Circuit Shorted to Battery	5E1-30
A014	Enable Relay Contacts or Fuse Open	5E1-32
A015	Enable Relay Contacts Shorted to Battery	5E1-34
A016	Enable Relay Coil Circuit Open	5E1-36
A017	Enable Relay Coil Circuit Shorted to Ground	5E1-38
A018	Enable Relay Coil Circuit Shorted to Battery or Coil Shorted	5E1-40
A021	Left Front Wheel Speed = 0 (1 of 2)	5E1-42
A022	Right Front Wheel Speed = 0 (1 of 2)	5E1-46
A023	Left Rear Wheel Speed = 0 (Non-Tubular Axle) (1 of 2)	5E1-50
A023	Left Rear Wheel Speed = 0 (Tubular Axle) (1 of 2)	5E1-52
A024	Right Rear Wheel Speed = 0 (Non-Tubular Axle) (1 of 2)	5E1-56
A024	Right Rear Wheel Speed = 0 (Tubular Axle) (1 of 2)	5E1-58
A025	Left Front Excessive Wheel Speed Variation (1 of 2)	5E1-62
A026	Right Front Excessive Wheel Speed Variation (1 of 2)	5E1-66
A027	Left Rear Excessive Wheel Speed Variation (Non-Tubular Axle) (1 of 2)	5E1-70
A027	Left Rear Excessive Wheel Speed Variation (Tubular Axle) (1 of 2)	5E1-72
A028	Right Rear Excessive Wheel Speed Variation (Non-Tubular Axle) (1 of 2)	5E1-76
A028	Right Rear Excessive Wheel Speed Variation (Tubular Axle) (1 of 2)	5E1-78
A036	Low System Voltage	5E1-82
A037	High System Voltage	5E1-84
A038	Left Front EMB Will Not Hold Motor	5E1-86
A041	Right Front EMB Will Not Hold Motor	5E1-88
A042	Rear Axle ESB Will Not Hold Motor	5E1-90
A044	Left Front Channel Will Not Move	5E1-92
A045	Right Front Channel Will Not Move	5E1-94
A046	Rear Axle Channel Will Not Move	5E1-96
A047	Left Front Motor Free Spins	5E1-98
A048	Right Front Motor Free Spins	5E1-100
A051	Rear Axle Motor Free Spins	5E1-102
A052	Left Front Channel in Release Too Long	5E1-104
A053	Right Front Channel In Release Too Long	5E1-106
A054	Rear Axle Channel in Release Too Long	5E1-108

Table 8-1 DIAGNOSTIC TROUBLE CODE AND SYMPTOM TABLE (continued) (Courtesy of Chevrolet Motor Division of General Motors Corporation)

DIAGNOSTIC TROUBLE CODE AND SYMPTOM TABLE		PAGE #
DIAGNOSTIC TROUBLE CODE	DESCRIPTION	
A055	Motor Driver Fault Detected	5E1-110
A056	Left Front Motor Circuit Open	5E1-112
A057	Left Front Motor Circuit Shorted to Ground	5E1-114
A058	Left Front Motor Circuit Shorted to Battery or Motor Shorted	5E1-116
A061	Right Front Motor Circuit Open	5E1-118
A062	Right Front Motor Circuit Shorted to Ground	5E1-120
A063	Right Front Motor Circuit Shorted to Battery or Motor Shorted	5E1-122
A064	Rear Axle Motor Circuit Open	5E1-124
A065	Rear Axle Motor Circuit Shorted to Ground	5E1-126
A066	Rear Axle Motor Circuit Shorted to Battery or Motor Shorted	5E1-128
A067	Left Front EMB Circuit Open or Shorted to Ground	5E1-130
A068	Left Front EMB Circuit Shorted to Battery or Driver Open	5E1-132
A071	Right Front EMB Circuit Open or Shorted to Ground	5E1-134
A072	Right Front EMB Circuit Shorted to Battery or Driver Open	5E1-136
A076	Left Front Solenoid Circuit Open or Shorted to Battery	5E1-138
A077	Left Front Solenoid Circuit Shorted to Ground or Driver Open	5E1-140
A078	Right Front Solenoid Circuit Open or Shorted to Battery	5E1-142
A081	Right Front Solenoid Circuit Shorted to Ground or Driver Open	5E1-144
A082	Calibration Memory Failure	5E1-146
A086	Red Brake Warning Lamp Activated by ABS	5E1-148
A087	Red Brake Warning Lamp Circuit Open	5E1-150
A088	Red Brake Warning Lamp Circuit Shorted to Battery	5E1-152
A091	Open Brake Switch Contacts During Deceleration	5E1-154
A092	Open Brake Switch Contacts When ABS Was Required	5E1-156
A093	DTCs A091 or A092 Set in Current or Previous Ignition Cycle	5E1-158
A094	Brake Switch Contacts Always Closed	5E1-160
A095	Brake Switch Circuit Open	5E1-162
A096	Brake Lamps Circuit Open	5E1-164

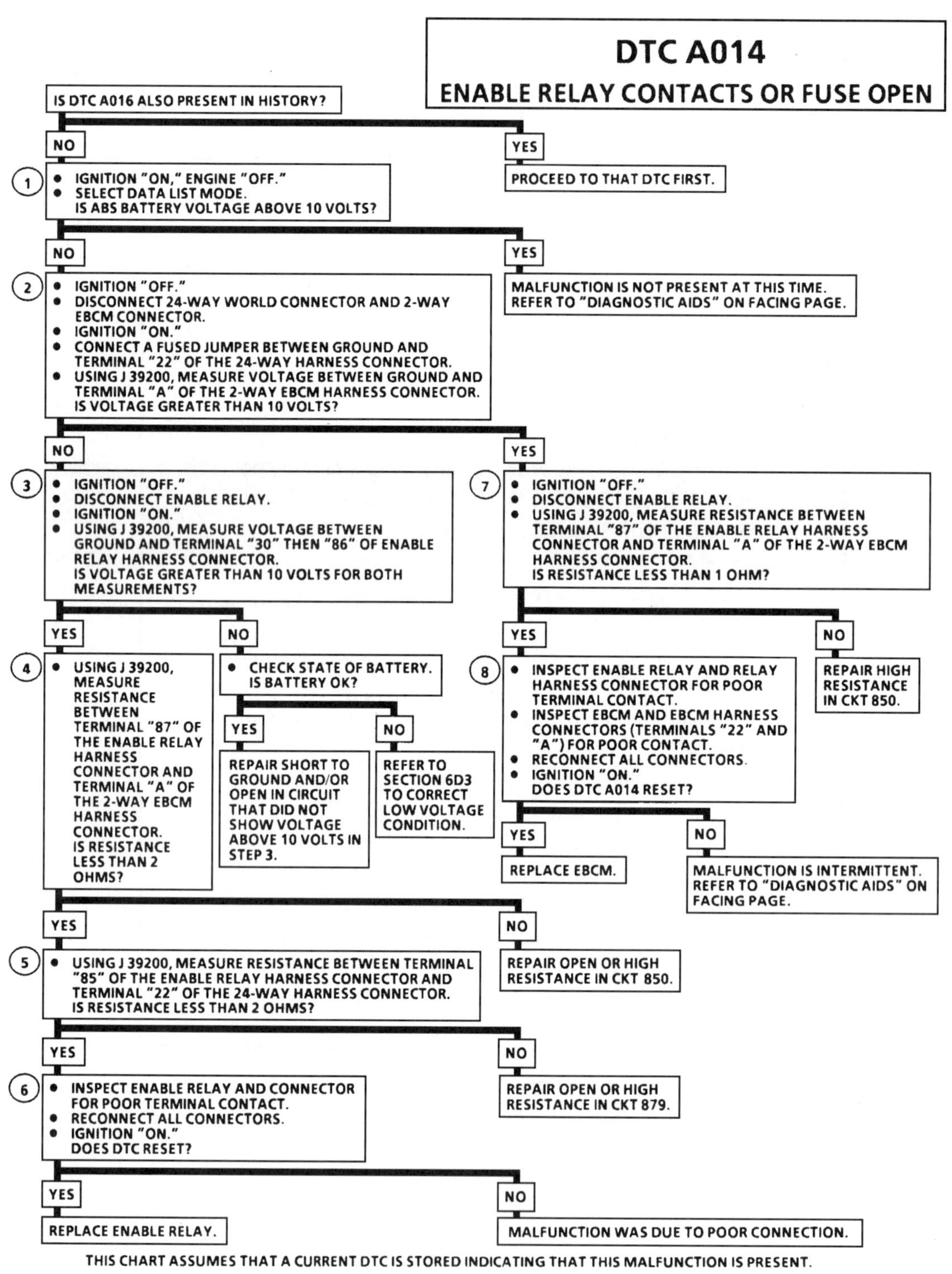

Figure 8-3 Typical digital trouble code (DTC) diagnostic flow chart showing step-by-step instructions for troubleshooting systems with a suspected open enable relay or fuse (Courtesy of Chevrolet Motor Division of General Motors Corporation)

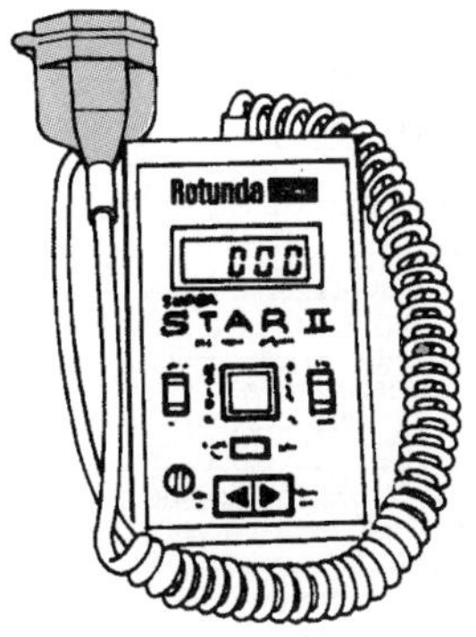

Rotunda Super Star II tester

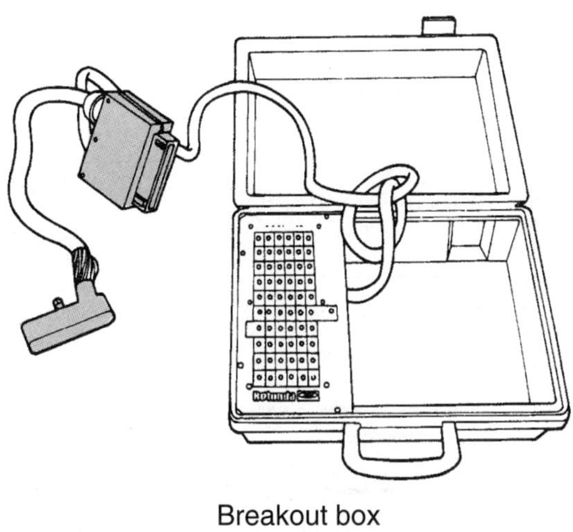

Breakout box

Figure 8-4 Ford's Rotunda Super Star II tester and EEC-IV breakout box with an ABS test adapter installed (Reprinted with the permission of Ford Motor Company)

Ford Motor Company provides a basic tester for reading out trouble codes. To perform function tests during many pinpoint test procedures, Ford's EEC-IV break-out box, which is used for electronic engine control troubleshooting, is plugged into the ABS control circuit through use of a special adapter (Figure 8-4).

General Motors' Tech-1 scan tool reads out digital trouble codes and also performs functional testing in a number of modes, including a snapshot mode, which captures ABS operating conditions during a test drive (Figures 8-5, 8-6, and 8-7). See Photo Sequence 12 for a typical procedure for troubleshooting a GM ABS-VI using a scan tool.

Honda's antilock brake checker can simulate all system functions and operating conditions, but it cannot be used to check system status during test drives. The functional ALB checker does not display numerical trouble codes. Instead, Honda uses the amber ABS system light and red brake light to flash out the digital trouble codes. On some of its models, the codes are flashed out by two LED lights located on the electronic control unit. As you can see, it is important to research the capabilities and proper use of the test equipment the vehicle manufacturer provides. Misuse of test equipment can be dangerous. For example, connecting test equipment during a test drive that is not designed for this use may lead to loss of braking ability.

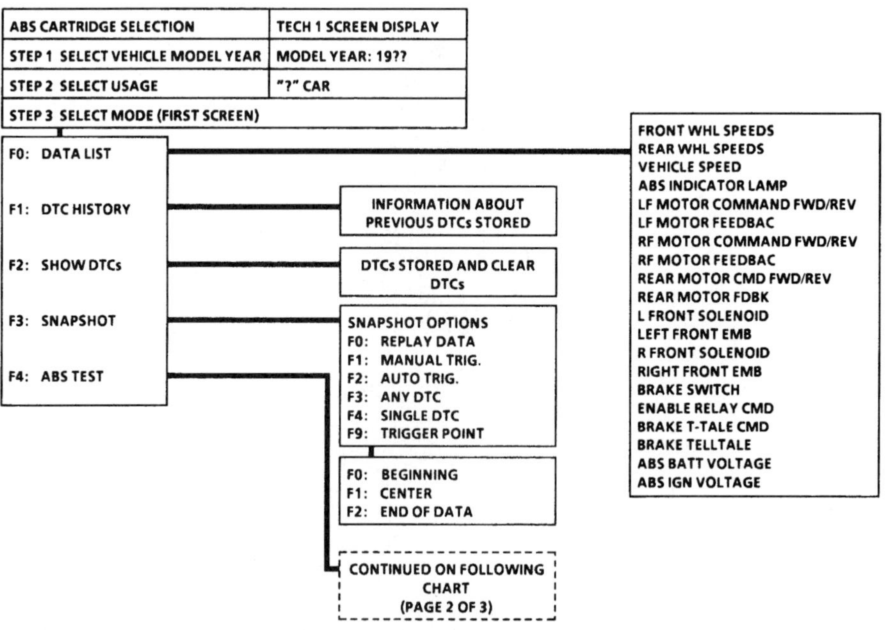

Figure 8-5 ABS-VI diagnostic functions possible using the Tech-1 scan tool (Courtesy of Chevrolet Motor Division of General Motors Corporation)

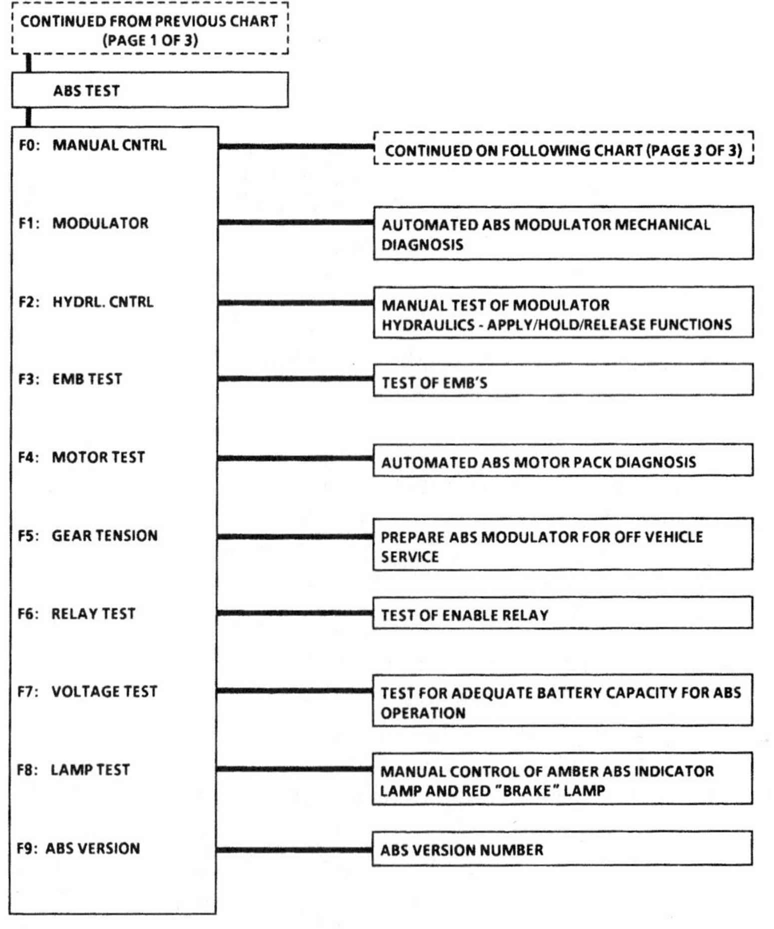

Figure 8-6 ABS-VI diagnostic functions possible using the Tech-1 scan tool (continued) (Courtesy of Chevrolet Motor Division of General Motors Corporation)

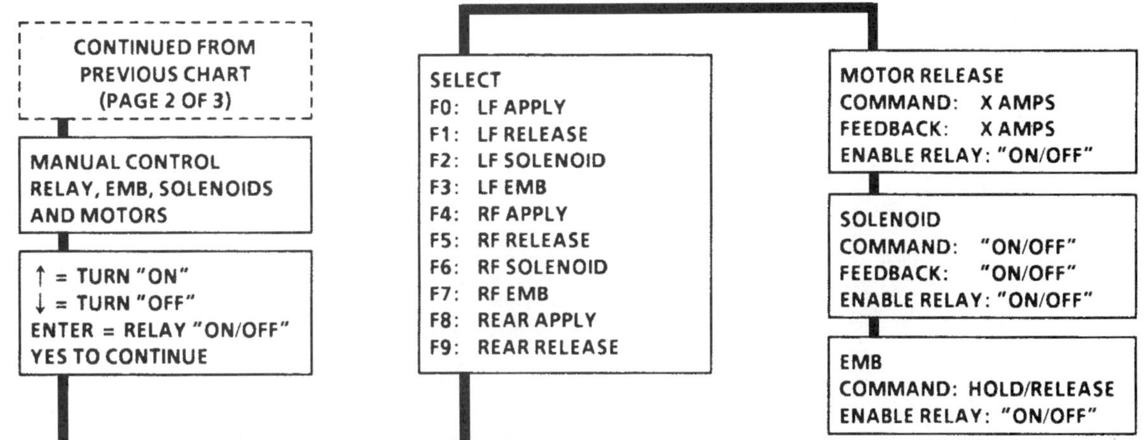

Figure 8-7 ABS-VI diagnostic functions possible using the Tech-1 scan tool (continued) (Courtesy of Chevrolet Motor Division of General Motors Corporation)

Photo Sequence 12
Troubleshooting a GM ABS-VI Using a Tech-1 Scan Tool

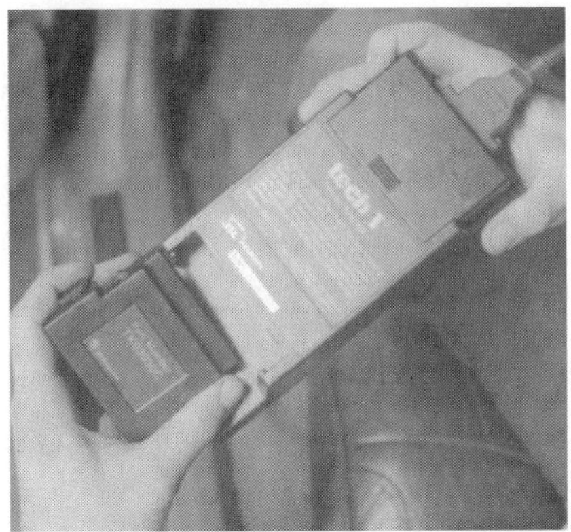

P12-1 Install the ABS cartridge into the scan tool.

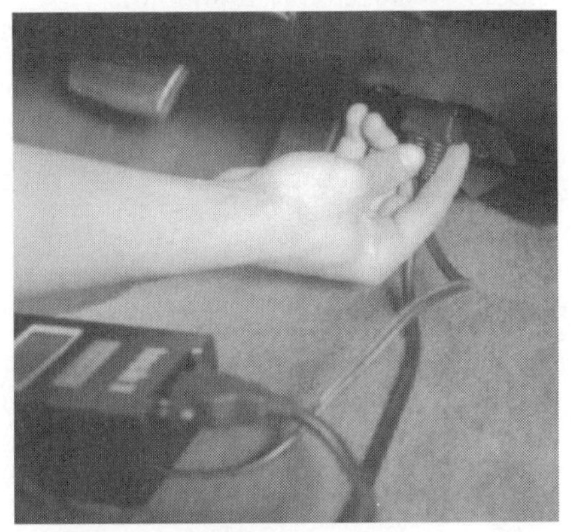

P12-2 With the engine and ignition switch in the OFF position, connect the scan tool to the data link connector (DLC).

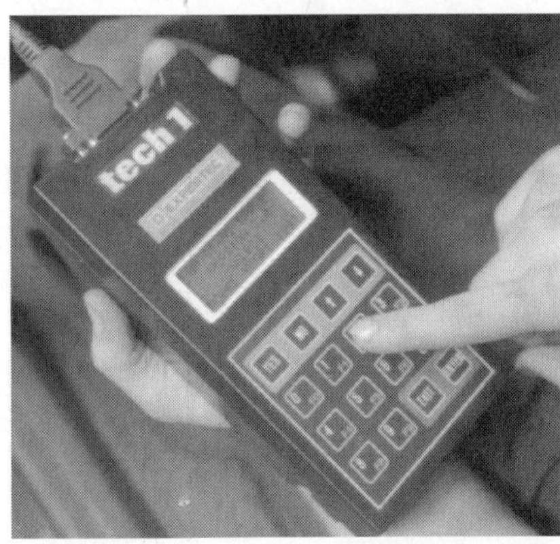

P12-3 Turn on the ignition, but leave the engine off.

P12-4 Select the vehicle model year.

P12-5 Select the usage (model type) of vehicle being serviced.

P12-6 Select the diagnostic trouble code mode to view any digital trouble codes occurring on the current ignition cycle.

Photo Sequence 12
Troubleshooting a GM ABS-VI Using a Tech-1 Scan Tool (continued)

P12-7 Read all current digital trouble codes (DTCs).

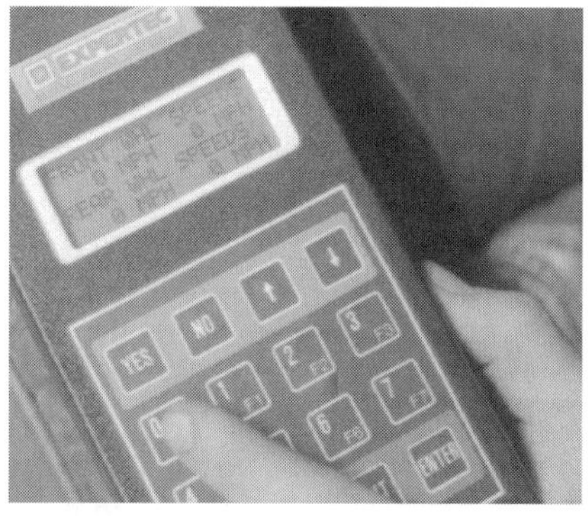

P12-8 Read all DTCs stored in the computer's memory. Note the last DTC to occur prior to the current ignition cycle. This is likely to be the problem that brought the customer to the shop.

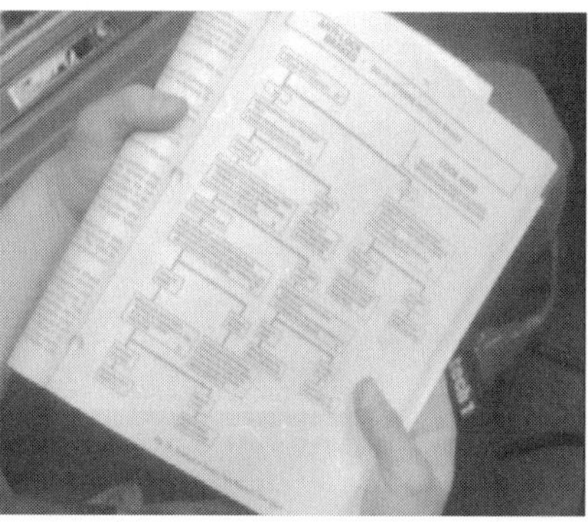

P12-9 Troubleshoot the specific DTC by following the specific diagnostic flow chart listed in the shop manual.

P12-10 If required, use the scan tool's data list mode to monitor system conditions such as wheel speed, the brake switch status, and other input and output signals.

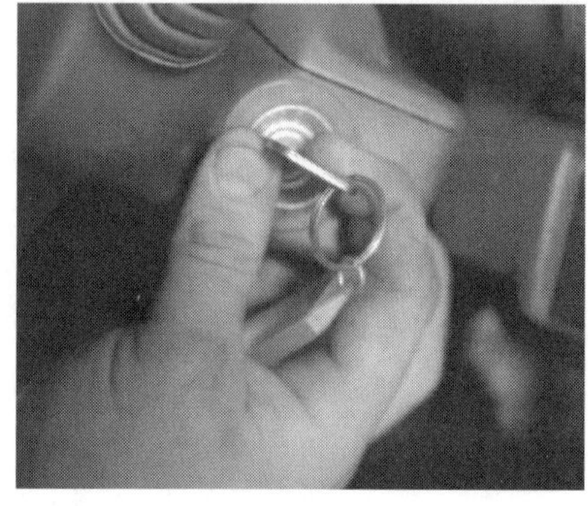

P12-11 If required, use the ABS Tests mode of the scan tool to manually operate ABS components such as the motor packs in the hydraulic modulator.

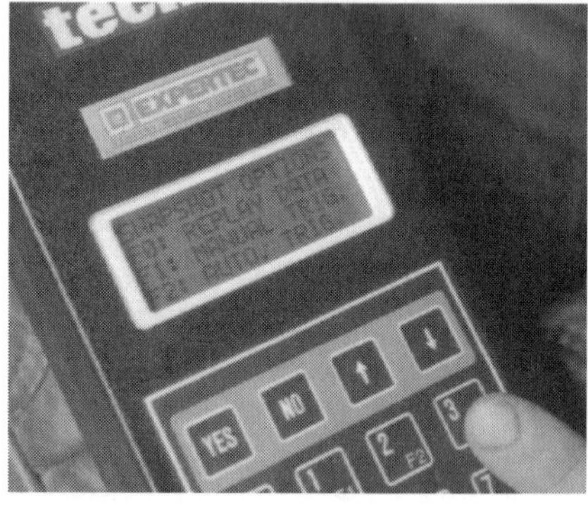

P12-12 If required, use the scan tool's ABS snapshot mode to capture a fault as it occurs on a controlled test drive.

P12-13 Once the problem has been isolated and repaired, clear all current and stored trouble codes.

P12-14 Always have the engine and the ignition off before disconnecting the scan tool from the DLC.

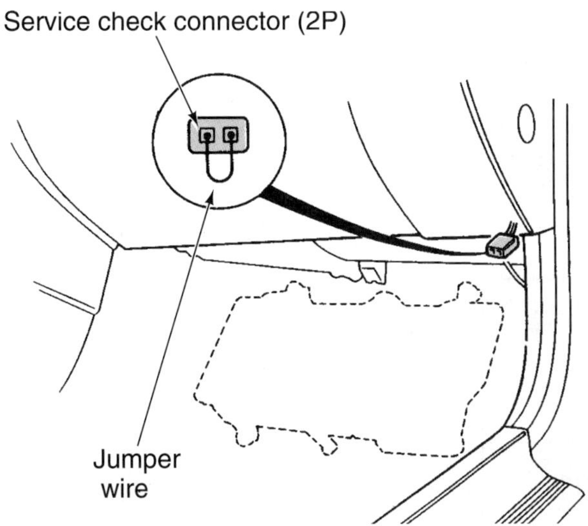

Figure 8-8 Making a jumper wire service check connection for accessing stored trouble codes (Courtesy of Honda Motor Co. Ltd.)

Accessing Trouble Codes Using the ABS Indicator Light

Following is a typical procedure for reading stored trouble codes in systems that use the ABS indicator light to flash out the code numbers.

1. Disconnect the service check connector from the connector cover and connect the two terminals of the service check connector with a jumper wire (Figure 8-8).
2. Turn on the ignition. Do not start the engine.
3. Watch the ABS indicator light and record its blinking frequency. Check the service manual for instructions on interpreting the flashing code. In this example, there are main and sub-code numbers separated by a one-second pause (Figure 8-9). This system can read out up to three separate trouble codes at one time. Each code number is separated by a five-second pause.
4. Record the main and sub-code numbers. Then refer to the applicable troubleshooting chart in the service manual (Figure 8-10).

CAUTION: Disconnect the jumper wire at the connector before starting the engine.

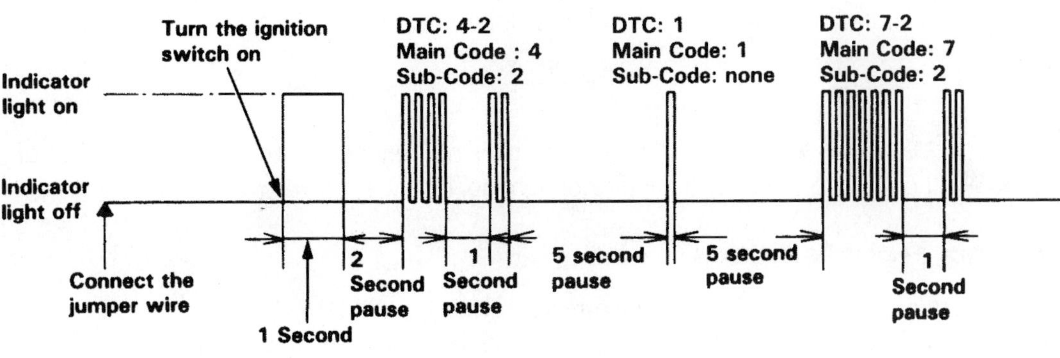

Figure 8-9 Reading main and sub codes of the flashing ABS indicator light during trouble code access sequence (Courtesy of Honda Motor Co. Ltd.)

<table>
<tr><th colspan="2">DIAGNOSTIC TROUBLE CODE (DTC)</th><th rowspan="2">PROBLEMATIC COMPONENT/ SYSTEM</th><th colspan="4">AFFECTED</th><th rowspan="2">PAGE</th><th rowspan="2">OTHER COMPONENT</th><th rowspan="2">PAGE</th></tr>
<tr><th>MAIN CODE</th><th>SUB CODE</th><th>FRONT RIGHT</th><th>FRONT LEFT</th><th>REAR RIGHT</th><th>REAR LEFT</th></tr>
<tr><td rowspan="5">1</td><td>—</td><td>Pump motor over-run</td><td>—</td><td>—</td><td>—</td><td>—</td><td>19-61</td><td>Solenoid
Pump motor
Pressure switch</td><td></td></tr>
<tr><td>2</td><td>Pump motor circuit problem</td><td>—</td><td>—</td><td>—</td><td>—</td><td>19-63</td><td>ABS Motor relay
ABS unit fuse
ABS Motor fuse</td><td>19-87</td></tr>
<tr><td>3</td><td>High pressure leakage</td><td>—</td><td>—</td><td>—</td><td>—</td><td>19-66</td><td>Solenoid</td><td></td></tr>
<tr><td>4</td><td>Pressure switch</td><td>—</td><td>—</td><td>—</td><td>—</td><td>19-67</td><td></td><td></td></tr>
<tr><td>8</td><td>Accumulator gas leakage</td><td>—</td><td>—</td><td>—</td><td>—</td><td>19-68</td><td>Pump motor</td><td></td></tr>
<tr><td>2</td><td>1</td><td>Parking brake switch-related problem</td><td>—</td><td>—</td><td>—</td><td>—</td><td>19-68</td><td>Brake fluid level switch
Brake system light</td><td></td></tr>
<tr><td rowspan="3">3</td><td>1</td><td rowspan="3">Pulser(s)</td><td>O</td><td></td><td></td><td></td><td rowspan="3">19-88</td><td rowspan="3">Wheel sensor installation</td><td rowspan="3"></td></tr>
<tr><td>2</td><td></td><td>O</td><td></td><td></td></tr>
<tr><td>4</td><td></td><td></td><td>O</td><td>O</td></tr>
<tr><td rowspan="4">4</td><td>1</td><td rowspan="4">Wheel sensor</td><td>O</td><td></td><td></td><td></td><td rowspan="4">19-69</td><td rowspan="4"></td><td rowspan="4"></td></tr>
<tr><td>2</td><td></td><td>O</td><td></td><td></td></tr>
<tr><td>4</td><td></td><td></td><td>O</td><td></td></tr>
<tr><td>8</td><td></td><td></td><td></td><td>O</td></tr>
<tr><td rowspan="3">5</td><td>—</td><td rowspan="3">Wheel sensor(s)</td><td></td><td></td><td>O</td><td>O</td><td rowspan="3">19-70</td><td rowspan="3">Modulator
Rear brake drag</td><td rowspan="3"></td></tr>
<tr><td>4</td><td></td><td></td><td>O</td><td></td></tr>
<tr><td>8</td><td></td><td></td><td></td><td>O</td></tr>
<tr><td rowspan="3">6</td><td>—</td><td rowspan="3">Fail-safe relay(s)</td><td colspan="2">O</td><td colspan="2">O</td><td rowspan="3">19-79</td><td rowspan="3"></td><td rowspan="3">19-105 (Function Test)</td></tr>
<tr><td>1</td><td colspan="2">O</td><td colspan="2"></td></tr>
<tr><td>4</td><td colspan="2"></td><td colspan="2">O</td></tr>
<tr><td rowspan="3">7</td><td>1</td><td rowspan="3">Solenoid related problem</td><td>O</td><td></td><td></td><td></td><td rowspan="3">19-76</td><td>ABS B3 fuse</td><td rowspan="3"></td></tr>
<tr><td>2</td><td></td><td>O</td><td></td><td></td><td>ABS B1 fuse
Front fail-safe relay</td></tr>
<tr><td>4</td><td></td><td></td><td>O</td><td>O</td><td>Rear fail-safe relay</td></tr>
</table>

Figure 8-10 Troubleshooting chart for flashed codes (Courtesy of Honda Motor Co. Ltd.)

SERVICE TIP: If you miscount the blinking frequency, turn the ignition switch off, then back on to cycle the indicator light again.

Erasing Trouble Codes

Once all system malfunctions have been corrected, clear the ABS DTCs. Codes cannot be erased until all codes have been retrieved, all faults have been corrected, and the vehicle has been driven above a set speed (usually 18 to 25 mph). It may be necessary to disconnect a fuse for several seconds to clear the codes on some systems. After service work is performed on the ABS system, repeat the previous test procedure to confirm that all codes have been erased.

Diagnosis of Chrysler Combined Antilock Brake and Traction Control System

Initialization and Dynamic Tests

Each time the ignition switch is turned on the controller antilock brake (CAB) performs an initialization test. During this test the CAB checks all the electrical components in the ABS and traction control system. The CAB momentarily cycles all the build pressure, decay pressure, isolation solenoids. If the driver has the brake pedal depressed during this test, pedal pulsations may be felt. This is a normal condition.

Each time the engine is started and the vehicle is accelerated to 5 to 10 mph (8 to 16 kmh), the CAB turns on the pump motor momentarily. When the CAB detects a fault during the initialization or dynamic tests, the CAB illuminates the amber ABS warning light and the amber traction control off light. Under this condition the CAB disables the ABS and traction control systems.

Diagnostic Mode

A compatible scan tester must be connected to the DLC under the instrument panel to enter the diagnostic mode (Figure 8-11). During the diagnostic mode both the red and amber brake warning lights flash. If a current fault is detected by the CAB, these lights remain illuminated without flashing. The vehicle speed must be below 10 mph (16 kmh) to enter the diagnostic mode. If the vehicle speed is above this value, the scan tester displays NO RESPONSE. The solenoid valves in the valve body cannot be actuated above 5 mph (8 kmh). When an attempt is made to energize the solenoid valves above this speed, a VEHICLE IN MOTION message is displayed.

Latching Faults. Many of the possible faults that may be detected by the CAB are latching faults. When the CAB detects a latching fault, the ABS and traction control systems are disabled and the amber ABS and traction control off warning lights are illuminated. These systems remain disabled until the ignition switch is cycled, even if the fault disappears. When the ignition switch is turned off and on, the systems are operational and the amber ABS and traction control off warning lights remain off until the CAB detects the fault again. The following are latching fault messages that may appear on the scan tester:

1. Main relay or power circuit failure
2. Pump motor circuit not operating properly
3. Pump motor running without command
4. Pedal travel sensor circuit

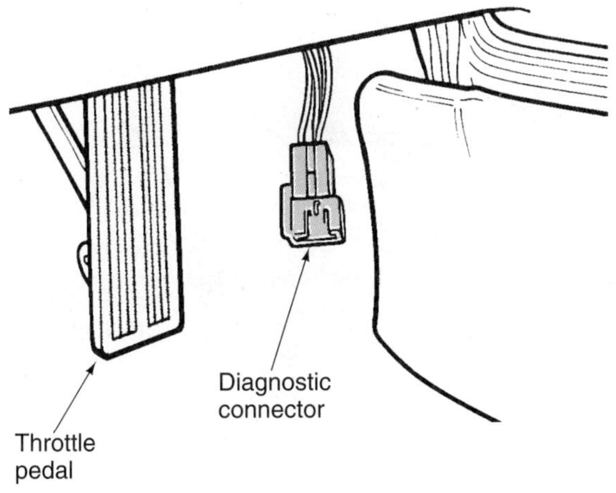

Figure 8-11 Chrysler ABS diagnostic connector location (Courtesy of Chrysler Corporation)

5. Solenoid valve fault (any one of 10 solenoids with traction control)
6. Fluid level switch #2 signal not being processed
7. Pressure switch or brake switch circuits
8. Wheel speed sensor circuit failure (any one of four sensors)
9. Signal missing, wheel speed sensor
10. Wheel speed sensor signals improper comparison
11. Wheel speed sensor continuity below 25 mph
12. Wheel speed sensor continuity above 25 mph

Non-Latching Faults. When the CAB detects a non-latching fault the amber ABS and traction control off warning lights are illuminated, and the ABS and traction control systems are disabled. If the defective condition disappears both warning lights go off, and the ABS and traction control systems are operational, but a DTC remains in the CAB memory. FLUID LEVEL SWITCH #2 OPEN is a non-latching fault.

Locked Fault. If the CAB detects a locked fault, the amber ABS and traction control off warning lights are illuminated and the ABS and traction control systems are disabled. These lights remain illuminated until the fault is erased from the CAB memory. HYDRAULIC FAILURE is a locked fault, and this fault can only be erased with the erase faults mode in a scan tester.

Noneraseable Fault. CONTROLLER FAILURE is a noneraseable fault. When this fault message is displayed on the scan tester, the CAB must be replaced.

Brake Bleeding. When bleeding the brakes on the combined ABS and traction control system, the first step is to bleed the brakes at each wheel with a pressure bleeder as explained later in this chapter. The second step in the brake bleeding procedure is to connect the scan tester to the ABS DLC under the instrument panel and use the scan tester to perform a hydraulic control unit bleed procedure. The third step is to repeat the brake bleeding procedure at each wheel using the pressure bleeder.

Since this system does not have a high-pressure accumulator, there is no special procedure for checking the brake fluid level in the master cylinder reservoir.

Ford Systems

Classroom Manual
Chapter 8, page 178

The trouble codes in most Ford ABS systems are retrieved with their Super Star Tester II, the NG5 Tester, or another compatible scan tester (Figure 8-12). The tester is connected to the test connector. The ignition is turned on and the codes are read and recorded.

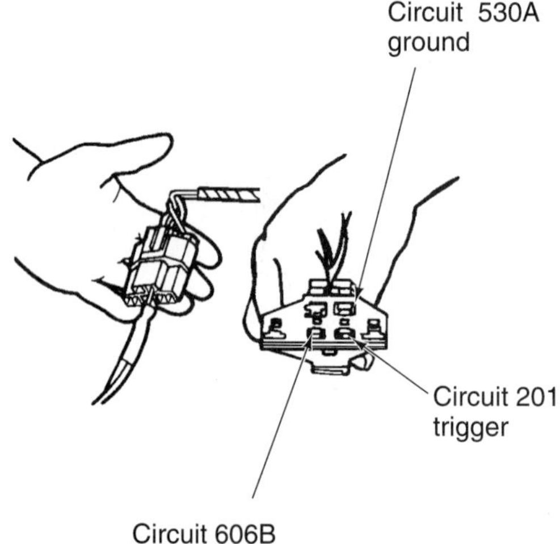

Figure 8-12 Star Tester connections (Reprinted with the permission of Ford Motor Company)

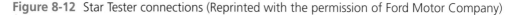

If the first code is in the 20s and no other codes are displayed, locate and repair the cause of this code before proceeding. No other codes can be output if a 20s fault exists. After that repair, continue with the test.

If only a code 10 or no codes at all are retrieved, checks with the breakout box must be completed as some of the possible faults within the system are not recognized by the control unit. The individual component tests that can be conducted with the breakout box and their specifications are shown (Figure 8-13).

Classroom Manual
Chapter 8, page 189

Ford trucks with RABS display trouble codes by causing the rear antilock light to flash. To retrieve these flashout codes, locate the RABS diagnostic connector and attach a jumper wire to it.

On-Board Self-Test Service Code Index

SERVICE CODE (COMPONENT)	PINPOINT TEST STEP
11 (Electronic Controller)	AA1
22 (Ref. Voltage of IFL)	BB1
23 (LH Front Outlet Valve)	BB3
24 (RH Front Inlet Valve)	BB4
25 (RH Front Outlet Valve)	BB5
26 (RH Rear Inlet Valve)	BB6
27 (RH Rear Outlet Valve)	BB7
28 (LH Rear Inlet Valve)	BB8
29 (LH Rear Outlet Valve)	BB9
31 (LH Front Sensor)	CC1
32 (RH Front Sensor)	CC6
33 (RH Rear Sensor)	CC11
34 (LH Rear Sensor)	CC16
35 (LH Front Sensor)	CC1
36 (RH Front Sensor)	CC6
37 (RH Rear Sensor)	CC11
38 (LH Rear Sensor)	CC16
41 (LH Front Sensor)	CC1
42 (RH Front Sensor)	CC6
43 (RH Rear Sensor)	CC11
44 (LH Rear Sensor)	CC16
51 (LH Front Outlet Valve)	DD1
52 (RH Front Outlet Valve)	DD3
53 (RH Rear Outlet Valve)	DD5
54 (LH Rear Outlet Valve)	DD7
55 (LH Front Sensor)	CC1
56 (RH Front Sensor)	CC6
57 (RH Rear Sensor)	CC11
58 (LH Rear Sensor)	CC16
61 (FLI Circuits)	EE1
62 (Travel Switch)	EE3
63 (Pump Motor Speed Sensor)	EE5
64 (Pump Motor Pressure)	EE8
71 (LH Front Sensor)	CC1
72 (RH Front Sensor)	CC6
73 (RH Rear Sensor)	CC11
74 (LH Rear Sensor)	CC16
75 (LH Front Sensor)	CC1
76 (RH Front Sensor)	CC6
77 (RH Rear Sensor)	CC11
78 (LH Rear Sensor)	CC16

Figure 8-13 Ford's on-board self-test service code index (Reprinted with the permission of Ford Motor Company)

FLASHOUT CODES CHART

CONDITION	ACTION TO TAKE
No Flashout Code	See Flashout Code 0
Yellow REAR ANTILOCK Light Flashes 1 Time This Code Should Not Occur	See Flashout Code 1
Yellow REAR ANTILOCK Light Flashes 2 Times Open Isolate Circuit	See Flashout Code 2
Yellow REAR ANTILOCK Light Flashes 3 Times Open Dump Circuit	See Flashout Code 3
Yellow REAR ANTILOCK Light Flashes 4 Times Red Brake Warning Light Illuminated RABS Valve Switch Closed	See Flashout Code 4
Yellow REAR ANTILOCK Light Flashes 5 Times System Dumps Too Many Times in 2WD (2WD and 4WD vehicles) Condition Occurs While Making Normal or Hard Stops. Rear Brake May Lock	See Flashout Code 5
Yellow REAR ANTILOCK Light Flashes 6 Times (Sensor Signal Rapidly Cuts In and Out) Condition Only Occurs While Driving	See Flashout Code 6
Yellow REAR ANTILOCK Light Flashes 7 Times No Isolate Valve Self Test	See Flashout Code 7
Yellow REAR ANTILOCK Light Flashes 8 Times No Dump Valve Self Test	See Flashout Code 8
Yellow REAR ANTILOCK Light Flashes 9 Times High Sensor Resistance	See Flashout Code 9
Yellow REAR ANTILOCK Light Flashes 10 Times Low Sensor Resistance	See Flashout Code 10
Yellow REAR ANTILOCK Light Flashes 11 Times Stop Lamp Switch Circuit Defective. Condition Indicated Only When Driving Above 35 mph	See Flashout Code 11
Yellow REAR ANTILOCK Light Flashes 12 Times Fluid Level Switch Grounded During a RABS Stop	See Flashout Code 12
Yellow REAR ANTILOCK Light Flashes 13 Times Speed Processor Check	See Flashout Code 13
Yellow REAR ANTILOCK Light Flashes 14 Times Program Check	See Flashout Code 14
Yellow REAR ANTILOCK Light Flashes 15 Times Memory Failure	See Flashout Code 15
Yellow REAR ANTILOCK Light Flashes 16 Times or More 16 or More Flashes Should Not Occur	See Flashout Code 16

NOTE: Refer to Obtaining the Flashout Code in this Section for procedure to obtain flashout code.

Figure 8-14 Ford's RABS flashout code chart (Reprinted with the permission of Ford Motor Company)

Momentarily touch an end of the jumper wire to ground. This should cause the rear antilock light to flash. The codes consist of short and long flashes. Count the flashes together. The codes begin with several short flashes and end with a long flash (Figure 8-14). A complete troubleshooting chart for Ford RABS is given (Figure 8-15).

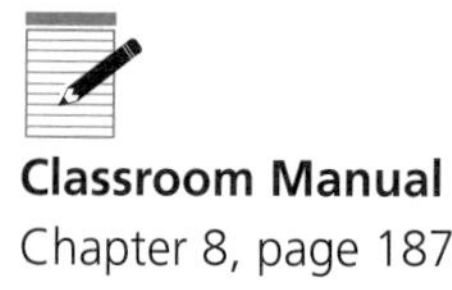

Classroom Manual
Chapter 8, page 187

General Motors' Systems

Most current model GM vehicles use the Delco-Moraine ABS-VI system. Some models also have a traction control system (TCS) incorporated into the ABS. The recommended procedure for diagnosing this system follows:

WARNING: When servicing the ABS or TCS system, the following steps should be followed in order. Failure to do this may result in the loss of important diagnostic data and may lead to difficulties and time-consuming procedures.

Troubleshooting Chart

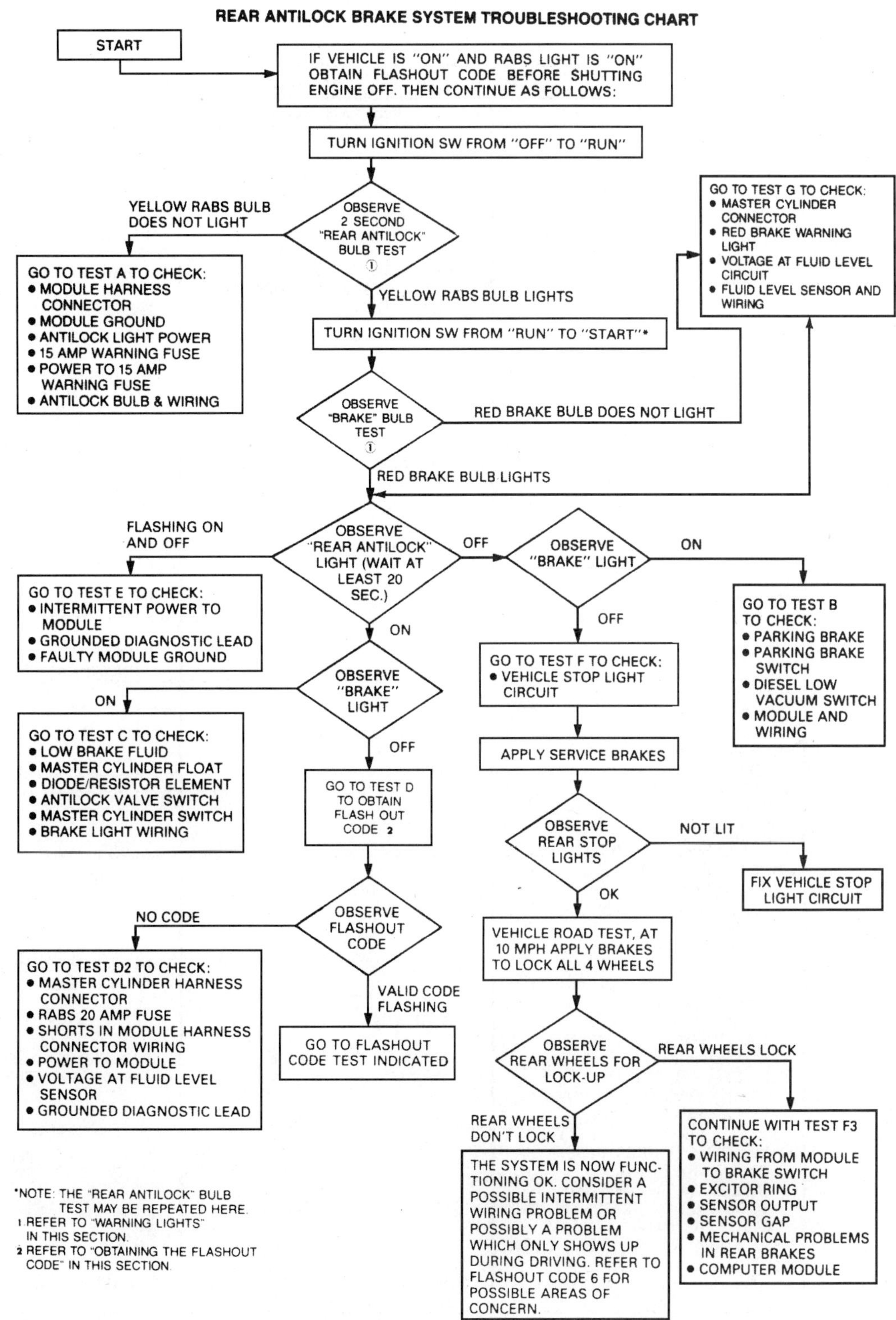

Figure 8-15 Complete troubleshooting chart for the Ford RABS system (Reprinted with the permission of Ford Motor Company)

Special Tool

Tech-1

GM's widely used ABS-VI system uses a lamp driver module. This module contains electronic components that receive a signal from the ECM that turns off the ABS indicator lamp. Unless the ECM commands the ABS indicator lamp off, it will remain on.

Classroom Manual

Chapter 8, page 187

1. Using the Tech-1 scan tool, read all current DTCs. Be certain to note them and do not clear them unless the procedures tell you to.
2. Visually inspect the following (Figure 8-16):
   a. Fluid reservoir for proper level
   b. Hydraulic modulator for any leaks or damage to the wiring

ABS/TCS VISUAL INSPECTION		
ITEM	**INSPECT FOR**	**CORRECTIVE ACTION**
PARKING BRAKE	- FULL RELEASE	- OPERATE MANUAL RELEASE LEVER TO VERIFY RELEASE - ADJUST CABLE OR REPAIR RELEASE SYSTEM AS REQUIRED
	- PROPER SWITCH FUNCTION WHEN NECESSARY, UNPLUG SWITCH CONNECTOR TO VERIFY	- REPAIR SWITCH AS REQUIRED - PROCEED TO ABS/TCS FUNCTIONAL TEST
MAJOR COMPONENTS - MASTER CYLINDER FLUID RESERVOIR	- LOW FLUID LEVEL	- DETERMINE CAUSE OF FLUID LOSS AND REPAIR. ADD FLUID TO MASTER CYLINDER RESERVOIR AS REQUIRED
- PRESSURE MODULATOR VALVE (PMV) ASSEMBLY	- LOW FLUID LEVEL IN PMV RESERVOIR	- DETERMINE CAUSE OF FLUID LOSS AND REPAIR. ADD FLUID TO MASTER CYLINDER RESERVOIR AS REQUIRED
	- EXTERNAL LEAKS	- REPAIR LEAKS AS REQUIRED
	- PROPER ASSEMBLY	- INSTALL OR POSITION COMPONENTS PROPERLY - PROCEED TO ABS/TCS FUNCTIONAL TEST
IP FUSE BLOCK - "ABS" FUSE - "CHIME" FUSE REAR FUSE BLOCK - "HZRD/STOP" FUSE RIGHT FRONT UNDERHOOD RELAY CENTER - "ABS MAIN" FUSE - "ABS PUMP MOTOR" FUSE	- OPEN (BLOWN FUSE) - PROPER ENGAGEMENT	- REPLACE - PROCEED TO ABS/TCS FUNCTIONAL TEST
CONNECTORS - MAIN RELAY - PUMP MOTOR RELAY - PMV FLUID LEVEL SWITCH - WHEEL SPEED SENSORS - ELECTRONIC BRAKE (AND TRACTION) CONTROL MODULE (EBCM/EBTCM) - PRESSURE MODULATOR VALVE (PMV) ASSEMBLY CONNECTORS: C1 - VALVE BLOCK CONNECTOR (ROUND) C2 - PUMP MOTOR CONNECTOR (GRY) - MASTER CYLINDER RESERVOIR - CRUISE/SHIFT INTERLOCK/BRAKE SWITCH - TCC/ANTILOCK BRAKE SWITCH - C101, C102, C414	- PROPER ENGAGEMENT - LOOSE WIRES OR TERMINALS - CORRODED OR BROKEN EYELETS	- PROPERLY ENGAGE CONNECTOR - REPAIR AS REQUIRED - PROCEED TO ABS/TCS FUNCTIONAL TEST
- GROUNDS - G201, G202, G104, G301	- CORRODED/LOOSE WIRES OR BROKEN EYELETS	- REPAIR AS NEEDED
WHEN NO PROBLEM IS NOTED IN VISUAL INSPECTION, PROCEED TO ABS/TCS FUNCTIONAL TEST		

Figure 8-16 GM ABS/TCS visual inspection chart (Courtesy of General Motors Corporation)

c. Brake parts of all four wheels; also check for proper operation of the base brake system
d. Wheel bearings for wear, damage, and looseness
e. Wheel sensors and their wiring for damage
f. Tires for wear patterns and excessive wear

3. If no DTCs are present or if the failure is intermittent and not reproducible, test drive the vehicle while using the snapshot feature of the Tech-1. Perform normal acceleration, stopping, and turning maneuvers. If this does not reproduce the malfunction, perform an ABS stop or TCS maneuver on a low-friction coefficient surface, such as gravel.
4. Once all system malfunctions have been corrected, clear the ABS/TCS DTCs.

There are no provisions for flash codes on this newer system; therefore, a Tech-1 or similar scan tool must be used.

Classroom Manual
Chapter 8, page 189

On some GM trucks with 4WAL, the trouble codes can be read with the Tech-1 scanner or by jumping terminal H to terminal A of the ALDL (Figure 8-17) and counting the "ANTILOCK" warning light flashes. The terminals must be jumped for a few seconds before the code will begin to flash. The light displays codes similar to the "SERVICE ENGINE SOON" light for the fuel and emissions system. The codes may be cleared with the Tech-1 scanner or by turning the ignition to the ON position and jumping across terminals H and A with a jumper wire for two seconds. Then remove the jumper wire for one second and repeat the grounding process for two seconds. The warning lights should be off and the codes erased.

The trouble codes on GM trucks with RWAL can be read in the same way as for 4WAL except the terminals need to be grounded for at least 20 seconds before codes will appear.

SERVICE TIP: Sometimes the first count sequence will be short; however, subsequent counts will be accurate.

Honda and Acura Systems

Honda provides an ALB checker to test their ABS systems after the trouble codes are retrieved. Disconnect the service connector from the connector cover under the dash prior to retrieving trouble codes (Figure 8-18). Connect the two terminals in the service connector with a jumper wire. After these terminals are connected, turn on the ignition switch, but do not start the engine. Count the

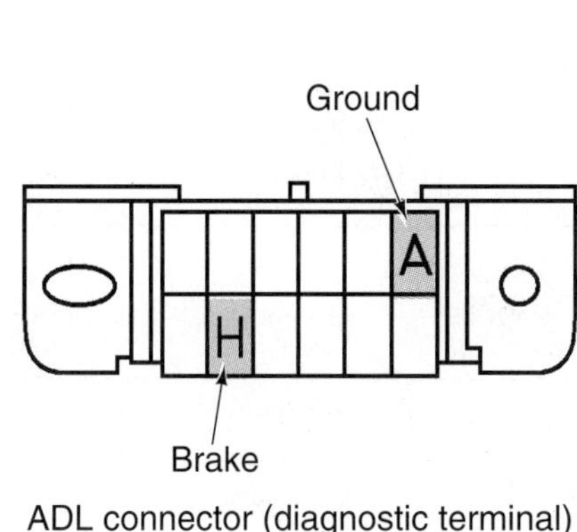

Figure 8-17 A GM assembly line diagnostic link (Courtesy of Chevrolet Motor Division of General Motors)

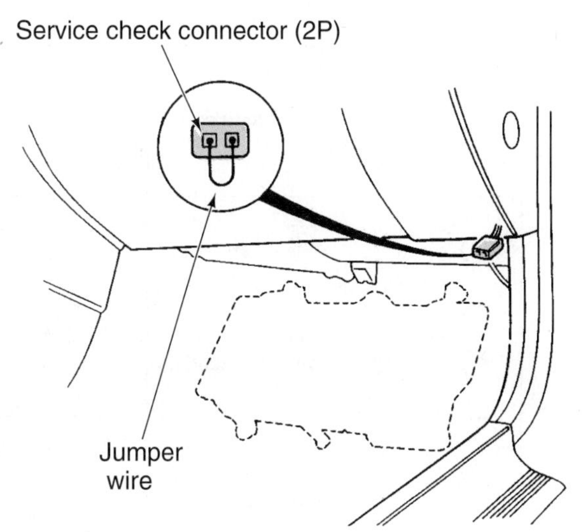

Figure 8-18 Honda's service check connector (Courtesy of Honda Motor Co. Ltd.)

flashes of the ABS warning light to obtain the trouble codes (Figure 8-19). The control unit has sufficient memory to store and display three trouble codes. To repeat the trouble code display, turn the ignition switch off and repeat the procedure. The trouble codes may be erased by disconnecting the ABS B2 fuse for at least three seconds.

WARNING: If the engine is started with the jumper wire connected to the service connector, the malfunction indicator light (MIL) will remain illuminated with the engine running.

The ALB checker simulates each ABS system function and operating mode to prove whether the system operation is normal or defective. Be sure the ABS warning light is not indicating an ABS system problem prior to using the ALB checker. When the ignition switch is turned on, the ABS

<table>
<tr><th colspan="2">DIAGNOSTIC TROUBLE CODE (DTC)</th><th rowspan="2">PROBLEMATIC COMPONENT/ SYSTEM</th><th colspan="4">AFFECTED</th><th rowspan="2">PAGE</th><th rowspan="2">OTHER COMPONENT</th><th rowspan="2">PAGE</th></tr>
<tr><th>MAIN CODE</th><th>SUB CODE</th><th>FRONT RIGHT</th><th>FRONT LEFT</th><th>REAR RIGHT</th><th>REAR LEFT</th></tr>
<tr><td>①</td><td>–</td><td>Pump motor over-run</td><td>–</td><td>–</td><td>–</td><td>–</td><td>19-61</td><td>Solenoid
Pump motor
Pressure switch</td><td></td></tr>
<tr><td></td><td>②</td><td>Pump motor circuit problem</td><td>–</td><td>–</td><td>–</td><td>–</td><td>19-63</td><td>ABS Motor relay
ABS unit fuse
ABS Motor fuse</td><td>19-87</td></tr>
<tr><td></td><td>③</td><td>High pressure leakage</td><td>–</td><td>–</td><td>–</td><td>–</td><td>19-66</td><td>Solenoid</td><td></td></tr>
<tr><td></td><td>④</td><td>Pressure switch</td><td>–</td><td>–</td><td>–</td><td>–</td><td>19-67</td><td></td><td></td></tr>
<tr><td></td><td>⑧</td><td>Accumulator gas leakage</td><td>–</td><td>–</td><td>–</td><td>–</td><td>19-68</td><td>Pump motor</td><td></td></tr>
<tr><td>②</td><td>①</td><td>Parking brake switch-related problem</td><td>–</td><td>–</td><td>–</td><td>–</td><td>19-68</td><td>Brake fluid level switch
Brake system light</td><td></td></tr>
<tr><td>③</td><td>①</td><td rowspan="3">Pulser(s)</td><td>O</td><td></td><td></td><td></td><td rowspan="3">19-88</td><td rowspan="3">Wheel sensor installation</td><td rowspan="3"></td></tr>
<tr><td></td><td>②</td><td></td><td>O</td><td></td><td></td></tr>
<tr><td></td><td>④</td><td></td><td></td><td>O</td><td>O</td></tr>
<tr><td>④</td><td>①</td><td rowspan="4">Wheel sensor</td><td>O</td><td></td><td></td><td></td><td rowspan="4">19-69</td><td rowspan="4"></td><td rowspan="4"></td></tr>
<tr><td></td><td>②</td><td></td><td>O</td><td></td><td></td></tr>
<tr><td></td><td>④</td><td></td><td></td><td>O</td><td></td></tr>
<tr><td></td><td>⑧</td><td></td><td></td><td></td><td>O</td></tr>
<tr><td>⑤</td><td>–</td><td rowspan="3">Wheel sensor(s)</td><td></td><td></td><td>O</td><td>O</td><td rowspan="3">19-70</td><td rowspan="3">Modulator
Rear brake drag</td><td rowspan="3"></td></tr>
<tr><td></td><td>④</td><td></td><td></td><td>O</td><td></td></tr>
<tr><td></td><td>⑧</td><td></td><td></td><td></td><td>O</td></tr>
<tr><td>⑥</td><td>–</td><td rowspan="3">Fail-safe relay(s)</td><td colspan="2">O</td><td colspan="2">O</td><td rowspan="3">19-79</td><td rowspan="3"></td><td rowspan="3">19-105
Function
Test)</td></tr>
<tr><td></td><td>①</td><td colspan="2">O</td><td colspan="2"></td></tr>
<tr><td></td><td>④</td><td colspan="2"></td><td colspan="2">O</td></tr>
<tr><td>⑦</td><td>①</td><td rowspan="3">Solenoid related problem</td><td>O</td><td></td><td></td><td></td><td rowspan="3">19-76</td><td>ABS B3 fuse</td><td rowspan="3"></td></tr>
<tr><td></td><td>②</td><td></td><td>O</td><td></td><td></td><td>ABS B1 fuse
Front fail-safe relay</td></tr>
<tr><td></td><td>④</td><td></td><td></td><td>O</td><td>O</td><td>Rear fail-safe relay</td></tr>
</table>

Figure 8-19 Honda's DTC code chart (Courtesy of Honda Motor Co. Ltd.)

warning light should be illuminated, and then go off after one second. Modes 1 through 5 in the ALB checker are used to verify proper ABS operation after one of these repair procedures:

1. A repair procedure involving suspension or body components that may have affected the wheel speed sensors or related wiring.
2. Replacement of components in the ABS system.
3. Replacement of the brake fluid in the ABS system or brake system bleeding.

Follow these steps when using the ALB tester:

1. Place the vehicle on a level shop floor with the wheels blocked. Place the transmission selector in park for automatic transmissions or neutral for manual transmissions.
2. With the ignition switch off, remove the cover from the ABS inspection connector under the passenger's seat.
3. Connect the ALB tester to the ABS inspection connector (Figure 8-20).
4. Start the engine and be sure the parking brake is released.
5. Turn the ALB tester mode selector switch to position #1 and push the start test switch. The test in progress light on the tester should be illuminated, and in a few seconds all four monitor lights on the tester should come on. The ABS warning light on the instrument panel should not come on.
6. Rotate the mode selector switch to position #2. Depress the brake pedal firmly and push the start test switch on the tester. The test in progress light should come on and the ABS warning light on the instrument panel should remain off. Kickback should be felt on the brake pedal.
7. Repeat step 6 with the mode selector switch in positions 3, 4, and 5. In each of these positions kickback should be felt on the brake pedal, and the test in progress light should come on while the ABS warning light remains off.

Explanation of Test Modes in ALB Tester. During test mode #1 a simulated driving speed signal of 0 mph to 113 mph (180 kmh) and then from this speed back to 0 mph is sent from each wheel speed sensor to the ABS control unit. There should be no pedal kickback in this mode.

In test mode #2 a driving signal from each wheel followed by a lockup signal from the left rear wheel is sent to the ABS control unit. There should be pedal kickback in this mode.

During test mode #3 a driving signal from each wheel followed by a lockup signal from the right rear wheel is sent to the ABS control unit. This lockup signal should result in pedal kickback.

Test mode #4 sends a driving signal from each wheel followed by a lockup signal from the left front wheel to the ABS control unit. There should be pedal kickback in this mode.

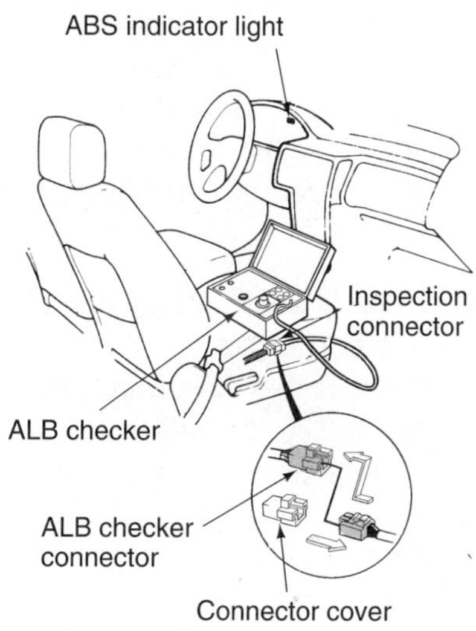

Figure 8-20 Connecting Honda's ALB checker (Courtesy of Honda Motor Co. Ltd.)

Test mode #5 sends a driving signal from each wheel followed by a lockup signal from the right front wheel to the ABS control unit. This mode should provide pedal kickback.

Interpretation of Test Results from ALB Tester. When the ABS warning light in the instrument panel is illuminated during any of the test modes, the ABS system has an electrical defect. Under this condition proceed with the DTC diagnosis. If there is no brake pedal kickback in modes 2 through 5, and the ABS warning light does not come on, check for air in the high pressure lines, a restricted high-pressure line, or a faulty modulator unit.

Wheel Speed Sensor Test with ALB Tester

The procedure for testing the wheel speed sensor signals with the ALB tester follows:

1. Leave the ALB tester connected to the ABS inspection connector.
2. Raise the vehicle on a lift.
3. Turn the ignition switch on, and position the ALB tester mode selector switch in the 0 position.
4. Rotate each wheel briskly by hand at one revolution per second. If the wheel speed sensor signal is satisfactory on the wheel being rotated, the monitor light flashes on the ALB tester for that wheel. When any monitor light does not flash, that wheel speed sensor signal is not received by the ABS control unit. If a wheel speed sensor signal is not provided during this test, check the sensor air gap, and check the sensor resistance with an ohmmeter. When these items are satisfactory, check the wires from the ABS control unit to the sensor.

Infiniti Systems

The J30 Infiniti ABS system also has a self-diagnostic mode. The testing and retrieval of trouble codes for this car model is similar to many other imported cars. To retrieve the codes, the self-diagnostic connector is grounded. The complete self-diagnostic procedure, as well as the location of the ABS warning light and the self-diagnosis connector is shown (Figure 8-21).

The DTC is represented by the number of times the warning light flashes. More than one DTC can be stored in the control unit's memory. When the codes are retrieved, the code of the most recent problem is displayed first.

Saturn Systems

Saturn DTCs are two-digit numbers that can range from 11 to 99 (Figure 8-22). When a malfunction is detected by the control unit, a code is set and the ABS warning light is lit. For 1991 through 1993 models, the warning light comes on solid to inform the driver that a problem has occurred and that the ABS has been disabled. If the warning flashes, this indicates that a problem has occurred but the ABS is still functional. For 1994 and newer vehicles, the warning lamp only flashes when certain codes are present. The codes that previously flashed the light do not turn the light on solid, rather they are held in memory to be retrieved with a scan tool. Only a scan tool can be used to monitor serial data from the control unit and retrieve and erase codes (Figure 8-23).

Relieving Accumulator Pressure

Classroom Manual
Chapter 8, page 166

Special Tools

Syringe

Clean rags

Some service operations require that the tubes, hoses, and fittings be disconnected. Many ABS systems use hydraulic pressures as high as 2,800 psi and an accumulator to store this pressurized fluid. Before disconnecting any lines or fittings, the accumulator must be fully depressurized. Following is a common method of depressurizing the ABS system. However, follow the service manual procedures for the vehicle you are working on.

1. Turn the ignition switch to the OFF position.
2. Pump the brake pedal between 25 and 50 times.
3. There should be a noticeable change in pedal feel when the accumulator is discharged.

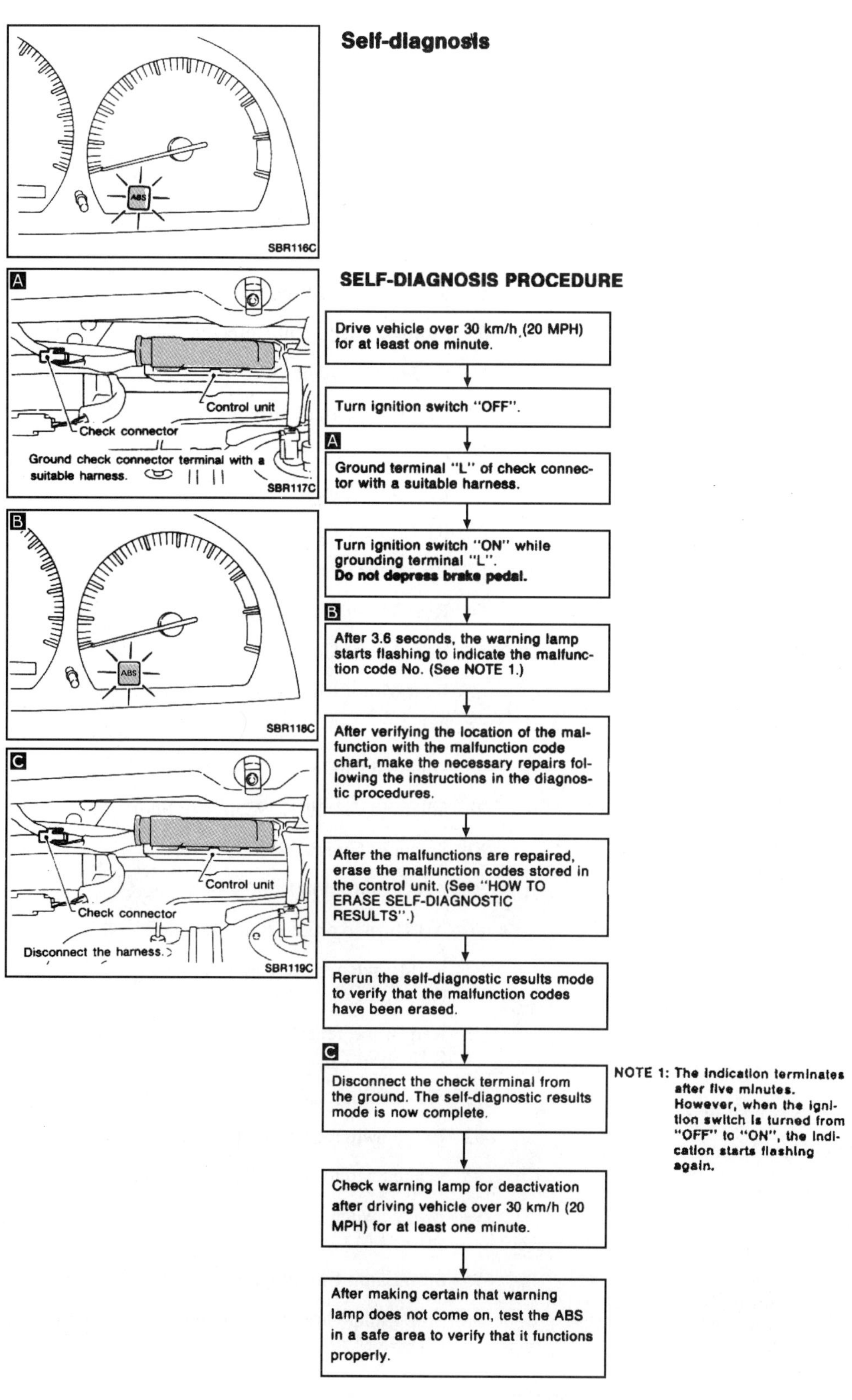

SBR116C
Self-diagnosis
A
Control unit
Check connector
Ground check connector terminal with a suitable harness.
SBR117C
SELF-DIAGNOSIS PROCEDURE
Drive vehicle over 30 km/h (20 MPH) for at least one minute.
Turn ignition switch "OFF".
A
Ground terminal "L" of check connector with a suitable harness.
B
SBR118C
Turn ignition switch "ON" while grounding terminal "L".
Do not depress brake pedal.
B
After 3.6 seconds, the warning lamp starts flashing to indicate the malfunction code No. (See NOTE 1.)
After verifying the location of the malfunction with the malfunction code chart, make the necessary repairs following the instructions in the diagnostic procedures.
C
Control unit
Check connector
Disconnect the harness.
SBR119C
After the malfunctions are repaired, erase the malfunction codes stored in the control unit. (See "HOW TO ERASE SELF-DIAGNOSTIC RESULTS".)
Rerun the self-diagnostic results mode to verify that the malfunction codes have been erased.
C
Disconnect the check terminal from the ground. The self-diagnostic results mode is now complete.
NOTE 1: The indication terminates after five minutes. However, when the ignition switch is turned from "OFF" to "ON", the indication starts flashing again.
Check warning lamp for deactivation after driving vehicle over 30 km/h (20 MPH) for at least one minute.
After making certain that warning lamp does not come on, test the ABS in a safe area to verify that it functions properly.

Figure 8-21 Self-diagnosis procedure for an Infiniti (Courtesy of Infiniti Motor Co.)

ABS DIAGNOSTIC CODE INDEX

Diagnostic codes for the *Antilock Brake System*.

I No Scan Data

11 ABS Telltale Circuit Open Or Grounded (91 - 92 MY)

12 ABS Telltale or Traction LED Fault (93 - 94 MY)

13 ABS Telltale Circuit Shorted to B+ (91 - 92 MY)

14 Switched Battery Circuit Open

15 Switched Battery Circuit Shorted to B+ (91 - 93 MY)

16 Enable Relay Coil Circuit Open

17 Enable Relay Coil Circuit Grounded (91 - 93 MY)

18 Enable Relay Coil Circuit Shorted to B+

21 LF Wheel Speed = 0 mph

22 RF Wheel Speed = 0 mph

23 LR Wheel Speed = 0 mph

24 RR Wheel Speed = 0 mph

25 LF Wheel Speed Acceleration fault

26 RF Wheel Speed Acceleration fault

27 LR Wheel Speed Acceleration fault

28 RR Wheel Speed Acceleration fault

31 Any 2 Wheel Speeds = 0 mph

36 System Voltage Low

37 ABS System Voltage High

38 Left Front ESB Does Not Hold Motor

41 Right Front ESB Does Not Hold Motor

42 Rear ESB Does Not Hold Motor

44 LF Motor Frozen

45 RF Motor Frozen

46 Rear Motor Frozen

47 LF Motor Circuit Current Low

48 RF Motor Circuit Current Low

51 Rear Motor Circuit Current Low

52 LF In Release Too Long

53 RF In Release Too Long

54 Rear In Release Too Long

55 Motor Circuit Fault Detected

56 LF Motor Circuit Open

57 LF Motor Circuit Grounded

58 LF Motor Circuit Shorted to B+

61 RF Motor Circuit Open

62 RF Motor Circuit Grounded

63 RF Motor Circuit Shorted to B+

64 Rear Motor Circuit Open

65 Rear Motor Circuit Grounded

66 Rear Motor Circuit Shorted To B+

76 Solenoid Circuit 1288 Open or Shorted to B+

77 Solenoid Circuit 1288 Grounded

78 Solenoid Circuit 1289 Open or Shorted to B+

81 Solenoid Circuit 1289 Grounded

82 ABS Calibration Fault

86 ABS turned on red *Brake* telltale

87 Red Brake Telltale Circuit Open (91 - 93 MY)

88 Red Brake Telltale Circuit Shorted to B+ (91 - 93 MY)

91 Brake Switch Circuit Open During Normal Stop

92 Brake Switch Circuit Open During ABS Stop

93 Brake Switch Circuit Open On Initialization

94 Brake Switch Circuit Always Closed (91 - 93 MY)

95 Stop Lamp Circuit Open

96 Stop Lamp Circuits Open or Grounds Open (91 - 93 MY)

Figure 8-22 Saturn's index to ABS DTCs (Courtesy of Saturn Co.)

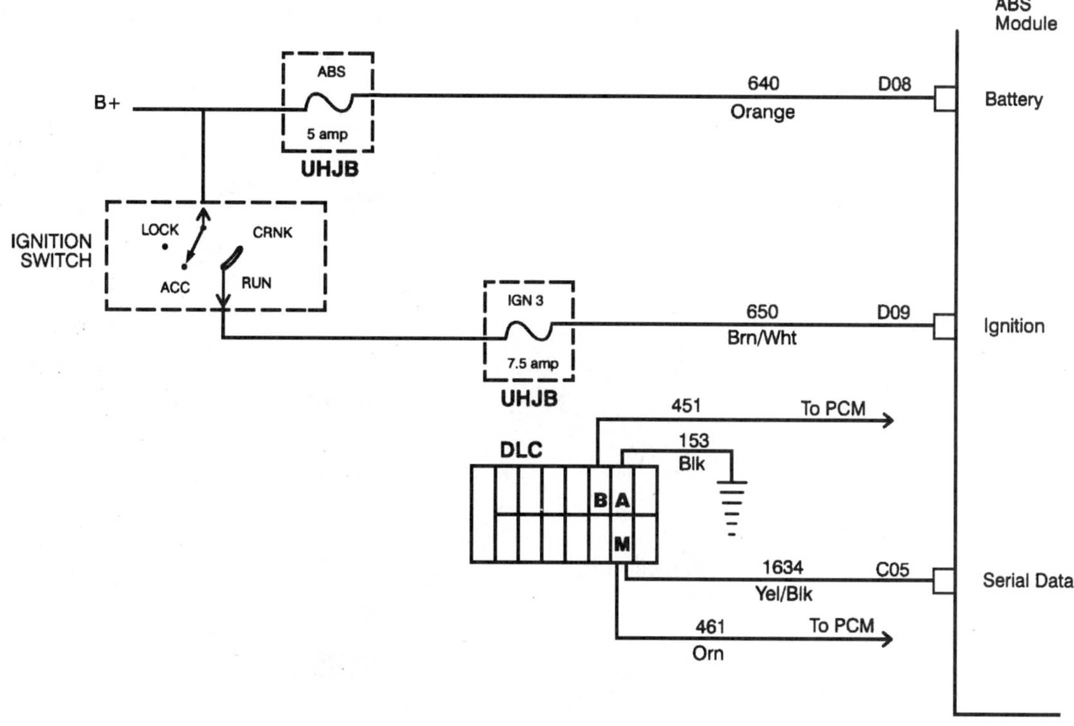

Figure 8-23 The control unit does not flash codes on Saturn ABS systems. To retrieve codes, a scan tool must be used to look at the serial data. (Courtesy of Saturn Co.)

Relieving Accumulator Line Pressure

Some manufacturers require the use of a special bleeder T-wrench to relieve pressure in the accumulator and associated lines before disassembling parts that contain highly pressurized brake fluid (Figure 8-24).

CAUTION: High-pressure fluid will squirt out if the tube that is shaded in Figure 8-24 is removed or if the solenoid head bolts are loosened.

To relieve pressure in these systems:

1. Remove the cap from the bleeder on the top of the power unit.
2. Install a special tool on the bleeder screw and turn it out slowly 90° to collect high-pressure fluid into the reservoir. Turn the special tool out one complete turn to drain the brake fluid thoroughly.
3. Retighten the bleeder screw and reinstall the cap. Discard the fluid collected.

To drain the brake fluid from the master cylinder, loosen the bleed screw and pump the brake pedal to drain the brake fluid from the master cylinder. To drain the brake fluid from the modulator reservoir, it can be sucked out through the top of the modulator reservoir with a syringe. It may also be drained through the pump joint after disconnecting the pump hose.

Testing Components with ABS Scan Tools

As mentioned earlier in this chapter, ABS scan tools and testers can often be used to monitor and/or trigger input and output signals in the ABS system. This allows you to confirm the presence

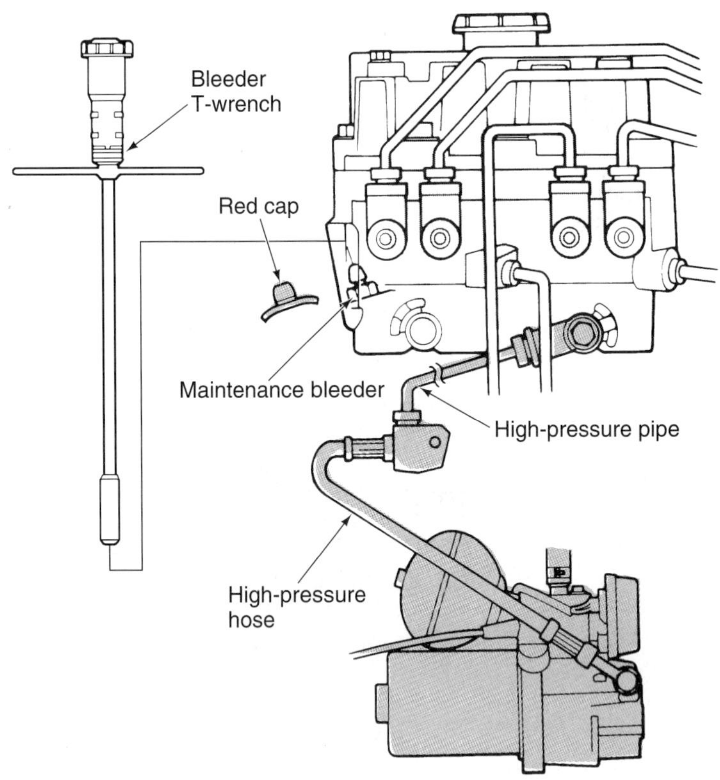

Figure 8-24 Relieving accumulator/line pressure using a bleeder T-wrench (Courtesy of Honda Motor Co. Ltd.)

of a suspected problem with an input sensor, switch, or output solenoid in the system. You can also check that the repair has been successful before driving the vehicle. Manual control of components and automated functional tests are also available when using many diagnostic testers. Details of many typical functions are contained in the following sections.

Solenoid Leak Test

This test checks for solenoid leaks in the hydraulic module. It is a typical example only. Perform the test on a level shop floor with the wheels securely blocked and the automatic transmission placed in Park.

Classroom Manual
Chapter 8, page 164

1. Disconnect the inspection connector from the connector cover and connect the inspection connector to the tester.
2. Remove the modulator reservoir filter, then fill the reservoir to the MAX level.

   NOTE: Do not reuse aerated brake fluid that has been bled from the power unit.
3. Bleed the high-pressure fluid from the maintenance bleeder connection with the special T-wrench tool (see Figure 8-24). Safely discard the fluid bled from the system.
4. Start the engine and release the parking brake.
5. Set the tester to the proper test mode and press the start test button. The ABS pump will begin to run.
6. With the pump running, place your finger over the top of the solenoid return tube in the modulator reservoir (Figure 8-25). If brake fluid movement is not felt in the return tube, the solenoids are not leaking. Reinstall the modulator reservoir filter and refill the fluid reservoir to the level marked MAX.

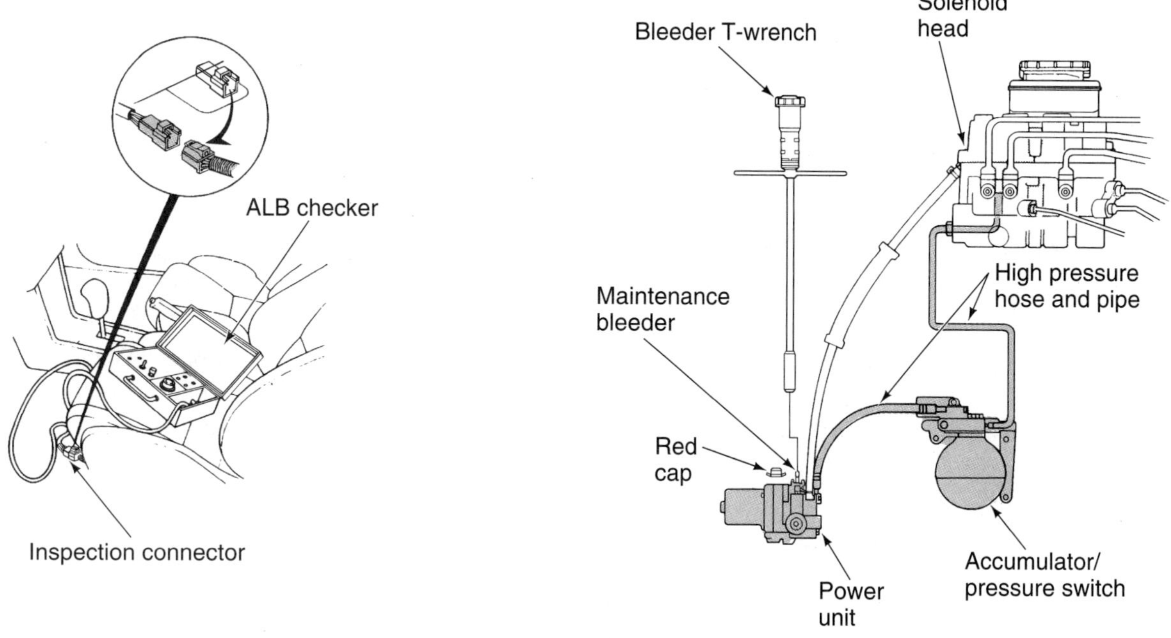

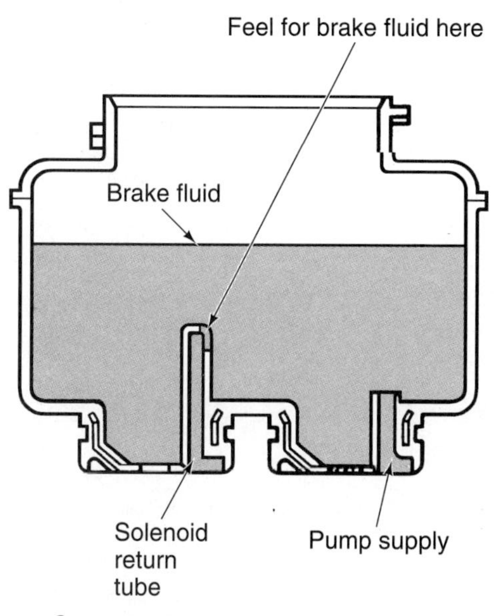

Figure 8-25 Performing solenoid leak test (Courtesy of Honda Motor Co. Ltd.)

7. If brake fluid movement is felt in the return tube, one of the solenoids is leaking. Bleed the high-pressure fluid from the maintenance bleeder with the special tool, then run through the ALB checker test modes specified in the shop manual. Bleed and test the module three or four times.
8. Now use the checker to activate the unit's pump and recheck for brake fluid coming from the return tube. If the solenoid leakage has stopped, reinstall the modulator reservoir filter and refill the reservoir to the MAX level. If one of the solenoids is leaking, the entire hydraulic module may need to be replaced. In some cases, it is possible to remove, inspect, and replace individual solenoids in the hydraulic module. The following sections outline one such procedure.

Solenoid Removal

1. Remove all of the brake fluid from the modulator reservoir.

CAUTION: Place the vehicle on level ground with the wheels blocked. Put the transmission in neutral for manual transmission models, and in park position for automatic transmission models.

2. Loosen the bleeder screw to relieve the high-pressure in the system.
3. Disconnect the inlet hose. Then remove the reservoir filter (Figure 8-26).
4. Unbolt the attaching screws and remove the modulator reservoir.
5. Remove the solenoid head from the cover.
6. Take off the solenoid cover. Then unbolt the solenoid set plate from the assembly.
7. Rotate the solenoid valves several times to loosen them. Then rotate the valves to align their tab with the notch on the set plate. Remove the solenoid valves with the set plate.

Inspection

1. Locate the inlet fitting on the solenoid valve. Connect a rubber hose to the fitting and apply compressed air to the solenoid valve through the hose (Figure 8-27).

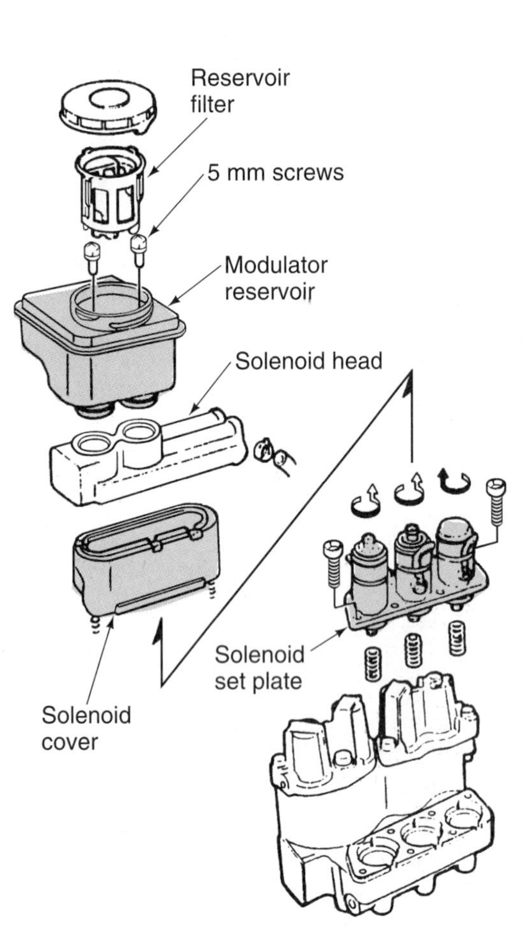

Figure 8-26 Disassembly of hydraulic modulator (Courtesy of Honda Motor Co. Ltd.)

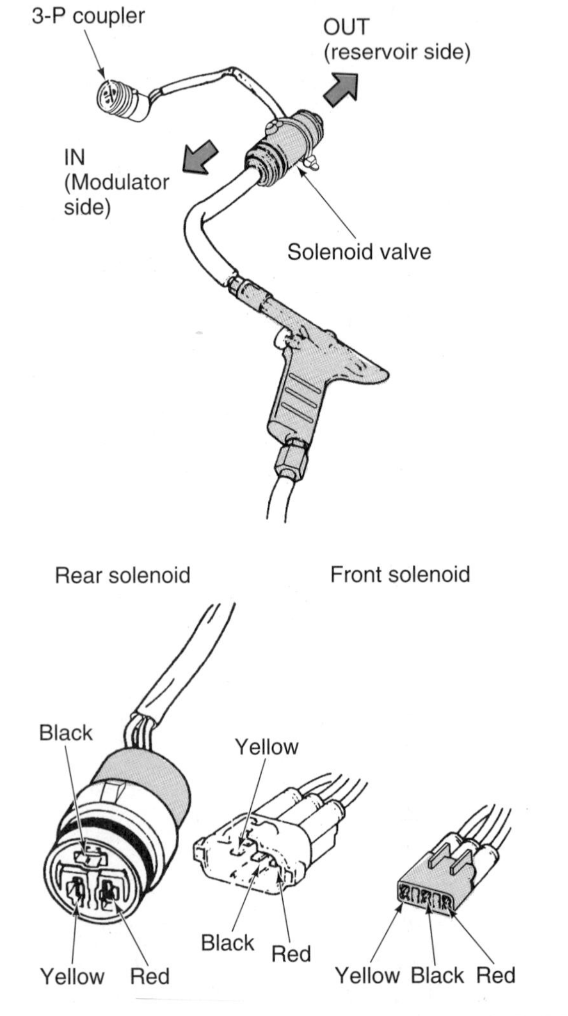

Figure 8-27 Testing solenoid operation using compressed air and a 12-V power source (Courtesy of Honda Motor Co. Ltd.)

2. Connect a 12-volt battery to the 3-P coupler terminals. Without voltage at the solenoid, there should be no airflow. When the battery is connected to the black and red terminals, air should flow through the valve. When the battery is connected to the black and yellow terminals, air should only flow into the valve.

Installing the Solenoids

1. Fill the modulator with clean brake fluid until the fluid level reaches the step in the solenoid mounting hole (Figure 8-28).
2. Coat the "O" rings with clean brake fluid and install it onto the solenoid valve.
3. Set the solenoid valves into the set plate.

WARNING: The solenoids are matched to a particular set plate opening. If the solenoid valves are interchanged when installed, the system will not work properly. Follow the markings to ensure correct installation.

4. Align the tab on the solenoid valve with the notch in the set plate, then rotate the valve 1/2 turn (Figure 8-29).
5. Install the solenoid adjust springs, solenoid valves, and set plate into the module body. Tighten the attaching bolts.
6. With the solenoid in place, install the solenoid cover, solenoid head, and modulator reservoir.
7. Install the reservoir filter and connect the low-pressure hose.

WARNING: Certain components of the ABS system are not intended to be serviced individually. Do not attempt to remove or disconnect these components. Only those components with approved removal and installation procedures in the manufacturer's service manual should be serviced.

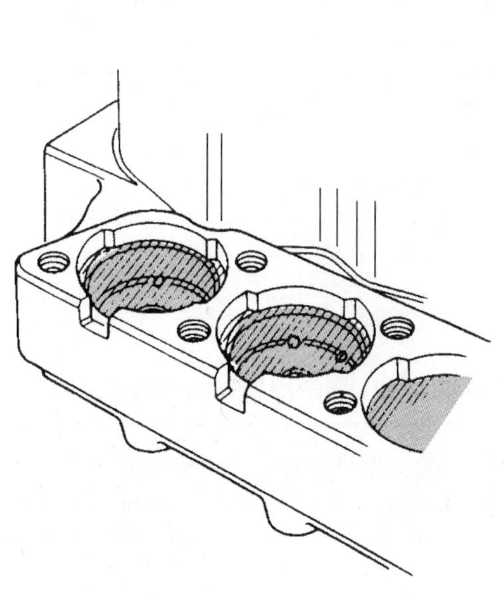

Figure 8-28 Filling the hydraulic module with brake fluid prior to installing the solenoid (Courtesy of Honda Motor Co. Ltd.)

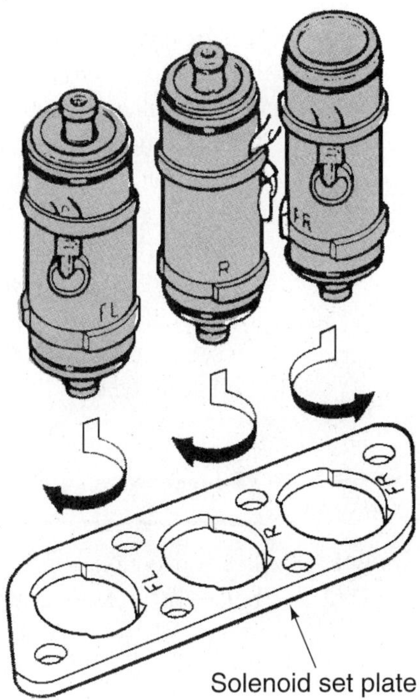

Figure 8-29 Solenoid installation into the modulator body (Courtesy of Honda Motor Co. Ltd.)

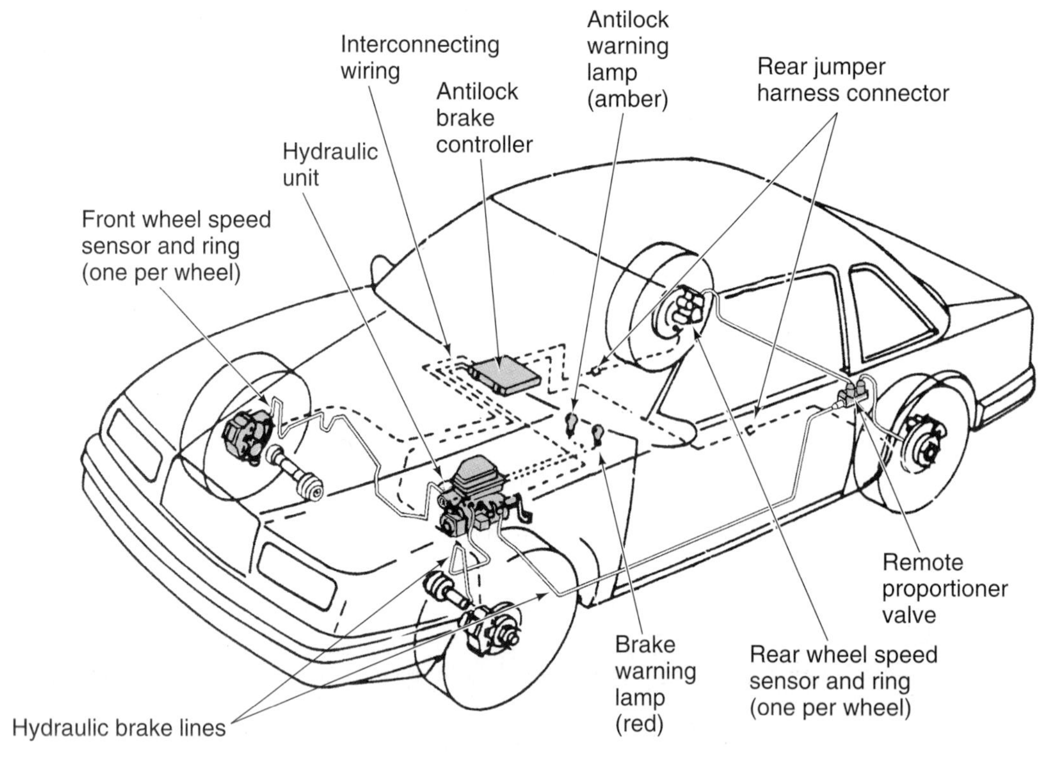

Figure 8-30 Components of a typical antilock braking system (Courtesy of General Motors Corporation)

Component Replacement

A typical antilock braking system (Figure 8-30) consists of a conventional hydraulic brake system (the base system) plus a number of antilock components. As described in Chapters 3 through 7, the base brake system consists of a vacuum power booster, master cylinder, front disc brakes, rear drum or disc brakes, interconnecting hydraulic tubing and hoses, a low-fluid sensor, and a red brake system warning light.

Antilock components are added to this base system to provide antilocking braking ability. Most ABS systems use the same operational principles, but the major components may be configured and/or named differently.

The electrical components of the ABS system are generally very stable. Common electrical system failures are usually caused by poor or broken connections. Other common faults can be caused by malfunction of the wheel speed sensors, pump and motor assembly, or the hydraulic module assembly.

Pump and Motor Assembly Replacement

Classroom Manual
Chapter 8, page 166

In a conventional braking system, the operator produces the hydraulic pressure needed to apply the brakes by pressing down on the brake pedal. A vacuum or hydraulic assist unit can help in generating application pressure. Pedal movement is passed on to pistons in the master cylinder that force the pressurized fluid out through the brake lines to the wheel assemblies.

Because the antilock brake system works independently of the conventional braking system, it needs its own method of generating hydraulic pressure. This is done through the use of an electric pump. Most systems use a single pump, but some, such as General Motors' ABS-VI system, use small electric motors in each wheel circuit to increase line pressure.

The pump normally consists of an electric motor, a filter, guide, piston rod, and cylinder body. Because the guide is positioned offset to the center of the motor's drive shaft, the rotation of the shaft and cylinder body produces a reciprocating motion in the piston rod. The reciprocating motion pressurizes the brake fluid. Pressurized brake fluid is then fed to the relief valve, accumulator, and hydraulic module. Pressures generated by the pump can be as high as 2,800 psi.

The service sample illustrated in Photo Sequence 13 is common to the GM Teves ABS system and is similar to other systems. However, it is intended only as an example and a basic guide. Refer to the manufacturer's service manual for the vehicle you are servicing for the exact procedure.

The pump may also be referred to as a motor pack or power unit. It may be a separate component or be integrated into the hydraulic module (GM).

CAUTION: Failure to follow the manufacturer's exact service procedures may result in brake system failure or personal injury. To install the replacement pump and motor assembly, reverse the procedure. Use clean brake fluid to lubricate the "O" rings before installing them. If any of the insulators are damaged or distorted, they must be replaced.

Photo Sequence 13
Pump and Motor Assembly Replacement

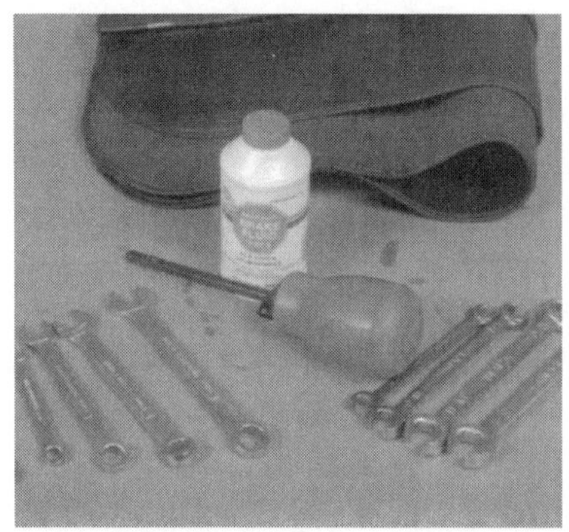

P13-1 To perform this task, you will need the following tools: fender covers, combination wrench set, line (flare) wrench set, syringe, and clean brake fluid.

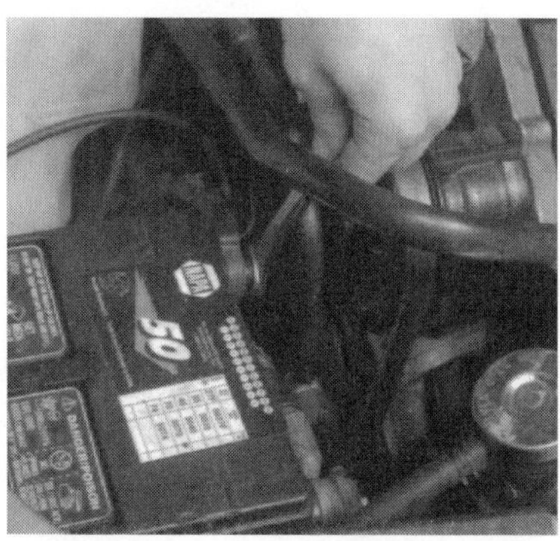

P13-2 Place the fender covers on the vehicle and disconnect the negative battery cable.

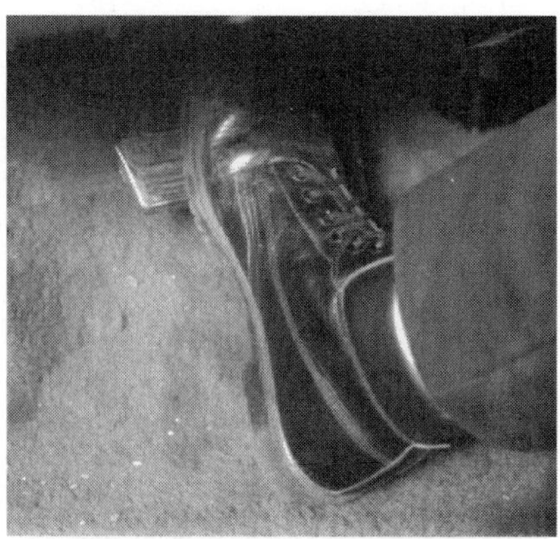

P13-3 Depressurize the accumulator by applying and releasing the brake pedal a minimum of 25 times. A noticeable change in pedal height and feel should occur when the accumulator is discharged.

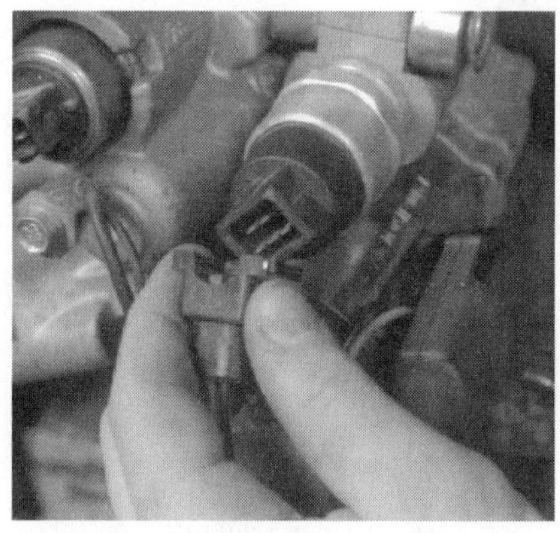

P13-4 Disconnect the electrical connector from the pressure switch and the electric motor.

P13-5 Use a clean syringe to remove the brake fluid from the reservoir.

P13-6 Unscrew the accumulator from the hydraulic module. Remove the "O"ring from the accumulator.

Photo Sequence 13
Pump and Motor Assembly Replacement (continued)

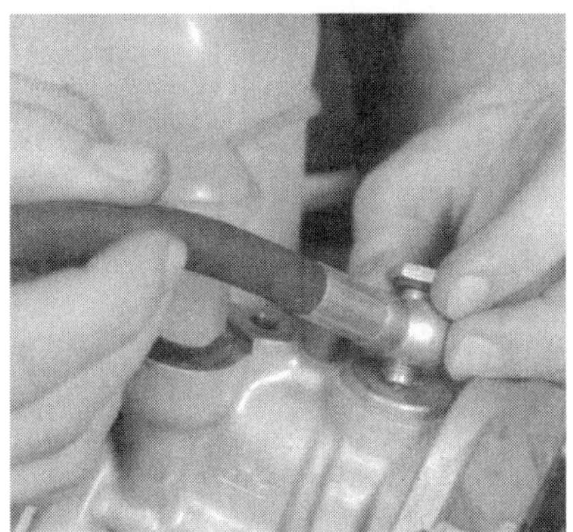

P13-7 Disconnect the high-pressure hose fitting connected to the pump.

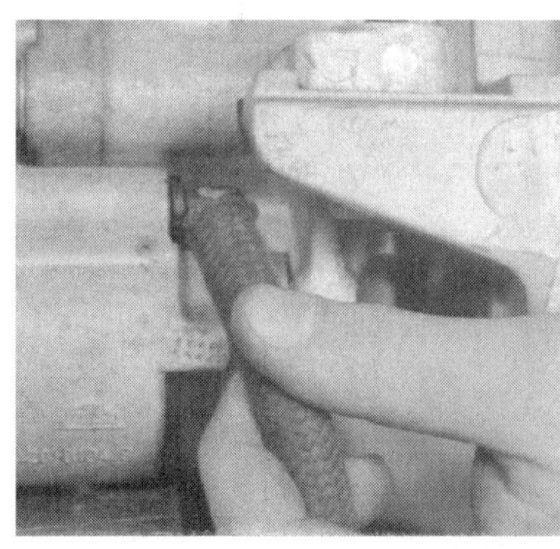

P13-8 Disconnect the wire clip. Then pull the return hose fitting out of the pump body.

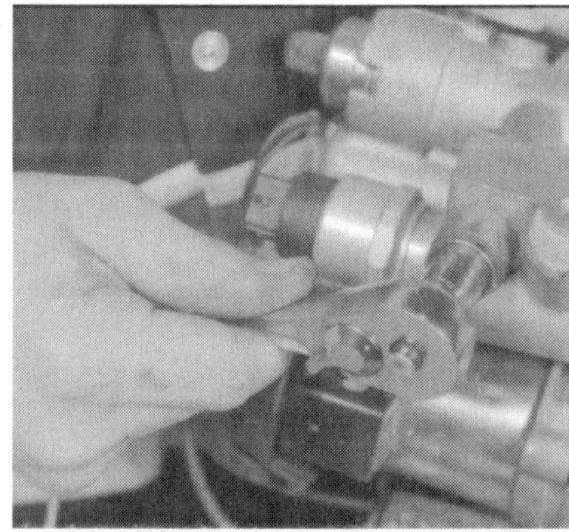

P13-9 Remove the bolt that attaches the pump and motor assembly to the hydraulic module.

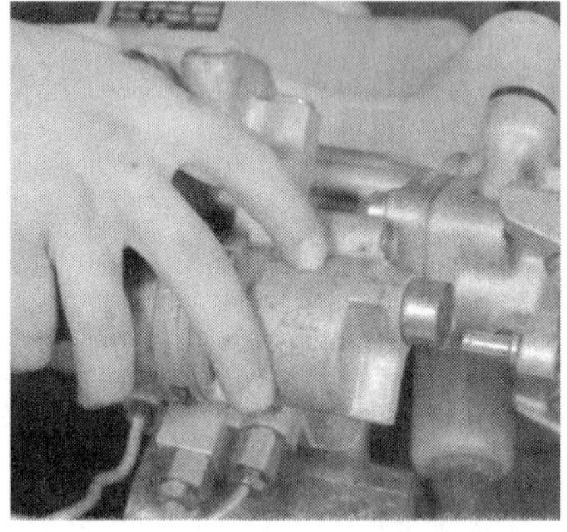

P13-10 Remove the pump and motor assembly by sliding it off the locating pin.

Classroom Manual
Chapter 8, page 164

The hydraulic module may also be called the modulator unit, hydraulic modulator, or brake pressure control valve block. In J1930 (OBD-II) terms, it is an Electronic Hydraulic Control Unit (EHCU).

Hydraulic Module (Valve Block) Assembly Replacement

The hydraulic module contains the electrical and mechanical components needed to modulate brake fluid pressure in each hydraulic brake circuit to the wheels controlled by the ABS system. These components include valves, solenoids, pistons, and various chambers and passageways (Figure 8-31). During normal driving, the ABS hydraulic circuits in the hydraulic module are bypassed and the base braking system controls the vehicle. But during hard braking, the ABS hydraulic circuits and the components in them are placed under the control of the electronic control unit.

WARNING: This service sample is similar to many systems. However, refer to the manufacturer's service manual for the system you are servicing.

To replace the hydraulic module (Figure 8-32), first disconnect the negative battery cable and then depressurize the accumulator.

1. Drain the brake fluid from the module.

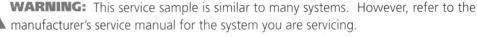

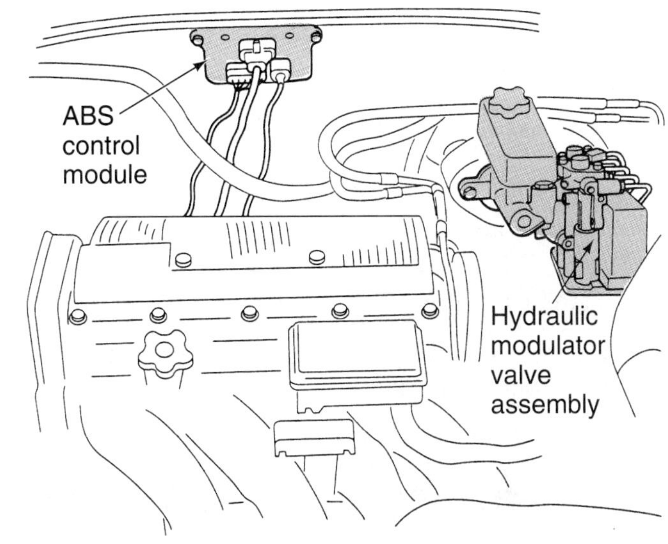

Figure 8-31 The location of the hydraulic modulator valve assembly for ABS-VI systems (Courtesy of General Motors Corporation)

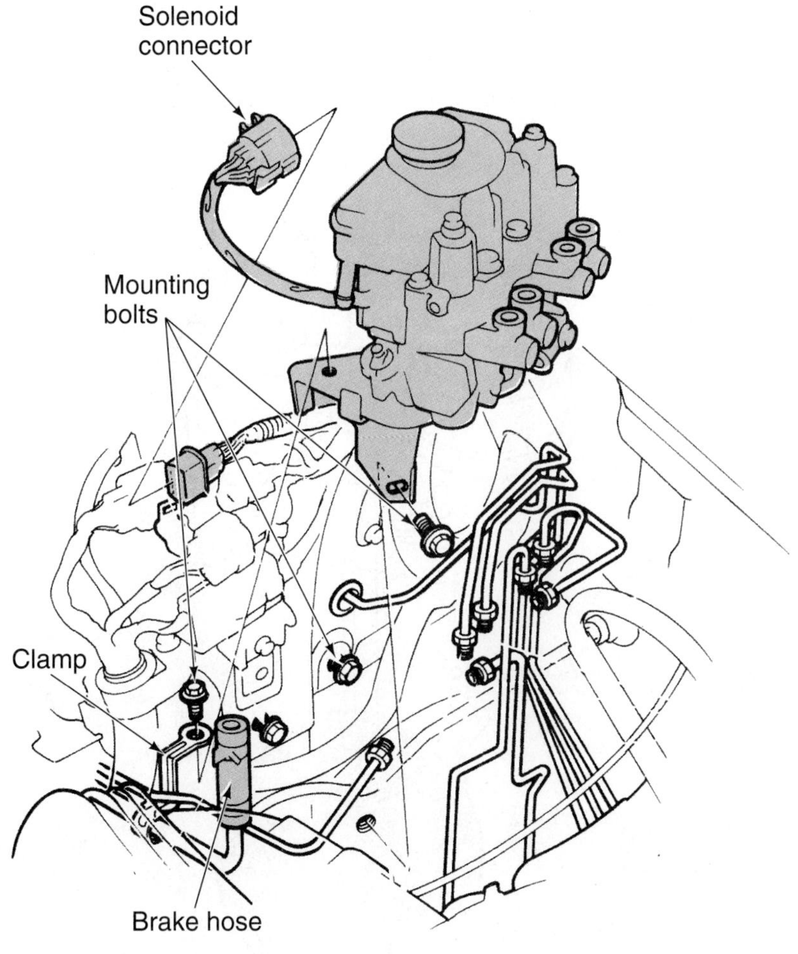

Figure 8-32 ABS hydraulic module connections and mounting details (Courtesy of Honda Motor Co. Ltd.)

2. Remove the components needed to access the module. These may include items such as the intake air duct and emission control box.
3. Disconnect the solenoid connector from the module.
4. Disconnect the brake tube lines from the module, taking care not to bend the pipes or damage the fittings. Disconnect the brake hose from the reservoir.
5. Remove any clamps and/or bolts securing the module in place and lift the module out.
6. Install by reversing the preceding steps. Check that all parts are free of dust and dirt before reassembly.
7. Bleed the system using the exact sequence recommended in the service manual.

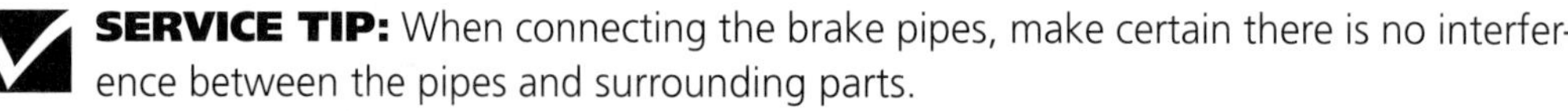

SERVICE TIP: When connecting the brake pipes, make certain there is no interference between the pipes and surrounding parts.

Accumulator/Pressure Switch Removal

Classroom Manual
Chapter 8, page 166

In some systems, the accumulator is remotely located from the hydraulic module.

Many ABS systems are equipped with an accumulator. The accumulator stores high-pressure brake fluid fed from the pump. When the antilock brake system operates, the accumulator helps the pump supply high-pressure brake fluid to the hydraulic module via the inlet side of the solenoid valve. To maintain high hydraulic pressures inside the accumulator, the accumulator is filled with a charge of nitrogen gas at high pressure. Follow the specific service manual instructions on handling and discharging the accumulator.

After the accumulator is discharged, drain the fluid. Then loosen and remove the retaining bolts. Remove the accumulator from the accumulator bracket (Figure 8-33).

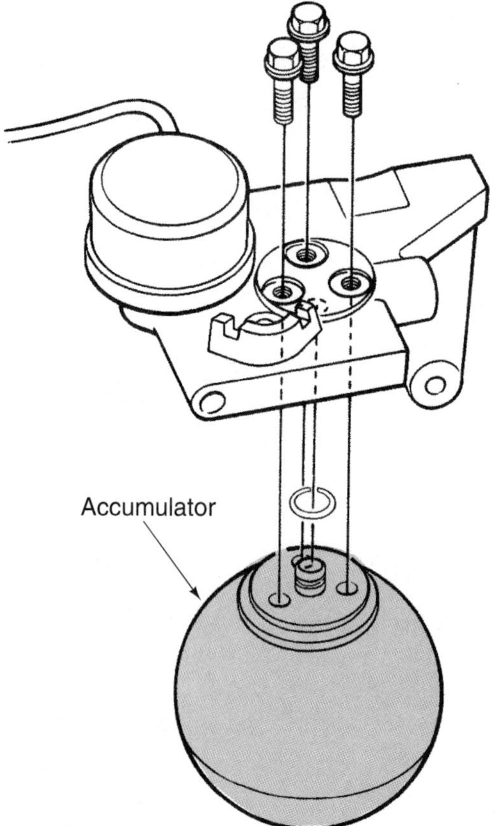

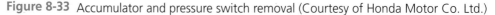

Figure 8-33 Accumulator and pressure switch removal (Courtesy of Honda Motor Co. Ltd.)

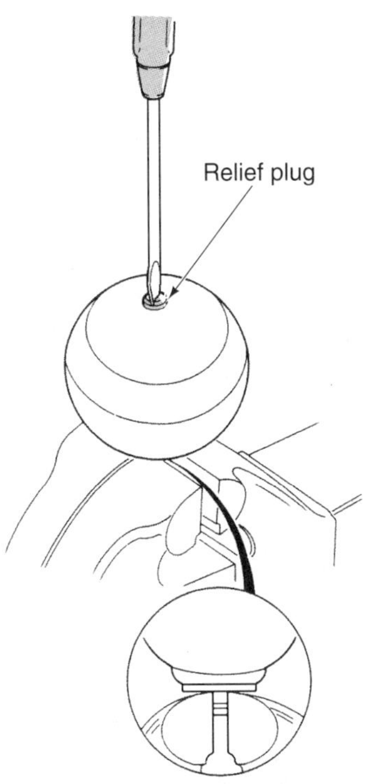

Figure 8-34 Accumulator disposal (Courtesy of Honda Motor Co. Ltd.)

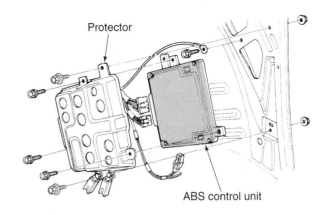

Figure 8-35 ABS electronic control unit mounting on inside of trunk wall (Courtesy of Honda Motor Co. Ltd.)

Accumulator Disposal

CAUTION: The accumulator contains high-pressure nitrogen gas. Do not puncture, expose to flame, or attempt to disassemble the accumulator. It may explode, resulting in serious personal injury. To safely dispose of the unit, follow the manufacturer's specific instructions for depressurizing and disposal. Often the accumulator is fitted with a relief plug that is loosened to allow the pressure to escape (Figure 8-34).

Electronic Control Unit Replacement

If the diagnostic trouble code troubleshooting flow chart recommends replacement of the electronic control unit, simply disconnect all electrical connectors and remove the mounting bolts holding the protective cover and unit in place (Figure 8-35). Handle the unit carefully. Remove any static charge from your body by touching a good ground before touching the ends of any connectors or terminals linked to the unit.

Classroom Manual
Chapter 8, page 162

After installing the unit, run the system self-check by turning the ignition switch to the on position.

SERVICE TIP: In many cases, the control unit's memory is cleared of all stored trouble code information when the mounting bolts are removed.

Relay Service

A wiring for a typical ABS system is shown (Figure 8-36). As you can see, various electrical relays are used to control the pump motor and/or to activate fail-safe systems. The ABS electronic control unit controls the operation of ABS power relays and pump/motor relays by grounding the relay circuit.

Classroom Manual
Chapter 8, page 162

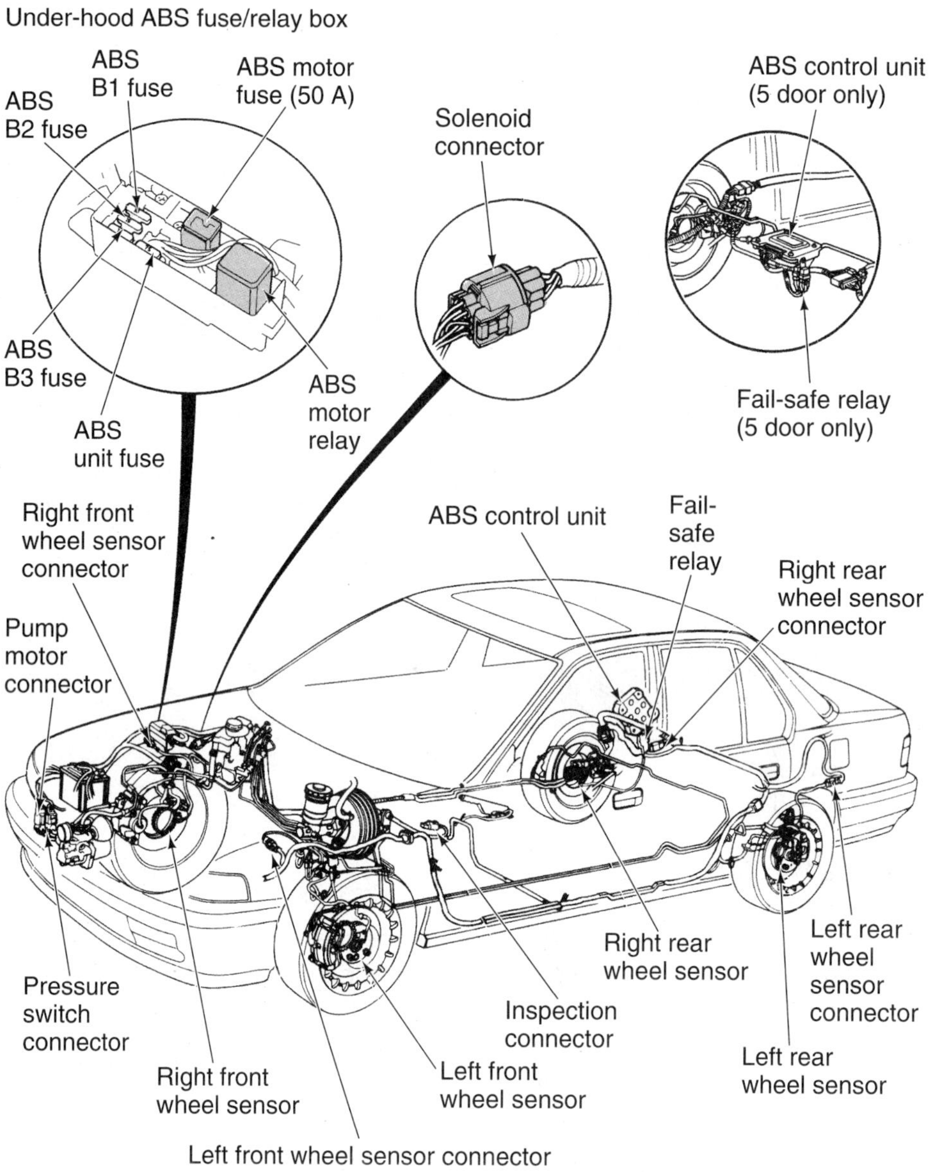

Figure 8-36 Wiring of a typical ABS system (Courtesy of Honda Motor Co. Ltd.)

Visually inspect all pump motor and fail-safe relays, checking for continuity as outlined in the service manual and troubleshooting flow charts.

Wheel Speed Sensor Service

Classroom Manual
Chapter 8, page 163

Visually inspect the wheel speed sensor rotors (pulsers) for chipped or damaged teeth. Use a feeler gauge to measure the air gap between the sensor and rotor all the way around while rotating the drive shaft or rear hub bearing unit by hand. If there is a specification on this gap, make certain it is within service manual specification (typically 0.02–0.04 in. or 0.4 mm–1.0 mm). If the gap exceeds specifications, the problem is likely a distorted knuckle that should be replaced.

Sensors are replaced by simply disconnecting the wiring at the sensor and unbolting the fasteners (Figure 8-37). Be careful not to twist the wiring cables or harness when installing the sensors. When possible, functionally test sensor input using the manufacturer's diagnostic tester.

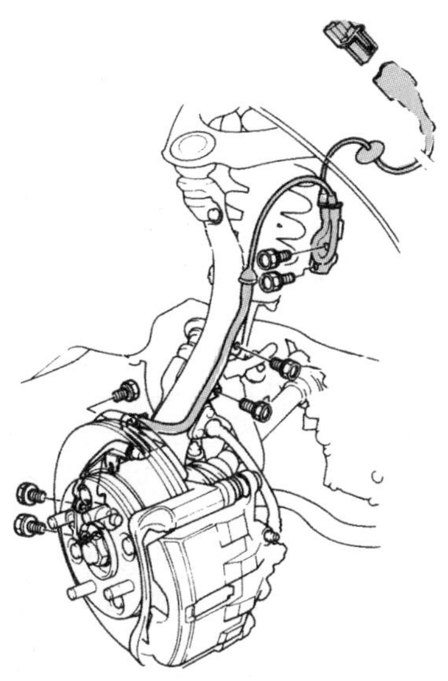

Figure 8-37 Front wheel sensor replacement (Courtesy of Honda Motor Co. Ltd.)

Wheel Speed Sensor Jumper Harness

On many GM vehicles, a jumper harness made of highly flexible twisted pair wiring is located between each wheel speed sensor and the main wiring harness (Figure 8-38). The rear jumper harnesses come together at a single connector at the rear body pass-through. This wiring exists because the main harness must connect to the suspension of the vehicle, thus the wiring in this area is subjected to the same motion as a spring or shock absorber. Consequently, any repair to this section of wiring will result in stiffening and eventual failure due to wire fracture. For this reason, the wheel speed sensor jumper harnesses are *not* repairable and must be replaced. Do *not* attempt to solder, splice, or crimp these harnesses as eventual failure will likely result.

Special Tool

Nonmagnetic feeler gauge

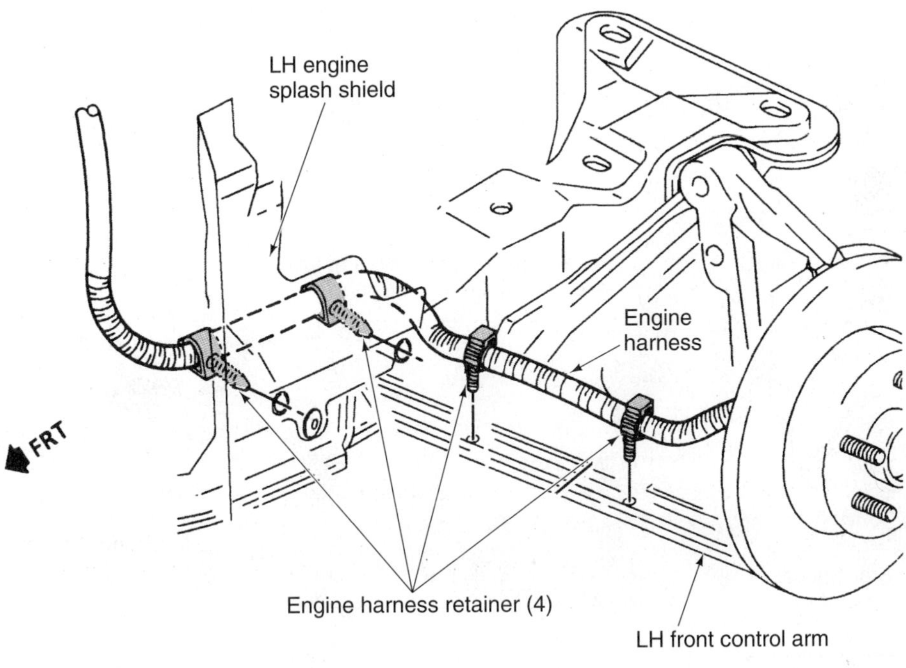

Figure 8-38 Wheel sensor jumper harness (Courtesy of Chevrolet Motor Division of General Motors Corporation)

Brake Pedal Switch

Classroom Manual
Chapter 8, page 168

The antilock brake pedal sensor switch is normally closed. When the brake pedal travel exceeds the antilock brake pedal sensor switch setting during an antilock stop, the antilock brake control module senses that the antilock brake pedal sensor switch is open and grounds the pump motor relay coil. This energizes the relay and turns the pump motor on. When the pump motor is running, the hydraulic reservoir is filled with high-pressure brake fluid, and the brake pedal is pushed up until the antilock brake pedal sensor switch closes.

When the antilock brake pedal sensor switch closes, the pump motor is turned off and the brake pedal drops some with each ABS control cycle until the antilock brake pedal sensor switch opens again and the pump motor is turned on again. This minimizes pedal feedback during ABS cycling.

If the antilock brake pedal sensor switch is not adjusted properly or is not connected to the circuit, it will result in objectionable pedal feel during ABS stops. Most concerns with the antilock brake pedal sensor switch or its installation will result in the pump motor running during the entire ABS stop. The brake pedal will become very firm, pushing the driver's foot up to an unusually high position.

To remove the switch, disconnect the wiring at the switch connector. Use a flat-blade screwdriver to carefully pry the connector locator from the brake pedal support (Figure 8-39). Unsnap the switch hook from the pin on the speed control dump valve adapter bracket. Use a needlenose pliers to squeeze the tabs on the switch mounting clip and push the clip through the hole in the pedal support bracket. Remove the sensor switch by feeding the wiring harness through the hole in the top of the pedal support bracket.

Adjustment

Adjust the ABS brake pedal sensor switch by fully depressing the switch plunger into its housing as shown (Figure 8-40).

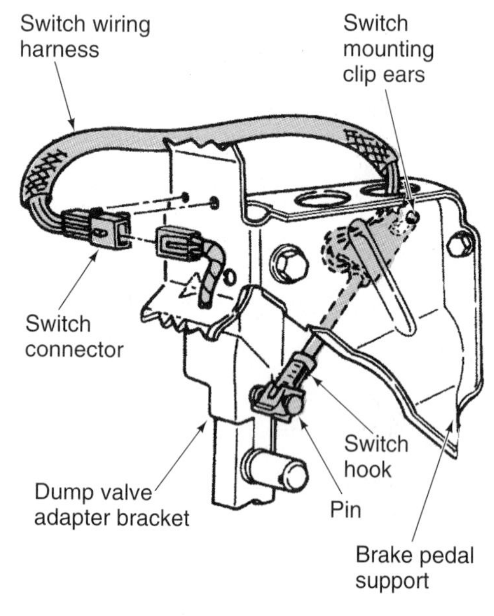

Figure 8-39 Brake pedal sensor switch mounting details (Reprinted with the permission of Ford Motor Company)

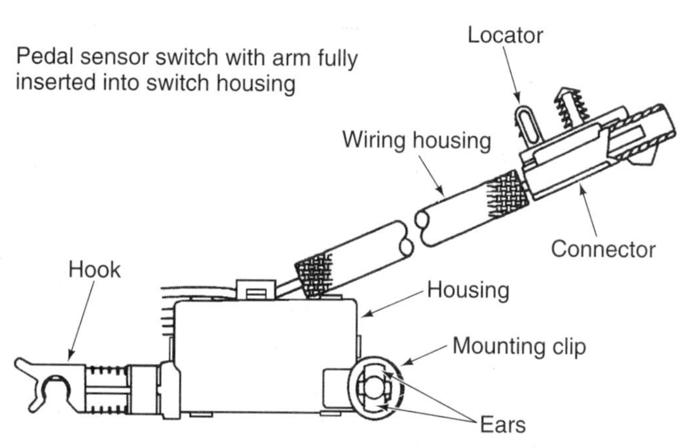

Figure 8-40 Brake pedal sensor switch with arm fully inserted into the switch housing (Reprinted with the permission of Ford Motor Company)

Slowly pull the arm out of the switch's housing until it is past the detent point. Press the brake pedal until the switch hook can be snapped onto the pin. Snap the hook onto the pin and pull the brake pedal back up to its normal unapplied position. This automatically positions the ABS sensor switch at the proper position.

SERVICE TIP: Any time the ABS brake pedal sensor switch is unhooked from the pin for any reason, it must be readjusted using the preceding procedure.

Bleeding Antilock Brake Systems

The front brakes of an antilock brake system can be bled in the conventional manner, with or without the accumulator being charged. However, bleeding the rear brakes requires a full-charged accumulator or a pressure bleeder attached to the reservoir cap opening, with a minimum of 35 psi.

Special Tools

Pressure bleeder

Clean rags

Catch can

Bleeding the ABS hydraulic circuits must be performed any time the braking system is opened, just as in the base brake system. Follow the same procedures for handling and storing brake fluid that were covered in Chapter 5.

WARNING: Use only DOT 3 and/or DOT 4 brake fluid from a clean, sealed container. Do not use fluid from an open container that may be contaminated with water. Do not use DOT 5 brake fluid in ABS systems.

Bleeding the ABS system requires removing all air from the pump and/or hydraulic module and high-pressure lines. This can be done in one of two ways, depending on the recommendation of the vehicle manufacturer. The first method requires the use of the diagnostic scan tool or other electronic breakout box bleeder adapter that can trigger pump operation. The pump purges the ABS lines of air, which is vented out through the brake fluid reservoir. Once the ABS circuits are cleared of air in this manner, the base brake system can then be pressure bled or manually bled using the standard methods covered in Chapter 5. If the ABS circuits are not bled in the recommended manner using the tester, air will be trapped in the hydraulic module, which will eventually lead to a spongy brake pedal condition. A typical ABS bleeding procedure using an ABS tester follows:

1. Park the vehicle on a level surface with the wheels blocked. Put the transmission in neutral for manual transmissions and in park for automatic transmissions.
2. Release the parking brake.
3. Disconnect the electronic system at the connection specified in the service manual and install the tester or bleeder adapter.
4. Fill the brake fluid reservoir that serves both the master cylinder and the hydraulic module to the maximum fill line.
5. Start the engine and allow it to idle for several minutes.
6. Some systems now require that you bleed off line pressure from a service fitting using a special bleeder adapter (see Figure 8-24). If this is required, refill the reservoir to replenish lost fluid.
7. Turn the selector on the tester to the proper mode for bleeding the system.
8. Firmly depress the brake pedal and activate the tester. You should feel a kickback on the brake pedal. Cycle the pump for the recommended time in all recommended test modes.
9. After bleeding the ABS system, refill the reservoir.
10. Bleed the base brake system as outlined in the service manual or in Chapter 5 of this shop manual.

SERVICE TIP: The Teves Mark IV must be bled as two independent systems. Bleeding must follow a three-step process.

1. Bleed the base brake system.

2. Bleed the hydraulic control unit (HCU) using the scan tool.
3. Repeat the bleeding of the base system.

SERVICE TIP: If a 4WAL's electronic hydraulic control unit (EHCU) needs to be bled, it should be bled after the master cylinder but before the calipers and wheel cylinders. Follow the procedure for doing this.

System Bleeding with Accumulator Charged

Once accumulator pressure is available to the system, the rear brakes can be held by opening the rear brake caliper bleeder screws for 10 seconds at a time while holding the brake pedal in the applied position with the ignition switch in the run position. Repeat until an air-free flow of brake fluid has been observed at each caliper, then close the bleeder screws. Pump the brake pedal several times to complete the bleeding procedure. Adjust the brake fluid level in the reservoir to the maximum level with a fully charged accumulator.

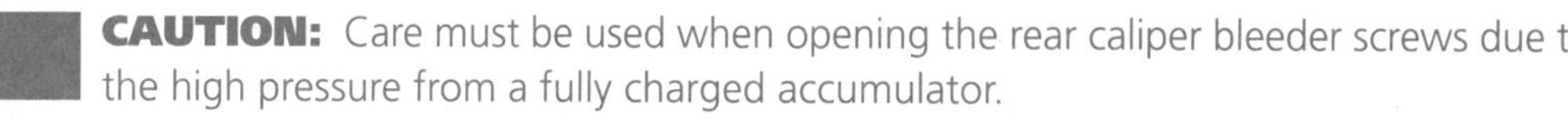
CAUTION: Care must be used when opening the rear caliper bleeder screws due to the high pressure from a fully charged accumulator.

System Bleeding with Pressure Bleeder

When using a pressure bleeder, it must be attached to the reservoir cap opening, and a minimum of 35 psi must be maintained on the system. With the brake pedal at rest and the ignition switch off, open the rear caliper bleeder screws for 10 seconds at a time. Once an air-free flow of brake fluid has been observed at each caliper, close the bleeder screws and place the ignition switch in the run position. Pump the brake pedal several times to complete the bleeder procedure. Siphon off the excess fluid in the reservoir to adjust the level to the maximum level with a fully charged accumulator.

Pressure or Manual Bleeding of ABS Systems

Not all ABS systems require a special scan tool or breakout box for system bleeding. Some systems are equipped with bleeder valves or recommend cracking open brake tube connections at certain points in the system while using pressure or manual bleeding methods similar to those outlined in Chapter 5.

SERVICE TIP: Before pressure bleeding the brakes on GM ABS-VI systems, perform the following procedure to ensure the front and rear displacement cylinder pistons are returned to the top-most position. Start the engine and allow it to run for at least 10 seconds (with your foot off the brake) to initialize the ABS. Make sure the ABS lamp is off after about 3 seconds. If the lamp remains on, a Tech-1 must be used to diagnose the system. If the lamp goes off and stays off, repeat this procedure one more time. Use a Tech-1 or T-100 scan tool to enter "manual control" function and "apply" the front and rear motors. The entire brake system can now be bled using either the pressure bleeding or manual bleeding procedures that follow.

SERVICE TIP: Only use diaphragm-type pressure bleeding equipment (Figure 8-41) with a rubber diaphragm between the air supply and the brake fluid. This prevents air, moisture, and other contaminants from entering the hydraulic system.

1. Clean the fluid reservoir cover and surrounding area, and remove the fluid reservoir cover.
2. Check the level of fluid. Add clean brake fluid if needed.

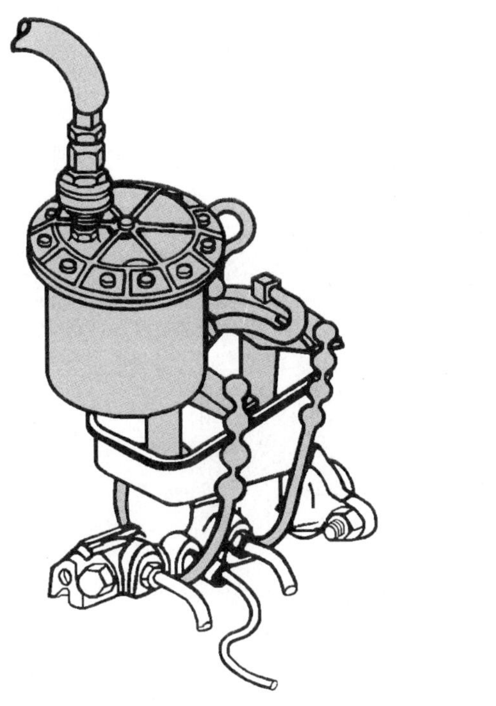

Figure 8-41 Pressure bleeder in place (Courtesy of Chevrolet Motor Division of General Motors Corporation)

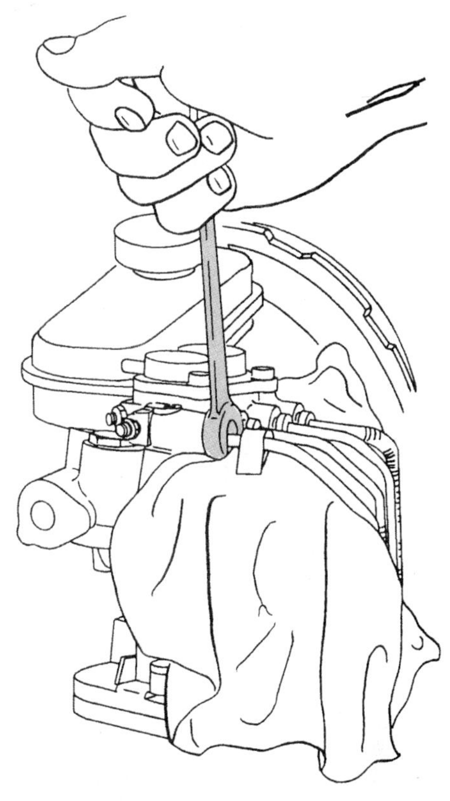

Figure 8-42 Pressure bleeding brake pipe connections (Courtesy of Chevrolet Motor Division of General Motors Corporation)

3. Connect the correct bleeder adapter to the fluid reservoir. Then connect the bleeder adapter to the pressure bleeding equipment.
4. Connect a clear plastic bleeder hose to the rearward bleeder valve. Submerge the opposite end of the bleeder hose in a clean container that is partially filled with brake fluid.
5. Adjust the pressure bleed equipment to 35-70 kPa (5-10 psi). Wait for approximately 30 seconds to be sure that there are no leaks. Set the pressure bleeder to 205-240 kPa (30-35 psi) and bleed the system.

Hydraulic Module/Master Cylinder Assembly

With all pressure bleeding equipment connected and pressurized, proceed as follows (Figure 8-42):

1. Slowly open the bleeder valve and allow fluid to flow until there is no trace of air in the fluid.
2. Close the valve, then reopen it until there are no air bubbles. Repeat this as often as needed.
3. Tighten the bleeder valve to manual specifications.
4. Install the bleeder hose onto the forward bleeder valve and repeat the above steps.

Brake Pipe Connections

With all pressure bleeding equipment connected and pressurized, proceed as follows (Figure 8-43):

CAUTION: Use a shop cloth to catch escaping brake fluid. Do not allow fluid to run down the motor pack and into the bottom or electrical connector.

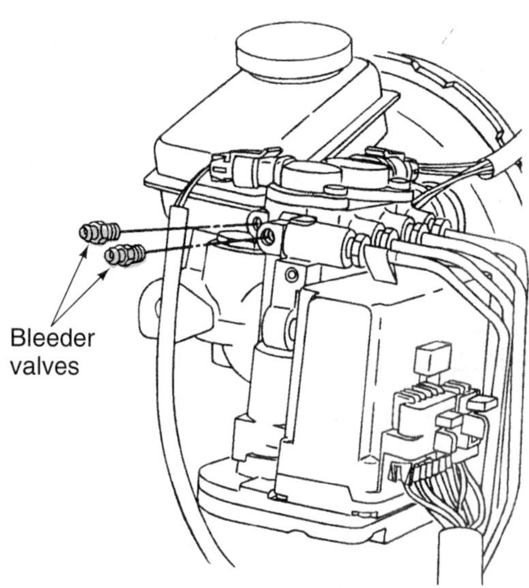

Figure 8-43 ABS-VI hydraulic modulator bleeder locations (Courtesy of Chevrolet Motor Division of General Motors Corporation)

1. Slightly open the forward brake pipe tube nut on the hydraulic module and check for air in the escaping fluid.
2. When the air flow stops, immediately tighten the tube nut to manual specifications.
3. Repeat steps 1 and 2 for the remaining three brake pipe connections, moving from front to rear.

Wheel Brakes

1. Raise the vehicle on the left or place it on jack stands.
2. When bleeding the wheel brakes, be careful to bleed in the recommended sequence.
3. Connect the bleeder hose to the bleeder valve and place the opposite hose end in a clean container that is partially filled with brake fluid.
4. Slowly open the bleeder valve and allow the fluid to flow.
5. Close the valve when fluid begins to flow without any air bubbles. Tap lightly with a rubber mallet to dislodge any trapped air bubbles.
6. Repeat this sequence on the left rear wheel brake.
7. Repeat also on the front wheel brakes.
8. Lower the vehicle.
9. Remove the bleeder adapter.
10. Check the fluid level in the reservoir and fill to the correct level using clean brake fluid, if necessary.
11. Install the fluid reservoir cover and diaphragm assembly.

Final Inspection

1. With the ignition on, apply the brake pedal with moderate force and hold. Note the pedal travel and feel.
2. If the pedal feels firm and constant and pedal travel is not excessive, start the engine. With the engine running, recheck the pedal travel. If it's still firm and not excessive, then go to step 1 under Brake Pipe Connections.

3. Road test the vehicle, making several normal (non-ABS) stops from a moderate speed to ensure proper brake system function.

SERVICE TIP: On GM ABS-VI systems, if the pedal feels soft or has excessive travel either initially or after the engine starts, use the Tech-1. Release, then apply the front and rear motors to ensure the pistons are in their upmost position. Now repeat the bleeding procedure.

Manual Bleeding

WARNING: Use only recommended brake fluids from a clean, sealed container. Do not use fluid from an open container that may be contaminated with water. Do not use DOT 5 brake fluid.

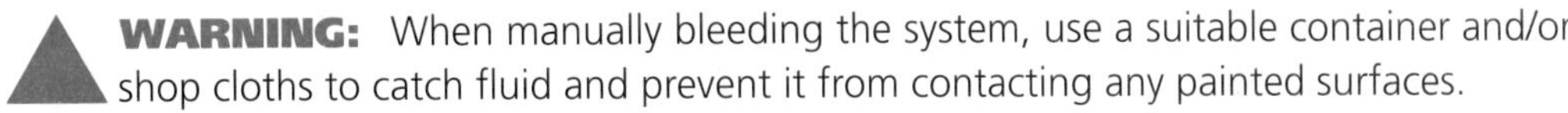

WARNING: When manually bleeding the system, use a suitable container and/or shop cloths to catch fluid and prevent it from contacting any painted surfaces.

1. Clean the fluid reservoir cover and surrounding area.
2. Remove/disconnect the fluid reservoir cover.
3. Inspect the fluid level in the reservoir and fill it to the correct level as needed.
4. Install the fluid reservoir cover.

Hydraulic Module/Master Cylinder Assembly

1. Prime the hydraulic motor/master cylinder assembly by attaching the bleeder hose to the rearward bleeder valve and submerging the opposite hose end in a clean container that is partially filled with brake fluid. Slowly open the rearward bleeder valve 1/2 to 3/4 turn. Depress the brake pedal and hold it until fluid begins to flow. Close the valve and release the brake pedal. Repeat the preceding procedure until no air bubbles are present. Repeat the procedure for the forward bleeder valve until fluid begins to flow.

WARNING: Fluid flowing from both modulator bleeder valves indicates the hydraulic module/master cylinder assembly is sufficiently full of fluid. However, the module may not be completely purged of air. To insure that it is, move to the wheel brakes and bleed them. This guarantees the lowest points in the system are completely free of air. At this point, the hydraulic module/master cylinder should be purged of any remaining air.

2. Remove/disconnect the fluid reservoir cover.
3. Inspect the fluid level in the reservoir and fill to the correct level if necessary.
4. Install/connect the fluid reservoir cover.
5. Raise the vehicle and suitably support it.
6. When bleeding the wheel brakes, it is important to use the sequence recommended in the shop manual. Connect the bleeder hose to the bleeder valve and place the opposite end of the hose in a clean container that is partially filled with brake fluid. Open the bleeder valve. Slowly depress the brake pedal. Close the valve and slowly release the brake pedal. Wait 5 seconds.
7. Repeat step 6, including the 5-second waiting period, until the brake pedal feels firm when partially depressed and no air bubbles are observed in the bleeder hose.
8. Repeat steps 6 and 7 on other wheels in the recommended sequence.

SERVICE TIP: Lightly tapping on the caliper casting with a rubber mallet will help free trapped air bubbles.

9. Remove/disconnect the fluid reservoir cover.
10. Inspect the fluid level in the reservoir and fill it to the correct level as needed.
11. Install/connect the fluid reservoir cover.
12. To bleed the ABS hydraulic module/master cylinder assembly, attach the bleeder hose to the rearward bleeder valve and submerge the opposite end of the hose in a clean container that is partially filled with brake fluid. Depress the brake pedal with moderate force. Slowly open the rearward bleeder valve 1/2 to 3/4 turn and allow fluid to flow. Close the valve and release the brake pedal. Wait 5 seconds and then repeat the preceding steps, including the 5-second wait, until all the air is purged from the system. Repeat this procedure for the forward bleeder valve until all air is purged from the system.
13. Remove/disconnect the fluid reservoir cover.
14. Inspect the fluid level in the reservoir and fill to the correct level as needed.
15. Install/connect the fluid reservoir cover.

Final Inspection

1. With the ignition on, apply the brake pedal with moderate force and hold. Note the pedal travel and feel.
2. If the pedal feels firm and constant and pedal travel is not excessive, start the engine. With the engine running, recheck the pedal travel. If the brake pedal is still firm and constant and pedal travel is not excessive, go to step 4.
3. If the pedal feels soft or has excessive travel either initially or after engine start, troubleshoot the system according to shop manual flow charts. On some GM systems you can use the Tech-1 tool to release, then apply the module motors 2 to 3 times and cycle solenoids 5 to 10 times. Be sure to apply the front and rear motors to ensure the pistons are in their upmost position. This may correct the problems associated with brake pedal travel and feel.
4. Road test the vehicle, making several normal (non-ABS) stops at a moderate speed to ensure proper brake system performance.

CASE STUDY

A customer brings his 1992 Chevrolet Lumina into the service department because the Brake Warning Light is on at all times. He is sure that the ABS doesn't work, because he read in the owner's manual that when the ABS light remains on, there is a problem with the system and the ABS is inoperative.

The technician verifies the complaint and finds the red brake light to be lit at all times. She further verifies that the vehicle is equipped with an ABS-VI system. Then she retrieves the ABS test procedures from the service manual. After the preliminary tests are completed, she uses a scan tool to retrieve any trouble codes from the system. No trouble codes are found and the Brake Warning Light remains on.

The technician consults a fellow certified technician. She summarizes the problem and the tests that have been conducted. The other technician inspects the car and identifies the major cause of the problem. The red Brake Warning Light is on, not the amber one. The amber light will light if a problem has affected the ABS. It will also light to indicate that the ABS is inoperative. The red light, however, will light when there is a problem of low brake fluid, a closed parking brake switch, or if there is a hydraulic problem in the base brake system.

Through a basic inspection, the technician found the brake fluid level to be low. Once the fluid level was corrected, the light went out. When troubleshooting a vehicle, it is important to understand the function of all warning and alert lights before jumping to conclusions.

Terms to Know

ABS brake sensor switch
ABS indicator light
Accumulator
Base brakes
Digital trouble code
Electronic control unit (ECU)
Hydraulic module
Motor pack assembly
Scan tool
Solenoid valve
Troubleshooting flow chart
Wheel speed sensor

ASE Style Review Questions

1. Antilock brake system operation is being discussed:
   Technician A says that the system works by rapidly releasing and applying hydraulic pressure to the disc or drum brake assemblies it controls.
   Technician B says that if the ABS system becomes inoperable, the base brake system still functions to stop the vehicle.
   Who is correct?
   A. A only **C.** Both A and B
   B. B only **D.** Neither A nor B

2. On/off control of the ABS system pump is being discussed:
   Technician A says the pump is often controlled by a pressure switch located in the master cylinder.
   Technician B says the pump is often controlled by a brake pedal sensor switch.
   Who is correct?
   A. A only **C.** Both A and B
   B. B only **D.** Neither A nor B

3. The amber ABS warning light is being discussed:
   Technician A says this light goes on when the system is performing its self-check whenever the engine is started.
   Technician B says this light goes on whenever the ABS system is activated during a hard stop.
   Who is correct?
   A. A only **C.** Both A and B
   B. B only **D.** Neither A nor B

4. Improper adjustment of the ABS system brake pedal sensor switch is being discussed:
   Technician A says most incorrect adjustments result in the pedal becoming very firm during ABS applications.
   Technician B says an incorrect adjustment can result in the ABS pump running during the entire ABS application.
   Who is correct?
   A. A only **C.** Both A and B
   B. B only **D.** Neither A nor B

5. A motor pack assembly on a GM ABS-VI system is leaking.
   Technician A installs the damaged external seal.
   Technician B replaces the entire motor pack assembly as a unit.
   Who is correct?
   A. A only **C.** Both A and B
   B. B only **D.** Neither A nor B

6. When determining if there is a problem with a given ABS component:
   Technician A performs the available component function tests using the system's scanning tool.
   Technician B performs a visual inspection of the component.
   Who is correct?
   A. A only **C.** Both A and B
   B. B only **D.** Neither A nor B

7. Accessing digital trouble codes from the ABS electronic control unit is being discussed:
Technician A says some systems read out the code numbers on the LED display of the service scan tool.
Technician B says some systems read out codes by flashing the amber ABS warning light and red brake system light in a given sequence.
Who is correct?
A. A only **C.** Both A and B
B. B only **D.** Neither A nor B

8. The flexible wiring harness between the wheel sensor and the main harness that is mounted to the suspension is damaged. Several wires have snapped.
Technician A replaces the entire wiring harness.
Technician B makes soldered repairs to the broken wires.
Who is correct?
A. A only **C.** Both A and B
B. B only **D.** Neither A nor B

9. Bleeding antilock braking systems is being discussed:
Technician A says cracking open the brake line connections at the hydraulic module and bleeding the air into a shop rag is the recommended procedure for some systems.
Technician B says pressure bleeding is required for bleeding antilock braking systems.
Who is correct?
A. A only **C.** Both A and B
B. B only **D.** Neither A nor B

10. Pressures inside an ABS are being discussed:
Technician A says these hydraulic pressures can reach dangerously high levels of up to 2,800 psi.
Technician B says the accumulator is a dangerous component if it is not removed and disposed of properly.
Who is correct?
A. A only **C.** Both A and B
B. B only **D.** Neither A nor B

Table 8-2 ASE TASK

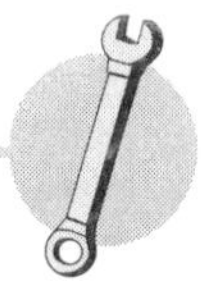

Diagnose the cause of antilock brake system problems.

Problem Area	Symptoms	Possible Causes	Classroom Manual	Shop Manual
HYDRAULICS	Low fluid level	1. Leak in the base brake hydraulic system	167	218
		2. Accumulator charge is low or is gone	166	228
		3. Leak in the ABS hydraulic circuit	164	248
	Pressure drops at the accumulator	1. Faulty pump/motor assembly	166	247
		2. Leak in the ABS hydraulic circuit	164	248
		3. Accumulator charge is low or is gone	166	228
		4. ABS operation on icy surface for an extended period of time		
	Pressure drops at the boost pressure and pressure differential switches	1. Master cylinder's primary piston is leaking	55	60
		2. Leak in the ABS hydraulic circuit	164	248
		3. Leak in base brake hydraulic system	167	218

Table 8-2 ASE TASK (continued)

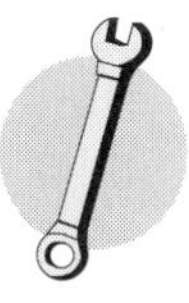

Diagnose the cause of antilock brake system problems.

Problem Area	Symptoms	Possible Causes	Classroom Manual	Shop Manual
HYDRAULICS	Pressure differential switch in proportional valve is actuated	1. Master cylinder's secondary piston is leaking	55	60
		2. Air in base brake hydraulic system	47	59
		3. Leak in base brake hydraulic system	167	218
ELECTRICAL SYSTEM OPERATION	Computer will not self-test	1. Parking brake was left engaged during test drive	—	219
		2. Low fluid level	167	55
		3. Leak in hydraulic system	47	59
		4. Problem within the system	159	219
	Low voltage to electronic control unit	1. High resistance in wiring	162	257
		2. Discharge battery	—	44
		3. Poor ground circuit	162	251
		4. Faulty ABS relay	168	251
	Pump/motor will not operate	1. Faulty relay	168	251
		2. Loss of voltage to motor	166	247
		3. Damaged pump/motor assembly	166	247
	Pump/motor stays running	1. Shorted relay	167	246
		2. Shorted motor switch	167	246
	No or improper solenoid operation	1. Faulty modulator/solenoid assembly	164	246
		2. Electrical problem in modulator/solenoid wiring harness	169	244
		3. Faulty wheel sensor(s)	163	252
	Wheel sensor fault	1. Sensor to tone wheel gap is incorrect	163	252
		2. Defective wire to sensor(s)	163	253
		3. Sensor and tone wheel are misaligned	163	252
		4. Open sensor	163	252
		5. Damaged tone wheel	163	252

Parking Brake Service

Upon completion and review of this chapter, you should be able to:

- ❑ Diagnose problems in the parking brake system.
- ❑ Inspect the parking brake system, including cables and parts for wear, rust, binding, and corrosion.
- ❑ Clean or replace all system parts as needed.
- ❑ Lubricate the parking brake system.
- ❑ Adjust calipers with integrated parking brakes.
- ❑ Adjust the parking brake assembly and check system operation.

System Description

Basic Tools

Basic mechanic's tool set

Hydraulic lift or safety stands

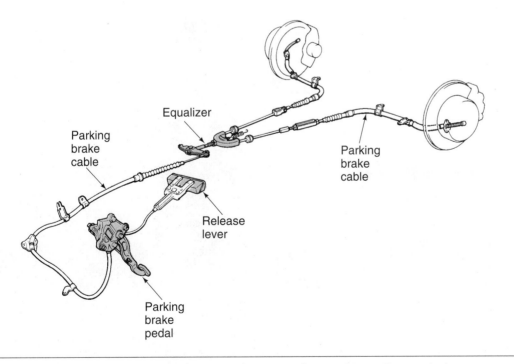

The job of the parking brake is to prevent the vehicle from rolling when it is parked. The parking brake is not part of the hydraulic braking system. It is a mechanical system using a lever assembly that connects through a cable to the rear drum or disc brake assembly. When the parking brake is applied, the brake shoes of a drum brake are forced out against the drum. In a rear disc brake, the caliper pistons are mechanically clamped in against the rotor.

The parking brake system consists of a ratchet-type foot pedal or hand lever assembly, a release lever, a front cable, an equalizer (also called an adjuster), right and left rear cables, and actuator mechanisms at the rear drum or disc brake assemblies. Brake cables are often protected by passing them through lengths of conduit. The cable and conduit are secured to the frame or underbody of the vehicle with clips and or mounting brackets. At points where the cable/conduit must pass through an opening in the frame or body, it is protected by a rubber grommet.

Foot-operated pedal parking brakes (Figure 9-1) are normally used on light trucks and larger passenger vehicles. They are mounted beneath the left-hand side of the dashboard assembly.

Equalizer
Parking brake cable
Parking brake cable
Release lever
Parking brake pedal

Figure 9-1 Typical parking brake system (Courtesy of Toyota Motor Corporation)

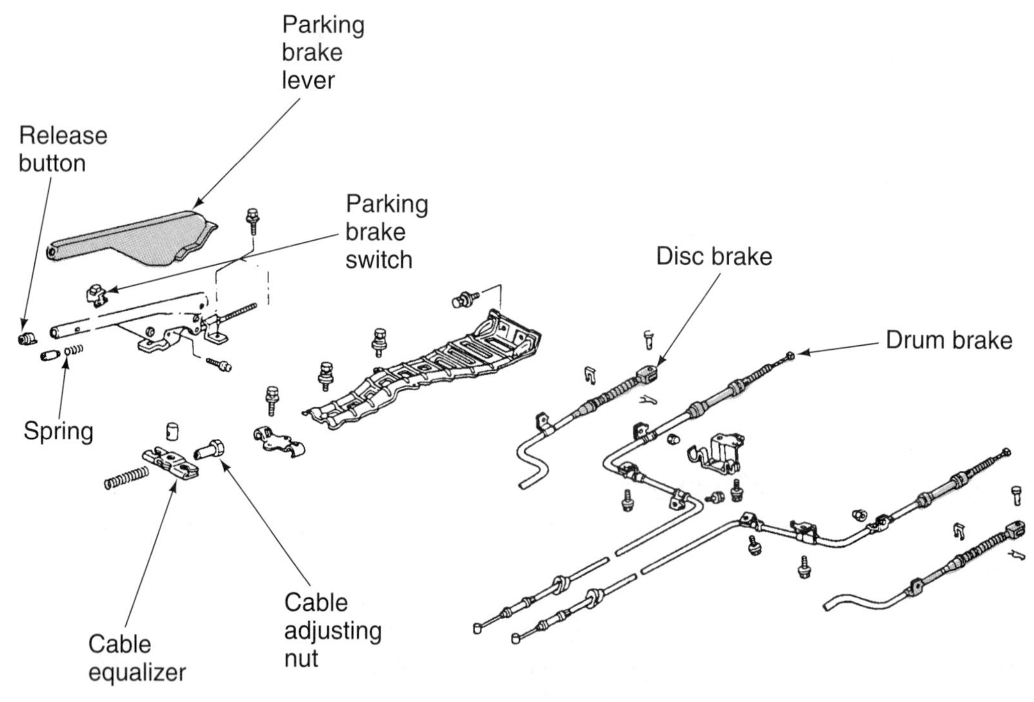

Figure 9-2 Components of console-mounted brake system. Note the difference in cable ends for drum and disc brakes. (Courtesy of Honda Motor Co., Inc.)

Hand-applied parking brakes are normally used on small to mid-sized cars and are mounted on the console between the front passenger seats (Figure 9-2).

Classroom Manual
Chapter 9, page 209

In pedal-type parking brakes, a pull-type hand release is located adjacent to the foot pedal (Figure 9-3). Hand-applied parking brakes are released by pressing in on a button on the lever tip. The lever can then be returned to the unset position.

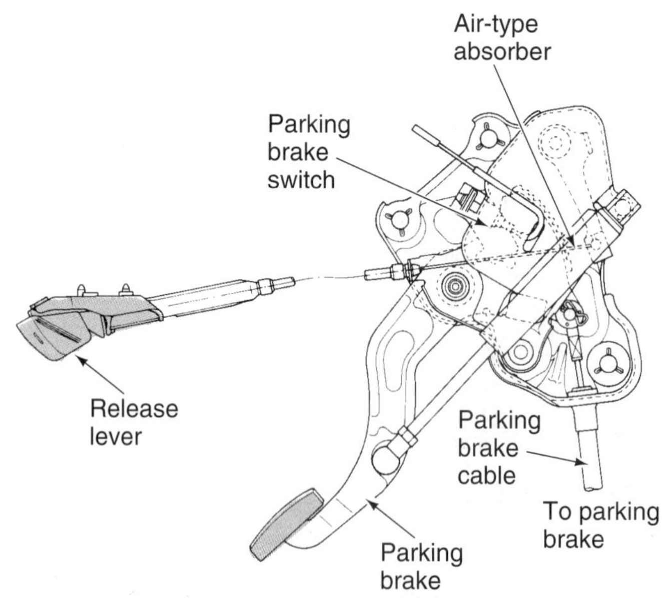

Figure 9-3 Parking brake pedal and manual release lever mechanism (Courtesy of Toyota Motor Corporation)

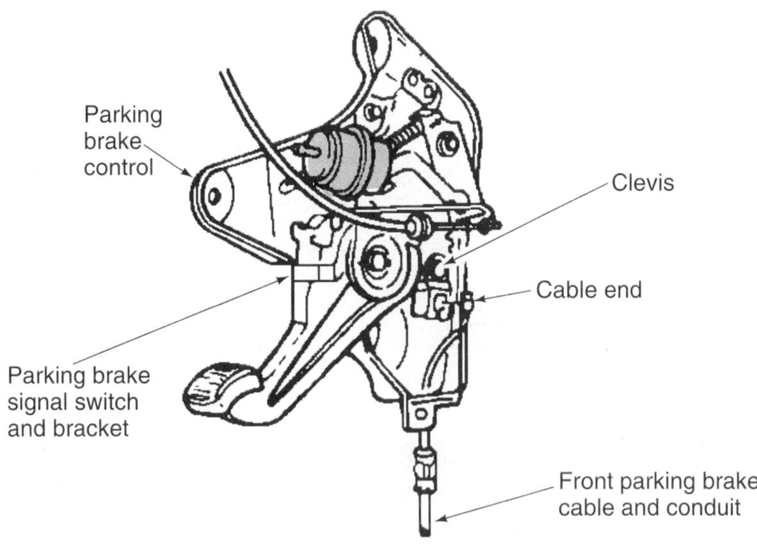

Figure 9-4 Parking brake mechanism with vacuum-powered release control motor (Reprinted with the permission of Ford Motor Company)

Automatic (Vacuum) Release Systems

On some pedal-applied systems, a vacuum-powered release control motor is added to the parking brake control (Figure 9-4). Typically, a rod connects the vacuum-actuated diaphragm inside the unit to the parking brake release handle. Hoses run from the power unit to the engine manifold. On the way to the manifold, the vacuum line is routed through a vacuum release switch located on the steering column or floor shift console. The vacuum release switch supplies engine vacuum to the parking brake release vacuum motor when the engine is running and the shift lever is placed in a forward gear. This vacuum provides the power the control motor uses to release the parking brake. The release switch vents parking brake release motor vacuum when the shift lever is placed in Park, Reverse, or Neutral. In a typical system, the release switch operates off a cam located on the shift tube. The cam opens and closes the switch between the Neutral and Drive positions. The vacuum release system can also be manually released by the operator at any time by simply pulling on the release lever as with an ordinary parking brake.

Parking Brake Indicator Light Switch

When the parking brake pedal is applied (Figure 9-4), it closes a switch, which completes an electric circuit to the brake indicator light in the instrument panel. The parking brake applied indicator light will then light when the ignition is turned on. The light goes out when the parking brake is released or the ignition is turned off. In some vehicles, this same indicator light may be used to alert the driver to problems in the antilock brake system.

On Canadian vehicles, applying the parking brake pedal also disables the daytime running lamps.

CUSTOMER CARE: The parking brake is not designed for use in place of the service brakes and should be applied only after the vehicle is brought to a complete stop, except in emergency. If you suspect a customer is using the parking brake for other than its intended job, remind the customer of the dangers of this practice and stress the importance of keeping the service brakes in good working order.

Operation

The starting point of a typical parking brake cable and lever system is the foot pedal or hand lever. This assembly is a variable ratio lever mechanism that multiplies the force exerted on the pedal or

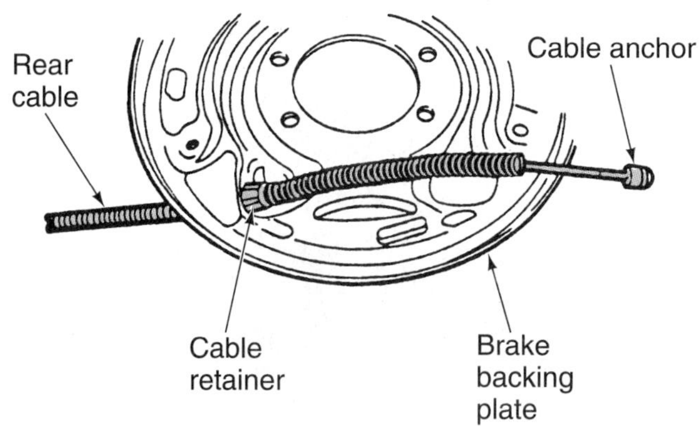
Rear
cable
Cable anchor
Cable
retainer
Brake
backing
plate

Figure 9-5 The parking brake cable enters the drum brake assembly through an opening in the drum backing plate. (Reprinted with the permission of Ford Motor Company)

hand lever by the operator. The cable connected to the pedal or hand lever mechanism travels a shorter distance, but the force it exerts on the equalizer is greater.

As the front cable assembly is pulled, tensile force from the front cable is transmitted through the car's brake cable system to the equalizer lever.

When the parking brake pedal is applied, the cables and equalizer exert a balanced pull on the parking brake levers of both rear brakes.

The equalizer lever multiplies the pulling force of the front cable and transfers it to the right and left rear cables. The equalizer also ensures equal pull on both rear cables. The equalizer functions by allowing the rear brake cables to slip slightly so as to balance out small differences in cable length or adjustment. This tension pulls the flexible steel cables attached to each of the rear brakes. The rear cable enters each rear brake through a conduit (Figure 9-5). The cable end engages the lower end of the parking brake lever. On drum brakes, this lever is hinged to the web of the secondary shoe and linked with the primary shoe by means of a strut. The lever and strut expand both shoes away from the anchor and wheel cylinder and into contact with the drum as the cable and lever are drawn forward. The shoe return springs reposition the shoes when the cable is slacked (Figure 9-6). On rear disc brake assemblies, pulling the lever forces the caliper pads against the disc brake rotor (Figure 9-7).

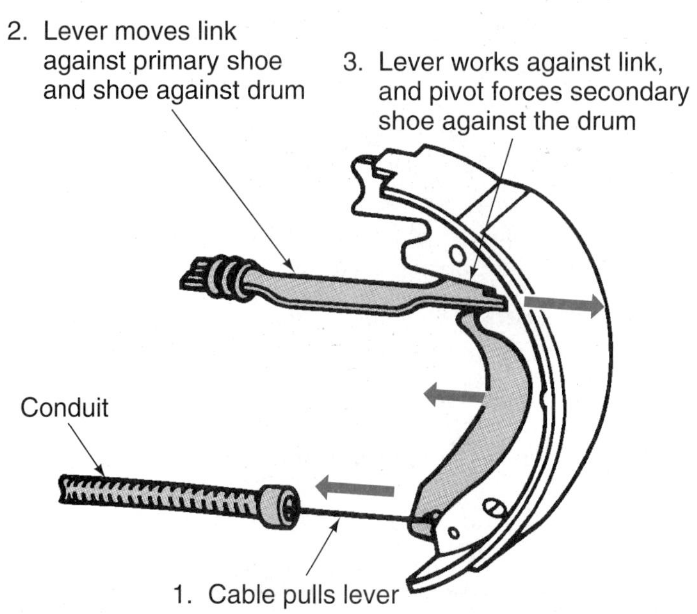
2. Lever moves link against primary shoe and shoe against drum
3. Lever works against link, and pivot forces secondary shoe against the drum
Conduit
1. Cable pulls lever

Figure 9-6 Application action of rear drum parking brake

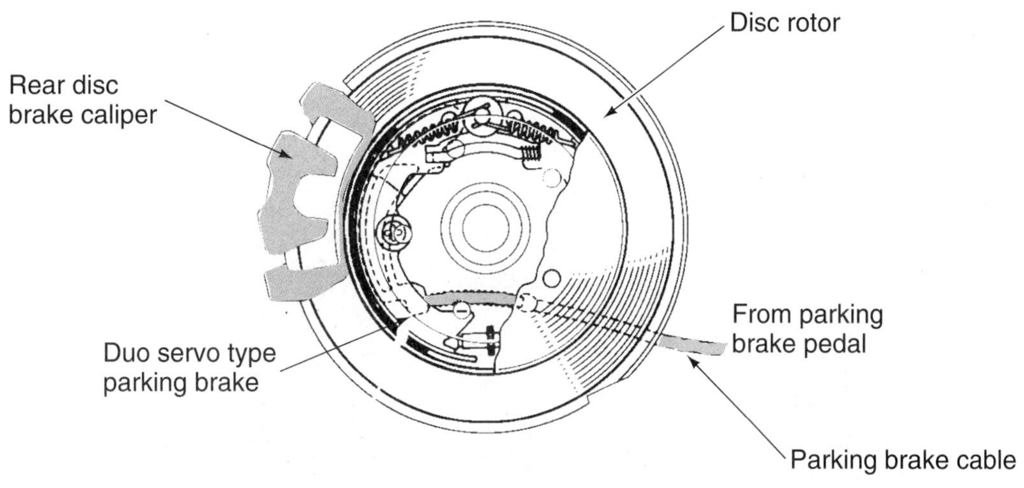

Figure 9-7 Parking brake cable connection at rear disc brake activation lever (Courtesy of Toyota Motor Corporation)

Checking Parking Brake Effectiveness

To check the operation of the parking brake system, raise the vehicle on the shop lift. Be sure the parking brake is fully released. Inspect for slack in the parking brake cables from the pedal to the equalizer and from the equalizer back to the rear brake assemblies. Next, check that there is no drag on the rear brakes when the wheels are turned. If cable slack or brake drag is present, adjustment is needed.

Look closely at the operation of the parking brake linkage as a helper applies then releases the parking brake pedal. The pedal should apply smoothly and return to its released position. The manual parking brake cables should move smoothly without any binding or slack. Protective conduit should be in good shape and the cables must not be frayed or corroded. If a problem is present, check the cable routings for kinks or binding. Clean and lubricate the parking brake assembly and noncoated metal cables with a brake lubricant. See Photo Sequence 14 for a typical inspection and adjustment procedure.

The parking brake cables of many newer vehicles have the wire strand coated with a plastic material. This plastic coating helps the cables slide smoothly against the nylon seals inside the conduit end fittings. It also protects the cable against corrosion damage.

You must remember that plastic-coated cables do not need periodic lubrication. You should also handle these cables extra carefully during service. Avoid contact with sharp-edged tools or sharp surfaces on the vehicle underbody. Damage to the plastic coating will impair the smooth operation of the system. It can also lead to corrosion.

Perform tests of the parking brake system and controls after making sure the linkage and manual controls operate properly. To check the parking brake system for holding effectiveness, perform the following steps:

1. Check the drum-to-lining clearance as outlined in Chapter 7, and adjust if needed.
2. Fully apply the parking brake pedal or hand lever. Raise and support the vehicle on the hoist.
3. Using both hands, attempt to turn a rear wheel forward. If the system is working properly, you should not be able to rotate the wheel forward. **NOTE:** You should be able to rotate the wheel backward during this test. This is normal.
4. Check for rear brake drag by releasing the parking brake and rotating the wheels forward. The rear wheels should turn freely. In some cases, light shoe-to-drum contact may be heard and is considered normal.

Some manufacturer's service manuals specify a number of clicks as a way of telling when the parking brake is properly applied for testing. Five to seven clicks is a common number.

Photo Sequence 14
Typical Procedure for Inspecting and Adjusting Rear Drum Parking Brake

P14-1 Proper adjustment of the parking brake begins with setting the parking brake to a near fully applied position.

P14-2 Raise the car on the hoist and make sure it is secure. The vehicle must be positioned so you can rotate the rear wheels freely. (If the parking brake is applied and adjusted properly, you should be unable to rotate the wheels.)

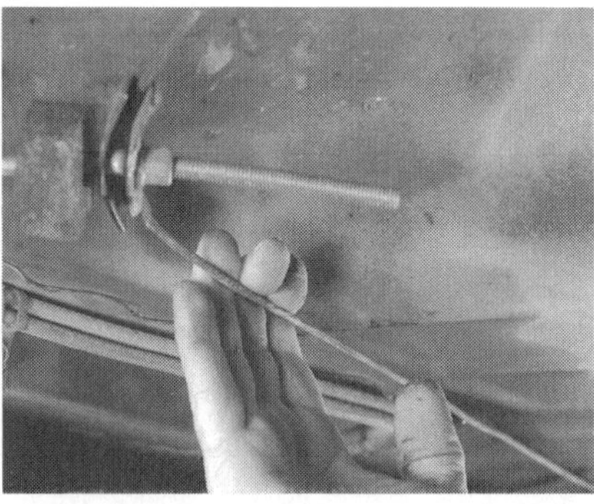

P14-3 Carefully inspect the entire length of the parking brake cable. Look for signs of fraying, breakage, and deterioration.

P14-4 Spray all exposed metal areas of the cable assembly with penetrating oil. This ensures a free-moving system.

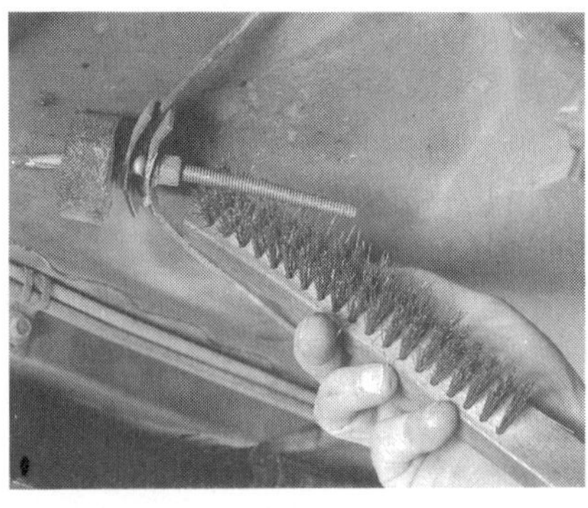

P14-5 Inspect the adjustment mechanism. Clean off the threaded areas and make sure the tightening nuts are not damaged.

P14-6 Loosen the adjustment lock nut. Adjust the parking brake by tightening the adjustment nut.

P14-7 When you can no longer turn the wheels by hand, stop tightening the adjusting nut.

P14-8 Lower the vehicle and release the parking brake.

P14-9 Raise the vehicle and rotate the wheels. If the wheels turn with only a slight drag, the parking brake is properly adjusted.

Photo Sequence 14
Typical Procedure for Inspecting and Adjusting Rear Drum Parking Brake (continued)

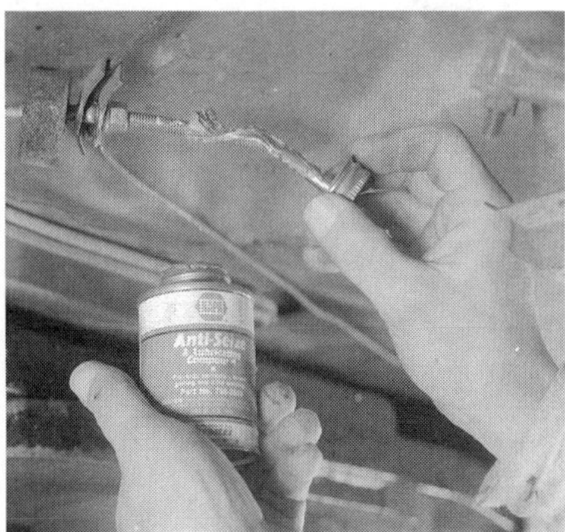

P14-10 After the proper adjustment is made, tighten the adjusting lock nut. Apply a coat of white grease to all contacting surfaces of the adjustment assembly.

Adjusting the Parking Brake Linkage

If the system fails the test outlined in the previous section, it must be adjusted. The parking brake must be adjusted anytime the parking brake cables have been replaced or disconnected, or if the brake holding ability is inadequate. Service manuals often describe the proper parking brake adjustment by the number of clicks heard as the pedal is pushed or the lever is pulled. For example, under heavy foot pressure, the pedal travel may be specified as less than 9 ratchet clicks or greater than 13 clicks. Before adjusting the parking brake, check the condition of the service brakes. They must be in good working order and adjusted properly. Incorrectly adjusted brakes will affect the parking brake adjustment.

To prevent damage to the threaded adjusting rod of the equalizer, clean the grease and dirt from the threads on either side of the adjusting nut before trying to turn the nut. Forcing the nut over dirty threads may damage the threads, lock up the nut, or strip the nut lands. You should also lubricate the threads of the adjusting rod before turning the nut.

Drum Brake Systems

If the rear drum brakes must be adjusted prior to adjusting the parking brake, start the engine and depress the brake pedal several times to set the self-adjusting drum brake. A typical parking brake linkage adjustment sequence for rear drum brake vehicles is shown as part of Photo Sequence 15. Begin by raising and supporting the rear axles on safety stands or by placing the vehicle on the hoist. Block the front wheels if safety stands are used. You must be able to turn the rear wheels by hand to test for proper linkage adjustment. Loosen the equalizer nut and set the parking brake pedal to the specified number of clicks listed in the service manual. This usually ranges from two to eight clicks, so check the manual.

Tighten the equalizer nut until the rear-wheels will not rotate without excessive force in a forward direction. Now loosen the equalizer nut until there is moderate drag when the rear wheels are rotated in a forward direction.

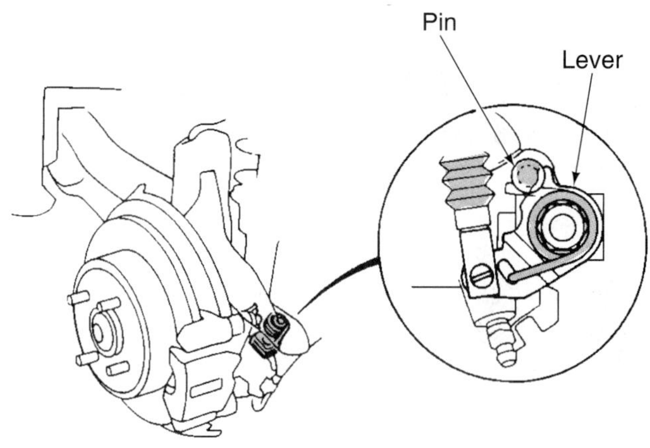

Figure 9-8 This type of rear disc brake parking system uses a lever/pin mechanism to set the brake. (Courtesy of Honda Motor Co., Ltd.)

Release the parking brake and rotate the rear wheels. There should be no brake drag in either direction if adjustment is correct. Lower the vehicle to the ground.

Rear Disc Brake Systems

After servicing a rear brake caliper, you must set the self-adjusting brakes before adjusting the brake pedal. Begin by loosening the parking brake adjusting nut. Then start the engine and depress the brake pedal several times to set the adjustment.

Once the brake calipers are set, raise the vehicle on the hoist. The rear disc brake parking brake for a mid-size import vehicle is shown (Figure 9-8). On this design, the lever of the rear brake caliper must contact the brake caliper pin.

To make the adjustment, pull the parking brake hand lever up one notch (click) and then tighten the adjusting nut until the rear wheels drag slightly when turned (Figure 9-9). Next, release the parking brake lever and check that the rear wheels do not drag when turned. Make any read-

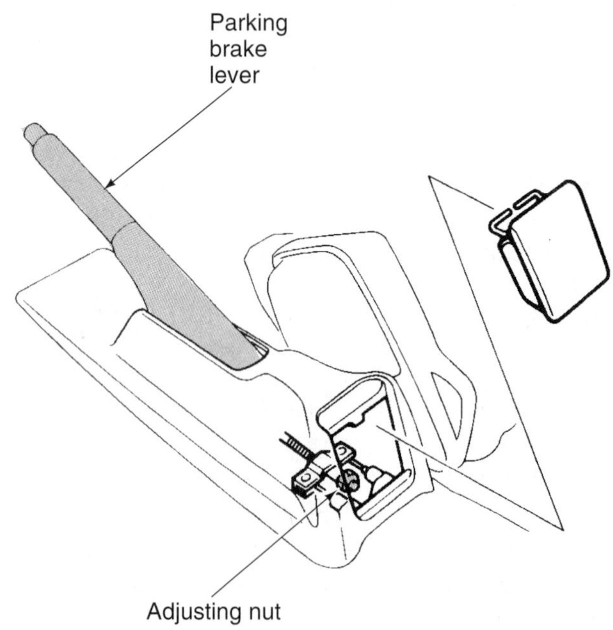

Figure 9-9 Adjusting nut location on hand lever-actuated rear disc brake parking brake system (Courtesy of Honda Motor Co., Ltd.)

justments necessary. When the equalizer is properly adjusted, the rear brakes should be fully applied when the parking brake lever is pulled up 5 to 10 clicks.

On a typical pedal-activated system found on larger domestic vehicles, the vehicle is raised on the hoist or safety stands. With the parking brake fully released, the adjusting nut is tightened against the rear parking brake cable adjuster/equalizer until there is a very slight (less than 1/16″) movement of either rear parking brake lever at the rear disc brake caliper. Once this movement is observed, firmly apply then release the parking brake pedal two times. Now lower the vehicle and test parking brake application.

Vacuum Release Parking Brake Tests

Special Tool

Vacuum gauge and connectors

On parking brakes equipped with vacuum release motors, sufficient vacuum must be available at the control motor. This vacuum specification, which can be found in the service manual, is usually around 10–12 in. Hg (35–40 kPa). System vacuum can be checked with a vacuum tester gauge.

Perform all vacuum system checks with the engine running at idle speed. Inspect the lines between all connecting points. Insufficient vacuum is normally caused by a loose hose connection or leak in the hose line. To detect leaks or bad connections in the parking brake vacuum hoses, listen for a hissing sound along the hose route. Make certain lines are not crossed with another hose, connected to the wrong connection, kinked close, or otherwise damaged.

WARNING: Never apply air pressure to the vacuum system of the parking brake release. Air pressure from the shop hose or other source will damage the actuator diaphragm in the parking brake release control motor.

The parking brake release control motor does not operate with the transaxle in Reverse.

To test the operation of the system, run the engine at idle with the transmission shifter in Neutral. Firmly set the parking brake. Move the shift control to D (drive) range and watch the parking brake pedal to see if it returns to its unapplied position when the parking brake releases. If the parking brake releases, then the parking vacuum control is working properly.

If the parking brake does not release, test for vacuum at the parking brake release vacuum hose that connects to the release control motor (see Figure 9-4). Remove the vacuum hose from the unit and use coupling connectors to tee the vacuum gauge into the line. Keep in mind that vacuum is sent to the control motor only when the transmission is in a forward Drive gear range. Normally, at least 10 in. Hg (35 kPa) of vacuum is required to run the parking brake release control motor. If minimum vacuum is not present, check for a damaged component and replace it.

Parking Brake Release Switch Replacement

A faulty parking brake release switch will prevent engine vacuum from reaching the acutator motor. If vacuum is present at the inlet to the switch but does not exit the switch when the shift control is in a forward gear, replace the switch. To replace a steering column-mounted parking brake release switch, begin by disconnecting the battery ground cable and the air bag backup power supply (if so equipped). Remove the necessary moldings from the instrument panel to access the steering column cover. Remove the steering column cover and any reinforcement piece that may be present (Figure 9-10).

Disconnect the PRNDL21 cable from the actuator housing and remove the retaining hardware that secures the steering column tube assembly in place. Carefully lower the steering column tube assembly and disconnect the vacuum hose extension assembly from the parking brake release switch (Figure 9-11). Remove the retaining screws that secure the parking brake release switch to the steering column tube assembly and remove the switch (Figure 9-12).

Begin installation of the new switch by shifting the system into neutral. Position the parking brake release switch over the column mounting bosses and push the switch against the turn signal cancel cam. Fasten the parking brake release switch with the retaining screws.

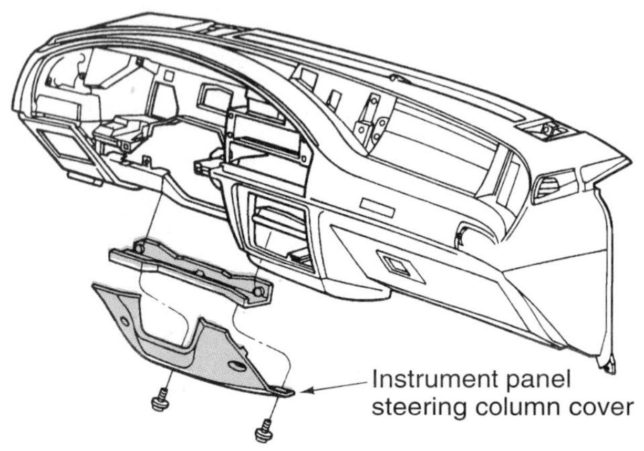

Figure 9-10 Removing the instrument panel steering column cover to access the column-mounted parking brake release switch (Reprinted with the permission of Ford Motor Company)

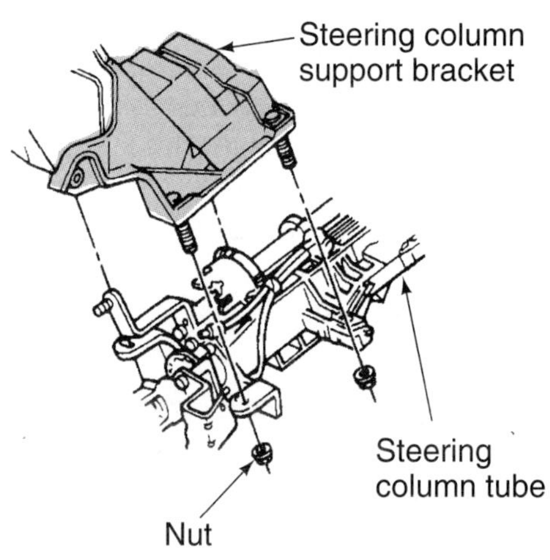

Figure 9-11 Removing the steering column tube assembly from its support bracket (Reprinted with the permission of Ford Motor Company)

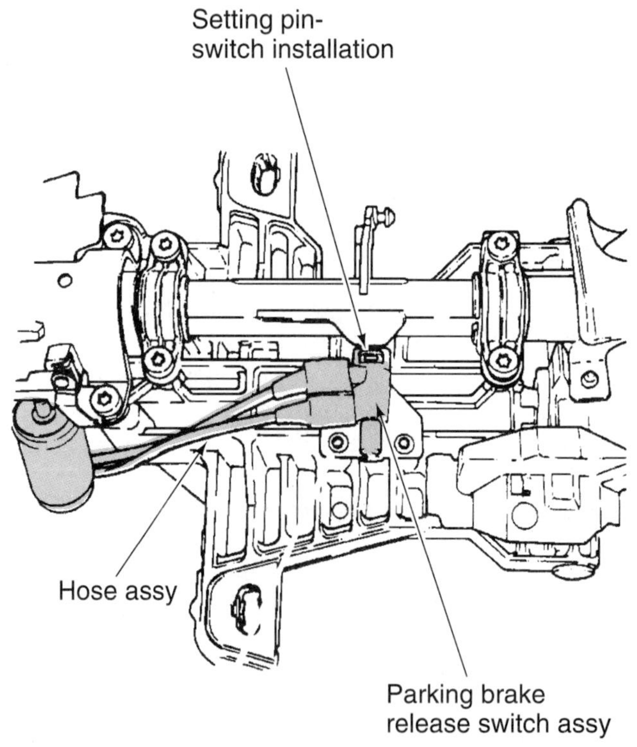

Figure 9-12 Vacuum hose connections at parking brake release switch (Reprinted with the permission of Ford Motor Company)

The new switch may be equipped with a plunger retainer that must be removed and discarded. Once this retainer is removed, connect the vacuum hose extension assembly to the parking brake release switch. Reassemble the steering column tube assembly, connect the PRNDL21 cable to the steering column, and reinstall the removed instrument panel trim and moldings. Complete the installation by reconnecting the air bag backup power supply and battery ground cable.

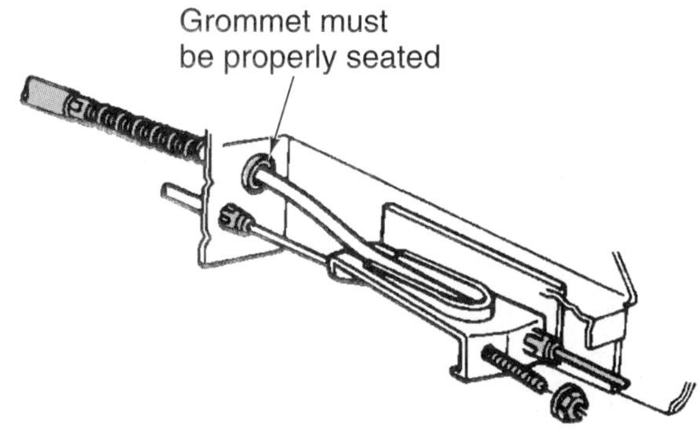
Grommet must
be properly seated

Figure 9-13 Backing off the adjusting nut from the equalizer. The nut does not have to be removed. (Reprinted with the permission of Ford Motor Company)

Parking Brake Control Service

To remove a typical pedal-actuated parking brake control mechanism, such as the one shown in Figure 9-4, begin by releasing the parking brake and raising the vehicle on the hoist. Back off the adjusting nut from the brake cable equalizer to remove all tension from the parking brake rear cables and conduits (Figure 9-13).

Once all tension is relieved, lower the vehicle to the ground. If the vehicle is equipped with a vacuum release system, disconnect the release vacuum hose from the release control motor. Next, disconnect the release cable from the parking brake release arm and remove the release cable grommet from the parking brake control.

On some light trucks, it is necessary to remove the dashboard trim panels to gain access to the brake pedal.

Disconnect the wiring connector running to the parking brake signal switch. Now remove the front parking brake cable and conduit end from the clevis at the parking brake control. On many systems, a push pin installed from the cowl side trim panel helps secure the cable. Remove the push pin and then remove the front parking brake cable and conduit retainer from the parking brake control. Use a properly sized box-end wrench to depress the retaining prongs. Remove the remaining bolts and hardware fastening the parking brake control to the trim panel and lift the parking brake control off of the vehicle.

To reinstall a typical unit, place the parking brake control in position and fit the front parking brake cable and conduit through its mounting hole. Install the pronged or other style retainer in place so that the retainer is securely locked into place. Now reconnect the control end of the cable into the clevis at the parking brake control. Reinstall all retaining bolts and push pin and hardware, tightening to torque specifications.

When so equipped, reconnect the brake release vacuum hose to the parking brake release control motor. Next, connect the release cable to the parking brake release arm and install the release cable grommet to the parking brake control. Reconnect the wiring connector to the parking brake signal switch and bracket. Complete the installation by raising the vehicle on the hoist and checking parking brake operation. Make any adjustment needed.

Release Handle and Cable Assembly Service

To replace the release handle and cable assembly (Figure 9-14), disconnect the release cable from the parking brake control release arm. It there is a grommet holding the cable in the parking brake control, remove it. Look under the instrument panel for a retaining clip or fastener that secures the cable and parking brake release handle to the instrument panel. On many vehicles, this fastener is a clip that can be carefully pryed off with a screwdriver. Remove the clip and pull the parking brake release handle and cable assembly out of the instrument panel.

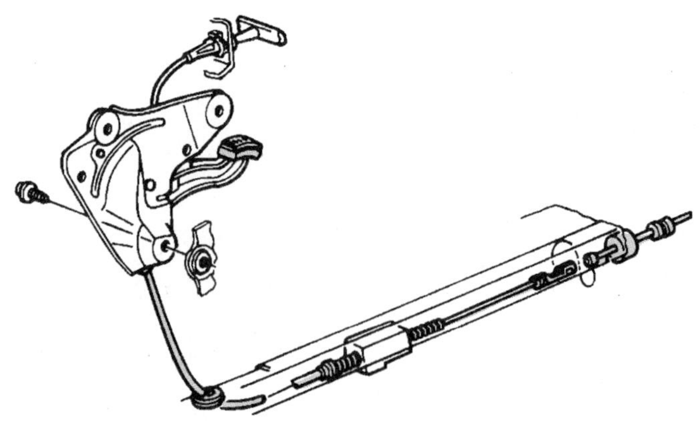

Figure 9-14 Parking brake release handle and control assembly (Reprinted with the permission of Ford Motor Company)

To reinstall the release handle, thread the cable and parking release handle assembly through the locating hole in the instrument panel. Secure the parking brake release handle to the instrument panel with the retainer clip. Connect the release cable to the parking brake control release arm and install the release cable grommet to the parking brake control.

Front Parking Brake Cable and Conduit Replacement

The cable that runs from the brake control mechanism (pedal or lever) to the equalizer/adjuster is known as the front cable. To remove and replace this cable, raise the vehicle on a hoist, and loosen the adjuster nut at the rear parking brake cable equalizer/adjuster.

Attach a piece of wire to the cable to aid in routing the cable through holes and tight spots.

Once the cable is backed off, lower the vehicle and disconnect the front parking brake cable and conduit from the parking brake control at the clevis. Using a properly sized box-end wrench, depress the conduit retaining prongs and remove the cable-end pronged fitting from the parking brake control.

Expose the front parking brake cable and conduit by removing the necessary instrument panel and pulling the carpet back. Raise the vehicle on the hoist and disconnect the front parking brake cable and conduit from the parking brake rear cable and conduit at the cable connector (Figure 9-15). Remove the front parking brake cable and conduit and push-in prong retainer from the cable bracket. Use a properly sized box-end wrench to depress the retaining prongs.

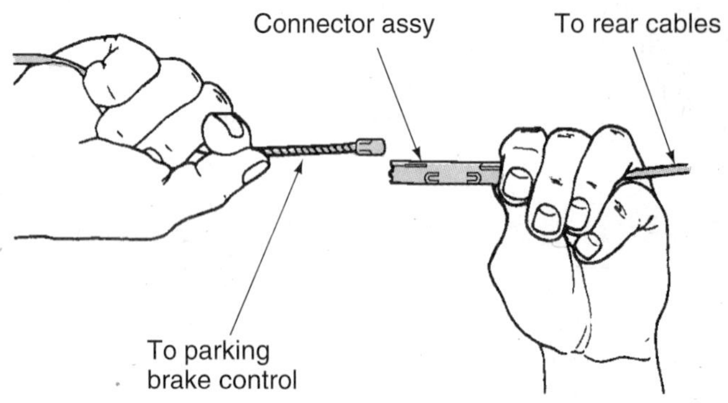

Figure 9-15 Disconnecting the parking brake cable at a connector assembly that links two cable segments together (Reprinted with the permission of Ford Motor Company)

To install the replacement cable, thread the front parking brake cable and conduit through the hole in the front floor pan. Install the grommet in place to secure the cable at the hole location. Position the front parking brake cable and conduit through the front cable bracket at the inner floor side member. Push the prong into the bracket. Seat the cable seal into the hole in the front floor pan. Connect the rear cable to the front cable at the connector. Lower the vehicle. Next, push the prong retainer into the parking brake control housing until the prongs are secure. Then connect the front parking brake cable and conduit to the parking brake control clevis. Reinstall the carpet and cowl side trim panel. Adjust the parking brake as described earlier in the chapter and check its operation.

The cable retaining fingers must be completely through the holes.

Rear Cable Replacement

Classroom Manual
Chapter 9, page 211

The vehicle has two rear parking brake cables—a left-hand cable to the left rear wheel and a right-hand cable to the right rear wheel. These rear cables are connected to the equalizer or adjuster. With the vehicle on a hoist, remove the parking brake cable adjusting nut. Next, disconnect the rear cable conduit end fitting from the front brake cable and conduit connector.

To remove the cable from the drum brake assembly, first remove the wheel and brake drum. Disconnect the end of the rear cable/conduit from the rear parking brake lever. Use the proper size offset box wrench or screwdriver to depress the conduit retaining prongs, and slide the cable end pronged fitting out through the hole in the support plate (Figure 9-16).

On a typical rear disc brake, the parking brake rear cable/conduit is secured at the rear disc brake assembly by a clip and/or pin that connects the end of the cable/conduit to the parking brake lever of the caliper assembly (Figure 9-17). Remove the clip and pin to free the cable. It may be necessary to remove grommets and pronged retainers before the rear cable can be freed.

SERVICE TIP: There may be slight differences in the routing and removal/installation of left-hand and right-hand rear parking brake cables. Refer to the vehicle service manual for details and specific instructions.

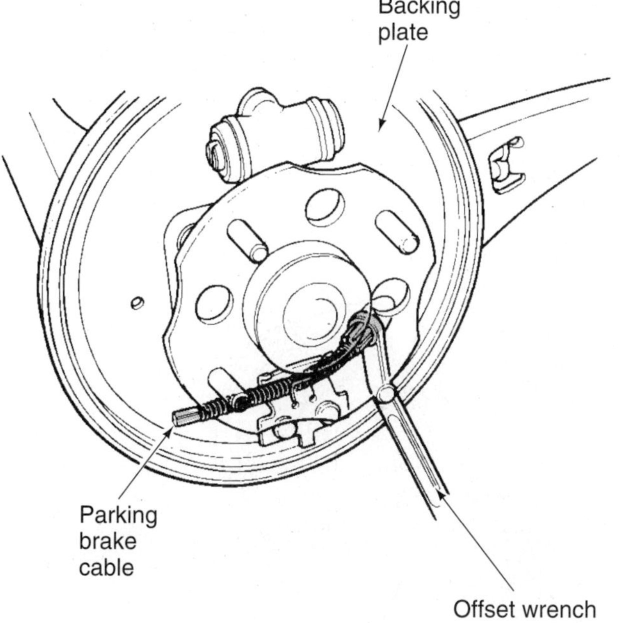

Figure 9-16 Disconnecting the parking brake cable at a rear drum brake assembly (Courtesy of Honda Motor Co., Inc.)

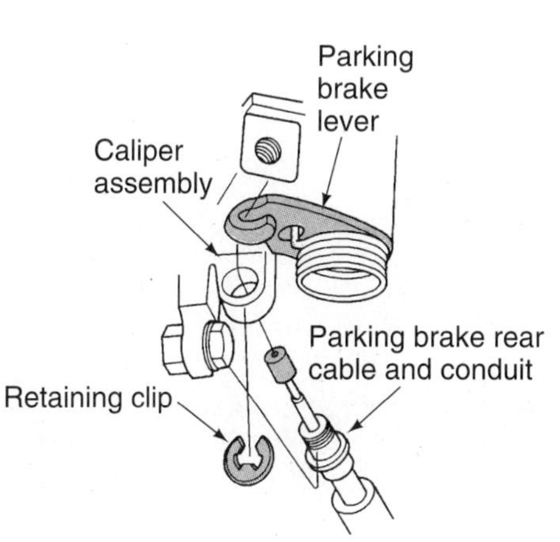

Figure 9-17 Disconnecting the parking brake cable at a rear disc brake assembly (Reprinted with the permission of Ford Motor Company)

To install a new rear cable, work the cable/conduit through the inner floor-side member bracket and the rear parking brake cable equalizer/adjuster. By sure the pronged retainer at the adjuster/equalizer firmly secures the cable to the adjuster. Install all grommets and clips used to keep the cable secure. Follow the original routing pattern.

On drum brakes, insert the cable/conduit end into the hole in the brake assembly support plate. Ensure that the retaining prongs lock the conduit in place where it passes through the backing plate. Hold the brake shoes in place on the backing plate and engage the brake cable into the parking brake lever. Install the brake drum and wheel assembly. Install the brake cable adjusting nut. Adjust the parking brake and lower the vehicle. Check for proper operation.

On disc brakes, insert the parking brake rear cable and conduit end into the rear disc brake caliper and install the retaining clip or pin that secures the cable to the lever.

Transmission or Drive Shaft Parking Brake

Not all parking brakes do their job by mechanically applying the drum or disc service brakes. Some motor home and large van parking brakes are transmission or drive shaft parking brakes, which are completely separate from the basic brake system. There are two types in use: the external band type and the internal shoe type.

The transmission or drive shaft parking brake system operates independently of the rear wheel brakes. These parking brakes are known in the trade as drive shaft brakes.

External Band Brake

The external band drive shaft parking brake is an older design not normally found on recent models. This brake consists of a drum attached to the drive shaft and an external band (Figure 9-18). The brake lining is attached to the inside of the band. When the brake is set, a lever mechanism squeezes the band against the drum. This keeps the drum, drive shaft, and rear wheels from turning.

An adjusting screw on the brake assembly is used to set a slight 0.020 to 0.030 in. clearance between the lining and the drum at all points around the external band. Once the band is properly

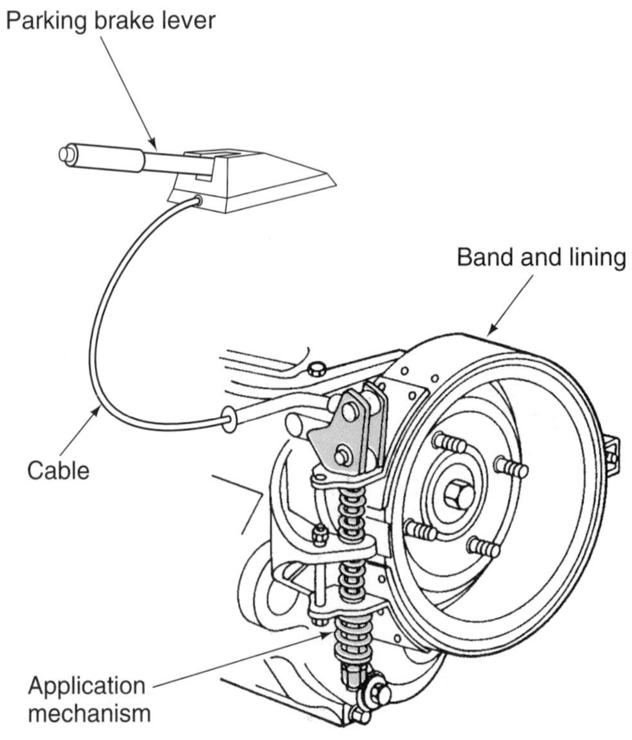

Parking brake lever
Band and lining
Cable
Application mechanism

Figure 9-18 External band drive shaft parking brake

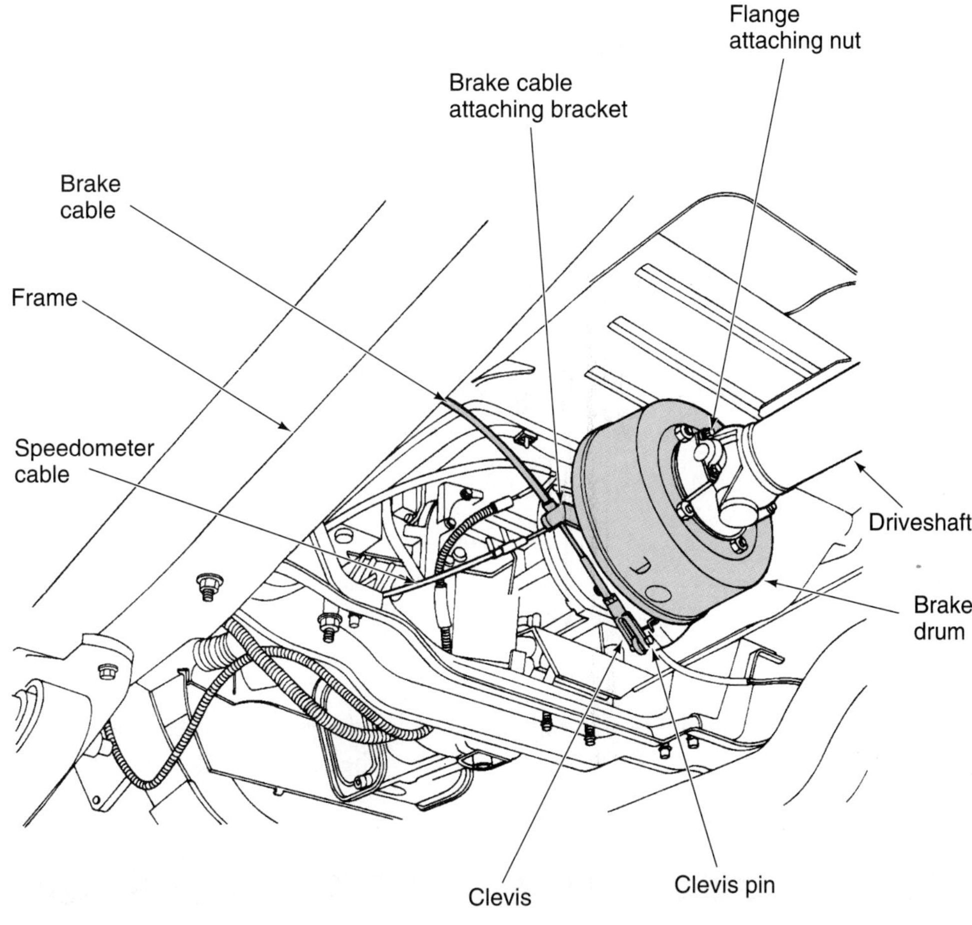

Figure 9-19 Internal shoe drive shaft parking brake (Reprinted with the permission of Ford Motor Company)

adjusted, all slack should be removed from the parking brake cable. To do this, loosen the clevis lock nut, remove the pin, and adjust the yoke on the threads until all slack is removed. Adjustment is correct when the clevis can just be reinstalled. Replace the pin and tighten the lock nut.

WARNING: Always check for proper band adjustment before adjusting the cable.

Internal Shoe

The internal shoe design is used on most vans and motor homes. It is similar in operation to the standard drum brake. The parking brake consists of a drum attached to the drive shaft and a shoe assembly at the rear of the transmission (Figure 9-19). The brake shoes are mechanically adjusted by a lever or a cam inside the assembly. To make the adjustment, place the transmission in Neutral and release the parking brake. Place the vehicle on the lift, supporting it properly. The drum should be rotatable by hand. Remove the adjusting screw cover and loosen the clamp bolt to free the cable adjusting nut. Back out the cable adjusting nut or slide the cable sleeve in the clamp (Figure 9-20). Expand the shoes by rotating the shoe adjusting nut until the lining drags on the drum. Next, back off the nut about one notch to provide 0.010 in. clearance between the linings and the drum. Check to be sure the shoulders on the nut are seated in the grooves in the sleeve. As the final step in servicing, remove all slack from the parking brake cable. This may be done by an adjusting nut located at the point where the cable enters the drum or by a cable adjusting nut

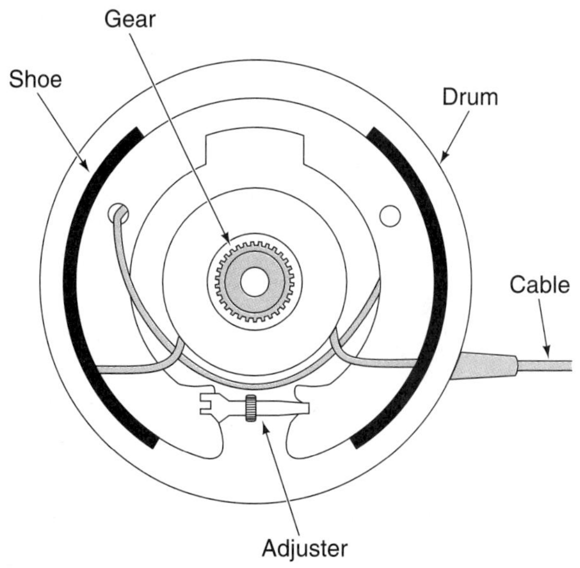

Figure 9-20 Adjusting an internal shoe drive shaft parking brake

located under the dash on the pedal control linkage. When properly adjusted, the hand lever or foot pedal should advance four to six notches to firmly set the parking brake. The lever or pedal should not advance more than one-half of the total available setting stroke.

CASE STUDY

A customer complains that his parking brake is not holding. He lives on a hill and has noticed that the vehicle has started to drift backward when he disengages the clutch to start the engine. The technician sets the parking brake. It applies firmly and releases smoothly. It may just be in need of a slight adjustment, but the technician decides a test drive may reveal additional problems. She is right. On the first hard braking application, the car nose dives. It is soon apparent to the technician that the front disc brakes are doing most of the braking on this vehicle. Back at the shop, she pulls the drum and inspects the brake assemblies. The shoes are severely worn and glazed. They are also out of adjustment. When the brakes are relined and adjusted, the parking brake system works fine.

Terms to Know

Adjuster
Conduit
Equalizer
Release mechanism
Vacuum control motor
Vacuum release switch

ASE Style Review Questions

1. *Technician A* says applying the parking brake disables the daytime running lights used on Canadian and some U.S. vehicles.
   Technician B says the parking brake applied indicator light may also be used to indicate problems in the vehicle's antilock brake system.
   Who is correct?
   A. A only **C.** Both A and B
   B. B only **D.** Neither A nor B

2. Vacuum-operated parking brake release systems are being discussed.
   Technician A says the vacuum release switch supplies engine vacuum to the parking brake release vacuum motor whenever the transaxle is placed in a forward gear.
   Technician B says the release switch vents the parking brake release motor when the transaxle is placed in Park, Reverse, or Neutral.
   Who is correct?
   A. A only **C.** Both A and B
   B. B only **D.** Neither A nor B

3. While lubricating the parking brake system:
   Technician A applies graphite to the contact areas of plastic-coated cables.
   Technician B applies a quality grease to metal cables.
   Who is correct?
   A. A only **C.** Both A and B
   B. B only **D.** Neither A nor B

4. During parking brake adjustment:
   Technician A checks and adjusts the drum-to-lining clearance before making any adjustment to the parking brake.
   Technician B fully applies the parking brake lever before making the adjustment.
   Who is correct?
   A. A only **C.** Both A and B
   B. B only **D.** Neither A nor B

5. A parking brake does not release.
   Technician A checks the drum-to-lining clearance for proper adjustment.
   Technician B checks for proper engine vacuum to the vacuum release system.
   Who is correct?
   A. A only **C.** Both A and B
   B. B only **D.** Neither A nor B

6. Parking brake vacuum release systems are being discussed.
   Technician A says the vacuum release switch is located on the parking brake foot pedal or hand lever mechanism.
   Technician B says the vacuum control motor is mechanically linked to the release handle by a metal rod.
   Who is correct?
   A. A only **C.** Both A and B
   B. B only **D.** Neither A nor B

7. When adjusting the parking brake linkage:
   Technician A cleans and lubricates the threads of the adjusting mechanism bolt to avoid damaging it.
   Technician B makes certain the rear wheels cannot be rotated forward with the parking brake fully applied.
   Who is correct?
   A. A only **C.** Both A and B
   B. B only **D.** Neither A nor B

8. When testing a vacuum control motor:
   Technician A checks for proper vacuum at the motor using a vacuum gauge installed with proper fittings.
   Technician B applies 5 to 10 psi of air pressure to the control motor to test the release action of the mechanical linkage.
   Who is correct?
   A. A only **C.** Both A and B
   B. B only **D.** Neither A nor B

9. Adjustment to rear disc brake parking brakes is being discussed.
Technician A says on some systems, the parking brake hand lever is pulled up one notch and the adjusting nut is tightened until the rear wheels drag slightly when turned.
Technician B says on a typical pedal-activated system with the parking brake fully released, the adjusting nut is tightened against the rear parking brake cable adjuster/equalizer until there is a very slight (less than 1/16″) movement of either rear parking brake lever at the rear disc brake caliper.
Who is correct?
   A. A only **C.** Both A and B
   B. B only **D.** Neither A nor B

10. *Technician A* says the cable running from the parking brake control mechanism to the equalizer/adjuster is commonly referred to as the front cable.
Technician B says there may be slight routing and length differences between the left and right rear cables in a parking brake system.
Who is correct?
   A. A only **C.** Both A and B
   B. B only **D.** Neither A nor B

Table 9-1 ASE TASK

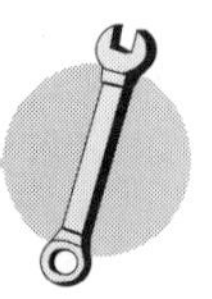

Diagnose problems with the parking brake system.

Problem Area	Symptoms	Possible Causes	Classroom Manual	Shop Manual
PARKING BRAKE FAILS TO APPLY	Vehicle rolls down incline when foot pedal or hand lever is fully applied.	1. Linkage adjusted too tight	211	271
		2. Binding linkage or control pedal/lever	212	269
		3. Poor lubrication	212	269
LEAKS OR BLOCKAGE IN VACUUM SUPPLY SYSTEM	Inoperative vacuum release system.	1. Loose hose connection at parking brake release control motor or at vacuum supply connection	210	273
		2. Leaks or kinks in line from vacuum motor to supply hose or line that connects to switch on steering column tube or shift console	210	273
		3. Crossed vacuum hoses or hoses connected to the wrong connection	210	274
		4. Pinched hoses	210	273
PARKING BRAKE FAILS TO RELEASE	Brakes drag when parking brake foot pedal or hand lever is fully released.	1. Insufficient or no vacuum reaching control motor (See Leaks in Vacuum Supply System above)	210	273
		2. Parking brake control mechanism or cables binding	209	275
		3. Transaxle/transmission not in Drive	212	269

Brake Electrical and Electronic Component Service

Upon completion and review of this chapter, you should be able to:

- ❑ Inspect, test, and replace the brake warning light, system switch, and wiring.
- ❑ Test and repair the parking brake indicator lights, switches, and wiring.
- ❑ Test, adjust, repair, or replace the brake stoplight switch and wiring.
- ❑ Inspect, test, and replace the master cylinder fluid level sensor.
- ❑ Inspect, test, and replace the disc brake pad wear sensor.
- ❑ Inspect, test, and replace the ABS warning light.

Basic Electrical Principles

Basic Tools

Basic mechanic's tool set

Digital multimeter

Test light

Continuity checker

As you learned in Chapter 8, antilock braking systems rely heavily on electronic actuation and control. But modern brake systems also use a number of simple electrical circuits to perform basic system status checks and safety functions. For example, an electrical circuit is energized if the parking brake is applied or if fluid level in the master cylinder reservoir falls below a certain point. An electrical circuit may also be energized if a pressure difference develops in the hydraulic brake lines. All of these circuits are tied into one or more brake system warning lights mounted on the instrument cluster. When a problem or certain condition exists, the warning light bulb illuminates and the driver is alerted.

Some vehicles equipped with electronic instrument panels flash a written message instead of illuminating a warning light. However, the basic circuit operation is essentially the same between the instrument panel and the warning circuit switch

Brake stoplight circuits to the rear taillight assemblies and center high mount brake lights are energized when the brake pedal is depressed. These lights warn other motorists that the brakes on the vehicle in front of them have been applied.

While most circuits involved in brake stoplight and warning light operation are simple circuits, the increased use of electrical equipment on modern vehicles may make them more difficult to find and trace throughout the vehicle. Always work from the service manual electrical schematics and follow the basic principles of electrical troubleshooting outlined in this chapter.

In a vehicle's electrical system, electrical power flows from a power source to a load device and then back to the source of power to form a complete circuit. In addition to the power source and loads, most automotive circuits contain circuit control components and protection components (Figure 10-1).

A normally open (NO) switch will not allow current flow when it is in its rest position. The contacts are open until they are acted upon by an outside force (such as a brake pedal being depressed) that forces them closed to complete the circuit.

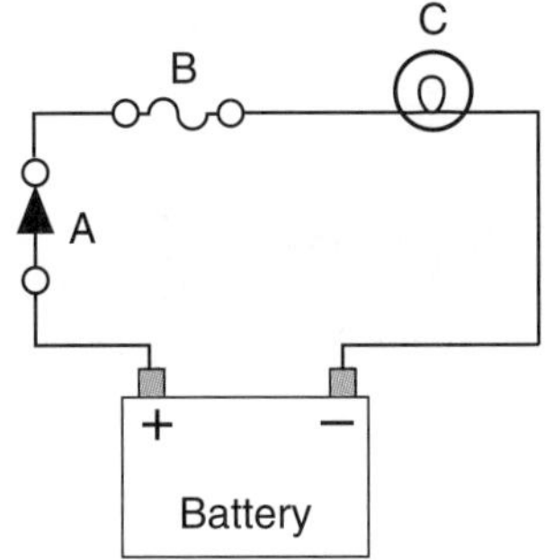

Figure 10-1 A simplified electric circuit including (A) a switch, (B) a fuse, and (C) a lamp

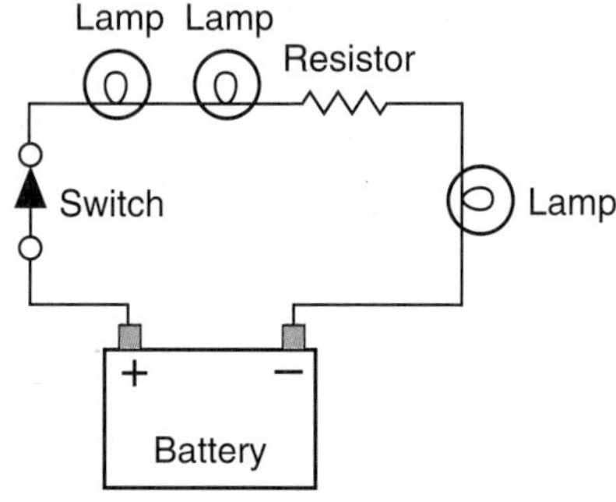

Figure 10-2 A simplified series circuit

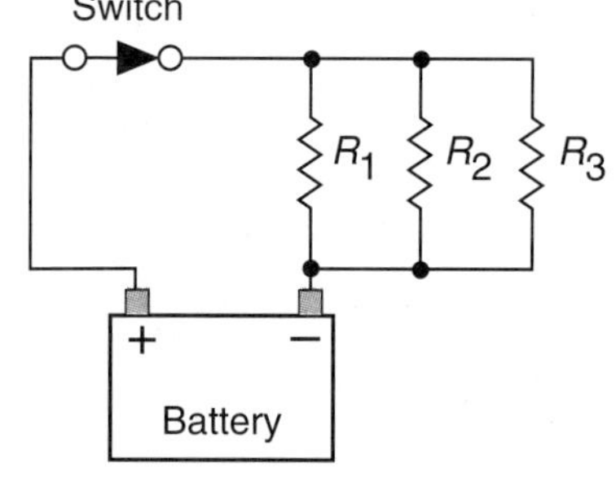

Figure 10-3 A parallel circuit with different resistances in each branch

The vehicle's battery and alternator are the power sources that provide power for all electrical circuits. Circuit protection devices include items such as fuses, circuit breakers, and fusible links. They provide overload protection for the circuit. Circuit controllers such as switches or relays are used to control the power flow within a circuit. They open and close the circuit. Circuit loads may be lamps, motors, or solenoids.

Electrical circuits can be set up as series circuits, parallel circuits, or series/parallel circuits.

Good continuity means there is very low resistance and the circuit is continuous. Very poor or no continuity shows there is high resistance in the circuit or that the circuit is open.

Series Circuit. In a series circuit, the electrical devices are connected to form one current path to and from the power source. In a series circuit, the voltage is shared equally by all the components in the circuit (Figure 10-2).

Parallel Circuit. In a parallel circuit, the electrical components are connected to form more than one current path to and from the power source. In a parallel circuit, the voltage is constant and equal for each current path (Figure 10-3).

Series/Parallel Circuits. A series/parallel circuit consists of a single current path and a circuit with more than one current path to and from the power source (Figure 10-4).

Circuit Malfunctions

Opened is the term used to mean that the circuit path for current flow is broken. A closed circuit means there are no breaks in the path and current will flow.

There are three electrical conditions that can result in a nonworking circuit: an open circuit, a short circuit, or a grounded circuit.

Open Circuit (Figure 10-5)**.** An open circuit occurs whenever there is a break (open) in the circuit. The break may be caused by a corroded connector, a broken wire, a wire that burned open from too much current, or a faulty component.

Short Circuit (Figure 10-6)**.** A short circuit happens when the current bypasses part of the normal circuit. This bypassing can be caused by wires touching, salt water found in or on the circuit switches or connectors, or solder melting and bridging conductors in a component.

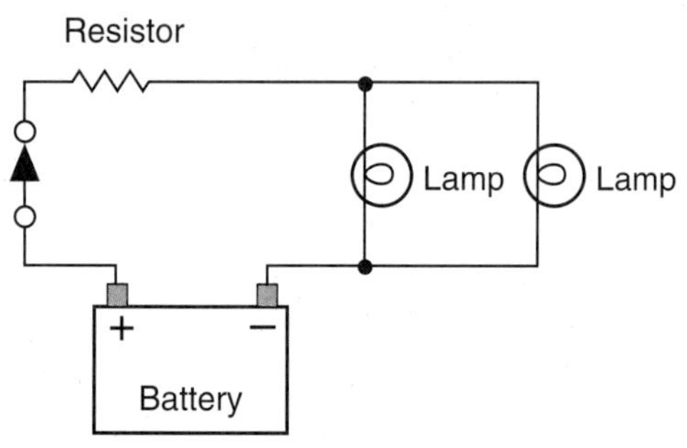

Figure 10-4 A series-parallel circuit

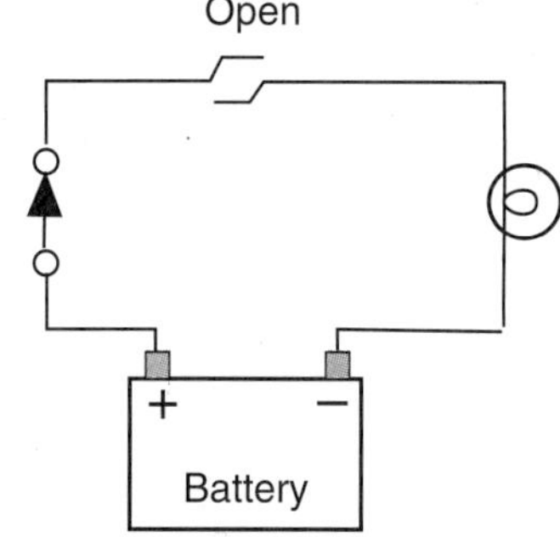

Figure 10-5 An open circuit stops all current flow

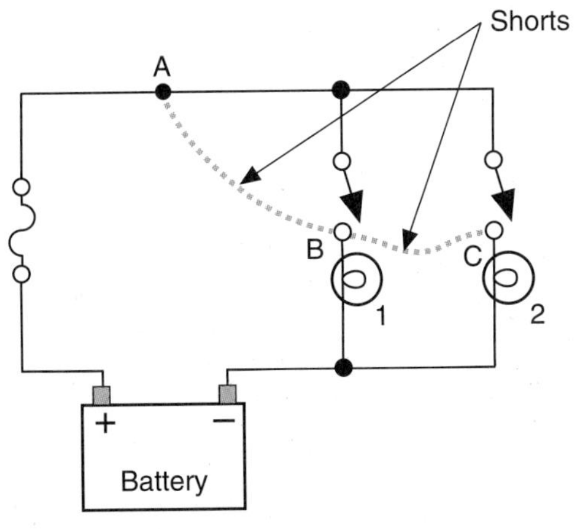

Figure 10-6 A short circuit can be a copper-to-copper contact between two adjacent wires.

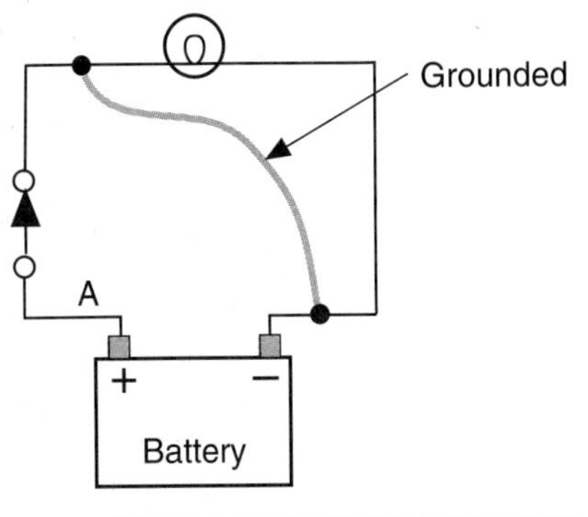

Figure 10-7 A grounded circuit

Grounded Circuit (Figure 10-7). A grounded circuit acts like a short circuit except that the current flows directly into a ground circuit that is not part of the original circuit. An unintentional ground circuit may be caused by a bare of broken wire rubbing against the vehicle's frame or body. A grounded circuit can also be caused by deposits of oil, dirt, or moisture around connections or terminals, which provide a good path to ground. Grounded circuits can draw excessive current from the battery and can be very damaging to components in the circuit. Testing for an unintentional ground circuit is illustrated (Figure 10-8). The circuit fuse is removed and a test lamp is installed across the fuse terminals. If the test lamp lights, there is an unwanted ground in the circuit.

The Process of Electrical Troubleshooting

A shorted circuit allows current to bypass part of the normal path.

To correctly isolate and repair an electrical problem, follow a logical troubleshooting procedure.

1. Verify the problem. Review the work order, operate the system, and list symptoms in order to check the accuracy and completeness of the owner's complaint. If the problem is intermittent, try to recreate the problem.

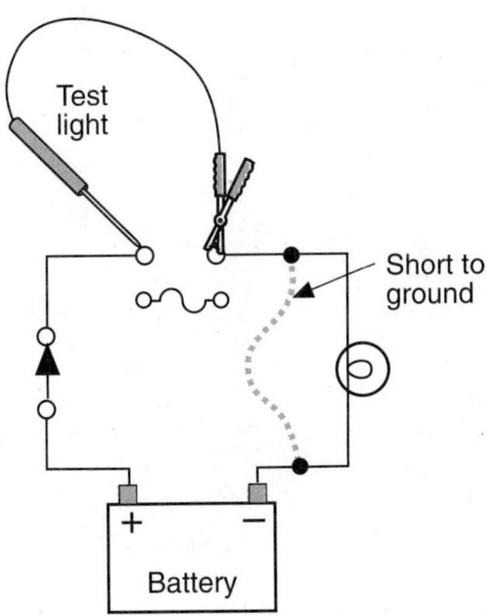

Figure 10-8 Testing for a grounded circuit by jumping across the removed fuse using a test lamp. If the lamp lights, there is an unwanted ground circuit.

2. Determine the possible cause(s) of failure. Refer to the circuit diagram for clues to the problem. Locating and identifying the circuit components may help determine where the problem is.

SERVICE TIP: Remember, shop manual circuit diagrams are designed to make it easy to identify common circuit problems, which will help narrow the problem to a specific area. If several circuits fail at the same time, check for a common power or ground connection. If part of a circuit fails, check the connections between the functioning areas of the circuit and the failed areas.

3. Identify the faulty circuit by studying the circuit diagram to determine circuit operation for the problem circuit. You should have enough information to narrow the failure to one component or one portion of the circuit.
4. Locate the failed component or element. Service manual procedures or diagnostic charts often detail a step-by-step approach to diagnosing a symptom. The test procedures are listed in numerical sequence and should be followed in that order. Each test step describes what must be done to the circuit, what test equipment to use, and where to connect the test equipment. It is important to remember that a problem in one system could result in a symptom in another system.
5. Make the repair and verify that the repair is complete by operating the system.

Classroom Manual
Chapter 10, page 234

Brake System Warning Light Circuits

The brake system warning light on most vehicles performs multiple warning functions; that is, several circuits are tied into the same warning light (Figure 10-9). Typically, this indicator light goes on when the parking brake is applied, when the brake fluid level is low, and as a circuit test while the engine is cranking. Battery voltage reaches the brake indicator lamp when the ignition switch is in "run," "bulb test," or "start." When any of the three switches connected to the brake indicator closes, the circuit is grounded and the indicator lamp lights.

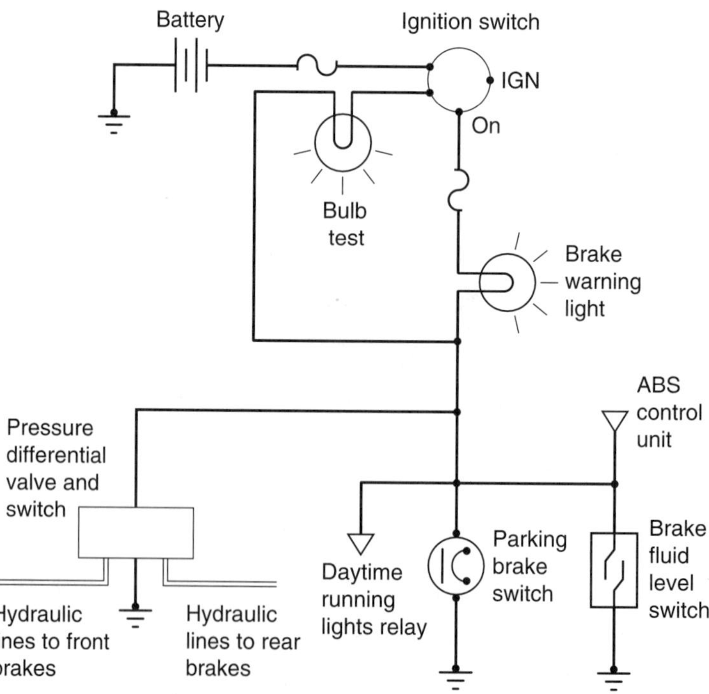

Figure 10-9 Simplified brake warning light circuit diagram

The ignition switch completes the circuit to ground when it is in the "bulb test" and "start" positions. The parking brake switch provides a ground when the park brake is applied.

On vehicles equipped with daytime running lights, a safety feature required by law in Canada and now being offered on American vehicles, the parking brake switch completes ground to the brake lamp via a diode within the daytime running light module.

The brake fluid level switch closes to light the brake indicator when the brake fluid in one of the two hydraulic brake systems falls below switch level. This can be caused by a leak in one of the brake lines. After repairing the problem, the switch can be reset to its open position by refilling the reservoir.

The electronic brake control module will also turn on the brake indicator if a problem causes the antilock brake system to revert to the base brake system with ABS control.

System Check

Tables 10-1 and 10-2 outline the basic checks and troubleshooting actions normally taken to pinpoint problems in the brake warning light system.

Table 10-1 PROPER WARNING LIGHT OPERATION

ACTION	NORMAL RESULTS
Release the parking brake, then turn the ignition switch slowly past the "run" position.	Brake warning light comes on.
Release the ignition switch to the "run" position.	Brake warning light does not light.
With the ignition switch in "run," apply the parking brake.	Brake warning light comes on.
Release the parking brake.	Brake warning light does not light.

Table 10-2 TROUBLESHOOTING THE WARNING LIGHT SYSTEM

CONDITION	POSSIBLE SOURCE	ACTION
Brake warning light on	• Low brake fluid level in system.	• Refill master cylinder reservoir, then check lines and all connections for leaks.
	• Leaking primary piston cup in master cylinder.	• Rebuild or replace the master cylinder.
	• Parking brake not fully released.	• Unfreeze control and adjust the parking brake cable/conduit. Replace any damaged components.
	• Wiring between parking brake switch and warning light inadvertently grounded.	• Repair the wiring.
	• Parking brake release switch grounded.	• Inspect the parking brake release switch.

Troubleshooting the Warning Light Circuit

Perform the basic checks for power and continuity before beginning more in-depth circuit diagnosis. The following checks are based on the circuit schematic shown (Figure 10-10). The corresponding troubleshooting flow chart is shown (Figure 10-11). As always, the service manual should be your first source of information on electrical troubleshooting. In this example, the manual outlines the following steps. Begin by checking fuse 9. Look at the lamp indicator with the ignition switch in "run" and the engine off. It should come on. Next, check ground 101 by operating the blower motor. If the vehicle is equipped with a daytime running lights module, set the parking brake to force the indicator to light. If the light does not come on, check circuit 33, circuit 1134, and

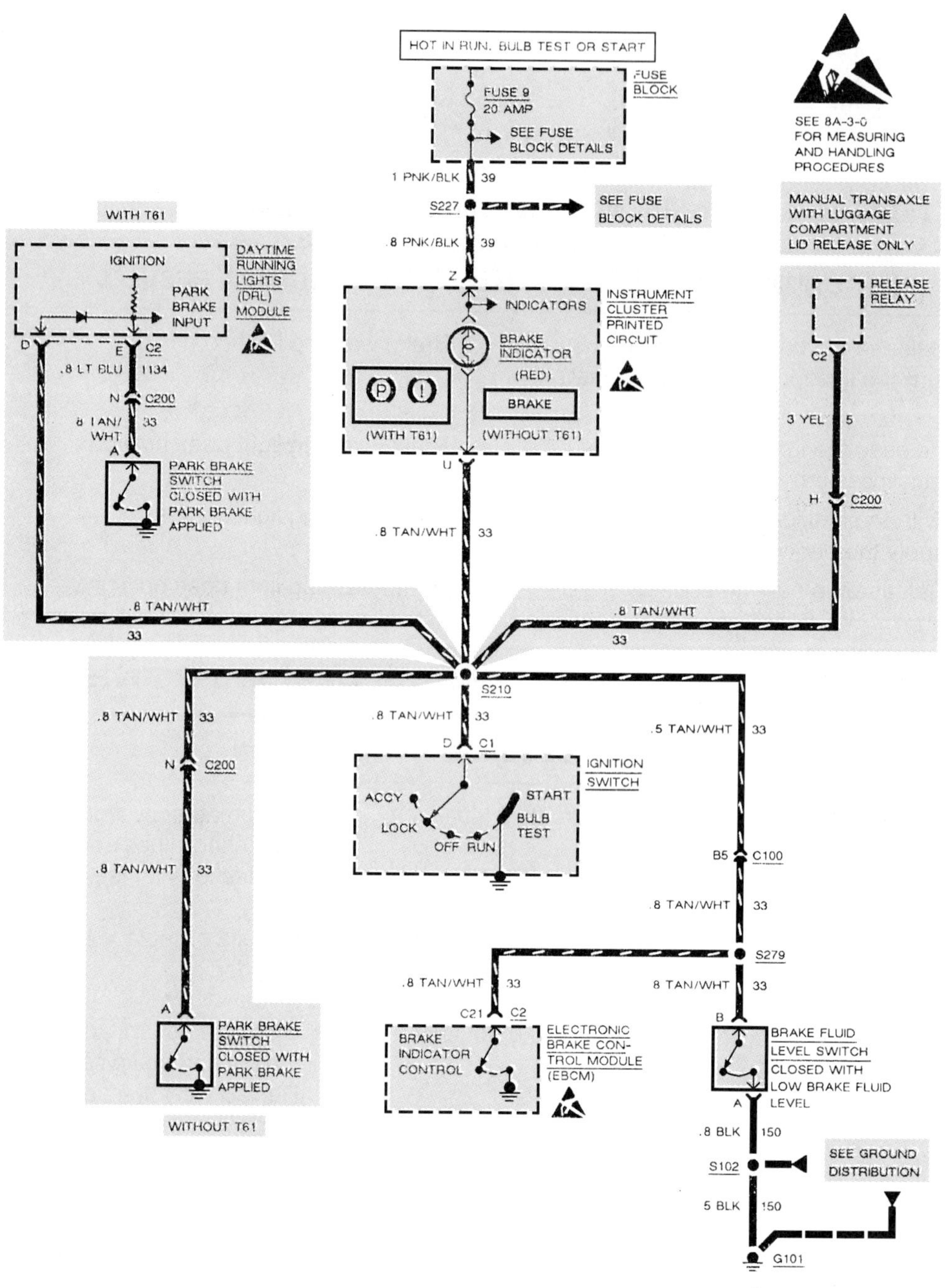

Figure 10-10 Brake warning light circuit schematic (Courtesy of Chevrolet Motor Division, General Motors Corporation)

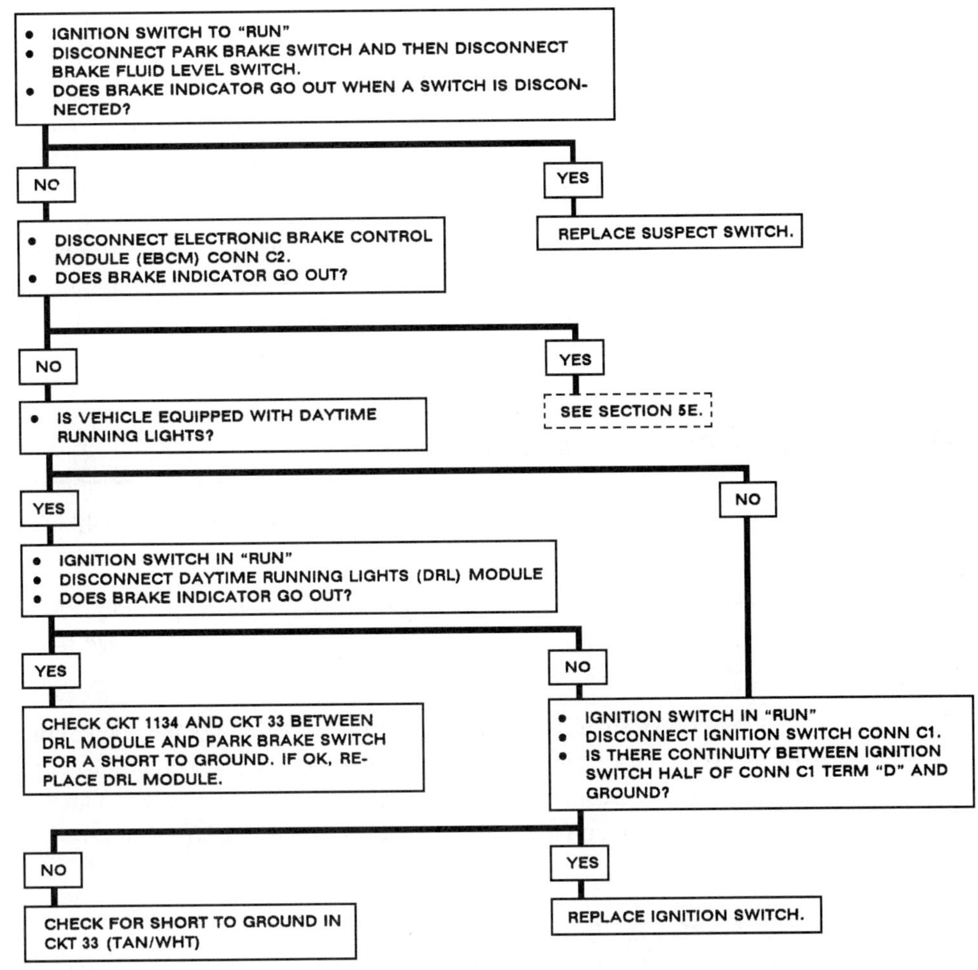

BRAKE INDICATOR REMAINS ON WITH IGNITION SWITCH IN "RUN" AND PARK BRAKE RELEASED
• IGNITION SWITCH TO "RUN"
• DISCONNECT PARK BRAKE SWITCH AND THEN DISCONNECT BRAKE FLUID LEVEL SWITCH.
• DOES BRAKE INDICATOR GO OUT WHEN A SWITCH IS DISCONNECTED?
NO
YES
REPLACE SUSPECT SWITCH.
• DISCONNECT ELECTRONIC BRAKE CONTROL MODULE (EBCM) CONN C2.
• DOES BRAKE INDICATOR GO OUT?
NO
YES
SEE SECTION 5E.
• IS VEHICLE EQUIPPED WITH DAYTIME RUNNING LIGHTS?
YES
NO
• IGNITION SWITCH IN "RUN"
• DISCONNECT DAYTIME RUNNING LIGHTS (DRL) MODULE
• DOES BRAKE INDICATOR GO OUT?
YES
NO
CHECK CKT 1134 AND CKT 33 BETWEEN DRL MODULE AND PARK BRAKE SWITCH FOR A SHORT TO GROUND. IF OK, REPLACE DRL MODULE.
• IGNITION SWITCH IN "RUN"
• DISCONNECT IGNITION SWITCH CONN C1.
• IS THERE CONTINUITY BETWEEN IGNITION SWITCH HALF OF CONN C1 TERM "D" AND GROUND?
NO
YES
CHECK FOR SHORT TO GROUND IN CKT 33 (TAN/WHT)
REPLACE IGNITION SWITCH.

Figure 10-11 Brake indicator light "ON" diagnostic chart (Courtesy of Chevrolet Motor Division, General Motors Corporation)

the parking brake switch. If these are ok, replace the daytime running lights module. If the brake indicator does not light for any reason, check the suspect switch and circuits 33 and 150 as shown in the diagram. Replace or repair as needed.

Testing the Warning Lamp

When checking the warning light bulb, it is often difficult to determine if the filament is good. If there is any doubt as to the bulb's serviceability, replace it with a known good one.

Keep in mind that it is highly unlikely that all of the warning lights on the instrument panel would fail at the same time. If the lights are not operating properly, check the fuse first. Next, check for voltage at the last common connection. If no voltage is present here, trace the circuit back to the battery. If voltage is found at the common connection, test each branch of the circuit in the same manner.

To test for a faulty warning lamp, turn the ignition switch to the "start" position. The lamp test circuit should light the warning lamp. If the light does not illuminate when the test circuit is powered, check the suspect sender switch. To do this, disconnect the sender switch lead and use a jumper wire to connect the sender switch lead to ground. With the ignition switch in the "run" position, the warning lamp should light. If the lamp is illuminated, replace the sending switch. If the light does not come on, the bulb is burned out or the wiring is damaged. Use your test light to confirm voltage is present at the sensor terminal connector. If there is voltage, this indicates that the problem is a bad bulb.

WARNING: Always check the manufacturer's service manual to confirm the location of the sensor switch that you are testing. Accidentally grounding a terminal of the vehicle's engine control module or ABS control module could result in damage to the computer. A warning light sensor switch usually has one lead. The computer sensor has two to four leads and is contained in a weather-pack connector.

If the problem is that the warning light stays on, test it in the following manner: Disconnect the lead to the sender switch. The light should now go out when you place the ignition switch in the "run" position. If it does not, there is a short to ground in the wiring from the sender switch to the lamp.

Parking Brake Switch Test

Classroom Manual
Chapter 10, page 223

With the ignition switch in "run" or "start" and the parking brake switch closed, the light goes on to remind the driver that the parking brake is applied.

SERVICE TIP: On vehicles with daytime running lights, if the parking brake switch is okay, but the brake warning system light does not work, do the input test for the daytime running lights relay.

To test a typical parking brake switch on a console-mounted parking brake lever, remove the console and disconnect the connector from the switch. With the brake lever pulled up, use a continuity checker to ensure continuity exists between the positive terminal and a good ground (Figure 10-12). Continuity should be broken when the brake lever is in the down position.

Brake Fluid Level Switch Test

With the ignition switch in "run" or "start" and the brake fluid level switch closed (float down), the brake system light goes on to alert the driver of a low fluid condition in the master cylinder reservoir. To test the float switch, remove the reservoir cap and ensure that the float moves up and down without sticking. If the float sticks or binds, replace the reservoir assembly (Figure 10-13).

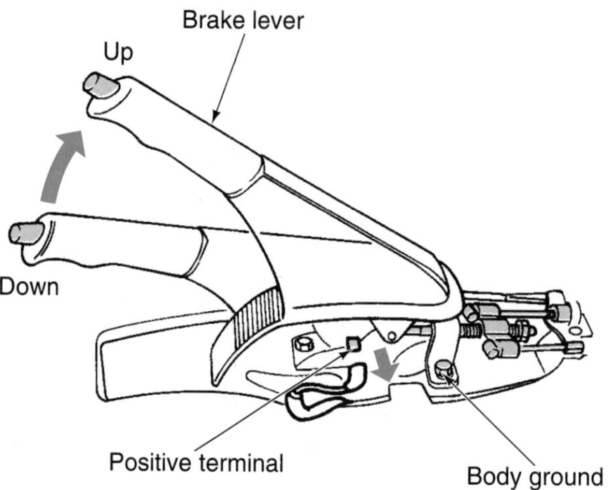

Figure 10-12 Testing the parking brake switch on a hand lever-activated parking brake (Courtesy of Honda Motor Co. Ltd.)

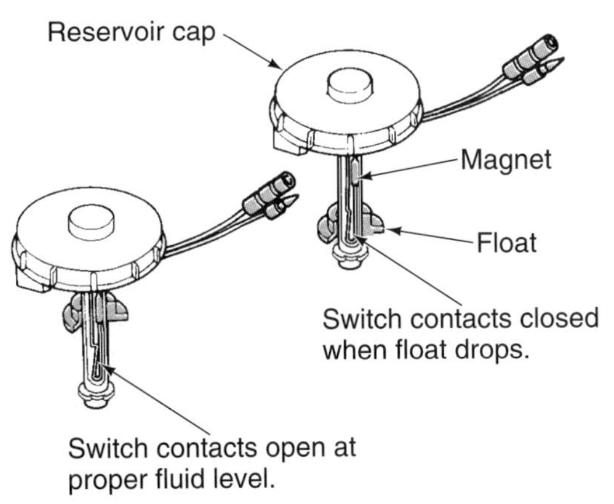

Figure 10-13 Performing a brake fluid level switch test (Courtesy of Honda Motor Co. Ltd.)

Figure 10-14 Removing a fluid level sensor from a master cylinder housing (Courtesy of General Motors Corporation)

There should be continuity when the float is down and no continuity when the float is up. Replace the reservoir cap if the proper test results are not obtained.

In many cases, the fluid level sensor mounts into the master cylinder. To test this type of fluid sensor, disconnect the electrical connector at the sensor and remove the sensor from the master cylinder (Figure 10-14). Once the sensor is removed, reconnect the electrical connector. Turn the ignition on and hold the sensor in the upright position. Within 30 seconds, the instrument panel warning light or the electronic instrument panel message "BRAKE FLUID LOW" should appear.

Next invert the sensor. The instrument panel message or the warning light should turn off immediately. This means that the sensor is working properly.

If the message does not appear or the light does not illuminate, remove the wiring from the sensor. Using a jumper wire, connect both sides of the plug. The instrument panel message should appear within 30 seconds or the light should go on. If the light or message is not activated, a problem exists with either the wiring or the instrument panel.

Testing the Combination Valve Warning Light Circuit

Some vehicles are equipped with combination valves. The pressure differential warning switch in a combination valve compares front and rear brake pressures from the master cylinder. If either the front or rear brakes malfunction, it energizes the warning indicator lamp on the instrument panel (Figure 10-15). The valve and the switch are designed so the switch "latches" in the warning position once a pressure difference is sensed. The only way to turn off the warning lamps is to fix the problem and apply enough pedal force to create about 450 psi of hydraulic line pressure.

The brake warning light electrical circuit between the pressure differential switch and the warning light should be checked prior to testing the pressure switch. To test the circuit, disconnect the lead wire from the pressure switch terminal on the combination valve. Connect a jumper wire from the lead to a ground on the engine or chassis. Turn the ignition key to ON. The brake warning light should come on. If the light does not come on, inspect and service the bulb, wiring, and connectors as required. If the warning light comes on with the jumper connected, turn the ignition switch to OFF and reconnect the lead wire to the pressure differential switch terminal.

To check the operation of the combination valve pressure differential switch, refer to the procedures outlined in Chapter 5.

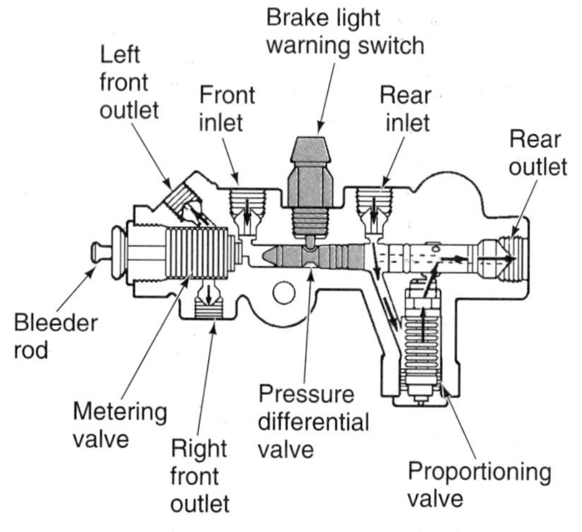

Figure 10-15 Pressure differential switch operation in a hydraulic system combination valve (Courtesy of Raybestos Division, Brake Systems Inc.)

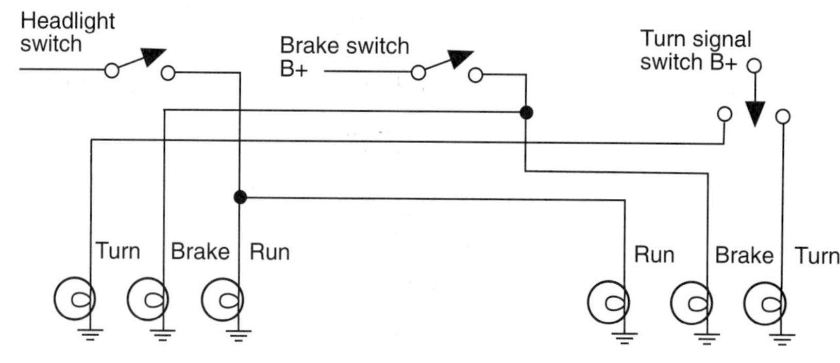

Figure 10-16 A three-bulb taillight circuit has individual control for each of the bulbs in the assembly.

Brake Stoplights

Studies indicate that locating a brake light at eye level significantly reduces rear-end collisions.

Because the turn signal switches used in a two-bulb system also control a portion of the operation of the brake lights, they have a complex system of contact points. Many brake light problems are caused by worn contact points in the turn signal switch.

Brake stoplights are used to warn others that the brakes of the vehicle have been applied and the vehicle is slowing down. Brake stoplights are included in the right and left taillight assembly. Beginning in 1986, the vehicle must also be equipped with a center high mounted stop lamp (CHMSL). This lamp is located on the center line of the vehicle no lower than 3 inches below the bottom of the rear window (6 inches on convertibles).

A three-bulb taillight contains bulbs for three separate vehicle functions: the running lights, the turn signal, and the brake stoplight. In a three-bulb taillight system, the brake stoplights are controlled directly by the brake light switch (Figure 10-16). In most vehicles, the brake light switch is attached to the brake pedal with a small clearance between the pedal arm pin and the eye of the pushrod (Figure 10-17). When the operator pushes in on the brake pedal, the switch plunger moves away from the pushrod. This closes the contact points, completing the circuit to energize the brake stoplights.

The brake light switch receives direct battery voltage through a fuse. This means the lights will operate even when the ignition is in the OFF position. Once the normally open switch is

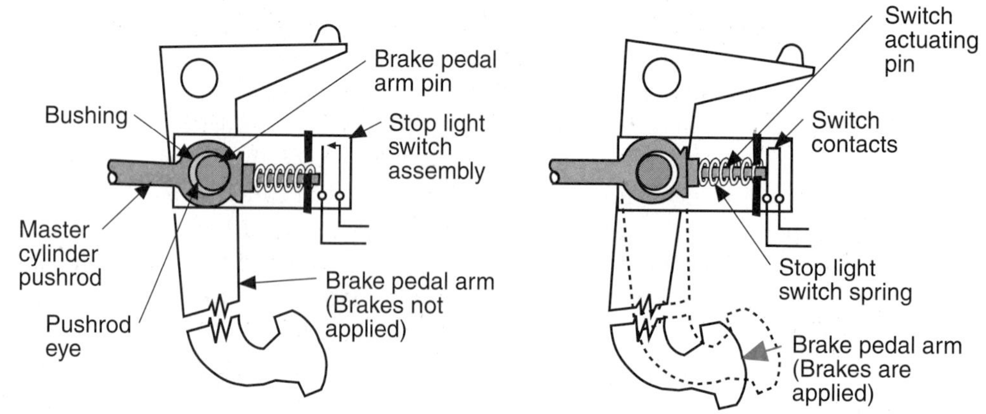

Figure 10-17 Typical brake pedal warning light switch operation (Reprinted with the permission of Ford Motor Company)

closed, voltage is applied to the brake lights. The lights on both sides of the vehicle and in the CHMSL are wired in parallel. The bulb is grounded to complete the circuit.

Brake Light Brake Switch Test

To troubleshoot a typical brake light circuit such as the one illustrated (Figure 10-18), first check for brake light illumination with the brake pedal applied. If one or both lights does not illuminate, check for a blown fuse in the fuse box. If the fuse is okay, check for a bad bulb. If the fuse and bulbs are all right, perform continuity checks on the brake switch. Disconnect the connector from the brake switch (Figure 10-19), depress the brake pedal, and check for continuity between the appropriate terminals. There should be continuity only when the brake pedal is depressed. If there is no continuity, replace the switch or adjust the pedal height and retest for continuity. If there is continuity, but the brake lights do not illuminate, troubleshoot the circuit for opens or a bad ground.

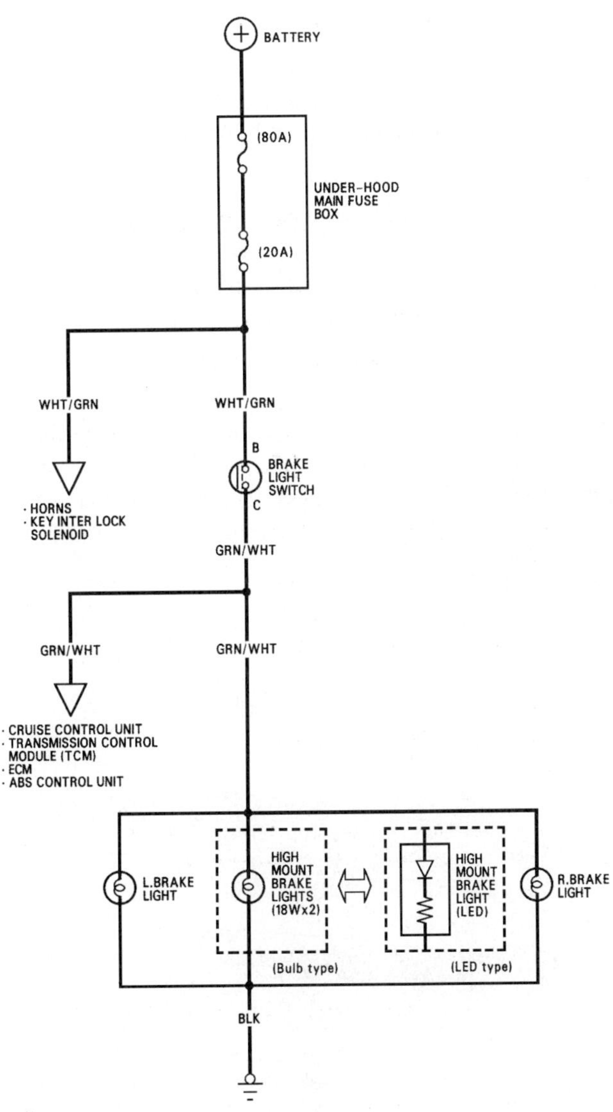

Figure 10-18 Circuit diagram for a typical high mount brake light circuit (Courtesy of Honda Motor Co. Ltd.)

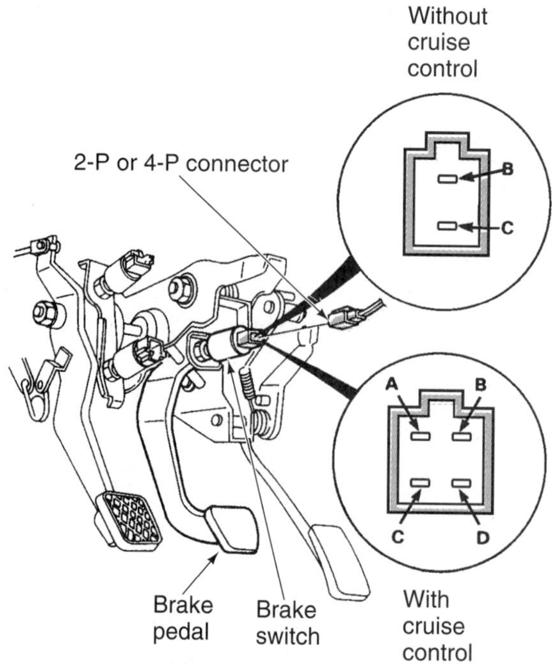

Figure 10-19 High mount brake light switch on brake pedal (Courtesy of Honda Motor Co. Ltd.)

Replacing CHMSL Brake Lights

CHMSL is the abbreviation for center high mounted stop light. It is also referred to as a collision avoidance light.

CHMSL are available in LED and standard bulb types. To replace a typical CHMSL LED-type brake light, remove the screws securing the light in place and carefully pry the light out of the tailgate spoiler (Figure 10-20). Disconnect the connectors from the light. Discard the old light and install the new one.

To replace a typical bulb-type CHMSL on a hatchback-type vehicle, open the hatch, then remove the screw that secures the lens cover in place (Figure 10-21). Next, remove the nuts that hold the light assembly to the hatch and disconnect the electrical connector. Turn the socket counterclockwise to remove the bulb. Installation is the reverse of disassembly.

CUSTOMER CARE: Clean the rear window glass before reinstalling the CHMSL assembly.

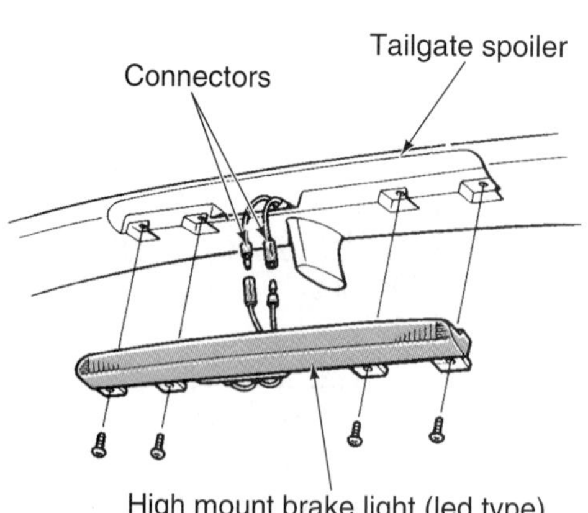

Figure 10-20 Removing an LED-type high mount brake light (Courtesy of Honda Motor Co. Ltd.)

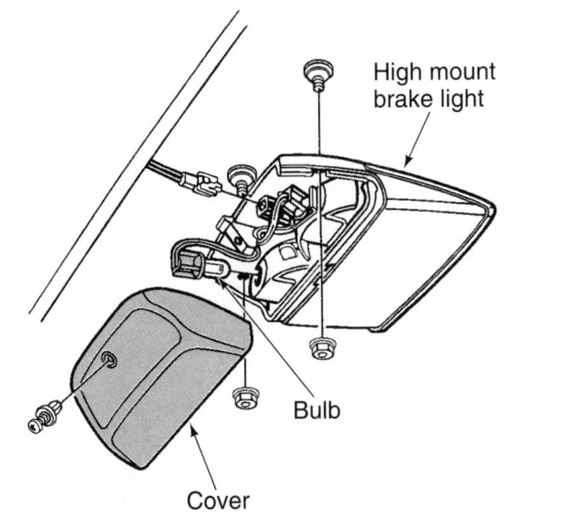

Figure 10-21 Bulb-type high mount brake light hatchback mounting and wiring (Courtesy of Honda Motor Co. Ltd.)

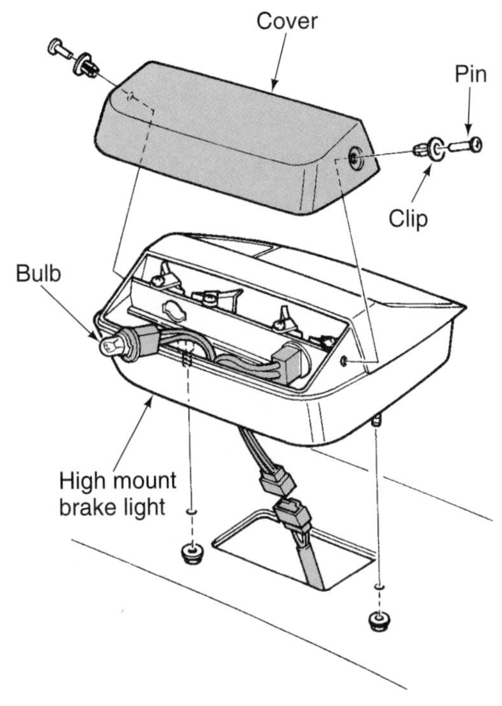

Figure 10-22 Bulb-type high mount brake light sedan mounting and wiring (Courtesy of Honda Motor Co. Ltd.)

A typical truck-mounted CHMSL assembly is shown (Figure 10-22). To remove this assembly, first open the trunk lid. Push the pin into the cover and then remove the clip. Remove the nuts that fasten the assembly to the trunk and then disconnect the electrical connector. Turn the socket 45° counterclockwise to remove the bulb. To reinstall the assembly, reverse the procedure. When attaching the clip to the cover, put the pin into the clip first, then push it into the cover.

Dual-Filament Bulb Taillight Assemblies

Many taillight assemblies use dual-filament bulbs (Figure 10-23) that perform two functions. The brake light circuit and the turn signal/hazard light circuit usually share a single dual-filament bulb, with the brake light circuit connected to the high-intensity filament contact.

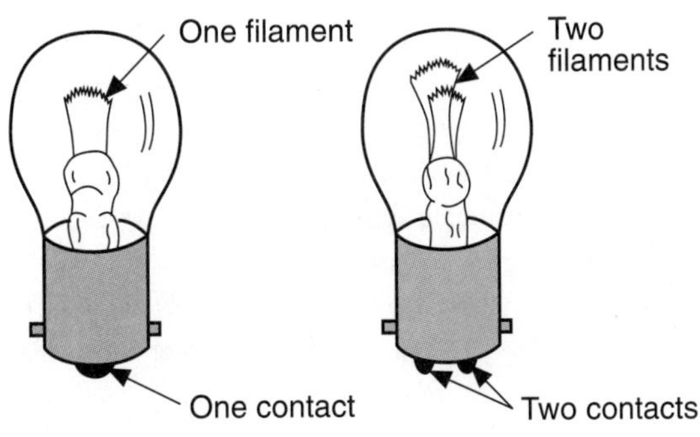

Figure 10-23 Examples of single- and dual-filament bulbs. These must never be used interchangeably.

Dual-filament bulbs are used to perform more than one function. They are not interchangeable with single-filament bulbs.

In this type of two-bulb circuit, the brake lights are wired through the turn signal and hazard switches (Figure 10-24). If neither turn signal is on, the current flows to both rear brake lights (Figure 10-25). If the left turn signal is on, current for the right brake light is sent to the lamp through

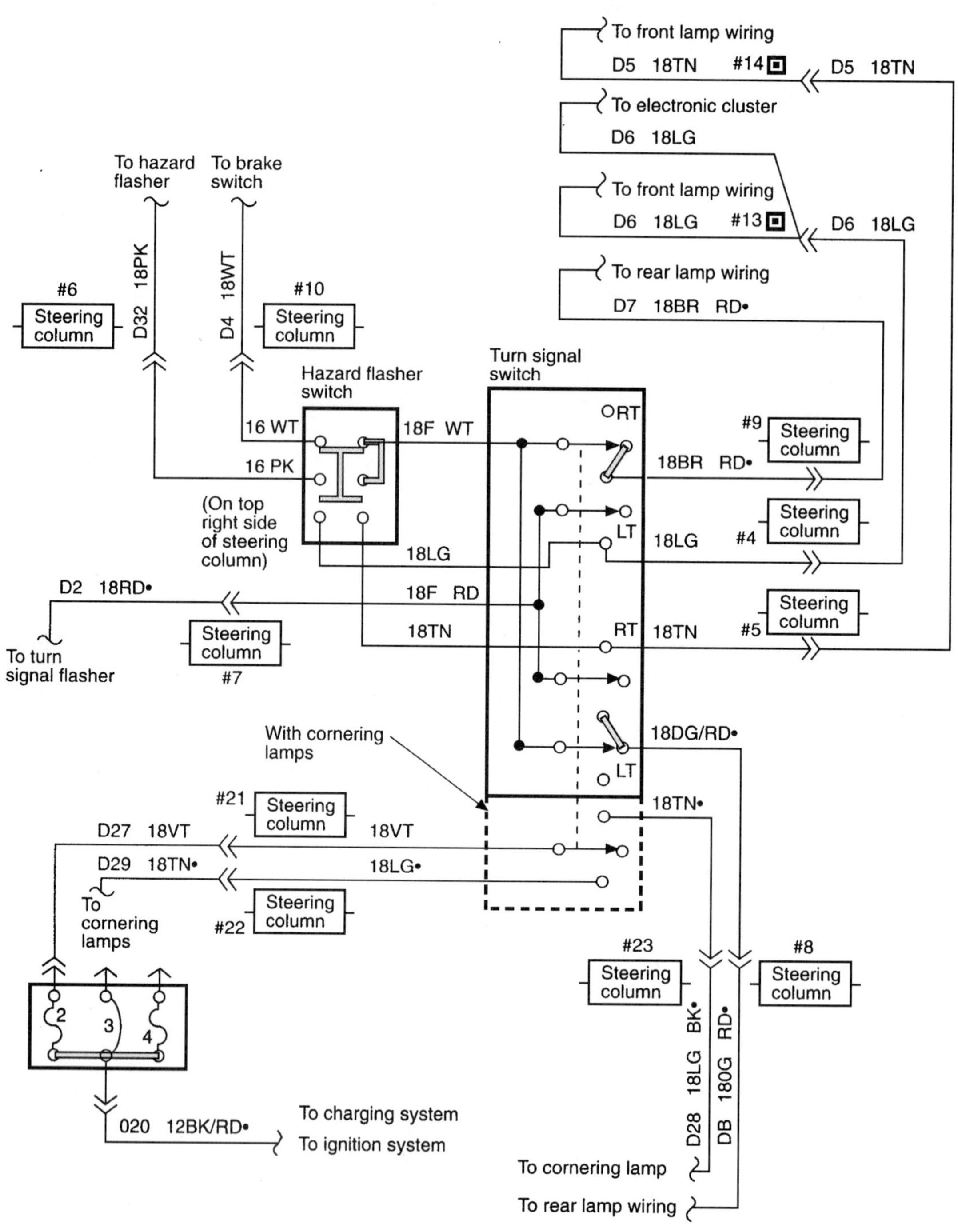

Figure 10-24 Turn signal switch used in a two-bulb taillight circuit (Courtesy of Chrysler Corporation)

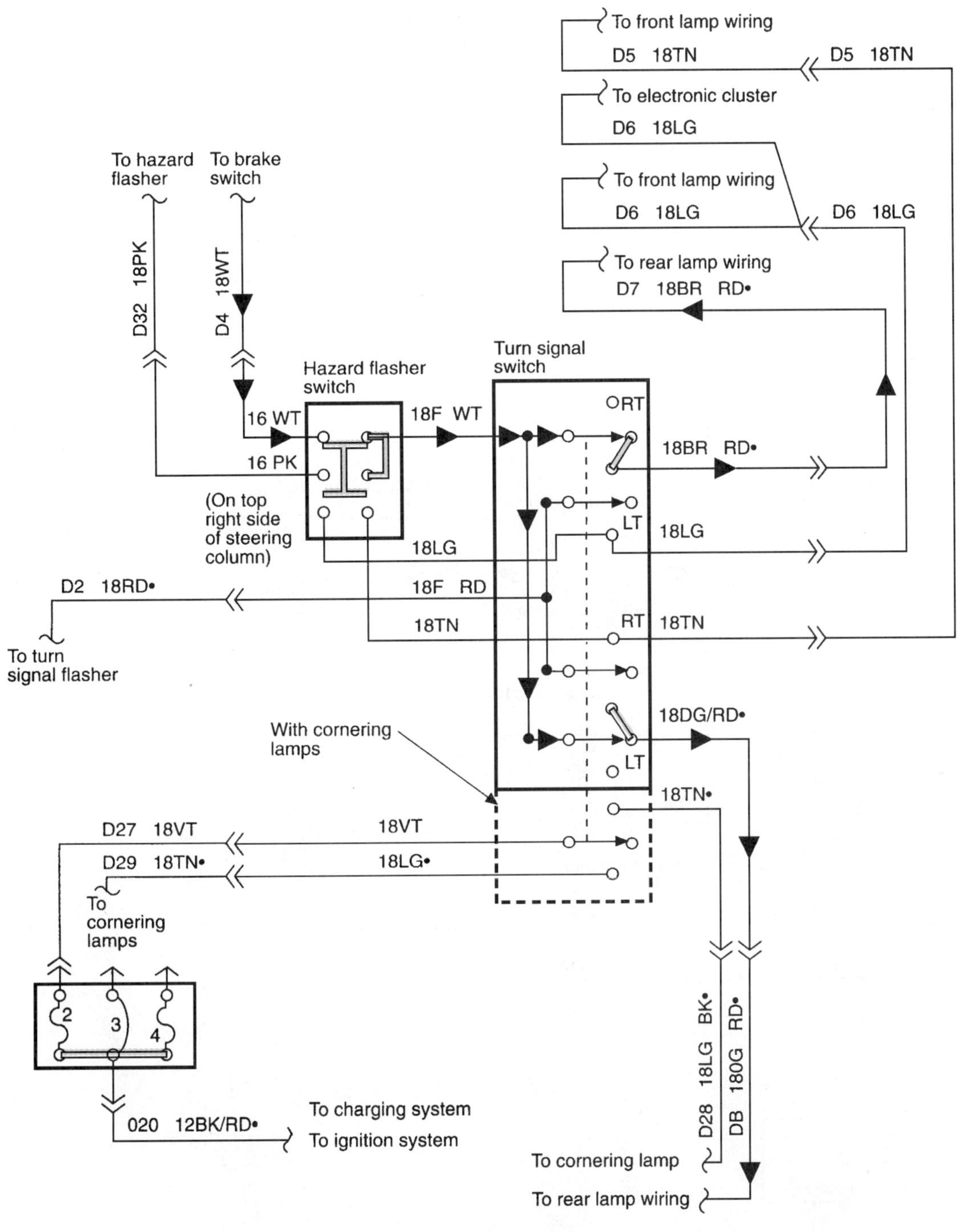

Figure 10-25 Brake light operation with the turn signals in the neutral position (Courtesy of Chrysler Corporation)

the turn signal switch and wire designated as 18BR/RD. The left brake light does not receive any voltage from the the brake switch because the turn signal switch opens that circuit (Figure 10-26).

In a two-bulb circuit, the CHMSL can be wired in one of two ways. The first way is to connect the brake light circuit between the brake light switch and the turn signal switch (Figure 10-27). However, this method increases the number of conductors needed in the harness. Because of this, most manufacturers prefer to install diodes in the conductors that are connected between the left and right side bulbs (Figure 10-28). If the brakes are applied when the turn signal switch is in its neutral position, the diodes allow voltage to flow to the CHMSL. When the turn signal switch is

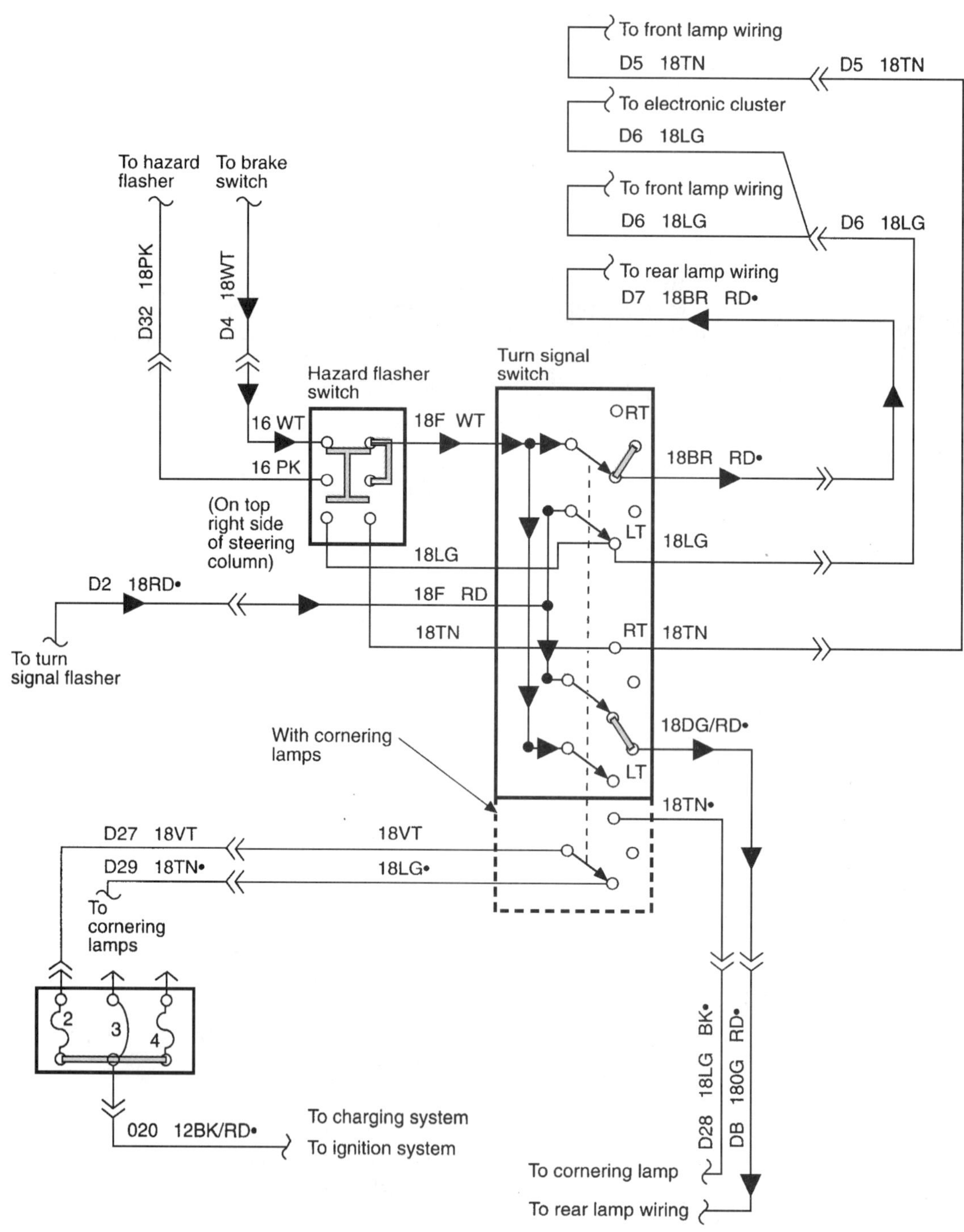

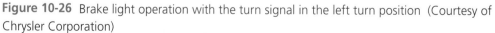

Figure 10-26 Brake light operation with the turn signal in the left turn position (Courtesy of Chrysler Corporation)

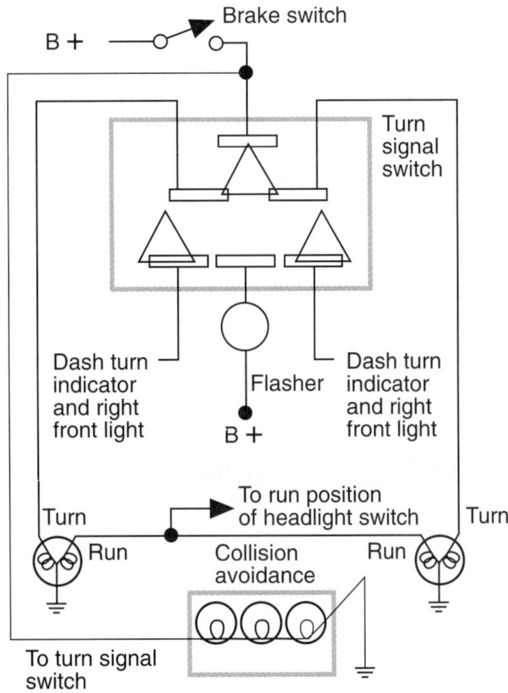

Figure 10-27 Wiring the CHMSL into the two-bulb circuit between the brake light switch and the turn signal

placed in the left turn position, the left light must receive a pulsating voltage from the flasher. However, the steady voltage being applied to the right brake light would cause the left light to illuminate continuously if the diode were not used. Diode 1 blocks the voltage from the right lamp, preventing it from reaching the left light. Diode 2 allows the voltage from the right brake light circuit to reach the CHMSL.

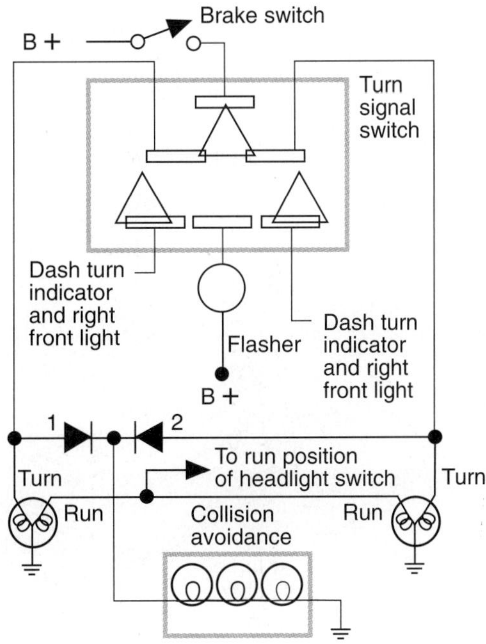

Figure 10-28 Two-bulb taillight circuit incorporating a CHMSL into the brake light system using diodes

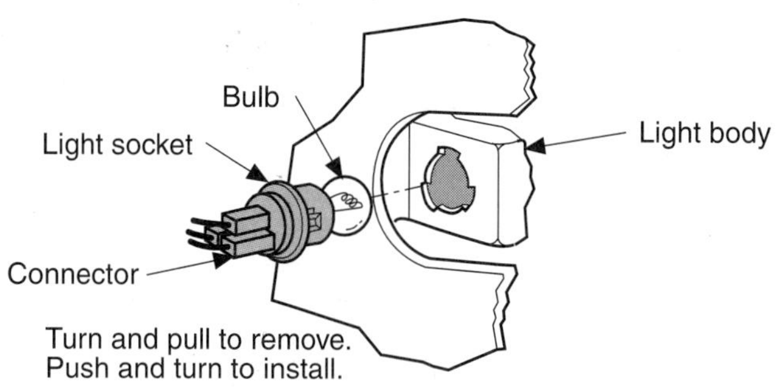

Figure 10-29 Bulb and socket removal from the taillight assembly (Reprinted with the permission of the Nissan Motor Corporation in USA)

Troubleshooting Taillight Assemblies

If all of the taillights do not operate, check the condition of the fuse in the circuit. If the fuse is good, use a voltmeter to test the brake light circuit. Check the brake light operation through the turn signal switch using a 12-volt test light to probe for voltage into and out of the switch. The ignition switch must be in the run position.

The taillight bulbs usually can be replaced without removing the lens assembly. Remove the bulb and socket by twisting the socket slightly and pulling it out of the lens assembly (Figure 10-29). Push in on the bulb and turn it. When the lugs align with the channels of the socket, pull the bulb out to remove it (Figure 10-30). The illustration shows how the taillight lens assembly mounts to the vehicle body (Figure 10-31). It is normally held in position by several attachment nuts.

Circuit Wiring Maintenance and Repair

All electrical connections must be kept clean and tight. Loose or corroded connections may cause a discharged battery, difficult starting, dim lamps, or possible damage to the alternator. Wires must be replaced if insulation becomes burned, cracked, or deteriorated.

Always use rosin flux solder to splice a wire or repair one that is frayed or broken, and use insulating tape to cover all splices or bare wires.

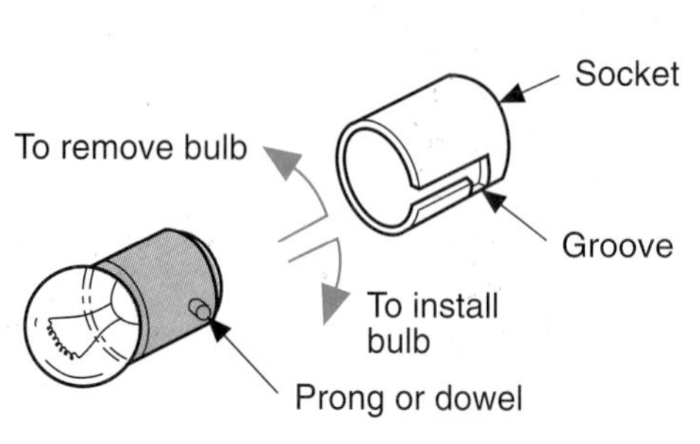

Figure 10-30 Removing the bulb from the socket (Reprinted with the permission of the Nissan Motor Corporation in USA)

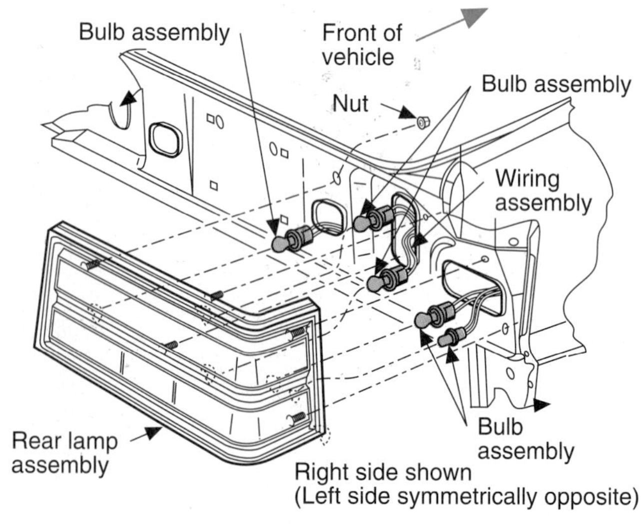

Figure 10-31 Taillight lens assembly (Reprinted with the permission of Ford Motor Company)

When replacing a wire, it is important that the correct size wire be used as shown on applicable wiring diagrams or in parts books. Each harness or wire must be held securely in place to prevent chafing or damage to the insulation due to vibration.

Wire size in a circuit is determined by the amount of current, the length of the circuit, and the voltage drop allowed. Wire size is specified using the metric gage. A metric gage describes the wire size directly in a cross-sectional area measured in square millimeters.

NOTE: Never replace a wire with one of a smaller size or replace a fusible link with a wire of larger size. Using the incorrect size could cause repeated failure and damage to the vehicle's electrical system.

Repairing Wire

Many of the electrical system repairs a technician performs involve replacing damaged conductors (wires). It is important to make these repairs in a way that does not increase the resistance in the original circuit or lead to shorts or grounds in the repaired area. Numerous methods are used to repair damaged wire, with many factors influencing the choice the technician must make. These include the type of repair required, accessibility of the wiring run, the type of conductor and size of wire needed, and the circuit requirements. The three most common repair methods are:

- ❑ Wrapping the damaged insulation with electrical tape
- ❑ Crimping the connections with a solderless connector
- ❑ Soldering splices

Copper Wire Repairs

Copper wire is the most popular type of wire used in automotive electrical systems. It is an excellent conductor, flexible, and easy to work with. However, the wire's protective insulation can be damaged or worn and small gauge wire can break due to stress or steady vibrations.

If the break or problem occurs in an inaccessible area, it is necessary to bypass part of the new wire. This involves cutting the wire before it enters and after it exits the inaccessible area and installing a replacement wire by using straps, hangers, and grommets as needed.

Crimping means to bend or deform by pinching a connector so that the wire connection is securely held in place.

Special Tools

Crimping tool

Electrical tape or heat shrink tubing

Solderless connector

Safety glasses

Fender covers

Crimping. Crimping solderless connections is an acceptable way of splicing wire that is not exposed to weather, dirt, corrosion, or excessive movement. To make a splice using a solderless connector:

1. Strip enough insulation from the end of the wire to allow it to completely penetrate the solderless connector.
2. Position the wire in the connector and crimp the connector (Figure 10-32). To ensure a good crimp, place the open area of the connector facing toward the anvil. Make certain you compress the wire under the crimp.

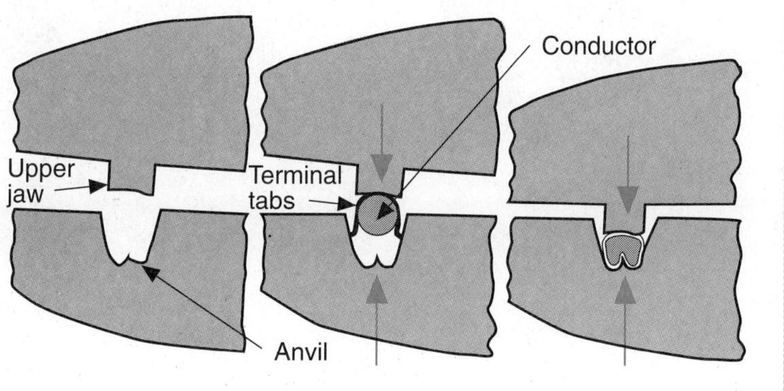

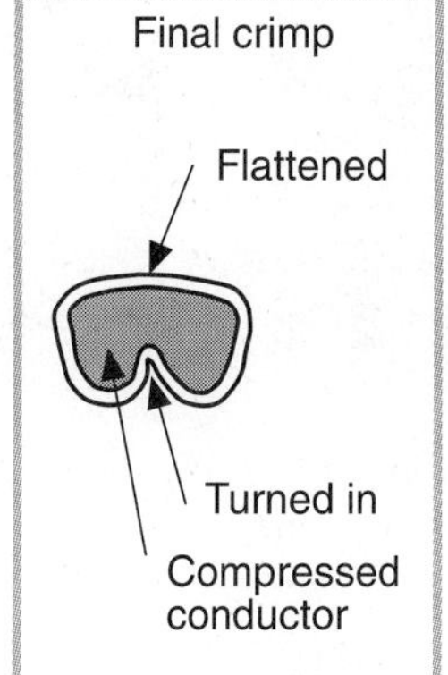

Figure 10-32 Crimping a solderless connection

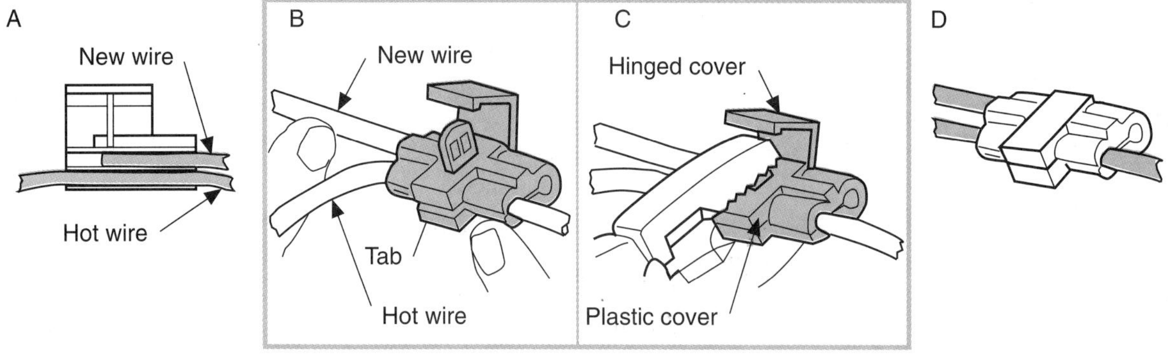

Figure 10-33 Using a tap connector to splice in another wire: (A) Place wires in position in the connector, (B) close the connector around the wires, (C) use pliers to force the tab into the conductors, and (D) close the hinged cover.

3. Insert the stripped end of the other wire into the connector and crimp in the same manner.
4. Wrap the connection in electrical tape or install a length of heat shrink tubing over the connection.

Heat shrink tubing is plastic tubing that shrinks in diameter when heated above a certain temperature.

The tap splice connector is another style of crimping connector. As shown (Figure 10-33), this connector allows for adding an additional circuit to an existing feed wire without stripping the wires.

Soldering

Soldering is preferred over crimping when splicing copper wires. When deciding where to cut the damaged wire, avoid points close to other splices or connections. As a rule, do not have two splices or connections within 1.5 in. (40 mm) of each other. Use a wire of the same size or larger than the wire being replaced, and always make the splice by soldering, follow the steps illustrated in Photo Sequence 15.

Photo Sequence 15
Typical Procedure for Soldering Copper Wire

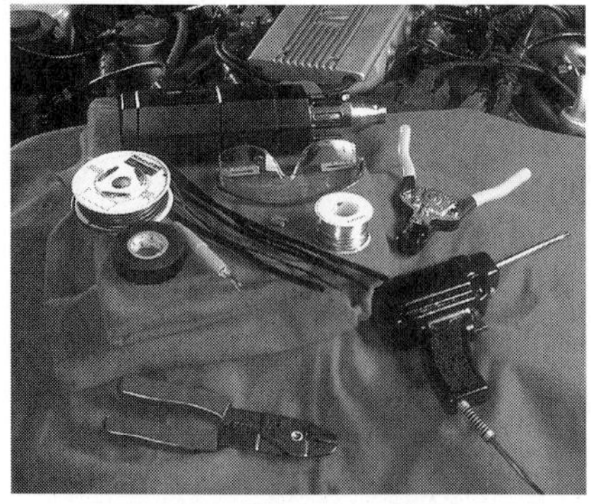

P15-1 Tools required to solder copper wire: 100-watt soldering iron, 60/40 rosin core solder, crimping tool, splice clip, heat shrink tubing, heating gun, safety glasses, sewing seam ripper, electrical tape, and fender covers.

P15-2 Place the fender covers over the fenders.

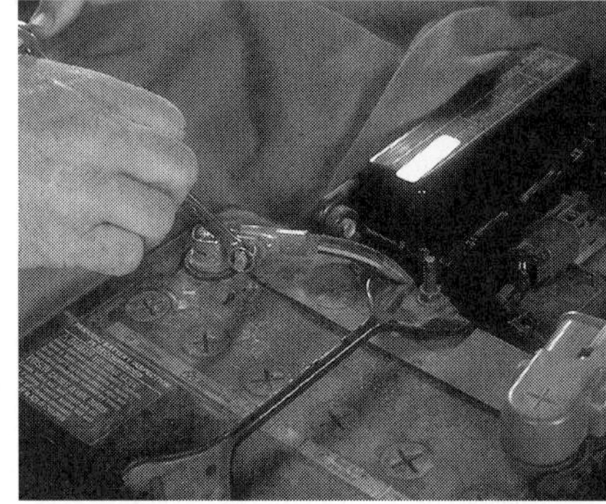

P15-3 Disconnect the fuse that powers the circuit being worked on. If the circuit is not protected by a fuse, disconnect the battery.

Photo Sequence 15
Typical Procedure for Soldering Copper Wire (continued)

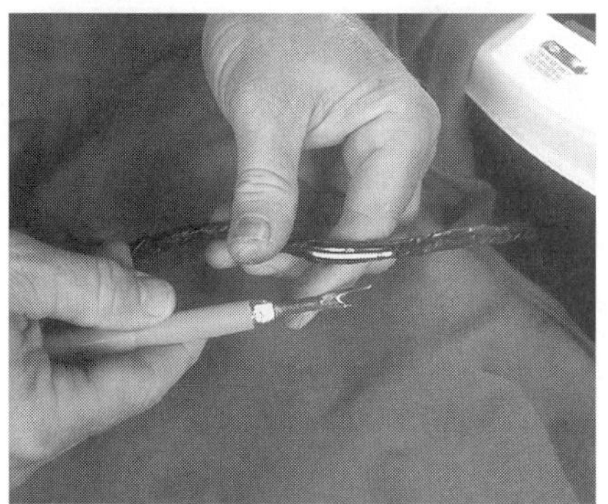

P15-4 If the wiring harness is taped, use a seam ripper to open the wiring harness.

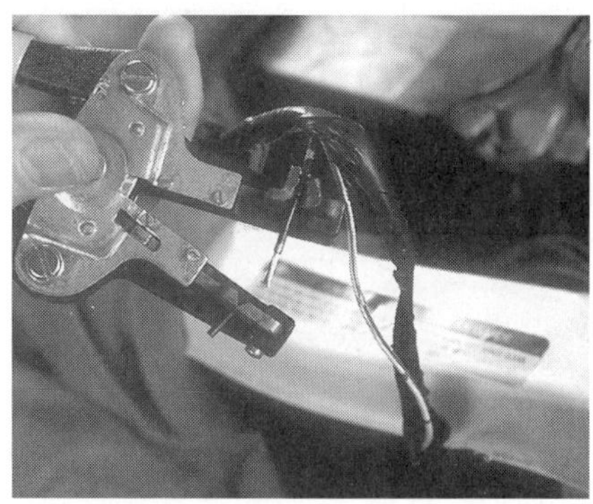

P15-5 Cut out the damaged wire using the cutters on the crimping tool.

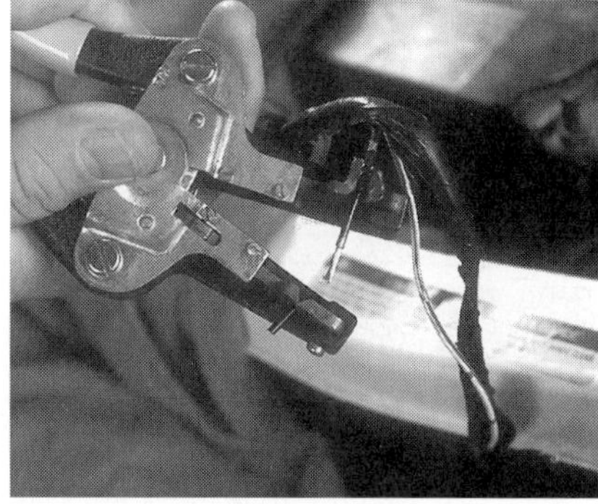

P15-6 Using a wire stripper, remove about one-half inch of the insulation from both wires. Be careful not to nick or cut any of the other wires in the area.

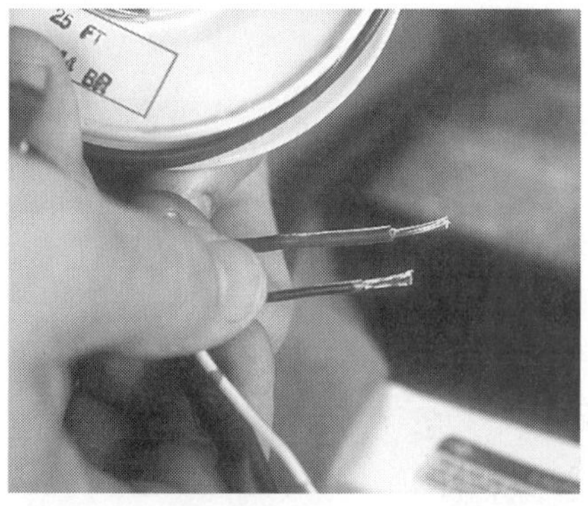

P15-7 Determine the correct gauge and length of replacement wire.

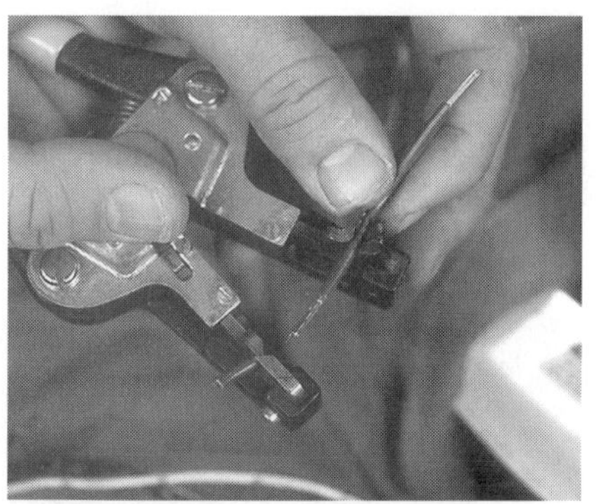

P15-8 Using the correct size stripper, remove one-half inch of insulation from each end of the replacement wire.

P15-9 Select the proper size splice clip to hold the splice.

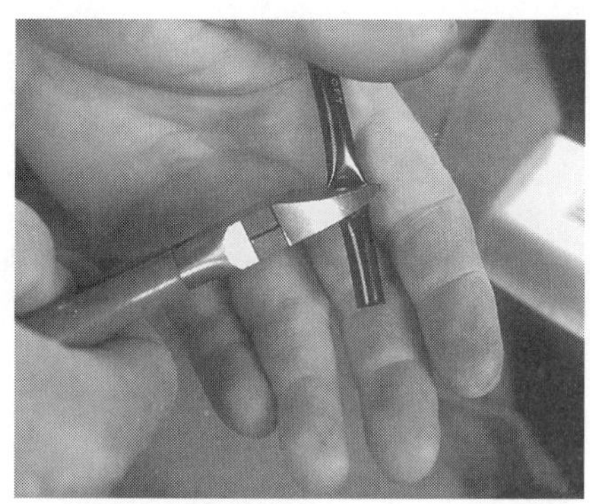

P15-10 Place the correct length and diameter of heat shrink tubing over the two ends of the wire.

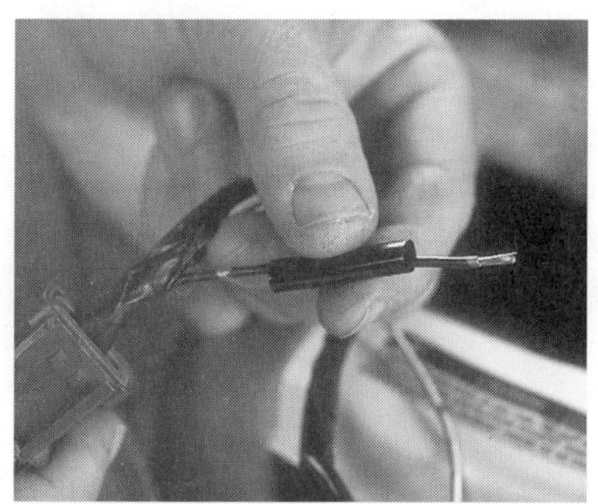

P15-11 Slide the tube far enough away so it is not exposed to the heat of the soldering iron.

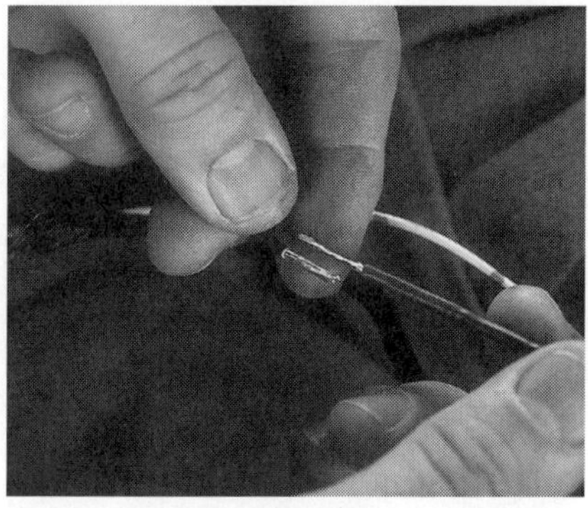

P15-12 Overlap the two splice ends and hold them in place with your thumb and forefinger.

Photo Sequence 15
Typical Procedure for Soldering Copper Wire (continued)

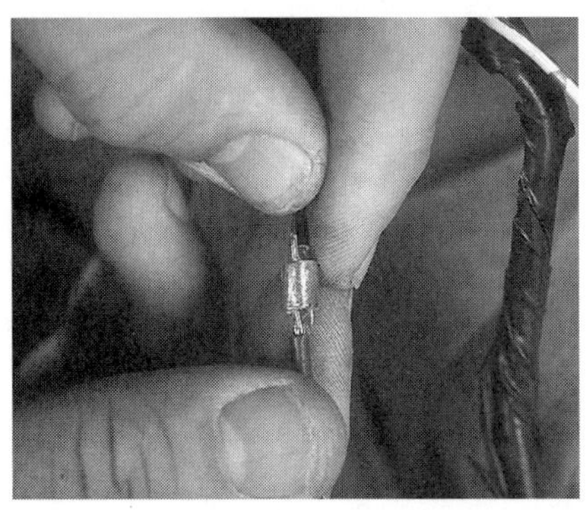

P15-13 Center the splice clip around the wires making sure the wires extend beyond the splice clip in both directions.

P15-14 Crimp the splice clip firmly in place. Crimp the clip at both ends of the repair.

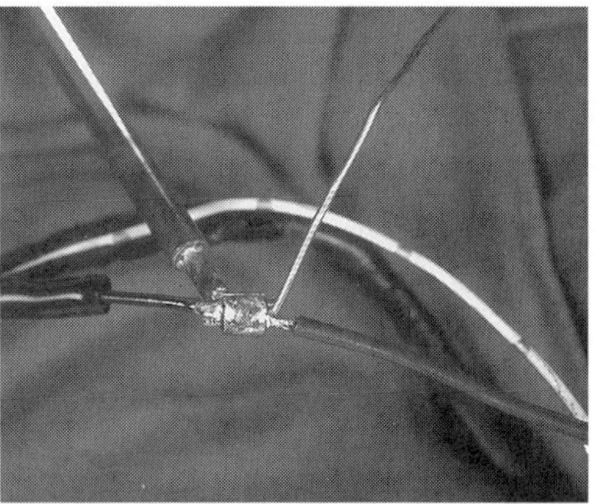

P15-15 Heat the splice clip with the soldering iron while applying solder to the opening in the back of the clip. Do not apply solder to the iron. The iron should be 180° away from the opening of the clip.

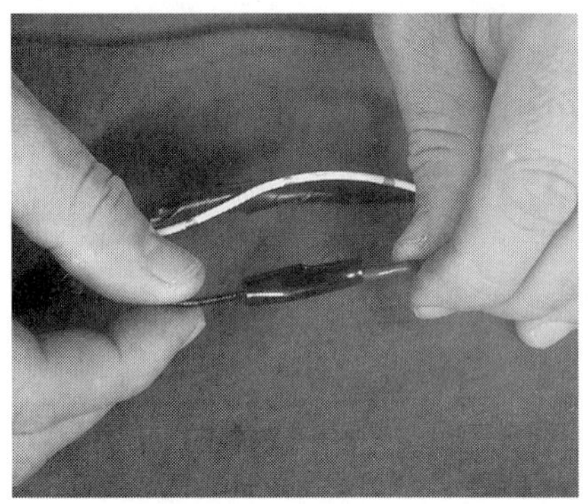

P15-16 After the solder cools, slide the heat shrink tube over the splice.

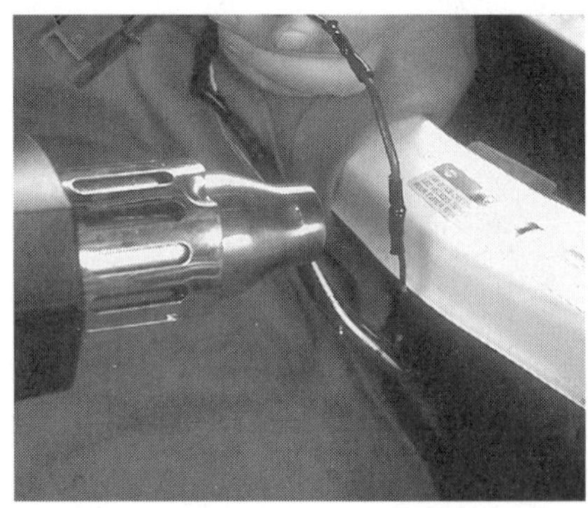

P15-17 Heat the tubing with the hot air gun until it shrinks around the splice. Do not overheat the tubing.

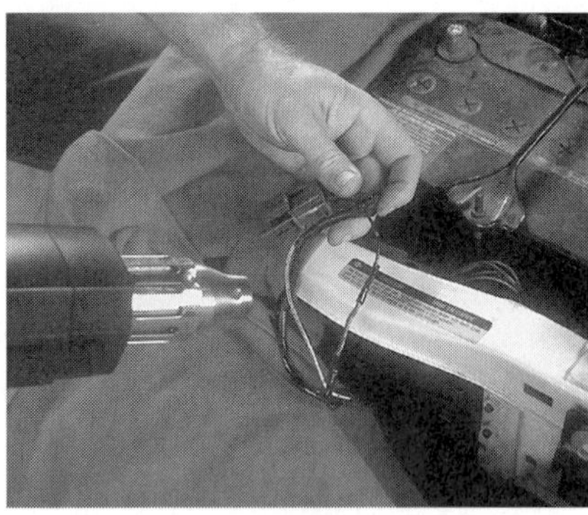

P15-18 Repeat for all other splices to complete the repair.

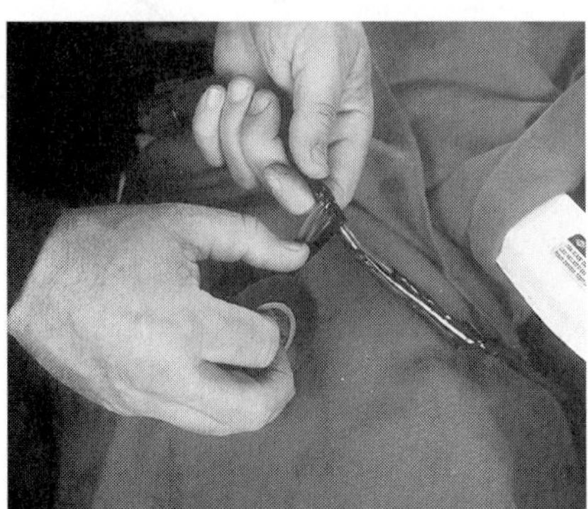

P15-19 Retape the wiring harness.

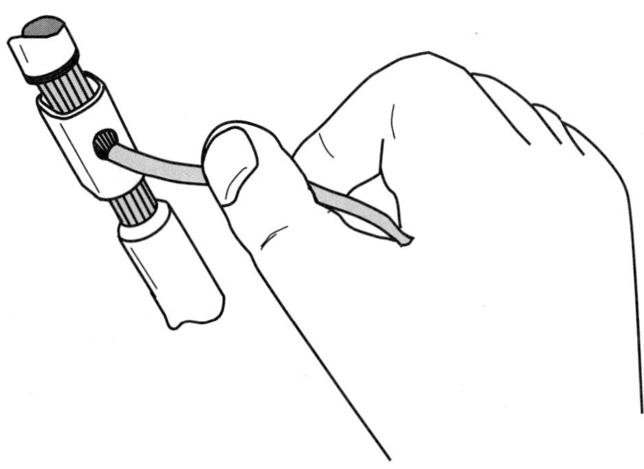

Figure 10-34 Using a splice clip to join wires (Courtesy of General Motors Corporation)

WARNING: Do not use acid core solder for electrical repairs. The acid will corrode the wire and increase its resistance.

A splice clip is a special connector used along with solder to create a good connection. The splice clip differs from a solderless connector in that it does not have insulation. The hole is used for applying the solder (Figure 10-34).

A second way of soldering wire together uses wire joints instead of splice clips. To use this method, begin by removing about one inch of insulation from the wires. Join the wires using one of the techniques illustrated (Figure 10-35). Heat the twisted connection with the soldering iron and apply solder to the strands of wire. As always, do not apply the solder directly to the iron. Heat the wire and allow the solder to flow onto it (Figure 10-36). Insulate the soldered connection using electrical tape or heat shrink tubing.

Special Tools

100-watt soldering iron

60/40 rosin core solder

Crimping tool

Splice clip

Heat shrink tubing

Heating gun

Safety glasses

Sewing seam ripper

Electrical tape

Fender covers

Repairing Aluminum Wire

WARNING: Do not attempt to solder aluminum wire. High heat will damage the wire.

Single-strand aluminum wire has been used by General Motors in limited applications. The aluminum wire uses a thick plastic insulator and is placed in a brown harness. Cut away the damaged

Special Tools

Crimping tool

Petroleum jelly

Safety glasses

Electrical tape or

Heat shrink tubing

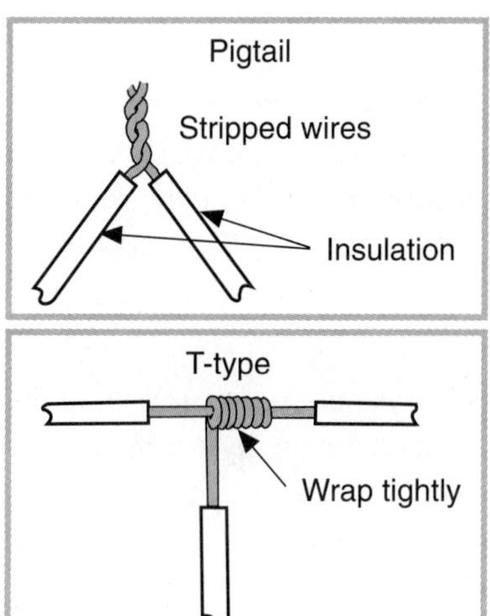

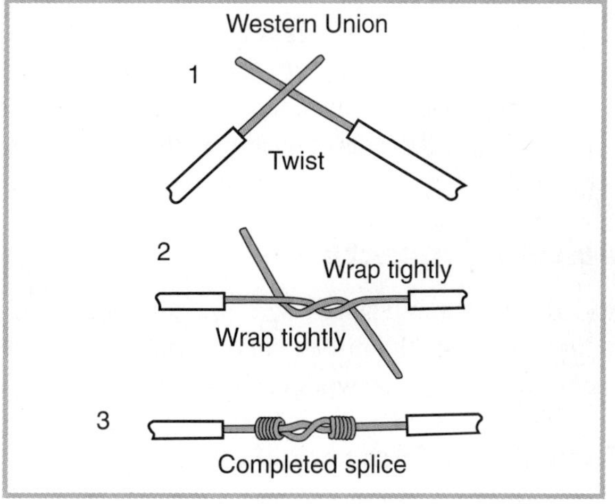

Figure 10-35 Methods of twist joining wire prior to soldering

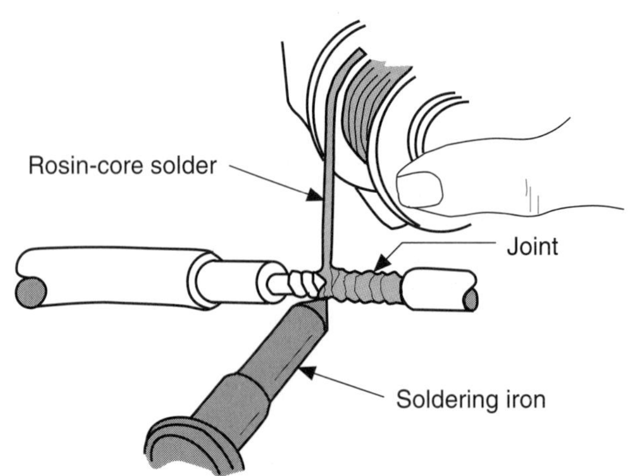

Figure 10-36 Heat the joint with the soldering iron and allow the heat to draw the solder into the joint. Do not apply solder directly to the iron.

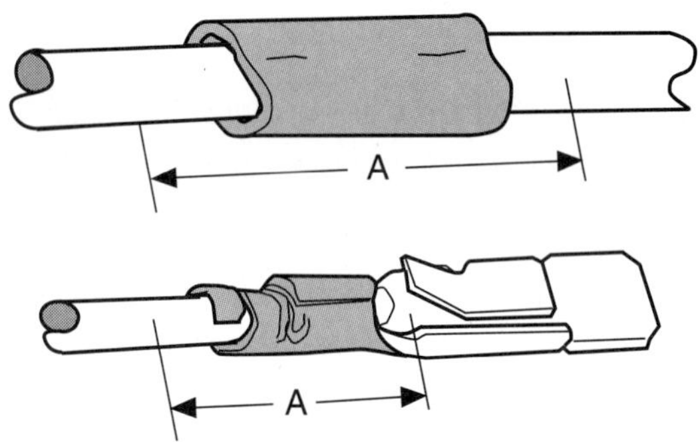

Figure 10-37 Joining aluminum wire. Apply petroleum jelly in the areas indicated. (Courtesy of General Motors Corporation)

section of wire and strip the insulation off of the last 1/4 inch of ends to be joined. Apply a good coat of petroleum jelly to the wires and connectors as shown (Figure 10-37). The jelly protects the wire against corrosion buildup. Crimp the connector using the technique covered earlier in the chapter. Insulate the splice using electrical tape or shrink tubing.

Repairing Connector Terminals

It is important to inspect the condition of the connectors when diagnosing the vehicle. Check connectors for cracks or signs of melting in the connector walls or contacts that are bent, scorched, corroded, or missing. If any of these problems are present, the connector should be replaced.

Molded Connectors

Molded connectors (Figure 10-38) are one-piece connectors that cannot be separated. If the connector is damaged, it must be cut out and a new connector spliced in wire by wire. This can be a time-consuming job, so work carefully around these connectors to avoid damaging them.

Hard-shell connectors have a hard plastic shell that holds the connecting terminals of separate wires.

Hard-Shell Connectors

Hard-shell connectors normally allow removing the terminals for repair. A pick or special tool is used to depress the locking tang of the connector (Figure 10-39). Pull the lead back to release the

Figure 10-38 Molded connectors are one-piece connectors that cannot be separated.

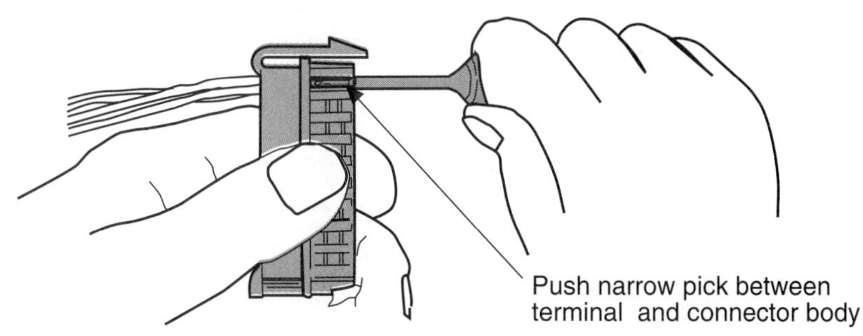
Push narrow pick between
terminal and connector body

Figure 10-39 Depress the locking tang to remove the terminal from the hard-shell connector. (Courtesy of General Motors Corporation)

locking tang from the connector. Remove the pick and pull the lead completely out of the connector. Make the repair to the terminal using the same procedure as for repairing copper or aluminum wire.

You must reform the terminal locking tank to ensure a good lock in the connector (Figure 10-40). Work with the pick to bend the lock tang back into its original shape. Insert the lead into the back of the connector. A noticeable "catch" should be felt when the lead is about half-way through the connector. Tug lightly to ensure the lead has caught inside the connector.

Special Tools

Pick

Crimping tool

Safety glasses

Fender cover

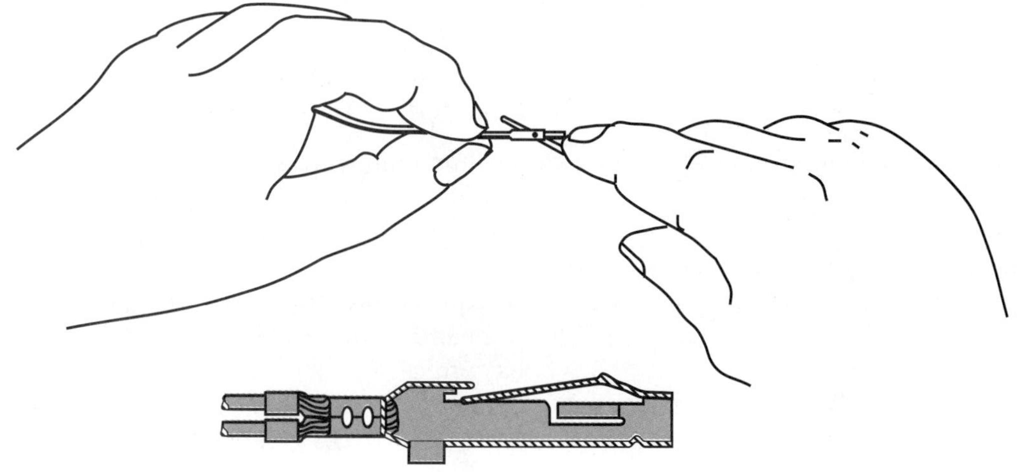

Figure 10-40 Reform the locking tang to its original position. (Courtesy of General Motors Corporation)

CASE STUDY

A minivan is brought to the shop with both rear stoplamps inoperative. The technician reviews the stoplamp circuit wiring diagram given in the shop manual to trace power through the circuit. To test for opens in the circuit, she first connects an unpowered test lamp from the proper terminal at the stoplamp switch connector to ground. The test lamp lights, indicating power is reaching the switch connector from the fuse block. She next connects the test lamp at the proper terminal at the stoplamp switch connector to ground and depresses the brake pedal. The test lamp operates, indicating the switch is operative. She continues by connecting the test lamp at the proper terminal on the turn signal connector to ground. The test lamp does not light, indicating an open exists between the stoplamp switch connector and the turn signal connector. The technician traces the wire between the two connectors and finds a break in the wire. She crimps, solders, and wraps the damaged wire. The stoplamps now operate.

Terms to Know

Brake stoplights
CHMSL
Continuity
Crimp
Diode
Dual-filament
Ground circuit
Hard-shell connector
Molder connector
Open circuit
Parallel circuit
Schematic or wiring diagram
Series circuit
Series-parallel circuit
Short circuit
Solderless connector
Splice clip

ASE Style Review Questions

1. *Technician A* says a broken or frayed wire can cause an unintentional grounded circuit. *Technician B* says dirt and grease buildup at terminals and connections can cause the same problem.
Who is correct?
A. A only **C.** Both A and B
B. B only **D.** Neither A nor B

2. *Technician A* says dual-filament light bulbs provide a backup filament in the bulb to double the life expectancy of the bulb.
Technician B says a dual-filament bulb serves two distinct functions, such as stoplights and turn signals.
Who is correct?
A. A only **C.** Both A and B
B. B only **D.** Neither A nor B

3. *Technician A* states that the brake stoplights in each taillight assembly and the center high mount stoplight assembly are wired as a series circuit.
Technician B says these components are wired in parallel.
Who is correct?
A. A only **C.** Both A and B
B. B only **D.** Neither A nor B

4. During aluminum wire repairs:
Technician A uses petroleum jelly on the wires to prevent corrosion and makes a crimped heat tubed sealed splice.
Technician B makes the repair with a twisted wire joint and 60/40 rosin core solder.
Who is correct?
A. A only **C.** Both A and B
B. B only **D.** Neither A nor B

5. Using schematics (wiring diagrams) is being discussed:
Technician A always traces the circuit out on the diagram before beginning work.
Technician B checks the diagram if visual inspection and tracing on the vehicle become difficult.
Who is correct?
A. A only **C.** Both A and B
B. B only **D.** Neither A nor B

6. Wire repair is being discussed:
Technician A avoids points close to other splices or connections when determining where to cut the damaged wire and stays at least 1.5 inches (40 mm) away from other splices or connectors.
Technician B uses a repair wire of the same size or larger than the wire being replaced.
Who is correct?
A. A only **C.** Both A and B
B. B only **D.** Neither A nor B

7. The brake system warning light is being discussed:
Technician A says the light can be activated by any number of switches such as the brake switch or the low fluid level switch in the master cylinder reservoir.
Technician B says the warning light should always light when the ignition is cranking.
Who is correct?
A. A only **C.** Both A and B
B. B only **D.** Neither A nor B

8. The brake warning light is being discussed:
Technician A says a ground between the parking brake switch and the lamp will cause it to burn continuously.
Technician B says that if the bulb does not light, replace it with a known good bulb before troubleshooting the entire circuit.
Who is correct?
A. A only **C.** Both A and B
B. B only **D.** Neither A nor B

9. Brake stoplights are being discussed:
Technician A says these lights receive fused voltage directly from the battery and will light with the ignition in the off position if the brake pedal is pressed.
Technician B says the lights are controlled by a normally open switch on the brake pedal.
Who is correct?
A. A only **C.** Both A and B
B. B only **D.** Neither A nor B

10. Wire splicing is being performed:
Technician A uses acid core solder on twisted copper wire joints.
Technician B wraps the splice in electrical tape or installs heat shrink tubing over the splice.
Who is correct?
A. A only **C.** Both A and B
B. B only **D.** Neither A nor B

APPENDIX A

Automotive Brake System Special Tool Suppliers

Baum Tools Unlimited
Longboat Key, FL

CTA Manufacturing Corp.
Carlstadt, NJ

Durston Manufacturing Co.
VIM Tools, LaVerne, CA

Goodson Auto Machine Shop Supplies
Winona, MN

Kent-Moore, Div. SPX Corp.
Roseville, MI

Lisle Corp.
Clarinda, IA

MATCO Tool Co.
Stow, OH

Mityvac/Neward
Cucamonga, CA

NAPA
Atlanta, GA

OTC, Div. SPX Corp.
Owatonna, MI

Snap-On Tools
Kenosha, WI

APPENDIX B

Metric Conversions

	to convert these	to these,	multiply by:
TEMPERATURE	Centigrade Degrees	Fahrenheit Degrees	1.8 then + 32
	Fahrenheit Degrees	Centigrade Degrees	0.556 after − 32
LENGTH	Millimeters	Inches	0.03937
	Inches	Millimeters	25.4
	Meters	Feet	3.28084
	Feet	Meters	0.3048
	Kilometers	Miles	0.62137
	Miles	Kilometers	1.60935
AREA	Square Centimeters	Square Inches	0.155
	Square Inches	Square Centimeters	6.45159
VOLUME	Cubic Centimeters	Cubic Inches	0.06103
	Cubic Inches	Cubic Centimeters	16.38703
	Cubic Centimeters	Liters	0.001
	Liters	Cubic Centimeters	1000
	Liters	Cubic Inches	61.025
	Cubic Inches	Liters	0.01639
	Liters	Quarts	1.05672
	Quarts	Liters	0.94633
	Liters	Pints	2.11344
	Pints	Liters	0.47317
	Liters	Ounces	33.81497
	Ounces	Liters	0.02957
WEIGHT	Grams	Ounces	0.03527
	Ounces	Grams	28.34953
	Kilograms	Pounds	2.20462
	Pounds	Kilograms	0.45359
WORK	Centimeter Kilograms	Inch-Pounds	0.8676
	Inch-Pounds	Centimeter-Kilograms	1.15262
	Meter Kilograms	Foot-Pounds	7.23301
	Foot-Pounds	Newton-Meters	1.3558
PRESSURE	Kilograms/Square Centimeter	Pounds/Square Inch	14.22334
	Pounds/Square Inch	Kilograms/Square Centimeter	0.07031
	Bar	Pounds/Square Inch	14.504
	Pounds/Square Inch	Bar	0.0689

GLOSSARY

Note: Terms are highlighted in color, followed by Spanish translation in bold.

Abrasion Wearing or rubbing away of a part.
Abrasión El desgaste o consumo por rozamiento de una parte.

ABS Antilock brake system.
ABS Un sistema de frenos antideslizante.

Acceleration An increase in velocity or speed.
Aceleración Un incremento en la velocidad.

Accumulator A device used in some antilock brake systems to maintain high pressure in the system. It is normally charged with nitrogen gas.
Acumulador Un dispositivo que se usa en algunos sistemas de frenos antideslizantes para mantener una alta presión en el sistema. Normalmente cargado con gas nitrógeno.

Air bleeding The removal of air from a hydraulic system.
Purgar con aire Quitar el aire de un sistema hidráulico.

Air brakes A braking system used on heavy trucks using air pressure as the power brake source.
Frenos de aire Un sistema de frenos utilizado en los camiones pesados que usa el presión de aire para suministrar los frenos de potencia.

Alignment An adjustment to a line or to bring into a line.
Alineación Un ajuste que se efectúa en una linea o alinear.

Anchor pin Anchoring pins on which the heel of the brake shoe rotates.
Perno de anclaje Las clavijas de anclaje en las cuales gira el extremo inferior de la zapata de freno.

Antifriction bearing A bearing designed to reduce friction. This type bearing normally uses ball or roller inserts to reduce the friction.
Cojinetes de antifricción Un cojinete diseñado con el fin de disminuir la fricción. Este tipo de cojinete suele incorporar una pieza inserta esférica o de rodillos para disminuir la fricción.

Antirattle spring A spring that holds parts in clutches and disc brakes together and keeps them from rattling.
Resorte antigolpeteo Un resorte que une las partes en los embragues o en los frenos de disco y los impide de rechinar.

Anti-seize Thread compound designed to keep threaded connections from damage due to rust or corrosion.
Antiagarrotamiento Un compuesto para filetes diseñado para protejer a las conecciones fileteados de los daños de la oxidación o la corrosión.

Arbor press A small, hand-operated shop press used when only a light force is required against a bearing, shaft, or other part.
Prensa para calar Una prensa de mano pequeña del taller que se puede usar en casos que requieren una fuerza ligera contra un cojinete, una flecha u otra parte.

Asbestos A material that was commonly used as a gasket material in places where temperatures were great. This material is being used less frequently today because of health hazards that are inherent to the material.
Amianto Una materia que se usó frecuentemente como materia de empaques en sitios en los cuales las temperaturas eran extremas. Esta materia se usa menos actualmente debido a los peligros al salud que se atribuyan a esta materia.

Axial Parallel to a shaft or bearing bore.
Axial Paralelo a una flecha o al taladro del cojinete.

Axis The center line of a rotating part, a symmetrical part, or a circular bore.
Eje La linea de quilla de una parte giratoria, una parte simétrica, o un taladro circular.

Axle The shaft or shafts of a machine upon which the wheels are mounted.
Semieje El eje o los ejes de una máquina sobre los cuales se montan las ruedas.

Backing plate A plate to which drum braking mechanisms are affixed.
Placa de respaldo Una placa a la cual se afija los mecanismos de un freno de tambor.

Ball bearing An antifriction bearing consisting of hardened inner and outer races with hardened steel balls that roll between the two races and support the load of the shaft.
Rodamiento de bolas Un cojinete de antifricción que consiste de pistas endurecidas interiores e exteriores que contienen bolas de acero endurecidos que ruedan entre las dos pistas, y sostiene la carga de la flecha.

Bearing The supporting part that reduces friction between a stationary and rotating part or between two moving parts.
Cojinete La parte portadora que reduce la fricción entre una parte fija y una parte giratoria o entre dos partes que muevan.

Bearing cage A spacer that keeps the balls or rollers in a bearing in proper position between the inner and outer races.
Jaula del cojinete Un espaciador que mantiene a las bolas o a los rodillos del cojinete en la posición correcta entre las pistas interiores e exteriores.

Bearing cone The inner race, rollers, and cage assembly of a tapered roller bearing. Cones and cups must always be replaced in matched sets.
Cono del cojinete La asamblea de la pista interior, los rodillos, y el jaula de un cojinete de rodillos cónico. Se debe siempre reemplazar a ambos partes de un par de conos del cojinete y los anillos exteriores a la vez.

Bearing cup The outer race of a tapered roller bearing or ball bearing.
Anillo exterior La pista exterior de un cojinete cónico de rodillas o de bolas.

Bearing race The surface upon which the rollers or balls of a bearing rotate. The outer race is the same thing as the cup, and the inner race is the one closest to the axle shaft.
Pista del cojinete La superficie sobre la cual rueden los rodillos o las bolas de un cojinete. La pista exterior es lo mismo que un anillo exterior, y la pista interior es la más cercana a la flecha del eje.

Bleeder valve A valve on the master cylinder, caliper, or wheel cylinder that allows air and fluid to be drained from the system.
Tornillo de purga Una válvula en el cilindro maestro, en la abrazadera, o en el cilindro de la rueda que permite que el aire y el líquido se vacíen del sistema.

Bolt torque The turning effort required to offset resistance as the bolt is being tightened.
Torsión del perno El esfuerzo de torsión que se requiere para compensar la resistencia del perno mientras que esté siendo apretado.

Booster A vacuum or hydraulic device attached to the master cylinder to ease operation and/or increase the effectiveness of a brake system.
Reforzador Un dispositivo de vacío o hidráulico conectado al cilindro maestro que facilita la operación y/o hace más eficaz un sistema de frenos.

Brake anchor Pivot pin on brake backing plate on which the shoe rests.
Anclaje de freno Una clavija de pivote en la placa de respaldo sobre la cual descanse la zapata.

Brake drag The continuous contact between pad or lining with the brake disc or drum.

Arrastre de los frenos Un contacto continuo entre la almohadilla o el forro con el disco o el tambor del freno.

Brake drum A bowl-shaped cast iron housing against which the brake shoes press to stop its rotation.

Tambor Una caja redonda de fierro colado contra la cual se aprietan las zapatas para detener su rotación.

Brake drum micrometer Measures the inside diameter of a brake drum.

Micrómetro del tambor Toma la medida del diámetro interior de un tambor de freno.

Brake fade Loss of braking effectiveness caused by excessive heat reducing the friction between the pad and disc or shoe and the drum.

Amortiguamiento del frenado La pérdida de la eficiencia de frenar debido al calor excesivo que reduce la fricción entre la almohadilla y el disco o la zapata y el tambor.

Brake fluid A hydraulic fluid used to transmit force through brake lines. Brake fluid must be noncorrosive to both the metal and rubber components of the brake system.

Líquido de freno Un líquido hidráulico que se usa para transmitir la fuerza por las líneas de freno. El líquido de freno no debe corroer los componentes de metal ni de hule que se encuentran en el sistema de frenos.

Brake flushing A procedure for cleaning a brake system by removing fluid and washing out sediment and condensation.

Purgar los frenos Un procedimiento para limpiar el sistema de frenos en que se vacia el líquido y se quitan los sedimentos y la condensación.

Brake grab Sudden and undesirable increase in braking force.

Amarro de freno Un incremento del acción de frenar repentino y no deseado.

Brake horsepower (bhp) Power delivered by the engine and available for driving the vehicle: bhp = torque × rpm/5,252.

Caballo indicado al freno (bhp) Potencia que provee el motor y que es disponible para el uso del vehículo: bhp = de par motor × rpm/5,252.

Brake lines Lines that carry brake fluid from the master cylinder to the wheels.

Líneas de freno Las líneas que llevan el líquido de freno del cilíndro maestro hasta las ruedas.

Brake lining Heat-resistant friction material that is pressed against the metal drum or disc to achieve braking force in a brake system.

Forro de frenos Una material resistente al calor que se oprime contra el tambor o disco de metal para ejecutar una fuerza de frenar en un sistema de frenos.

Brake pads The parts of a disc brake system that hold the linings.

Almohadilla s de freno Las partes en un sistema de frenos de disco que sostienen los forros.

Brake pedal Pedal pushed upon to activate the master cylinder.

Pedal de freno El pedal que se oprime para accionar el cilíndro maestro.

Brake shoe The metal assembly onto which the frictional lining is attached for drum brake systems.

Zapata de freno La asamblea metálica a la cual se afija el forro de fricción para el sistema de freno de tambor.

Brinnelling Rough lines worn across a bearing race or shaft due to impact loading, vibration, or inadequate lubrication.

Efecto brinel Lineas ásperas que aparecen en las pistas de un cojinete o en las flechas debido al choque de carga, la vibración, o falta de lubricación.

Burnish To smooth or polish by the use of a sliding tool under pressure.

Bruñir Tersar o pulir por medio de una herramienta deslizando bajo presión.

Burr A feather edge of metal left on a part being cut with a file or other cutting tool.

Rebaba Una lima espada de metal que permanece en una parte que ha sido cortado con una lima u otro herramienta de cortar.

Bushing A cylindrical lining used as a bearing assembly made of steel, brass, bronze, nylon, or plastic.

Buje Un forro cilíndrico que se usa como una asamblea de cojinete que puede ser hecho del acero, del latón, del bronze, del nylon, o del plástico.

Cage A spacer used to keep the balls or rollers in proper relation to one another. In a constant-velocity joint, the cage is an open metal framework that surrounds the balls to hold them in position.

Jaula Una espaciador que mantiene una relación correcta entre los rodillos o las bolas. En una junta de velocidad constante, la jaula es un armazón abierto de metal que rodea a las bolas para mantenerlas en posición.

Caliper Major component of a disc brake system. Houses the piston(s) and supports the brake pads.

Mordaza Un componente principal del sistema de frenos de disco. Contiene el pistón (los pistones) y sostiene las almohadillas.

Castellate Formed to resemble a castle battlement, as in a castellated nut.

Acanalado De una forma que parece a las almenas de un castillo (véa la palabra en inglés), tal como una tuerca con entallas.

Castellated nut A nut with six raised portions or notches through which a cotter pin can be inserted to secure the nut.

Tuerca con entallas Una tuerca que tiene seis porciones elevadas o muescas por los cuales se puede insertar un pasador de chaveta para retener a la tuerca.

C-clip A C-shaped clip used to retain the drive axles in some rear axle assemblies.

Grapa de C Una grapa en forma de C que retiene a las flechas motrices en algunas asambleas de ejes traseras.

Chamfer A bevel or taper at the edge of a hole or a gear tooth.

Chaflán Un bisél o cono en el borde de un hoyo o un diente del engranaje.

Chase To straighten up or repair damaged threads.

Embutir Enderezar o reparar a los filetes dañados.

Chassis The vehicle frame, suspension, and running gear. On FWD cars, it includes the control arms, struts, springs, trailing arms, sway bars, shocks, steering knuckles, and frame. The drive shafts, constant-velocity joints, and transaxle are not part of the chassis or suspension.

Chasis El armazón de un vehículo, la suspensión, y el engranaje de marcha. En los coches de FWD, incluye los brazos de mando, los postes, los resortes (chapas), los brazos traseros, las estabilizadoras, las articulaciones de la dirección y el armazón. Los árboles de mando, las juntas de velocidad constante, y la flecha impulsora no son partes del chasis ni de la suspensión.

Circlip A split steel snap ring that fits into a groove to hold various parts in place. Circlips are often used on the ends of FWD drive shafts to retain the constant-velocity joints.

Grapa circular Un seguro partido circular de acero que se coloca en una ranura para posicionar a varias partes. Las grapas circulares se suelen usar en las extremidades de los árboles de mando en FWD para retener las juntas de velocidad constante.

Clearance The space allowed between two parts, such as between a journal and a bearing.

Holgura El espacio permitido entre dos partes, tal como entre un muñon y un cojinete.

Coefficient of friction The ratio of the force resisting motion between two surfaces in contact to the force holding the two surfaces in contact.

Coeficiente de la fricción La relación entre la fuerza que resiste al movimiento entre dos superficies que tocan y la fuerza que mantiene en contacto a éstas dos superficies.

Compensating port An opening in a master cylinder that permits fluid to return to the reservoir.

Abertura de compensación Un orificio en el cilindro maestro que permite que el líquido regresa al recipiente.

Compound A mixture of two or more ingredients.

Compuesto Una combinación de dos ingredientes o más.

Concentric Two or more circles having a common center.

Concéntrico Dos círculos o más que comparten un centro común.

Constant-velocity joint (also called CV joint) A flexible coupling between two shafts that permits each shaft to maintain the same driving or driven speed regardless of operating angle, allowing for a smooth transfer of power. The constant-velocity joint consists of an inner and outer housing with balls in between, or a tripod and yoke assembly.

Junta de velocidad constante (también llamado junta CV) Un acoplador flexible entre dos flechas que permite que cada flechamantenga la velocidad de propulsión o arrastre sin importar el ángulo de operación, efectuando una transferencia lisa del poder. La junta de velocidad constante consiste de una caja interior e exterior entre los cuales se encuentran bolas, o de un conjunto de trípode y yugo.

Contraction A reduction in mass or dimension: the opposite of expansion.

Contracción Una reducción en la masa o en la dimensión: el opuesto de la expansión.

Control arm A suspension component that links the vehicle frame to the steering knuckle or axle housing and acts as a hinge to allow up-and-down wheel motions. The front control arms are attached to the frame with bushings and bolts and are connected to the steering knuckles with ball joints. The rear control arms attach to the frame with bushings and bolts and are welded or bolted to the rear axle or wheel hubs.

Brazo de mando Un componente de la suspención que une el armazón del vehículo al articulación de dirección o a la caja del eje y que se porta como una bisagra para permitir los movimientos verticales de las ruedas. Los brazos de mando delanteros se conectan al armazón por medio de pernos y bujes y se conectan al articulación de dirección por medio de los articulaciones esféricos. Los brazos de mando traseros se conectan al armazón por medio de pernos y bujes y son soldados o empernados al eje trasero o a los cubos de las ruedas.

Control valve The component beneath the master cylinder that contains the hydraulic controls for the brake system.

Válvula de control El componente en la parte inferior del cilindro maestro que contiene los controles hidráulicos para el sistema de frenos.

Corrode To eat away gradually as if by gnawing, especially by chemical action.

Corroer Roído poco a poco, primariamente por acción químico.

Corrosion Chemical action, usually by an acid, that eats away (decomposes) a metal.

Corrosión Un acción químico, regularmente de un ácido, que corroe (descompone) un metal.

Cotter pin A type of fastener made from soft steel in the form of a split pin, that can be inserted into a drilled hole. The split ends are spread to lock the pin in position.

Pasador de chaveta Un tipo de fijación hecho de acero blando en forma de una chaveta, que se puede insertar en un hueco tallado. Las extremidades partidas se despliegen para asegurar la posición de la chaveta.

Counterclockwise rotation Rotating in the opposite direction of the hands on a clock.

Rotación en sentido inverso Girando en el sentido opuesto de las manecillas de un reloj.

Coupling A connecting means for transferring movement from one part to another; may be mechanical, hydraulic, or electrical.

Acoplador Un método de conexión que transfere el movimiento de una parte a otra; puede ser mecánico, hidráulico, o eléctrico.

Deflection Bending or movement away from normal due to loading.

Desviación Curvación o movimiento fuera de lo normal debido a la carga.

Degree A unit of measurement equal to 1/360th of a circle.

Grado Una uneda de medida que iguala al 1/360 parte de un círculo.

Density Compactness; relative mass of matter in a given volume.

Densidad La firmeza; una cantidad relativa de la materia que ocupa a un volumen dado.

Depth gauge Based on a micrometer, this gauge is used to measure the depth of holes, slots, and keyways.

Galga de profundidades Principalmente como un micrómetro, esta galga se usa para medir la profundidad de los hoyos, las ranuras y las mortajas.

Diagnosis A systematic study of a machine or machine parts to determine the cause of improper performance or failure.

Diagnóstico Un estudio sistemático de una máquina o las partes de una máquina con el fín de determinar la causa de una falla o de una operación irregular.

Dial indicator A measuring instrument with the readings indicated on a dial rather than on a thimble as on a micrometer.

Indicador de carátula Un instrumento de medida cuyo indicador es en forma de muestra en contraste al casquillo de un micrómetro.

Disc brake A brake design in which the member attached to the wheel is a metal disc and braking force is applied by two brake pads that are squeezed against the disc by the caliper.

Frenos de disco Un diseño de frenos en el cual el miembro conectado a la rueda es un disco de metal y la fuerza del frenado se aplica por medio de la mordaza que aprieta dos almohadillas de freno contra el disco.

Disc runout A measurement of how much a brake disc wobbles from side to side as it rotates.

Corrimiento del disco Una medida del movimiento oscilatorio de un disco de un lado al otro mientras que gira.

Distortion A warpage or change in form from the original shape.

Distorción El abarquillamiento o un cambio en la forma original.

Dowel A metal pin attached to one object which, when inserted into a hole in another object, ensures proper alignment.

Espiga Una clavija de metal que se fija a un objeto que, al insertarla en el hoyo de otro objeto, asegura una alineación correcta.

Dowel pin A pin inserted in matching holes in two parts to maintain those parts in fixed relation to one another.

Clavija de espiga Una clavija que se inserte en los hoyos alineados de dos partes para mantener a ésos dos partes en una relación fija el uno al otro.

Drum brake A brake design in which the component attached to the wheel is shaped like a drum. Shoes press against the inside of the drum to provide braking action.

Frenos de tambor Un diseño de frenos en el cual el componente conectado a la rueda tiene la forma de un tambor. La zapatas oprimen contra la superficie interior para efectuar la acción de frenar.

Dry friction The friction between two dry solids.

Fricción seca Fricción entre dos sólidos secos.

Dual master cylinder A master cylinder consisting of two separate sections, one for the rear brakes and the other for the front brakes.

Cilindro maestro dual Un cilindro maestro que consiste de dos secciones distinctas, una para los frenos traseros y la otra para los frenos delanteros.

Dual-servo-action brakes Both brake shoes are self-energizing; that is, they tend to multiply braking forces as they are applied.

Freno duo-servo Ambas zapatas son autoenergéticas; quiere decir, suelen multiplicar la fuerza de frenar al aplicarlas.

Dynamic In motion.

Dinámico En movimiento.

Eccentric One circle within another circle wherein the circles do not have the same center or a circle is mounted off center. On FWD cars, front-end camber adjustments are accomplished by turning an eccentric cam bolt that mounts the strut to the steering knuckle.

Excéntrico Se dice de dos círculos, el uno dentro del otro, que no comparten el mismo centro o de un círculo ubicado descentrado. En los coches FWD, los ajustes de la inclinación se efectuan por medio de un perno excéntrico que fija el poste sobre el articulación de dirección.

Efficiency The ratio between the power of an effect and the power expended to produce the effect; the ratio between an actual result and the theoretically possible result.

Eficiencia La relación entre la potencia de un efecto y la potencia que se gasta para producir el efecto; la relación entre un resultado actual y el resultado que es una posibilidad teórica.

Elastomer Any rubber-like plastic or synthetic material used to make bellows, bushings, and seals.

Elastómero Cualquiera materia plástic parecida al hule o una materia sintética que se utiliza para fabricar a los fuelles, los bujes y las juntas.

Endplay The amount of axial or end-to-end movement in a shaft due to clearance in the bearings.

Juego de las extremidades La cantidad del movimiento axial o del movimiento de extremidad a extremidad en una flecha debido a la holgura que se deja en los cojinetes.

Face The front surface of an object.

Cara La superficie delantera de un objeto.

Fatigue The buildup of natural stress forces in a metal part that eventually causes it to break. Stress results from bending and loading the material.

Fatiga El incremento de tensiones e esfuerzos normales en una parte de metal que eventualmente causan una quebradura. Los esfuerzos resultan de la carga impuesta y del doblamiento de la materia.

Feeler gauge A metal strip or blade finished accurately with regard to thickness used for measuring the clearance between two parts; such gauges ordinarily come in a set of different blades graduated in thickness by increments of 0.001 inch.

Calibrador de lainas Una lámina o hoja de metal que ha sido acabado precisamente con respecto a su espesor que se usa para medir la holgura entre dos partes; estas galgas típicamente vienen en un conjunto de varias espesores graduados desde el 0.001 de una pulgada.

Fiber composites A mixture of metallic threads along with a resin form a composite offering weight and cost reduction, long-term durability, and fatigue life. Fiberglass is a fiber composite.

Compuestos de fibra Una mezcla de hilos metálicos con una resina que formen un compuesto ofreciendo una reducción en costo y peso, y una durabilidad y utilidad de largo plazo. La fibra de vidrio es una fibra compuesta.

Fit The contact between two machined surfaces.

Ajuste El contacto entre dos superficies maquinadas.

Fixed caliper A disc brake caliper that has pistons on both sides of the rotor. It is rigidly fixed to the suspension.

Mordaza fija Una mordaza de freno de disco que tiene pistones en ambos lados del rotor. Se fija rígidamente a la suspención.

Floating caliper A disc brake caliper that has one piston. The caliper is free to move on pins in response to the piston pressing against the rotor.

Mordaza flotante Una mordaza de freno de disco que tiene un pistón. La mordaza mueve libremente sobre clavijas respondiendo al pistón oprimiendo contra el rotor.

Foot-pound (or ft-lb) This is a measure of the amount of energy or work required to lift 1 pound a distance of 1 foot.

Libra-pie (o lb.p.) Una medida de la cantidad de energía o fuerza que requiere mover una libra a una distancia de un pie.

Force Any push or pull exerted on an object; measured in pounds and ounces, or in Newtons (N) in the metric system.

Fuerza Cualquier acción empujado o jalado que se efectua en un objeto; se mide en pies y onzas, o en Newtones (N) en el sistema métrico.

Four wheel drive On a vehicle, driving axles at both front and rear, so that all four wheels can be driven.

Tracción a cuatro ruedas En un vehículo, se trata de los ejes de dirección fronteras y traseras, para que cada una de las ruedas puede impulsar.

Frame The main understructure of the vehicle to which everything else is attached. Most FWD cars have only a subframe for the front suspension and drive train. The body serves as the frame for the rear suspension.

Armazón La estructura principal del vehículo al cual todo se conecta. La mayoría de los coches FWD sólo tiene un bastidor auxiliar para la suspensión delantera y el tren de propulsión. La carrocería del coche sirve de chassis par la suspensión trasera.

Friction The resistance to motion between two bodies in contact with each other.

Fricción La resistencia al movimiento entre dos cuerpos que estan en contacto.

Front wheel drive (FWD) The vehicle has all drive train components located at the front.

Tracción de las ruedas delanteras (FWD) El vehículo tiene todos los componentes del tren de propulsión en la parte delantera.

FWD Abbreviation for front wheel drive.

FWD Abreviación de tracción de las ruedas delanteras.

Galling Wear caused by metal-to-metal contact in the absence of adequate lubrication. Metal is transferred from one surface to the other, leaving behind a pitted or scaled appearance.

Desgaste por fricción El desgaste causado por el contacto de metal a metal en la ausencia de lubricación adecuada. El metal se transfere de una superficie a la otra, causando una aparencia agujerado o con depósitos.

Gasket A layer of material, usually made of cork, paper, plastic, composition, metal, or a combination of these, placed between two parts to make a tight seal.

Empaque Una capa de una materia, normalmente hecho del corcho, del papel, del plástico, de la materia compuesta, del metal, o de cualquier combinación de éstos, que se coloca entre dos partes para formar un sello impermeable.

Gasket cement A liquid adhesive material or sealer used to install gaskets.

Mastique para empaques Una substancia líquida adhesiva o una substancia impermeable que se usa para instalar a los empaques.

Glazing An extremely shiny condition of a metal surface, such as the bore of a cylinder.
Vidriado Una condición altamente luciente de una superficie de un metal, tal como el taladro de un cilindro.

Grind To finish or polish a surface by means of an abrasive wheel.
Amolar Acabar o pulir a una superficie por medio de una muela para pulverizar.

Half shaft Either of the two drive shafts that connect the transaxle to the wheel hubs in FWD cars. Half-shafts have constant-velocity joints attached to each end to allow for suspension motions and steering. The shafts may be of solid or tubular steel and may be of different lengths. Balance is not critical, as half-shafts turn at roughly one-third the speed of RWD drive shafts.
Semieje Cualquier de dos ejes o flechas de mando que conectan el transeje a los cubos de las ruedas en los coches de FWD. Los semiejes tienen juntas de velocidad continua conectado a cada extremo para permitir los movimientos de la suspención y la dirección. Los ejes pueden fabricarse del acero sólido o tubular y pueden variar en longitud. El balance no es crítico, puesto que los semiejes giran una tercera parte de la velocidad de los ejes de mando de RWD.

Hard pedal A loss in braking efficiency so that an excessive amount of pressure is needed to actuate the brakes.
Pedal de freno duro Una pérdida de la eficiencia del frenado en que se requiere una cantidad excesiva de presión para activar los frenos.

Heat treatment Heating, followed by fast cooling, to harden metal.
Tratamiento térmico Calentamiento, seguido por un enfriamiento rápido, para endurecer a un metal.

Honing A process whereby an abrasive material is used to smoothen a surface. Honing is a common operation in preparing cylinder walls for the installation of pistons.
Rectificación Un proceso por el cual una material abrasiva se usa para tersar una superficie. Rectificar es una operación comun en preparar los interiores de los cilindros para la instalación de los pistones.

Horsepower A measure of mechanical power, or the rate at which work is done. One horsepower equals 33,000 ft-lbs (foot-pounds) of work per minute. It is the power necessary to raise 33,000 pounds a distance of 1 foot in 1 minute.
Caballo de fuerza Una medida de fuerza mecánica, o el régimen en el cual se efectua el trabajo. Un caballo de fuerza iguala a 33,000 lb.p. (libras pie) de trabajo por minuto. Es la fuerza requerida para transportar a 33,000 libras una distancia de 1 pie en 1 minuto.

Hub The center part of a wheel, to which the wheel is attached.
Cubo La parte central de una rueda, a la cual se monta la rueda.

Hydraulic booster A power brake booster operated by hydraulic pressure from the power steering pump.
Hidrorreforzador Un reforzador de freno de potencia que opera con la presión hidráulica de la bomba de la dirección hidráulica.

Hydraulic fluid reservoir A part of a master cylinder assembly that holds reserve fluid.
Recipiente del líquido hidráulico Una parte de la asamblea del cilindro maestro que contiene el líquido en reserva.

Hydraulic press A piece of shop equipment that develops a heavy force by use of a hydraulic piston-and-jack assembly.
Prensa hidráulica Una herramienta del taller que provee una fuerza grande por medio de una asamblea de gato con un pistón hidráulico.

Hydraulic pressure Pressure exerted through the medium of a liquid.
Presión hidráulica La presión esforzada por medio de un líquido.

ID Inside diameter.
DI Diámetro Interior.

Idle Engine speed when the accelerator pedal is fully released and there is no load on the engine.
Marcha lenta La velocidad del motor cuando el pedal accelerador esta completamente desembragada y no hay carga en el motor.

Increments Series of regular additions from small to large.
Incremento Una serie de agregaciones regulares de pequeña a grande.

Index To orient two parts by marking them. During reassembly the parts are arranged so the index marks are next to each other. Used to preserve the orientation between balanced parts.
Índice Orientar a dos partes marcándolas. Al montarlas, las partes se colocan para que las marcas de índice estén alineadas. Se usan los índices para preservar la orientación de las partes balanceadas.

Inner bearing race The inner part of a bearing assembly on which the rolling elements, ball or roller, rotate.
Pista interior de un cojinete La parte interior de una asamblea de cojinetes en la cual ruedan las bolas o los rodillos.

Integral Built into, as part of the whole.
Íntegro Contenido, como una parte del total.

Jam nut A second nut tightened against a primary nut to prevent it from working loose. Used on inner and outer tie-rod adjustment nuts and on many pinion-bearing adjustment nuts.
Contra tuerca Una tuerca secundaria que se aprieta contra una tuerca primaria para prevenir que ésta se afloja. Se emplean en las tuercas de ajustes interiores e exteriores para las barras de acoplamiento y también en muchas de las tuercas de ajuste de portapiñones.

Key A small block inserted between the shaft and hub to prevent circumferential movement.
Chaveta Un tope pequeño que se meta entre la flecha y el cubo para prevenir un movimiento circunferencial.

Keyway A groove or slot cut to permit the insertion of a key.
Ranura de chaveta Un corte de ranura o mortaja que permite insertar una chaveta.

Knock A heavy metallic sound usually caused by a loose or worn bearing.
Golpe Un sonido metálico fuerte que suele ser causado por un cojinete suelto o gastado.

Knuckle The part of the suspension that supports the wheel hub and serves as the steering pivot. The bottom of the knuckle is attached to the lower control arm with a ball joint, and the upper portion is usually bolted to the strut.
Articulación La parte de la suspensión que sostiene al cubo de la rueda y sirve como punto pivote de dirección. La parte inferior de la articulación se une al brazo de mando inferior por medio de una articulación esférica, y la parte superior suele ser empernado al poste.

Knurl To indent or roughen a finished surface.
Moletear Indentar or desbastar a una superficie acabada.

Linkage Any series of rods, yokes, and levers, and so on, used to transmit motion from one unit to another.
Biela Cualquiera serie de barras, yugos, palancas, y todo lo demás, que se usa para transferir los movimientos de una unedad a otra.

Locknut A second nut turned down on a holding nut to prevent loosening.
Contra tuerca Una tuerca segundaria apretada contra una tuerca de sostén para prevenir que ésta se afloja.

Lock pin Used in some ball sockets (inner tie-rod end) to keep the connecting nuts from working loose. Also used on some lower ball joints to hold the tapered stud in the steering knuckle.

Clavija de cerrojo Se usan en algunas rótulas (las extremidades interiores de la barra de acoplamiento) para prevenir que se aflojan las tuercas de conexión. También se emplean en algunas juntas esféricas inferiores para retener al perno cónico en la articulación de dirección.

Lockplates Metal tabs bent around nuts or bolt heads.

Placa de cerrojo Chavetas de metal que se doblan alrededor de las tuercas o las cabezas de los pernos.

Lockwasher A type of washer that, when placed under the head of a bolt or nut, prevents the bolt or nut from working loose.

Arandela de freno Un tipo de arandela que, al colocarse bajo la cabeza de un perno, previene que el perno o la tuerca se aflojan.

Lubricant Any material, usually a petroleum product such as grease or oil, that is placed between two moving parts to reduce friction.

Lubricante Cualquier substancia, normalmente un producto de petróleo como la grasa o el aceite, que se coloca entre dos partes en movimiento para reducir la fricción.

Lug nut The nuts that fasten the wheels to the axle hub or brake rotor. Missing lug nuts should always be replaced. Overtightening can cause warpage of the brake rotor in some cases.

Tuerca de las ruedas Las tuercas que sujetan las ruedas al cubo de flecha o al rotor de los frenos. Las tuercas de las ruedas que se pierden siempre deben reemplazarse. Si se aprietan demasiado puede causar una deformación en el rotor del freno en algunos casos.

Master cylinder The liquid-filled cylinder in the hydraulic brake system or clutch where hydraulic pressure is developed when the driver depresses a foot pedal.

Cilindro maestro El cilindro lleno de líquido en el sistema de frenos hidráulico o en el embrague en el cual la presión hidráulica se presenta cuando el conductor comprime un pedal bajo su pie.

Metering valve A component that momentarily delays the application of front disc brakes until the rear drum brakes begin to move. Helps to provide balanced braking.

Válvula de medición Un componente que retrasa momentáneamente la aplicación de los frenos de disco delanteras hasta que comienzan a moverse los frenos de tambor traseros. Ayuda en proporcionar el enfrenado más equilibrado.

Micrometer A precision measuring device used to measure small bores, diameters, and thicknesses. Also called a mike.

Micrómetro Un dispositivo de medida precisa que se emplea a medir los taladros pequeños y los espesores. También se llama un mike (mayk).

Misalignment When bearings are not on the same center line.

Desalineamineto Cuando los cojinetes no comparten la misma linea central.

Mounts Made of rubber to insulate vibrations and noise while they support a power train part, such as engine or transmission mounts.

Monturas Hecho de hule para insular a las vibraciones y a los ruidos mientras que sujetan una parte del tren de propulsión, tal como las monturas del motor o las monturas de la transmisión.

Neoprene A synthetic rubber that is not affected by the various chemicals that are harmful to natural rubber.

Neoprene Un hule sintético que no se afecta por los varios productos químicos que pueden dañar al hule natural.

Newton-meter (N·m) Metric measurement of torque or twisting force.

Metro-Newton (N·m) Una medida métrica de la fuerza de torsión.

Nut A removable fastener used with a bolt to lock pieces together; made by threading a hole through the center of a piece of metal that has been shaped to a standard size.

Tuerca Un retén removable que se usa con un perno o tuerca para unir a dos piezas; se fabrica al filetear un hoyo taladrado en un pedazo de metal que se ha formado a un tamaño especificado.

Oil seal A seal placed around a rotating shaft or other moving part to prevent leakage of oil.

Empaque de aceite Un empaque que se coloca alrededor de una flecha giratoria para prevenir el goteo de aceite.

O-ring A type of sealing ring, usually made of rubber or a rubber-like material. In use, the O-ring is compressed into a groove to provide the sealing action.

Anillo en O Un tipo de sello anular, suele ser hecho de hule o de una materia parecida al hule. Al usarse, el anillo en O se comprime en una ranura para proveer un sello.

Outer bearing race The outer part of a bearing assembly on which the balls or rollers rotate.

Pista exterior de un cojinete La parte exterior de una asamblea de cojinetes en la cual ruedan las bolas o los rodillos.

Out-of-round Wear of a round hole or shaft that when viewed from an end will appear egg-shaped.

Defecto de circularidad Desgaste de un taladro o de una flecha circular, que al verse de una extremidad, tendrá una forma asimétrica, como la de un huevo.

Oxidation Burning or combustion; the combining of a material with oxygen. Rusting is slow oxidation, and combustion is rapid oxidation.

Oxidación Quemando o la combustión; la combinación de una materia con el oxígeno. El orín es una oxidación lenta, la combustión es la oxidación rápida.

Parallel The quality of two items being the same distance from each other at all points; usually applied to lines and, in automotive work, to machined surfaces.

Paralelo La calidad de dos artículos que mantienen la misma distancia el uno al otro en cada punto; suele aplicarse a las líneas y, en el trabajo automotívo, a las superficies acabadas a máquina.

Parking brake A mechanically operated brake used to hold the vehicle when it is parked.

Freno de estacionamiento (emergencia) Un freno que se opera a mano para sostener al vehículo al estacionarse.

Parking brake strut A bar between the brake shoes. When the parking lever is actuated, the parking brake strut pushes the leading brake shoe into the drum.

Poste del freno de estacionamiento Una barra entre las zapatas de freno. Al actuarse la palanca de estacionamiento, el poste del freno de estacionamiento empuje la zapata delantera contra el tambor.

Pascal's law The law of fluid motion.

Ley de Pascal La ley del movimiento del fluido.

Piston seal The seal fitted to the disc brake pistons. It provides the return motion to the piston as well as sealing in the brake fluid.

Sello del pistón El sello que se ajusta a los pistones de los frenos de disco. Provee el movimiento de regreso al pistón mientras que previene una fuga del líquido de freno.

Pitch The number of threads per inch on any threaded part.

Paso El número de filetes por pulgada de cualquier parte fileteada.

Pivot A pin or shaft upon which another part rests or turns.

Pivote Una chaveta o una flecha que sostiene a otra parte o sirve como un punto para girar.

Power booster Used to increase pedal pressure applied to a brake master cylinder.

Reforzador Se usa para incrementar la presión del pedal aplicado al cilindro maestro del freno.

Power brakes A brake system that employs vacuum or hydraulics to assist the driver in producing braking force.

Frenos de potencia Un sistema de frenos que emplea un vacío o las hidráulicas para asistir al conductor en efectuar una fuerza de enfrenado.

Preload A load applied to a part during assembly so as to maintain critical tolerances when the operating load is applied later.

Carga previa Una carga aplicada a una parte durante la asamblea para asegurar sus tolerancias críticas antes de que se le aplica la carga de la operación.

Press fit Forcing a part into an opening that is slightly smaller than the part itself to make a solid fit.

Ajustamiento a presión Forzar a una parte en una apertura que es de un tamaño más pequeño de la parte para asegurar un ajustamiento sólido.

Pressure Force per unit area, or force divided by area. Usually measured in pounds per square inch (psi) or in kilopascals (kPa) in the metric system.

Presión La fuerza por unedad de una area, o la fuerza divida por la area. Suele medirse en libras por pulgada cuadrada ($lb/pulg^2$) o en kilopascales (kPa) en el sistema métrico.

Pressure bleeding Pressure bleeding uses air to pressurize brake fluid in order to force air out of the hydraulic brake system.

Purga con presión En purgar con presión uno usa el aire para sobrecomprimir el líquido de freno así forzando el aire fuera del sistema hidráulico de frenos.

Pressure differential valve Used in dual brake systems to sense unequal hydraulic pressure between the front and rear brakes.

Válvula del diferencial de presión Usado en los sistemas de frenos dobles para sentir una presión desigual entre los frenos delanteros y traseros.

Primary shoe When the car is moving forward, the shoe facing the front of the car is the leading or primary shoe.

Zapata primaria Al moverse hacia frente el coche, la zapata en la dirección hacia la parte delantera del coche es la zapata de guía o primaria.

Proportioning valve This valve regulates the hydraulic pressure in the rear brake system. It is located between the inlet and outlet ports of the rear system in the control valve. It allows equal pressure to be applied to both the front and rear brakes until a particular pressure is obtained.

Válvula dosificadora Esta válvula regula la presión hidráulica en el sistema de frenos trasero. Se ubica entre las aberturas de entrada y salida del sistema trasero en la válvula de control. Permite que una presión equilibrada se aplica a ambos los frenos delanteros y traseros hasta que se obtiene una presión específica.

psi Abbreviation for pounds per square inch, a measurement of pressure.

$Lb/pulg^2$ Una abreviación de libras por pulgada cuadrada, una medida de la presión.

Puller Generally, a shop tool used to separate two closely fitted parts without damage. Often contains a screw, or several screws, which can be turned to apply a gradual force.

Extractor Generalmente, una herramienta del taller que sirve para separar a dos partes apretadas sin incurrir daños. Suele tener una tuerca o varias tuercas, que se pueden girar para aplicar una fuerza gradual.

Pulsation To move or beat with rhythmic impulses.

Pulsación Moverse o batir con impulsos rítmicos.

Pulsing pedal A condition where the brake pedal moves up and down when it is applied. Normally due to an unparallel brake rotor.

Pedal pulsante Una condición en la cual el pedal de freno se mueve hacia arriba y abajo al aplicarse. Normalmente se debe a un rotor de freno que esta fuera de paralelo.

Race A channel in the inner or outer ring of an antifriction bearing in which the balls or rollers roll.

Pista Un canal en el anillo interior o exterior de un cojinete antifricción en el cual ruedan las bolas o los rodillos.

Ratio The relation or proportion that one number bears to another.

Relación La correlación o proporción de un número con respeto a otro.

Reamer A round metal-cutting tool with a series of sharp cutting edges; enlarges a hole when turned inside it.

Escariador Una herramienta redonda para cortar a los metales que tiene una seria de rebordes mordaces agudos; al girarse en un agujero lo agranda.

Rear wheel drive A term associated with a vehicle where the engine is mounted at the front of the vehicle and the driving axle and driving wheels are mounted at the rear of the vehicle.

Tracción trasera Un término que se asocia con un vehículo en el cual el motor se ubica en la parte delantera y el eje propulsor y las ruedas propulsores se encuentran en la parte trasera del vehículo.

Rivet A headed pin used for uniting two or more pieces by passing the shank through a hole in each piece, and securing it by forming a head on the opposite end.

Remache Una clavija con cabeza que sirve para unir a dos piezas o más al pasar el vástago por un hoyo en cada pieza y asegurarlo por formar una cabeza en el extremo opuesto.

Roller bearing An inner and outer race upon which hardened steel rollers operate.

Cojinete de rodillos Una pista interior y exterior en la cual operan los rodillos hecho de acero endurecido.

Rollers Round steel bearings that can be used as the locking element in an overrunning clutch or as the rolling element in an antifriction bearing.

Rodillos Articulaciones redondos de acero que pueden servir como un elemento de enclavamiento en un embrague de sobremarcha o como el elemento que rueda en un cojinete antifricción.

RPM Abbreviation for revolutions per minute, a measure of rotational speed.

RPM Abreviación de revoluciones por minuto, una medida de la velocidad rotativa.

RTV sealer Room-temperature vulcanizing gasket material that cures at room temperature; a plastic paste squeezed from a tube to form a gasket of any shape.

Sellador RTV Una materia vulcanizante de empaque que cura en temperaturas del ambiente; una pasta plástica exprimida de un tubo para formar un empaque de cualquiera forma.

Runout Deviation of the specified normal travel of an object; the amount of deviation or wobble a shaft or wheel has as it rotates. Runout is measured with a dial indicator.

Corrimiento Una desviación de la carrera normal e especificada de un objeto. La cantidad de desviación o vacilación de una flecha o una rueda mientras que gira. El corrimiento se mide con un indicador de carátula.

RWD Abbreviation for rear wheel drive.

RWD Abreviación de tracción trasera.

SAE Society of Automotive Engineers.

SAE La Sociedad de Ingenieros Automotrices.

Score A scratch, ridge, or groove marring a finished surface.

Entalladura Una raya, una arruga o una ranura que desfigure a una superficie acabada.

Scuffing A type of wear in which there is a transfer of material between parts moving against each other; shows up as pits or grooves in the mating surfaces.

Erosión Un tipo de desgaste en el cual hay una tranferencia de una materia entre las partes que estan en contacto mientras que muevan; se manifiesta como hoyitos o muescas en las superficies apareadas.

Seal A material, shaped around a shaft, used to close off the operating compartment of the shaft, preventing oil leakage.

Sello Una materia, formado alrededor de una flecha, que sella el compartimiento operativo de la flecha, previniendo el goteo de aceite.

Sealer A thick, tacky compound, usually spread with a brush, that may be used as a gasket or sealant to seal small openings or surface irregularities.

Sellador Un compuesto pegajoso y espeso, comúnmente aplicado con una brocha, que puede usarse como un empaque o un obturador para sellar a las aperturas pequeñas o a las irregularidades de la superficie.

Seat A surface, usually machined, upon which another part rests or seats; for example, the surface upon which a valve face rests.

Asiento Una superficie, comúnmente maquinada, sobre la cual yace o se asienta otra parte; por ejemplo, la superficie sobre la cual yace la cara de la válvula.

Secondary shoe When the car is moving forward, the shoe facing the rear is the trailing or secondary shoe.

Zapata secundaria Al moverse hacia frente el coche, la zapata en la dirección hacia la parte trasera del coche es la zapata seguidora o secundaria.

Self-energizing The increase in friction contact between the toe of the brake shoe caused by the drum rotation tending to pull the shoe into the drum.

Autoenergético El incremento del contacto frotativo entre la parte superior del freno producido por la rotación del tambor que tiene una tendencia a jalar a la zapata hacia el tambor.

Sliding fit Where sufficient clearance has been allowed between the shaft and journal to allow free-running without overheating.

Ajuste corredera Donde se ha dejado una holgura suficiente entre la flecha y el muñon para permitir una marcha libre sin sobrecalentamiento.

Snap ring Split spring-type ring located in an internal or external groove to retain a part.

Anillo de seguridad Un anillo partido tipo resorte que se coloca en una muesca interior o exterior para retener a una parte.

Spalling A condition where the material of a bearing surface breaks away from the base metal.

Escamación Una condición en la cual una materia de la superficie de un rodamiento se separa del metal base.

Spline Slot or groove cut in a shaft or bore; a splined shaft onto which a hub, wheel, or gear with matching splines in its bore is assembled so that the two must turn together.

Acanaladura (espárrago) Una muesca o ranura cortada en una flecha o en un taladro; una flecha acanalada en la cual se asamblea un cubo, una rueda,o un engranaje, que tiene un acanaladura pareja en el taladro de manera de que las dos deben girar juntos.

Spongy pedal A condition where the brake pedal does not give firm resistance to foot pressure. Normally caused by air in the hydraulic system.

Resistencia esponjosa Una condición en la cual el pedal de freno no ofrece una resistencia firme a la presión del pie. Suele ser causado por la presencia del aire en el sistema hidráulico.

Spring A device that changes shape when it is stretched or compressed, but returns to its original shape when the force is removed; the component of the automotive suspension system that absorbs road shocks by flexing and twisting.

Resorte Un dispositivo que cambia de forma al ser estirado o comprimido, pero que recupera su forma original al levantarse la fuerza; es un componente del sistema de suspensión automotívo que absorba los choques del camino al doblarse y torcerse.

Spring retainer A steel plate designed to hold a coil or several coil springs in place.

Retén de resorte Una chapa de acero diseñado a sostener en su posición a un resorte helicoidal o más.

Squeak A high-pitched noise of short duration.

Chillido Un ruido agudo de poca duración.

Squeal A continuous high-pitched noise.

Alarido Un ruido agudo continuo.

Star-adjuster Star-shaped rotor used as an adjustment device in drum brakes.

Ajustador de estrella y tornillo Un rotor en forma de estrella que se usa como dispositivo de ajuste en los frenos de tambor.

Stress The force to which a material, mechanism, or component is subjected.

Esfuerzo La fuerza a la cual se somete a una materia, un mecanísmo o un componente.

Tap To cut threads in a hole with a tapered, fluted, threaded tool.

Roscar con macho Cortar las roscas en un agujero con una herramienta cónica, acanalada y fileteada.

Temper To change the physical characteristics of a metal by applying heat.

Templar Cambiar las características físicas de un metal mediante una aplicación del calor.

Tension Effort that elongates or "stretches" a material.

Tensión Un esfuerzo que alarga o "estira" a una materia.

Thickness gauge Strips of metal made to an exact thickness, used to measure clearances between parts.

Calibre de espesores Las tiras del metal que se han fabricado a un espesor exacto, sirven para medir las holguras entre las partes.

Thread chaser A device, similar to a die, that is used to clean threads.

Peine de roscar Un dispositivo, parecido a una terraja, que sirve para limpiar a las roscas.

Threaded insert A threaded coil that is used to restore the original thread size to a hole with damaged threads.

Pieza inserta roscada Una bobina roscada que sirve para restaurar a su tamaño original una rosca dañada.

Thrust load A load that pushes or reacts through the bearing in a direction parallel to the shaft.

Carga de empuje Una carga que empuja o reacciona por el cojinete en una dirección paralelo a la flecha.

Thrust washer A washer designed to take up end thrust and prevent excessive endplay.

Arandela de empuje Una arandela diseñada para rellenar a la holgura de la extremidad y prevenir demasiado juego en la extremidad.

Tolerance A permissible variation between the two extremes of a specification or dimension.

Tolerancia Una variación permisible entre dos extremos de una especifcación o de un dimensión.

Torque A twisting motion, usually measured in ft-lbs (N•m).

Torsión Un movimiento giratorio, suele medirse en lb.p. (N•m).

Traction The gripping action between the tire tread and the road's surface.

Tracción La acción de agarrar entre la cara de la rueda y la superficie del camino.

Vacuum Any pressure lower than atmospheric pressure.

Vacío Cualquier presión que es más baja la presión atmosférica.

Vacuum brake booster A diaphragm-type booster that uses manifold vacuum and atmospheric pressure for its power.

Reforzador de vacío Un reforzador de tipo de diafragma que usa el vacío del colector y la presión atmosférica para operar.

Vehicle identification number (VIN) The number assigned to each vehicle by its manufacturer, primarily for registration and identification purposes.

Número de identificacíon del vehículo (VIN) El número asignado a cada vehículo por su fabricante, primariamente con el propósito de la registración y la identificación.

Vibration A quivering, trembling motion felt in the vehicle at different speed ranges.

Vibración Un movimiento de estremecer o temblar que se siente en el vehículo en varios intervalos de velocidad.

Viscosity The resistance to flow exhibited by a liquid. A thick oil has greater viscosity than a thin oil.

Viscosidad La resistencia al flujo que manifiesta un líquido. Un aceite espeso tiene una viscosidad mayor que un aceite ligero.

Wheel A disc or spokes with a hub at the center that revolves around an axle, and a rim around the outside on which the tire is mounted.

Rueda Un disco o rayo que tiene en su centro un cubo que gira alrededor de un eje, y tiene un rim alrededor de su exterior en la cual se monta el neumático.

Wheel cylinder A mechanism located at each wheel in a drum brake system. It uses hydraulic pressure to force the brake shoes against the drum to stop the wheel from turning.

Cilindro de la rueda Un mecanismo ubicado en cada rueda de un sistema de frenos de tambor. Utiliza la presión hidráulica para forzar las zapatas contra el tambor asi previniendo que gira la rueda.

INDEX

Note: Page numbers in bold print reference non-text material.